# Fundamentals of Satellite Remote Sensing

## An Environmental Approach

## Second Edition

# Fundamentals of Satellite Remote Sensing

## An Environmental Approach

## Second Edition

Emilio Chuvieco

CRC Press
Taylor & Francis Group
Boca Raton London New York

CRC Press is an imprint of the
Taylor & Francis Group, an **informa** business

CRC Press
Taylor & Francis Group
6000 Broken Sound Parkway NW, Suite 300
Boca Raton, FL 33487-2742



Printed on acid-free paper
Version Date: 20160114

International Standard Book Number-13: 978-1-4987-2805-8 (Hardback)



**Visit the Taylor & Francis Web site at**
**http://www.taylorandfrancis.com**

**and the CRC Press Web site at**
**http://www.crcpress.com**

*To all mothers…*

*Because they chose to give us life*

# Contents

# Preface

The information being collected by sensors on board Earth observation satellites has revolutionized our knowledge of the planet's environment. The growing concern about the present environmental problems and future sustainability of our planet has spurred great interest in monitoring the main variables of the Earth system. We need to gain a better understanding of the biogeochemical cycles to enhance our efforts to mitigate or adapt to future changes. Satellite remote sensing is a particularly valuable tool to collect critical information about the state of the environment. The impacts of climate change, deforestation, desertification, fire, floods, and other natural and man-made disturbances require an up-to-date, spatially comprehensive, and global source of geographical data. Satellite Earth observations provide such global, recurrent, and comprehensive views of the many dynamic processes that are affecting the resources, and habitability of our planet. Human beings have demonstrated their capacity to dramatically alter Earth's natural systems with both beneficial and negative consequences. To make more informed decisions concerning our changing environment, a better understanding of the tools used in environmental monitoring becomes more and more relevant. Watching how our planet changes from space may not be sufficient to modify our current way of living, but it will certainly help us to manage our resources more wisely and move toward more sustainable policies. Albert Einstein said that information is not knowledge, and certainly good information does not necessarily provide good knowledge, but we can also affirm that bad information implies bad knowledge and therefore good knowledge requires good information, because access to accurate and updated information is the basis for making rational decisions.

This book is a modest attempt to help students and professionals become more familiar with the remote sensing technology. Many books have already been published on this topic, which I obviously do not try to amend. The scope of this book relies on focusing on only satellite remote sensing systems and provide an environmental orientation to illustrate the different interpretation techniques. A basic knowledge of aerial photography is assumed, and the main effort here is dedicated to digital interpretation. However, Chapter 5 on visual analysis of satellite images offers a synergic use of visual and digital interpretation techniques. Chapter 8 on accuracy assessment and the connection between remote sensing and geographic information systems will help the reader extend the interpretation of satellite images to a more operational, applications-oriented framework.

This book was initially published in Spanish in 1990 and was updated periodically in the same language. This edition is based on the first English version published in 2010 with the help of Professor Alfredo Huete. In addition to translating all the material, we have tried to adapt the contents to a more international audience. More than half of the figures of this edition are new compared to the former. All are now offered in full color, so the reader can better appreciate the quality of information provided by Earth observation satellite missions. Formulas are referenced using parentheses and are numbered sequentially, preceded by the number of the chapter, in the same way as the figures. We have illustrated the different interpretation techniques using a set of sample images acquired from different ecosystems and at different spatial resolutions.

We thank the publishing company for their support in the process, and the valuable suggestions of colleagues and students who have helped to shape the text in a more pedagogical way.

All figures are available as powerpoint slides from the CRC Press website: http://www.crcpress.com/product/isbn/9781498728058. I hope this addition will help students and professors to improve the educational content of the book.

Writing a textbook is really a difficult task that is highly under-appreciated by current academic evaluation criteria. Our main interest in carrying out this task was to help students and colleagues discover a fascinating world, as observed from above Earth, and to facilitate their learning of concepts and tools so that they can make a more effective use of satellite data for the benefit of humankind.

**Emilio Chuvieco**

# Author

**Emilio Chuvieco** is a professor of geography and director of the Environmental Ethics chair at the University of Alcalá, Spain, where he coordinates the master's and PhD programs in geographic information technologies. He also leads the "Environmental Remote Sensing Research Group." He was a visiting professor at the University of California at Berkeley and Santa Barbara, the Canadian Remote Sensing Center, Cambridge University, and the University of Maryland. He has advised 35 PhD dissertations, has been a principal investigator of 27 research projects, is an author of 26 books and 330 scientific papers and book chapters, and is a former president of the Spanish Remote Sensing Society and the Geographic Information Technologies group of the Association of Spanish Geographers. Professor Chuvieco is a corresponding member of the Spanish Academy of Sciences since 2004 and a member of the GOFC-GOLD Fire implementation team. Currently, he is science leader of the Fire Disturbance project within the European Space Agency's Climate Change Initiative program. He is co-editor-in-chief of the journal *Remote Sensing of Environment.*

# 1 Introduction

## 1.1 DEFINITION AND OBJECTIVES

One of the earliest dreams of humans has been to observe Earth and its landscapes from a bird's perspective. This was only possible with the advent of balloons, gliders, and airplanes. A key driving force in our quest to fly above the ground has been to find new perspectives from which to observe Earth's diverse landscapes. Our view of Earth is quite limited when we are confined to the ground. Our desire to survey entire landscapes, mountain ranges, volcanoes, hurricanes, rivers, and ice fields has been evident since the beginning of aeronautics and now forms the foundation of space-based Earth observation (EO). Today, as a result of rapid technological advances, we routinely survey our planet's surface from different platforms: low-altitude unmanned aerial vehicles (UAVs), airplanes, and satellites. The surveillance of Earth's terrestrial landscapes, oceans, and ice sheets constitutes the main goal of *remote sensing* techniques. The term "remote sensing" was first utilized in the early 1960s to describe any means of observing the Earth from afar, particularly as applied to aerial photography, the main sensor used at that time. In a broader context, remote sensing activities include a wide range of aspects, from the physical basis to obtain information from a distance, to the operation of platforms carrying out the sensor system, to the data acquisition, storage, and interpretation. Finally, the remotely collected data are converted to relevant information, which is provided to a vast variety of potential end users: farmers, foresters, fishers, journalists, glaciologists, ecologists, geographers, etc. Nowadays, satellite images have become widely accessible by ordinary people, who use them within well-known images or map services, as basis for planning tourist routes or finding shops and services.

In this book, we will mainly deal with EO from spaceborne imaging sensors. There are also a multitude of airborne sensing systems (aerial photography, videography, lidar, and radar [or radio detection and ranging] systems) that play an important role in remote sensing of the Earth's surface. Airborne remote sensing is particularly valuable in local studies, as well as in the validation of satellite data within research and development. However, we will focus on satellite systems, as they provide worldwide coverage with an ample variation of spatial, spectral, and temporal detail, which make them suitable for a wide range of studies. This book is targeted mainly at environmental scientists, including biologists, geologists, ecologists, geographers, foresters, agronomists, pedologists, oceanographers, and cartographers. We will mostly deal with terrestrial applications, but some examples of atmospheric and oceanographic uses of remote sensing data will be considered as well.

Remote sensing may be more formally defined as the acquisition of information about the state and condition of an object through sensors that do not touch it.

A remote observation requires some kind of energy interaction between the target and the sensor. The sensor-detected signal may be solar energy (from the Sun) that is reflected from the Earth's surface or it may be self-emitted energy from the surface itself. In addition to passive receivers, other sensors produce their own energy pulses and therefore are able to observe the Earth's surface regardless of solar conditions. The radiant energy signal that is detected and measured by the satellite sensor is then either stored in memory on board the satellite or transmitted to a ground receiving station for later interpretation.

Remote sensing also includes the analysis and interpretation of the acquired data and imagery. For environmental scientists, this is the most important aspect of remote sensing since the main value of this technique is to provide relevant information for monitoring Earth's resources. Extracting relevant information requires a good understanding of the physical basis and the acquisition process, as well as a solid knowledge of the algorithms used to process the original data.

In summary, remote sensing includes the following six components (Figure 1.1):

1. An energy source, which produces the electromagnetic radiation that interacts between the sensor and the surface. The most important source of energy is the Sun, as it illuminates and heats the Earth.

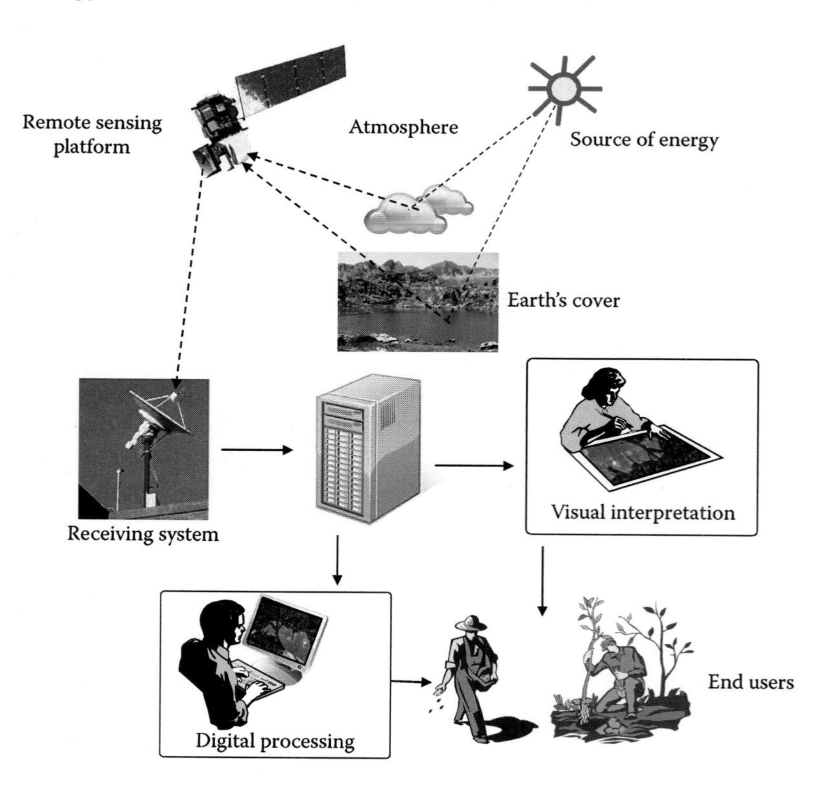


**FIGURE 1.1** Illustration of the main components associated with remote sensing activities.

2. Earth's surface, consisting of vegetation, soils, water, rocks, snow, ice, and human structures. These surfaces receive the incident energy from the source (1) and, as a result of physical and chemical interaction with the incoming energy, reflect and emit a part of that energy back toward the satellite sensor. Part or the whole energy pulse may be filtered by the atmosphere, depending on its gas and particulate matter concentrations.
3. Sensor and platform. The sensor is the instrument measuring and recording the energy coming from the surface. The platform provides the major services for sensor operation, such as attitude and orbit control, power supply, and communications with the ground receiving system. Ordinarily an EO satellite includes different sensors, depending on its main mission. Meteorological satellites commonly include sensors to detect atmospheric humidity, temperature, albedo, ozone, or aerosol concentrations.
4. The ground receiving system collects the raw digital data measured by the sensor, stores the data, and formats them appropriately. Some basic preprocessing corrections are performed by the ground system, prior to the distribution of the imagery.
5. The analyst, who converts the processed image data into thematic information of interest, using visual and/or digital techniques.
6. The user community, which utilizes the information extracted from the original data for a wide variety of applications.

We can illustrate the six components by using the human eye as an example. The eye is a sensor (3) that *sees* the reflected sunlight (1) from the various objects observed (2). The received electromagnetic signal is transmitted to the brain (4), which then generates an image of what is observed. The observer (5) further analyzes and interprets the image and, as the end user (6), applies this knowledge toward making appropriate decisions concerning his or her own behavior. Human vision is a sophisticated and complete remote sensing system, providing great spatial and color detail.

Despite the sophisticated nature of human eyesight, our personal remote sensing capabilities have several important limitations. Our eyes are restricted to a narrow portion of the electromagnetic spectrum, known as the "visible region." Other forms of electromagnetic energy, such as infrared, microwave, and ultraviolet radiation, cannot be seen with our eyes. In addition, our vision is dependent on a source of external energy, and we are unable to see if objects are not illuminated by the Sun or a source of artificial light. On the other hand, we have a limited perspective constrained by our own height, which restricts our observing capabilities to a limited, and often oblique, point of view. Remote sensing techniques make it possible for us to overcome these limitations through the use of devices that provide information from nonvisible radiations, at different perspectives, and regardless of solar conditions. These new capabilities have not only greatly expanded our knowledge about the Earth's environment but also provided increasing monitoring capabilities, to detect relevant changes. Remote sensing, in conjunction with parallel developments in geographical information systems (GISs), global positioning systems (GPS), and other ground data collection systems, now provides vast amounts of information about the land surface, to improve our understanding of the Earth system and better contribute to preserving it.

The effective utilization of such enormous quantities of data is accomplished with the growing availability of computer processors, which perform the tedious and time-consuming tasks of data handling and transformations to enable quick and efficient user interpretations and problem solving. Digital analysis of satellite data extends our capability to integrate spatial information, by combining information from different sources (soil or land cover maps, terrain models, climatic data, etc.). By enhancing our analytical capacity, we are able to focus more on data interpretation, problem solving, and making appropriate management decisions, tasks in which human intelligence is irreplaceable.

In this book, the study of remote sensing systems focused around the six components described earlier: energy sources and interactions with terrestrial surfaces (Chapter 2); sensors and platforms (Chapter 3); analysis of the acquired data (Chapters 4 through 7), including validation of the results (Chapter 8); and connection to end users within a GIS framework (Chapter 9). In the first part of this book, we cover the basic processes that allow for the acquisition of spaceborne imagery, including the physical principles of energy transmission and image acquisition; optical, thermal, and microwave radiant energy interactions on the Earth's surface (Chapter 2); and an overview of some of the main satellite observation systems (Chapter 3). The remaining chapters focus on visual and digital image analysis and interpretation techniques and their applications to science and management (Chapters 4 through 8). The final chapter is devoted to the integration of remote sensing with GIS for environmental analysis (Chapter 9). This first chapter also presents an overview of remote sensing with discussion on its foreseeable developments and some remarks on legal issues and international debate concerning EO across national boundaries.

## 1.2 HISTORICAL BACKGROUND

Remote sensing, as an applied tool and methodology, has evolved historically in parallel with other technological advancements, such as the improvements in optics, sensor electronics, satellite platforms, transmission systems, and computer data processing. These developments have resulted in enormous progress in the quantity, quality, and diversity of data available to the scientific community. A summary of milestones in remote sensing observation is included in Figure 1.2.

The first remote sensing acquisition can be traced back to the mid-1800s, along with the development of aerial photography. In 1839, the first ever photos were taken in France by Daguerre, Talbot, and Niepce, and by 1840 the French began using photos to produce topographic maps. In 1858, the first aerial photos were taken from a height of 80 m over Bievre, France, by Gaspard Félix Tournachon using cameras mounted on a hot air balloon. The first balloon photographies used for urban planning were acquired by James Wallace Black in 1860 over the city of Boston. The first attempts to use the new perspective provided by aerial platforms in military reconnaissance occurred in 1861, during the American Civil War, when Thaddeus Lowe

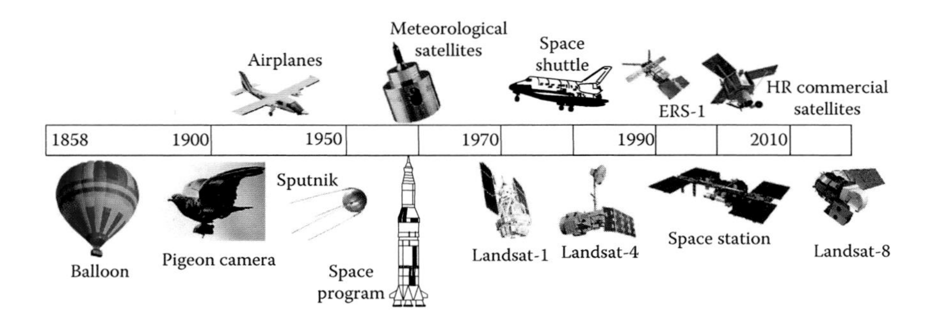


**FIGURE 1.2** Historical development of remote sensing systems.

was appointed Chief of the Union Army Balloon Corps by President Abraham Lincoln. In the 1880s, the British used kites to obtain aerial photography, and in the early 1900s, carrier pigeons were able to fly as more advanced, smaller, and lighter cameras were developed. The great San Francisco earthquake of 1906 was captured on film using a panoramic camera mounted 600 m above San Francisco Bay and supported by a string of kites.

The next major milestone in remote sensing occurred in 1909, when Wilbur Wright shot the first photographs over Italy from an airplane, establishing a new era of observations from airborne platforms. By 1915 and during World War I, the British Royal Air Force was collecting aerial reconnaissance photos with cameras designed specifically for aircraft surveying (Brookes 1975). In 1930, the first aerial multispectral photographs were collected by Krinov and colleagues in Russia. The following year, the first near-infrared film was developed and tested by Kodak Research Laboratories.

In the 1940s, the military made significant advancements in the development and use of color infrared films, which were used to distinguish real vegetation from camouflaged targets that were painted green to look like vegetation. The greatest developments in aerial reconnaissance and photo interpretation were made during World War II. Other significant advancements were made with thermal scanners and imaging radar systems, which create images by focused radar beams that scan across an area.

In the late 1940s and early 1950s, improved navigation systems gave way to the first space-based sensor devices. Early experiments with the V-2, Aerobee, and Viking rockets recorded panoramic images of the Southwest United States from altitudes of 100–250 km. The first space-based photo was taken on March 7, 1947, at about 200 km above New Mexico while testing captured German V-2 rockets (Figure 1.3). In 1950, scientist Otto Berg pieced together several photos from one of these tests into a mosaic of a large tropical storm over Brownsville, Texas.

One of the main strategic objectives of the Cold War era was the space race. The launch of Sputnik by the Soviet Union in 1957 was the start of a long series of civil and military missions that enabled us to explore not only our planet but also the moon and neighboring planets. In April 1960, NASA launched the Television Infrared Observational Satellites (TIROS-1) that began relaying continental views

**FIGURE 1.3** This photo was taken by an automatic K-12 camera, using black-and-white infrared film, from a Viking sounding rocket that reached a height of 227 km. This scene spans parts of New Mexico, Arizona, Nevada, California, and northwest Mexico (upper Gulf of California on the left). (From http://earthobservatory.nasa.gov/Features/Observing/.)

of global cloud patterns (Figure 1.4). These repetitive images enabled a deeper understanding of atmospheric conditions and circulation patterns and provided early warnings of serious natural catastrophes.

In the 1960s, several manned missions were launched to explore our Solar System and to reach the Moon, but these missions also acquired valuable images of our planet. Astronauts provided us with breathtaking and scientifically useful images of the Earth during the Mercury and Apollo programs. The scientific value of exploring our planet from space became particularly evident in the last Mercury mission, in May 1963, in which NASA astronaut Gordon Cooper took a series of spectacular photographs with a handheld Hasselblad camera over many regions inaccessible to Western scientists, such as the terrain of central Tibet. These photographs provided valuable information about the geology, hydrology, and vegetation of remote regions and provided the impetus for the creation of space programs devoted to the study of the planet Earth.

The first mission that officially incorporated photography to investigate potential geological and meteorological applications of space technology was the Gemini Titan mission in 1965. The Apollo 6 and Apollo 7 missions in 1968 also had strong photographic components (Figure 1.5), and in 1969, Apollo 9 carried the first multispectral remote sensing package, which consisted of four Hasselblad cameras with different filters.

NASA, encouraged by the success of these early EOs, introduced digital technologies in remote sensing in the late 1960s. Digital imaging sensors developed during the Ranger, Surveyor, and Lunar Orbiter Probes provided key experience that aided the development of the first Earth Resources Technology Satellite (ERTS),

**FIGURE 1.4** Scientist giving a vibration test to Television Infrared Observation Satellite (TIROS). The first TIROS was operational for 78 days. (From http://grin.hq.nasa.gov/ABSTRACTS/GPN-2003-00028.html.)

launched on July 23, 1972. ERTS-1 was later renamed Landsat 1 with the launching of a second satellite in 1975, which provided the first detailed, high-resolution, multispectral images of the entire land surface of our planet.

The Landsat series continues today, providing a continuous and consistent high-quality 40+ year dataset of Earth's land surface and its changes over time. This has been one of the most successful outcomes of remote sensing, resulting in a wide range of civilian applications from the assessment of natural resources to the monitoring of natural disasters such as flooding, drought, fires, volcanic eruptions, and hurricanes. The Landsat missions were also accompanied by other NASA environmental observation projects, such as Skylab, a manned space laboratory in 1973; Seasat, an oceanographic satellite in 1978; and the Heat Capacity Mapping Mission (HCMM) for thermal investigations in 1978.

As a result of the Landsat series, the decade of the 1980s saw an increasing interest by the international scientific community in developing spaceborne remote sensing systems. The most outstanding developments began with the launching of the

**FIGURE 1.5** Photograph taken by the Apollo 7 over Lake Chad (central Africa) in 1968. (From http://earthobservatory.nasa.gov/.)

first *Systeme Pour l'Observation de la Terre* (SPOT) satellite in early 1986 by the French *Centre National d'Etudes Spatiales* (CNES). The Japanese launched their first Marine Observation Satellite (MOS-1) in February 1987, while the Indian Remote Sensing (IRS-1) satellite was launched in March 1988 and the European Remote Sensing satellite (ERS-1) in 1991.

More recently, remote sensing missions have increased exponentially, as will be discussed in more detail in Chapter 3. Benefiting from the experience and technical knowledge gained from the earlier launched missions, as well as the declining costs in the design of sensors and platforms, Russia, Canada, Germany, the United Kingdom, Brazil, Argentina, China, South Korea, Taiwan, and Israel, to mention some, have launched Earth-observing satellites for environmental and natural resource monitoring purposes. This trend has increased the worldwide availability and diversity of remotely sensed data while reducing acquisition costs for the end users.

Many of the civilian uses of remote sensing have been the result of research performed for military applications. The delicate affair of the capture of a U-2 spy plane by the Soviets in 1960 led to the U.S. administration to make a great investment in developing military reconnaissance satellites, much less vulnerable than airplanes to detection and eventual bringing down. In 1995, some of the photos taken by those military satellites were declassified, covering the period between 1960 and 1972, including high-resolution photos (from 2.5 to 7.5 m) from the Corona, Lanyard, and Gambit mission. In 2002, other missions were declassified covering the KH-7 and KH-9 missions with high-resolution photos acquired up to 1980. These photos are accessible through the United States Geological Survey (USGS) long-term archive facility (https://lta.cr.usgs.gov/products_overview, last access November 2015).

The private sector has also played an important role in promoting and advancing remote sensing capabilities, through the creation of consortiums that have particularly promoted very-high-spatial-resolution satellite projects. The first commercial satellites were launched in the late 1990s, the most famous being the IKONOS satellite launched by Space Imaging; the EROS A1, launched in 2000 by ImageSat International; and Quickbird, launched in October 2001 by Digital Globe. The role of the private commercial sector has greatly increased in the last decade, particularly related to high-spatial-resolution applications, including homeland security and strategic observations, real state, tourism, location of services, and commodities. This trend has extended traditional applications and users of satellite observation (Table 1.1). For instance, current satellite missions providing images with spatial resolutions in the range of 0.5–2.5 m are very useful to monitor crop conditions, urban growing, or large public works. After natural or human disasters, they are also particularly useful to provide a quick assessment to help urgent recovery actions (Figure 1.6). More recently, these high-resolution images have provided critical information for humanitarian activities, such as supervising conditions in refugee camps or detect impacts of guerrillas in remote areas. A recent report from Amnesty International in Nigeria on terrorist activities as detected by satellite images is particularly striking (Figure 1.7).

Recent trends point to a growing availability of satellite missions aiming to generate operational services, where remotely sensed data are routinely used. This is already the case of meteorological forecast where climate models are closely linked to diverse EO products (cloud temperature and height, moisture conditions, rainfall, wind fields, etc.). New EO missions include public access policies that facilitate an extended use of the acquired data. In addition to the raw and calibrated data, some of the recent programs include the development of corrected reflectances or temperatures and even final products, thus facilitating the use of remotely sensed data even to nonexperts. The best example is the Moderate-Resolution Imaging Spectroradiometer (MODIS) program, which has been able to derive 45 standard products from the raw data acquired by the Terra and Aqua NASA satellite (see Chapter 3 for further explanations). The quick Internet connections further facilitate the distribution of data, enabling the operation of near-real-time services.

## 1.3 INTERNATIONAL SPACE LAW

Satellite remote sensing systems collect images across the world without regard to political boundaries. This can lead to the violation of the *political air space* of a country and to the disclosure of its strategic information, which may be exploited by another country. Because of this, there has been much concern and discussion of the political, economic, and environmental consequences of space remote sensing activities.

The first discussions on the legal aspects of aerial surveillance took place in the 1950s soon after the launch of the first spy satellites. At an international conference in Geneva in 1955, President Eisenhower proposed an open skies policy, which would allow free and mutual observations between the United States and the Soviet Union. The concept behind this failed proposal was to ease tensions and slow the arms race by allowing each nation to conduct aerial reconnaissance on the military resources of the other (Leghorn and Herken 2001). The treaty was revived again in 1989 by President George H.W. Bush and negotiated by the members of NATO and

**TABLE 1.1**
**Comparison of Traditional and New Users of Remote Sensing Image Data**

| Traditional Users | New Users |
|---|---|
| Governments | News media information providers |
| Civil planners (mapping, land management, disaster response, etc.) | Electronic media organizations |
| | Print media |
| Armed forces | Trade journals |
| Intelligence services | |
| Scientific centers | |
| State and local governments | |
| Multinational Organizations | NGOs |
| United Nations (UN) agencies (e.g., UN Special Commission, UN High Commissioner Refugees) | Environmental policy |
| | Arms control, nonproliferation, and disarmament |
| Global change research programs | Regional conflict resolution |
| Regional centers (e.g., Western European Union Satellite Centre) | Humanitarian relief |
| | Human rights |
| | Biodiversity, deforestation, etc. |
| Business | Academia and research organizations |
| Resource extraction (e.g., oil, gas) | Media studies departments |
| Resource management (e.g., forestry) | Security policy studies departments |
| | Archaeology departments |
| | Transportation studies |
| | Agricultural and ecosystem studies |
| Academia and research organizations | Business |
| • Geography and geology departments | • Utilities (e.g., telecommunications, pipelines) |
| • Remote sensing programs | • Insurance firms (e.g., hazard assessments) |
| • Environmental studies | • Precision agriculture |
| | Environmental impact assessments |
| Remote sensing industry | Customers |
| Aerial photography firms | Real estate |
| Satellite imagery data providers | Individuals interested in images of their homes or cultural attractions |
| Value-added, image processing firms | |
| GIS firms | |
| Remote sensing professional organizations | |

*Source:* Baker, J.C., New users and established experts: Bridging the knowledge gap in interpreting commercial satellite imagery, in: J.C. Baker, K.M. O'Connell, and R.A. Williamson, eds., *Commercial Observation Satellites: At the Leading Edge of Global Transparency*, RAND–ASPRS, Santa Monica, CA, 2001, pp. 533–557.

the Warsaw Pact. It was signed in Helsinki, Finland, on March 24, 1992, and put in force on January 1, 2002, after being adopted by 26 nations. The treaty is the most wide-ranging international effort to date to promote openness and transparency of military forces and activities.

The treaty enables all participants, regardless of size, a direct role in gathering information about military forces and activities of concern to them. All Open Skies

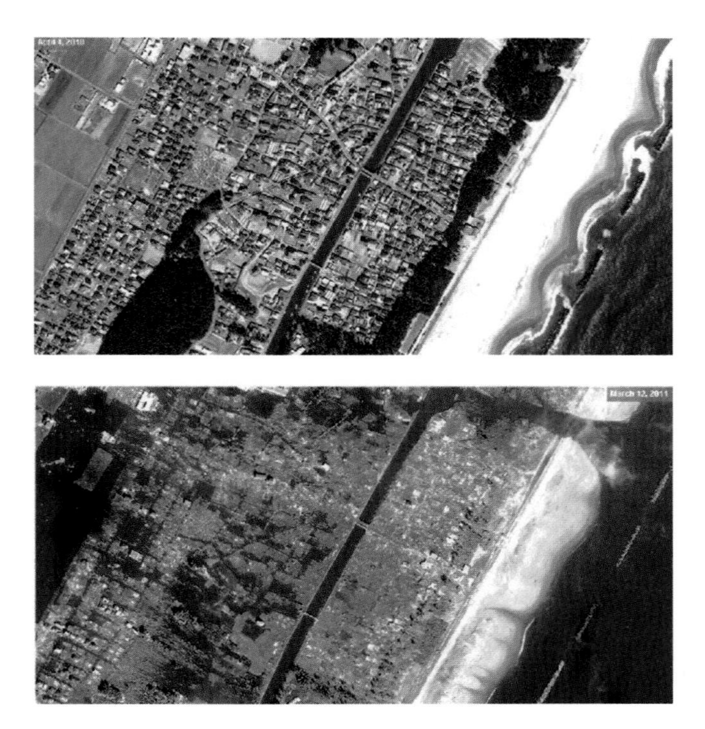

**FIGURE 1.6** GeoEye images acquired before and after the Japanese tsunami of 2011. The images were displayed by the *New York Times* web page. (From http://www.nytimes.com/interactive/2011/03/13/world/asia/satellite-photos-japan-before-and-after-tsunami.html?module=Search&mabReward=relbias%3Ar%2C{%222%22%3A%22RI%3A12%22}.)

Treaty aircraft and sensors must pass specific certification procedures and only treaty-permitted sensors can be installed and launched. Open Skies aircraft may have video, optical panoramic, and framing cameras for daylight photography, infrared line scanners for day/night capability, and synthetic aperture radar for day/night all-weather capability. Collected imagery from Open Skies missions is made available to any participant state willing to pay the costs of reproduction. The treaty provides that at the request of any participant state, the observing state will provide it a copy of the data collected during a mission over the observed state.

The United Nations Committee on the Peaceful Uses of Outer Space (part of the UN Office for Outer Space Affairs [UNOOSA]) is the primary international forum for the development of laws and principles governing outer space. One of its primary roles has been to develop basic legal principles on space observations to avoid tensions between observer and observed nations. Many developing countries have expressed the need to control the distribution of satellite images taken over their territory. Some countries, such as Brazil and Argentina, believe that the nation that owns a satellite system must first ask permission to remotely observe another country and that by no means should the data be handed over to third countries. In 1982, at the United Nations UNISPACE Congress in Vienna, an agreement was

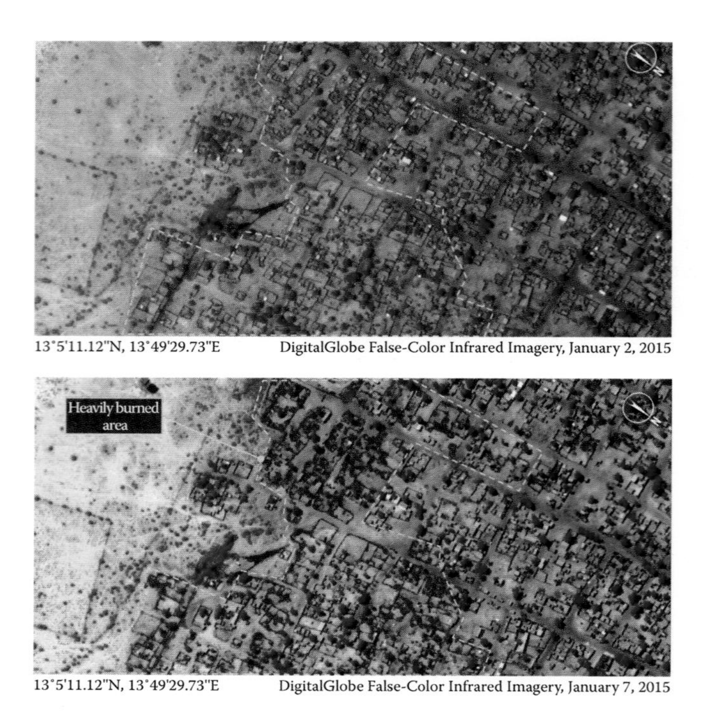


**FIGURE 1.7** Images acquired by Digital Globe satellites showing the impacts of guerrilla attacks on civilians in the north of Nigeria. (Acquired for Amnesty International. http://blog.amnestyusa.org/africa/the-story-behind-the-nigeria-satellite-images/, last access November 2015.)

reached to facilitate the unrestricted access to imagery obtained over each nation and to require authorization from a country before disclosing imagery acquired within it to third countries (O.T.A. 1984). Russia and India proposed a 50 m spatial resolution limit of space-based observations to preserve sovereign military security, while the United States, the United Kingdom, and Japan were opposed to any restrictions whatsoever on data observation and distribution.

The United Nations sponsored the Treaty on Principles Governing the Activities of States in the Exploration and Use of Outer Space, including the Moon and Other Celestial Bodies, also known as the Outer Space Treaty, which was adopted by the General Assembly and entered into force on October 10, 1967. Ninety-nine states have ratified, and an additional twenty-six have signed this treaty as of January 2008. The Outer Space Treaty provides the basic framework on international space law and guarantees that exploration and use of outer space shall be carried out for the benefit and in the interests of all countries. Additional principles governing the use of remote sensing included in this treaty are as follows:

- Remote sensing shall promote the protection of Earth's natural environment.
- Remote sensing shall promote the protection of humankind from natural disasters.

- The observed state shall have access to all primary and processed data acquired over its territory on a nondiscriminatory basis and on reasonable cost terms.
- The observed state shall also have access to the available analyzed information concerning the territory under its jurisdiction, taking particularly into account the needs and interests of the developing countries.
- States carrying out remote sensing activities shall promote international cooperation in these activities. To this end, they shall make available to other states opportunities for participation therein. Such participation shall be based in each case on equitable and mutually acceptable terms.
- Furthermore, a state must inform both the Secretary-General of the United Nations, as well as the interested nations that request it, of the remote sensing programs that will be developed.

The growth and expansion of remote sensing activities to many different countries and the increasing role of the commercial sector have lowered international concerns considerably. However, transferring space technology to developing countries has not been very effective, an unfortunate consequence considering that they are most in need of information about their natural resources. During the *Third U.N. Conference* (*UNISPACE III*) in Vienna in 1999, there were still concerns on the part of some countries (e.g., India and Israel) about the availability of remote sensing data over their territories; however, the main concerns were in reducing the costs of satellite imagery and not in controlling data availability (Florini and Dehqanzada 2001).

International coordination of satellite remote sensing is done through a dedicated committee (the Committee on Earth Observation Satellites [CEOS]: http://www.ceos.org), which promotes exchange of data to optimize societal impacts of EO missions. Working groups within CEOS include one dedicated to quick assessment of natural hazards through a unified system of space data acquisition and delivery to countries affected by major disasters (International Charter on Space and Major Disasters). Other active groups are the calibration and validation, information system and services, and coordination of satellite constellations.

## 1.4 BENEFITS OF ENVIRONMENTAL MONITORING FROM SATELLITE SENSORS

Satellite remote sensing has several advantages over other conventional methods of environmental sampling and monitoring, such as in situ field measurements or ground sensors. Spaceborne sensors provide a global view over the Earth's surface, at periodic times and including nonvisible regions of the electromagnetic spectrum. Data are accessible quickly and can be easily connected to other spatial databases, thus making a suitable information source for spatial data integration. More details on these characteristics follow.

### 1.4.1 Global Coverage

Satellite data are acquired by a platform that has a stable orbit around the Earth. Therefore, EO sensors make it possible to acquire consistent and repetitive imagery of the entire planet, including areas that are fairly remote and normally inaccessible, such as polar, mountain, desert, and forest areas. For example, remote sensing has been found to be highly useful in mapping and monitoring ice sheets, a particularly relevant topic for water mass studies in the context of Arctic warming (Figure 1.8). Other examples include the use of remote sensing in detecting remote forest fires, observing uncontrolled oil spills, and assessing damage from tsunamis or floods.

The global coverage provided by satellites is particularly useful in monitoring and understanding the dynamic processes affecting our environment. There is much concern over the many stresses placed on the environment, such as global warming, reduction in biologic diversity, depletion of freshwater, and land degradation and desertification. Tropical glaciers in the Peruvian Andes and the famous snows of Mount Kilimanjaro in East Africa are retreating rapidly and are predicted to disappear over the next two decades. By some estimates, as much as 40% of the Earth's land surface has been permanently transformed by human action, with significant consequences for biodiversity, nutrient cycling, soils, and climate.

Many of these environmental issues are most effectively addressed with a holistic planetary approach, for which global datasets, models, and information systems are needed. Assembling global databases is extremely difficult and time consuming and is often plagued by disparate data sources compiled with different criteria and formats. Country- or region-specific datasets must be merged and compiled to generate complete global datasets and images of the areas under study (Unninayar 1988).

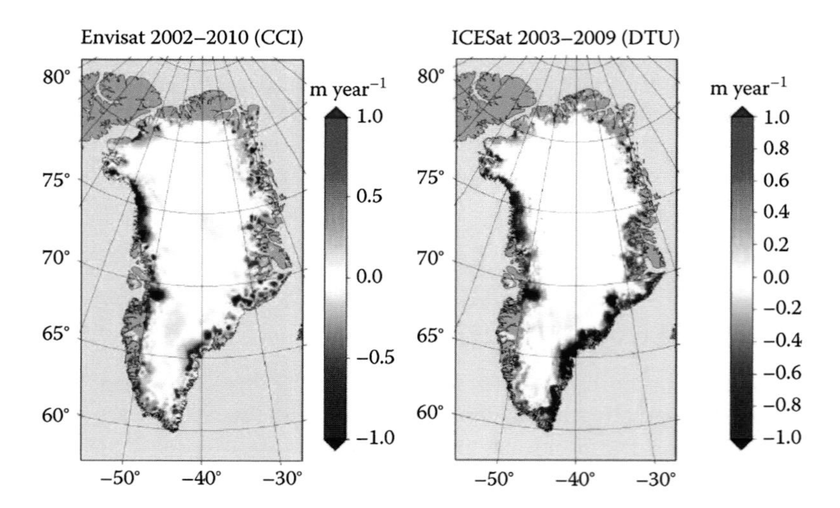


FIGURE 1.8 Overall changes in the ice sheet of Greenland as observed by Envisat and the ICESat satellites. (Courtesy of Rene Forsberg.)

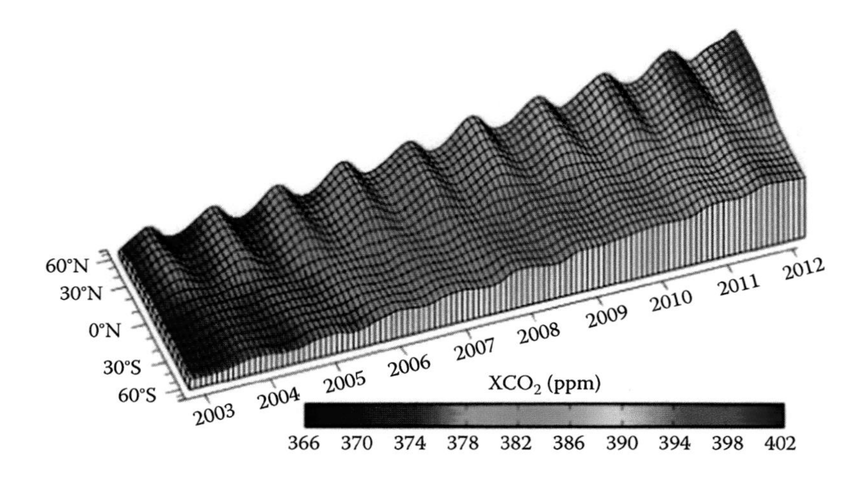


**FIGURE 1.9** Global trends in $CO_2$ concentrations as measured by the SCIAMACHY sensor on board the Envisat. (Courtesy of Michael Buchwitz, University of Bremen, Bremen, Germany, http://www.esa-ghg-cci.org/?q=node/115.)

EO satellite offers a quick and consistent source of uniformly collected data, from the same instrument and platform, with complete coverage of the planet for global studies.

The Global Climate Observing System (GCOS) program (http://www.wmo.int/pages/prog/gcos/), part of the World Meteorological Organization (WMO), UNESCO, ICSU, and the UN Environmental Agency, has defined a set of 50 Essential Climate Variables that can be retrieved and monitored by remote sensing systems. They include atmospheric (over land, sea, and ice), oceanic, and terrestrial parameters. The most actively monitored are temperature, water vapor, precipitation, radiation budget, aerosols, greenhouse gases (Figure 1.9), ozone, sea level, sea ice, ocean color, snow cover, glaciers, ice sheets, albedo, land cover, fire disturbance, soil moisture, and leaf area index.

### 1.4.2 Synoptic View

Satellite sensors are located far from the Earth's surface, observing large areas and thus providing a synoptic view of landscape features. A standard 1:18,000 aerial photography covers an area of approximately 16 km$^2$. At a 1:30,000 scale, the area coverage increases up to 49 km$^2$. By comparison, a single Landsat TM image captures 34,000 km$^2$ and NOAA's Advanced Very High Resolution Radiometer (AVHRR) image covers up to 9 million km$^2$. Satellite images are able to observe and depict phenomena that would be nearly impossible using the very local perspective of aerial photos. Large geologic features such as faults, fractures, and lithological contacts are more easily detected using satellite imagery, which can help locate mineral resources (Short and Blair 1986). It is also worth noting that satellite images cover such vast areas in a very short time and therefore the data are comparable throughout space, whereas conventional aerial photography requires mosaicking images acquired at different flight times.

### 1.4.3 Multiscale Observations

Satellite-based sensors, as will be discussed in more detail later, have a wide range of orbital altitudes, optics, and acquisition techniques. Consequently, the imagery acquired can be at very fine resolutions (fine level of detail) of 1 m or less with very narrow coverage swaths, or the images may have much larger swaths and cover entire continents at very coarse resolutions (>1 km) (Figure 1.10).

The management and monitoring of Earth's natural resources require spatial data at various scales, because the impacts of driving factors are often scale dependent (Ehleringer and Field 1993). For example, soil moisture may be more strongly related to local topography over short distances and to regional precipitation patterns over longer distances. Remote sensing may be the only feasible means of providing spatially distributed data such as land use patterns, topography, and seasonal hydrologic and vegetation parameters at multiple scales and on consistent and timely bases.

**FIGURE 1.10** Satellite images from the U.S. East Coast acquired by different sensors. (Images extracted from Google Earth.)

### 1.4.4 Observations over the Nonvisible Regions of the Spectrum

Satellite sensors are capable of acquiring data over various portions of the electromagnetic spectrum that cannot be sensed by the human eye or conventional photography. The ultraviolet, near-infrared, shortwave infrared, thermal infrared, and microwave portions of the spectrum provide valuable information over critical environmental variables. For example, the thermal infrared portion of the spectrum allows us to study the spatial distribution of sea surface temperatures and marine currents, as well as water stress in crops. The ultraviolet radiation is critical to monitor the ozone layer. The middle infrared region is ideal to sense $CO_2$ concentrations and the microwave radiation is more sensitive to estimate soil moisture and snow cover.

### 1.4.5 Repeat Observation

The orbital characteristics of most satellite sensors enable repetitive coverage of the same area of Earth's surface on a regular basis with a uniform method of observation. The repeat cycle of the various satellite sensor systems varies from 15 min to nearly a month. This characteristic makes remote sensing ideal for multitemporal studies, from seasonal observations over an annual growing season to interannual observations depicting land surface changes (see Section 7.3). Such periodic observations are vital given the highly dynamic nature of many environmental phenomena. There are many examples of multitemporal applications of remote sensing, including desertification studies (Figure 1.11), drought and flooding patterns, snow cover melting, deforestation assessments, and meteorological phenomena.

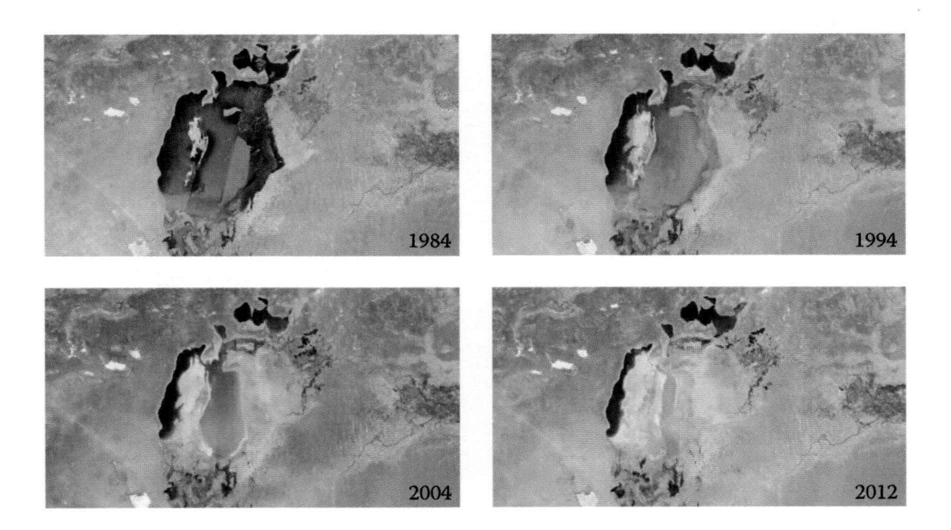


FIGURE 1.11 Multitemporal Landsat images acquired over the Aral Sea in Kazakhstan. (Images available from https://earthengine.google.org.)

### 1.4.6 Immediate Transmission

Nowadays, all remote sensing systems record image data in digital format, facilitating a real-time transmission to the ground receiving station and eventually to the end user. This is particularly relevant when dealing with disasters and natural hazards that require quick access to imagery (as it is the case of the International Disaster Charter: https://www.disasterscharter.org/web/guest/home).

Traditionally, only meteorological sensors had direct transmission to the end user, as the signal was not codified and could be acquired by a relatively cheap receiving antenna. Currently, images from global sensors can still be acquired freely, in the case of MODIS, for instance, using a more sophisticated antenna. In addition, the Land Rapid Response system provides near-real-time data of some products, such as the active fire information that are received by email within hours of satellite acquisition. For most medium- and high-spatial-resolution sensors, images can be obtained fairly quickly using fast Internet connections, but ordinarily not yet in real time as the signal is usually coded. These sensor systems transmit images when they are within the coverage area of the antenna or otherwise record onboard for later transmission. The user obtains the image with a certain time delay, due to calibration and preprocessing of the data.

### 1.4.7 Digital Format

As most images are now in digital format, the integration of satellite-derived information with other sources of spatial data is relatively straightforward. Computer-assisted visualizations of image data are also possible, as in the generation of 3D views by combining satellite imagery with digital elevation models (Figure 1.12). These views can be created from different angles and under different simulated conditions.

FIGURE 1.12 The Strait of Gibraltar in three dimensions, combining terrain data retrieved from the SRTM (see Chapter 3) mission with a pseudo-natural color image acquired by the Landsat TM in 1987. (From http://earthobservatory.nasa.gov/.)

In summary, there are many advantages to the use of remote sensing data for the study and monitoring of Earth's landscapes. However, this does not imply that EO satellites can be used to retrieve any environmental variable. Remote sensing also has significant limitations related to the available spatial, spectral, and temporal resolutions. The images may not be recorded at the required spatial detail or not frequent enough to cope with the user requirements. Furthermore, persistent cloud cover may notably reduce useful observations of optical sensors, thus severely restricting observations in cloudy areas. Radar observations are an alternative as they are independent of cloud coverage, but they are much less frequent than optical observations.

On the other hand, current remote sensing techniques may not be sensitive enough to detect the variable of interest. For instance, subsurface soil moisture or deep water temperature is unlikely sensed by EO systems, which receive radiation mostly from the upper layers of the surface.

In summary, EO satellite should be viewed as a complementary tool to other natural resource and environmental monitoring techniques, such as soil or plant samples, meteorological sensors, or flux towers, which in any case are useful to validate estimations provided by satellite instruments.

## 1.5 SOURCES OF INFORMATION ON REMOTE SENSING DATA

Although remote sensing is fairly new, much has been written on this subject, including papers read at numerous symposia and conferences, scientific journals, manuals, and technical reports. There are various reference guides and series periodicals dedicated to this field, including *Earth Resources: A Continuing Bibliography with Indexes*, a series published by NASA, and *Geographical Abstracts, G: Remote Sensing, Photogrammetry and Cartography*, published by Geo Abstracts. Also, there are reference databases specializing in remote sensing, such as the Canadian Remote Sensing Online Retrieval System (RESORS) and the English GEOBASE.

Some of the most important congresses are those organized by the professional societies dedicated to the development of remote sensing. These organizations include the International Society for Photogrammetry and Remote Sensing (ISPRS), the IEEE Geoscience and Remote Sensing Society (GRSS), the American Society of Photogrammetry and Remote Sensing (ASPRS), the Remote Sensing and Photogrammetry Society (RSPS), the International Society for Optical Engineering (SPIE), and the European Association of Remote Sensing Laboratories (EARSeL).

The main scientific journals dealing with EO studies ranked by the impact factor are the following: *Remote Sensing of Environment*, *IEEE Transactions on Geoscience and Remote Sensing*, *ISPRS Journal of Photogrammetry and Remote Sensing*, *IJ Applied Earth Observation and Geoinformation*, *Remote Sensing*, *IEEE Applied Earth Observations and Remote Sensing*, *Photogrammetric Engineering and Remote Sensing*, *IEEE Geoscience and Remote Sensing Letters*, *International Journal of Remote Sensing*, *Remote Sensing Letters*, *Canadian Journal of Remote Sensing*, *GIScience & Remote Sensing*, *The Photogrammetric Record*, and *Journal of Applied Remote Sensing*. There are a large number of textbooks dedicated to remote sensing, spanning a range of orientations and scopes. These can be classified in two broad groups: those intended for a general audience and those written for specific application fields (hydrology, oceanography, land use, geology, etc.).

The first group of books can be further classified into those that provide coverage of all the main topics of remote sensing acquisition and interpretation, or those books that focus on specific topics, such as land cover classification, or the physical principles of image acquisition. The general textbook group is the most common (Asrar 1989; Barret and Curtis 1999; Campbell and Wyme 2011; Chen 1985; Conway 1997; Couzy 1981; Cracknell and Hayes 1991; Curran 1985; Danson and Plummer 1995; Drury 1998; Elachi 1987; Estes and Lenger 1974; Gibson and Power 2000a,b; Harper 1983; Harris 1987; Holz 1973; Hord 1986; Howard 1991; Jensen 1996, 2000; Lillesand and Kiefer 2000; Lira 1987; Lo 1986; Mather 1998; McCloy 1995; Quattrochi and Goodchild 1997; Rees 1999; Richards and Xia 1999; Robin 1998; Schanda 1976; Schowengerdt 2007; Short 1982; Slater 1980; Swain and Davis 1978; Szekielda 1988; Thomas et al. 1987; Townshend 1981; Verbyla 1995; Weng 2012; Wilkie and Finn 1996). Special categories of general textbooks are those centered on specific sensors, such as radar (Allan 1983; Henderson and Lewis 1998; Trevett 1986) or lidar systems (Beraldin et al. 2010; Fujii and Fukuchi 2005; Vosselman and Klein 2010; Vosselman and Maas 2010).

Another group of textbooks are those oriented toward specific applications: urban areas (Au 1993; Lulla 1993), pedology and geology (Drury 1998; Mulders 1987; Rencz and Ryerson 1999; Short and Blair 1986; Metternicht 2008), hydrology (Gower 1994; Hall and Martinec 1985; Robinson 1985), oceanography (Gower 1994; Robinson 1985), climatology (Barret 1974; Carleton 1991; Conway 1997), natural vegetation (Achard and Hansen 2012; Franklin 2001; Frohn 1998; Hobbs and Mooney 1990; Howard 1991), forest fires (Ahern et al. 2001; Chuvieco 2003, 2009), archaeology (Lasaponara and Masini 2012), and global change studies (Chuvieco 2008; Chuvieco et al. 2010; Purkis and Klemas 2011).

Access to satellite images is becoming easier, largely due to wide distribution through the Internet. For images acquired by NASA or other American programs, the USGS EarthExplorer (https://earthexplorer.usgs.gov/) and GloVis (http://glovis.usgs.gov/) (Figure 1.13) servers are the most accessible portals to download EO images. Many other space agencies and companies also provide free data, including the European Space Agency and the Japanese Space Agency, JAXA.

Many organizations provide free, downloadable versions of images in standard formats. Web searches provide the best guide, as the number and location of sites are evolving rapidly. At the time of writing this chapter (December 2014), the most interesting references for accessing visual satellite data are NASA Visible Earth (http://visibleearth.nasa.gov/), which includes sample images from all NASA missions, and the Google Earth Engine database (https://earthengine.google.org/#intro), which includes a full Landsat historical archive to observe changes in the Earth landscapes for the last 30 years. A few years earlier from this initiative, the UN Environmental Program compiled a change detection atlas of Earth landscapes (UNEP: http://na.unep.net/atlas/).

Several books have been published in the last two decades compiling satellite images of different terrestrial landscapes. They provide an interesting source of information on the potentials of EO data, as well as a historical reference for

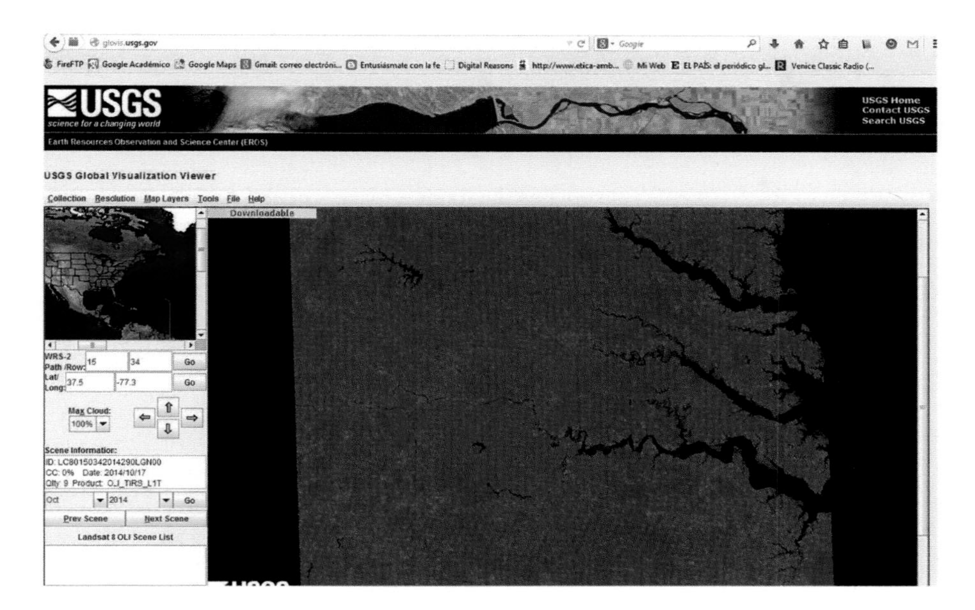


FIGURE 1.13 Visualization tool of GloVis to download images from different Earth observation sensor systems. (From http://glovis.usgs.gov/.)

analyzing land changes. Among the most relevant works are Mission to Earth (NASA 1976), Earth Watch (Sheffield 1981), Man on Earth (Sheffield 1983), Images of Earth (Francis and Jones 1984), Looking at Earth (Strain and Engle 1993), and the Satellite Atlas of the World (National Geographic 1999). Some may be now outdated.

## 1.6 REVIEW QUESTIONS

1. Identify a component that is not required for a remote sensing system:
   a. Source of energy
   b. Image processing equipment
   c. Platform
   d. Sensor
2. Which of these properties favors satellite versus airborne remote sensing?
   a. Spatial resolution
   b. Cost
   c. Temporal resolution
   d. Stereoscopic view
3. Which of these properties favors airborne over satellite remote sensing?
   a. Spatial resolution
   b. Spectral resolution
   c. Temporal resolution
   d. Stereoscopic view

4. Which of the following satellite programs is managed by a private company?
   a. Landsat
   b. GeoEye
   c. SPOT
   d. IRS
5. Which of the following sensors has the longest historical archive?
   a. NOAA-AVHRR
   b. Terra MODIS
   c. SPOT HRV
   d. ERS SAR

# 2 Physical Principles of Remote Sensing

Remote sensing has a strong physical basis, as it implies collecting electromagnetic (EM) signals coming from objects with different physical and chemical properties. This chapter explores the physical processes that make this interaction possible. Since the primary audience of this book comprises environmental scientists, we will just present the basis of these principles, focusing on how EM energy interacts with both the Earth's surface and atmosphere. Several textbooks have been written for those interested in more detailed physical explanation (Elachi 1987; Slater 1980), which would unnecessarily complicate the presentation of this introductory text. A better understanding of EM energy interactions with the Earth's surface will greatly benefit the interpretation of remote sensing imagery and the effective use of the acquired data for retrieving environmental information.

## 2.1 FUNDAMENTALS OF REMOTE SENSING SIGNALS

Remote sensing means that something is sensed remotely, that is, from a certain distance. This implies that there is a certain interaction between the object and the sensor that captures the former. For example, our eyes are sensors that can see a tree by deciphering the visible radiant energy that is reflected by the tree, by reflecting either sunlight or another illumination source. Our eyes would not be able to sense or see the tree in total darkness, since they are not sensitive to other sources of energy coming from the tree, such as thermal radiation. Both reflected and emitted energy from the tree are closely linked to its chemical, biological, and physical properties; for instance, the number and position of its leaves and branches, or their pigment or water content. These components impact different types of EM energy. For instance, the pigment status affects the blue and red regions of the visible (VIS) spectrum, while water content has more impacts on the shortwave infrared (SWIR) and thermal infrared (TIR) bands. Similarly, reflected or emitted radiation from the Earth's surface provides critical information on the properties of soils, ice, snow, water, vegetation, and rocks. The goal of remote sensing is to understand how EM energy interacts with the surface so that we may better extract relevant information from the images.

When we observe a tree, our eyes are sensitive only to the light the tree reflects. We can also use artificial sensors that are able to detect other sources of energy; for instance, thermal cameras that detect plant temperature. We could also use sensors with their own source of energy, such as laser or microwave (MW) pulses, which *illuminate* the tree and detect the return energy afterward. Thus, in remote sensing, we may consider three ways of *sensing* information about an object: by reflection, by emission, and by combined emission–reflection (Figure 2.1). The first one is the most

**FIGURE 2.1** Main types of radiation processes in remote sensing: (i) reflection, (ii) emission, and (iii) emission–reflection.

common because it utilizes sunlight, the main source of energy on Earth. The Sun illuminates the surface, which in turn reflects a portion of this energy back to space depending on the type and composition of cover present on the surface. The reflected EM energy is detected by the satellite sensor, which then records and transmits this signal to a receiving station.

Remote sensing observations may also be based on the emitted energy from the Earth's surface (Figure 2.1, ii), wherein the sensor detects energy coming from the surface itself. Since it does not directly depend on the Sun, this observation can be performed both during day and night. All objects warmer than absolute zero (0 K) emit energy, and the hotter the object, the higher the radiant energy it produces. The Sun is the hottest object, and it emits very large amounts of energy. Warm objects (a fire, lava, hot water) emit energy at longer wavelengths that we can sense with our skin, but not with our eyes.

Finally, we can also base remote sensing observations on active sensors, so named as they have their own source of energy. They are able to send pulses to the target objects and record later their reflection to characterize those objects. The most common active technique in the early days of remote sensing was radio detection and ranging (radar), working with MW energy, but in the last decades light detection and ranging (lidar) sensors have also become very popular. They work with polarized VIS or near-infrared (NIR) light and are extensively used these days to measure distances.

We will review how these different sensors work in the next chapter. This one focuses on how the signal is generated and interacts with atmospheric components and surface covers in its way down to the Earth and reflected/emitted up to the sensor.

The properties of EM radiation may be explained by two seemingly contradictory theories of light: wave theory (Huygens, Maxwell) and quantum theory (Planck, Einstein). According to wave theory, EM radiation is a form of energy derived from oscillating magnetic and electrostatic fields that are mutually orthogonal to each other and to the direction of propagation (Figure 2.2). EM energy is transmitted from one place to another following a harmonic and continuous model with a constant velocity, $c = 3 \times 10^8$ m s$^{-1}$ (speed of light). The properties of this energy can be described according to its wavelength ($\lambda$) and frequency ($\nu$), which are related by

$$c = \lambda \nu \tag{2.1}$$

where

$\lambda$ is the wavelength or distance between two successive peaks (usually in micrometers, 1 μm = $10^{-6}$ m; or nanometers, 1 nm = $10^{-9}$ m)

$\nu$ is the frequency, or number of cycles that pass over a fixed point per unit of time (in hertz, cycles s$^{-1}$)

As Equation 2.1 shows, the frequency of light is inversely proportional to its wavelength, such that the greater the wavelength, the smaller the frequency and vice versa.

The quantum theory of light describes radiation as a succession of discrete packets of energy known as photons or quanta, with mass equal to zero. The amount of energy transported by a photon is proportional to its frequency:

$$Q = h\nu \tag{2.2}$$

where

$Q$ is the radiant energy of a photon (in joules, J)

$\nu$ is the frequency

$h$ is Planck's constant ($6.626 \times 10^{-34}$ J s)

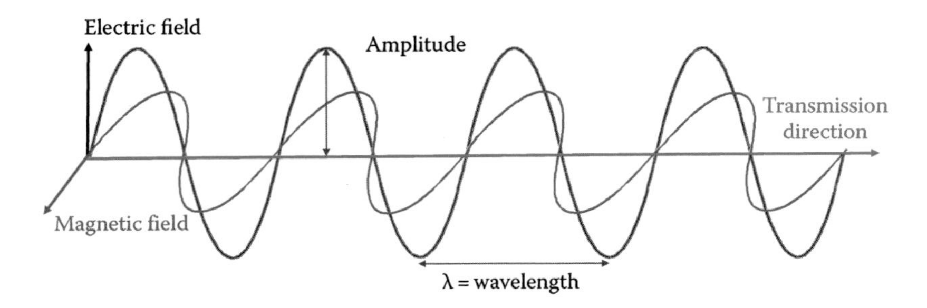


**FIGURE 2.2** The oscillating electric and magnetic components of electromagnetic radiation propagation (the wave theory of light).

Combining this with Equation 2.1 results in

$$Q = h\left(\frac{c}{\lambda}\right) \tag{2.3}$$

And, therefore, the greater the wavelength, or the smaller the frequency, the lower the energy content and vice versa. This implies that it is more difficult to detect longer than shortwave energy radiations, as the former have lower energy and require more sensitive means of detection.

## 2.2 ELECTROMAGNETIC SPECTRUM

Since radiation sources are very diverse and therefore EM radiations vary from very small to very long wavelengths, most textbooks tend to classify them in certain groups of wavelengths of frequencies that are finally organized in the so-called EM spectrum (Figure 2.3). It includes a continuous range of wavelengths or frequencies, but commonly several spectral regions or bands are identified, with particular radiation properties. Although most spectral bands are referred to in length units, MWs are commonly expressed in frequency units (gigahertz, GHz = $10^9$ Hz).

The shortest wavelengths, with the highest radiation energy, are gamma rays and x-rays, whose wavelengths range from $10^{-9}$ to $10^{-3}$ μm (or $10^{-15}$ to $10^{-9}$ m) and which are commonly used in astronomical observation and medical applications, respectively. The longest wavelengths are used for telecommunications, radio, and television, with wavelengths in the range of $10^8$–$10^{10}$ μm (or 100–10,000 m).

The spectral regions most commonly used in remote sensing observation are the following:

1. The VIS region (0.4–0.7 μm). It covers the spectral wavelengths that our eyes are capable of sensing and at which the Sun's energy is the highest. The VIS region can be further divided into the three primary colors: blue (0.4–0.5 μm), green (0.5–0.6 μm), and red (0.6–0.7 μm) (Figure 2.3).

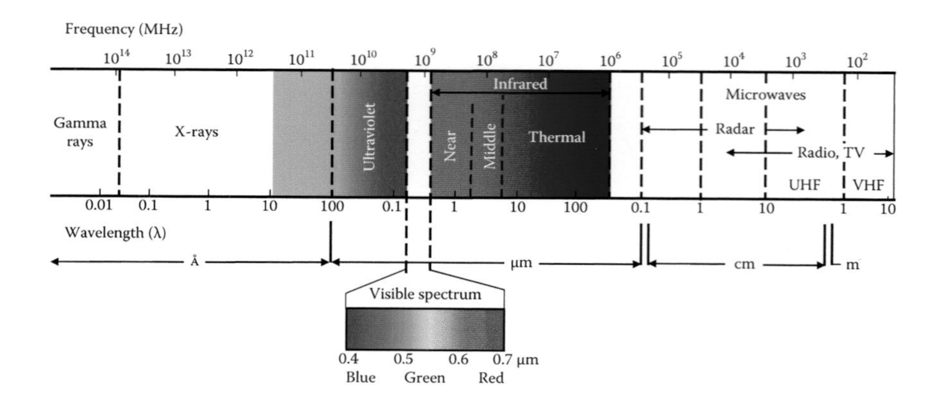


FIGURE 2.3 Major spectral bands within the electromagnetic spectrum.

2. The NIR region (0.7–1.2 μm). This portion of the spectrum lies just beyond the human eye's perception capability and is sometimes known as the reflective infrared or photographic infrared because part of this spectral region (0.7–0.9 μm) can be detected with special films. The NIR is of special interest because of its sensitivity to determine plant health status.
3. The mid-infrared region (MIR, 1.2–8 μm). This spectral region lies between the NIR and TIR regions. From 1.2 to 2.5 μm, the influence of the Sun's energy is still very relevant, and this band is commonly referred to as the SWIR region. This region provides the best estimations of the moisture content of soil and vegetation. From 3 to 8 μm, the signal becomes a continuous mixture of solar-reflected and a surface-emitted energy, becoming the more relevant emitted component as the wavelengths become longer. The 3–5 μm interval is particularly useful for detecting high-temperature sources, such as volcanoes or forest fires.
4. The thermal infrared region (TIR from 8 to 14 μm). This is the emitted energy from the Earth's surface that is commonly used to map surface temperatures. The thermal region has been widely used in detecting vegetation evapotranspiration (ET), ice and cloud properties, urban heat effects, and rock discrimination.
5. The microwave (MW) region, for radiations larger than 1 cm. This spectral region is where the imaging radar systems work. Its main advantage is the very low atmospheric absorption, which enables us to "see" through clouds. MW radiation can also penetrate forest canopies to various depths and are very useful in soil moisture and surface roughness analyses.

The EM energy signals received by a sensor across these different spectral regions vary with land cover type and with the biophysical and biochemical properties of the surface components. In the following sections, we examine how the various components of the terrestrial surface behave over these spectral regions. However, first we introduce some of the basic energy concepts and units of measurement used in remote sensing in order to gain a better understanding of the characteristic properties of spaceborne measurements.

## 2.3 TERMS AND UNITS OF MEASUREMENT

As mentioned earlier, a remote observation requires an energy source and a sensor that can detect the EM energy leaving the Earth's surface toward the field of view of the sensor. The EM energy of interest has a certain intensity, spectral composition, and direction (i.e., the energy may be directed toward or away from the surface). Here we describe the units commonly used in remote sensing applications. The precise formulas for each of these energy terms are included in Table 2.1 (Curran 1985; Elachi 1987; Rees 1999; Slater 1980):

1. Radiant energy ($Q$), measured in joules (J), is the most basic energy unit and refers to the total energy radiated in all directions away or toward a surface.

**TABLE 2.1**
**Radiometric Quantities Commonly Used in Remote Sensing**

| Concept | Symbol | Equation | Measured Unit |
|---|---|---|---|
| Radiant energy | $Q$ | — | Joules (J) |
| Radiant flux | $\phi$ | $\delta\phi/\delta t$ | Watts (W) |
| Exitance | $M$ | $\delta\phi/\delta A$ | W m$^{-2}$ |
| Irradiance | $E$ | $\delta\phi/\delta A$ | W m$^{-2}$ |
| Radiant intensity | $I$ | $\delta\phi/\delta\Omega$ | W sr$^{-1}$ |
| Radiance | $L$ | $\delta I/\delta A \cos\theta$ | W m$^{-2}$ sr$^{-1}$ |
| Spectral radiance | $L_\lambda$ | $\delta L/\delta_\lambda$ | W m$^{-2}$ sr$^{-1}$μm$^{-1}$ |
| Emissivity | $\varepsilon$ | $M/M_n$ | Unitless |
| Reflectance | $\rho$ | $\phi_r/\phi_i$ | Unitless |
| Absorptance | $\alpha$ | $\phi_a/\phi_i$ | Unitless |
| Transmittance | $\tau$ | $\phi_t/\phi_i$ | Unitless |

*Note:* sr, steradian, measure of the solid angle; μm, micrometer or micron ($10^{-6}$ m); $M_n$, exitance of a black body at the same temperature; $\phi_i$, incident flux; $\phi_r$, reflected flux; $\phi_a$, absorbed flux; $\phi_t$, transmitted flux; $\theta$, angle formed by the energy flux direction and the normal.

2. Radiant flux ($\phi$), measured in watts (W), is the number of joules per second (J s$^{-1}$ = W) and represents the rate of energy transfer in all directions per unit of time.
3. Radiant flux density is the rate of energy transfer per unit area measured in watts per square meter (W m$^{-2}$).
4. Radiant exitance or emittance ($M$) is the radiant flux density *leaving the surface* in all directions per unit area and per unit time (W m$^{-2}$).
5. Radiant irradiance ($E$) is the radiant flux density *incident upon* the surface per unit area and per unit time (W m$^{-2}$). It is the same concept as the radiant exitance, but in this case it refers to the energy arriving at the surface rather than leaving the surface.
6. Radiant intensity ($I$) is the total energy leaving the surface per unit time and within a unit solid angle ($\Omega$). The solid angle is a 3D angle that refers to the area of transmitted energy that a surface subtends and is measured in steradians (Figure 2.4). Radiant intensity is thus measured in watts per steradian (W sr$^{-1}$).
7. Radiance ($L$) is the total energy exiting in a certain direction per unit area and solid angle of measurement. It is the most fundamental term in remote sensing since it describes exactly what the sensor measures. Radiance is expressed in watts per square meter per steradian (W m$^{-2}$ sr$^{-1}$).

The foregoing energy terms may also be expressed in terms of wavelength basis and have the prefix *spectral* applied to them, such as spectral radiance or spectral irradiance. For example, the term *spectral radiance*, $L_\lambda$, refers to the energy output from a unit area, unit solid angle, and unit wavelength, or $L_\lambda = \text{W m}^{-2}\text{ sr}^{-1}\,\lambda^{-1}$. Similarly, *spectral irradiance*, $E_\lambda$, refers to the energy incident upon a surface per unit wavelength.

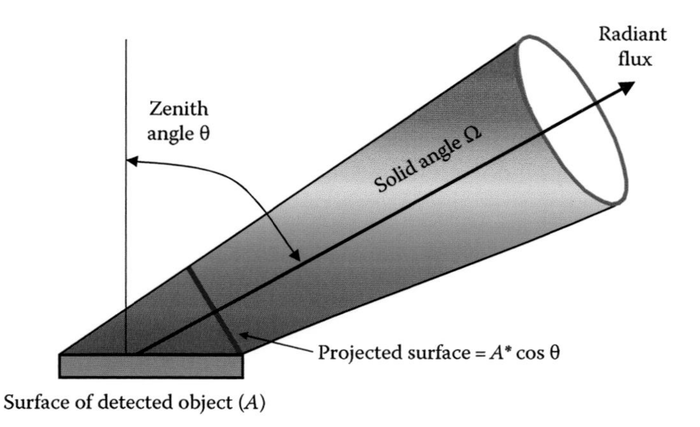


FIGURE 2.4 Solid angle and radiant energy changes with zenith angle.

There is also a series of dimensionless energy terms, varying from 0 to 1, that are widely used in characterizing the spectral properties of the Earth's surface:

8. Emissivity (ε): This is the relationship between the radiant exitance of a surface ($M$) relative to that of a perfect emitter at the same temperature ($M_n$). A perfect emitter is also known as a blackbody and has emissivity 1. Natural materials, on the other hand, are imperfect emitters with emissivity values ranging from 0 to <1. Emissivity values over different wavelengths are useful in characterizing materials.
9. Reflectance (ρ): This is the relationship between the energy reflected by a surface and the energy incident upon that surface.
10. Absorptance (α): This is the relationship between the energy absorbed by the surface and the energy incident upon that surface.
11. Transmittance (τ): This is the relationship between the energy transmitted through a surface and the energy incident upon that surface.

These unitless terms can also have *spectral* added to them, as in spectral reflectance. There are a few useful relationships that are derived from the aforementioned energy terms. The term *albedo* is the ratio of all outgoing energy to the incident energy for a given surface area. More specifically, albedo is the ratio of exitance ($M$) to irradiance ($E$) over all solar reflective, or shortwave, wavelengths and is the equivalent of hemispherical reflectance, that is, the reflectance integrated over all directions (Equation 2.4). Albedo is a fundamental variable in energy balance studies, climate modeling, and soil degradation studies. Spectral albedo refers to the exitance divided by the irradiance for a specific spectral band (Equation 2.5):

$$\text{Albedo} = \rho_{\text{hemispherical}} = \frac{M}{E} \tag{2.4}$$

$$\text{Spectral albedo} = \rho_{\text{hemispherical},\lambda} = \frac{M_\lambda}{E_\lambda} \tag{2.5}$$

It is important to note that a satellite sensor does not measure outgoing, hemispherical energy over all directions, but instead measures the directional radiance ($L_\lambda$) from only over a narrow angular field of view (Figure 2.4). The spectral directional radiance ($L_\lambda$) is related to the hemispherical spectral exitance ($M_\lambda$) as follows (Slater 1980):

$$M_\lambda = \pi L_\lambda \tag{2.6}$$

Similarly, the surface reflectances derived from satellite measurements are directional reflectances, as they refer to a specific measurement geometry between the satellite sensor and the Sun, relative to the surface. The relationship between the spectral radiance values received by a satellite sensor ($L_{sen,\lambda}$) and the surface spectral reflectance ($\rho_\lambda$) becomes

$$\rho_\lambda = \frac{\pi L_\lambda}{E_{0,\lambda}} \tag{2.7}$$

where $E_{0,\lambda}$ is the solar irradiance arriving at the surface. We will further comment in Section 6.7.3.4 the impact of varying the observation geometry on the retrieval of ground reflectance.

## 2.4 ELECTROMAGNETIC RADIATION LAWS

There is a set of physical laws that govern the behavior and characteristics of EM radiation. We saw in Equation 2.3 that the energy content of EM radiation varies inversely with wavelength. The spectral distribution of EM radiation emitted by a blackbody (a perfect emitter) can be characterized by Planck's radiation law as follows:

$$M_{n,\lambda} = \frac{c_1}{\lambda^5 (e^{(c_2/\lambda T)} - 1)} \tag{2.8}$$

where

- $M_{n,\lambda}$ (W m$^{-2}$ µm$^{-1}$) indicates the radiant spectral exitance at a certain wavelength ($\lambda$ in µm)
- $c_1$ and $c_2$ are constants ($c_1 = 3.741 \times 10^8$ W m$^{-2}$ µm$^4$ and $c_2 = 1.438 \times 10^4$ µm K)
- $T$ is the absolute temperature (K)

This equation describes the spectral exitance distribution of a blackbody at a certain temperature as a smooth curve with a single maximum (Figure 2.5). Planck's equation indicates that any object hotter than absolute zero (−273°C) emits radiant energy and that the energy increases in proportion to its temperature. As shown in

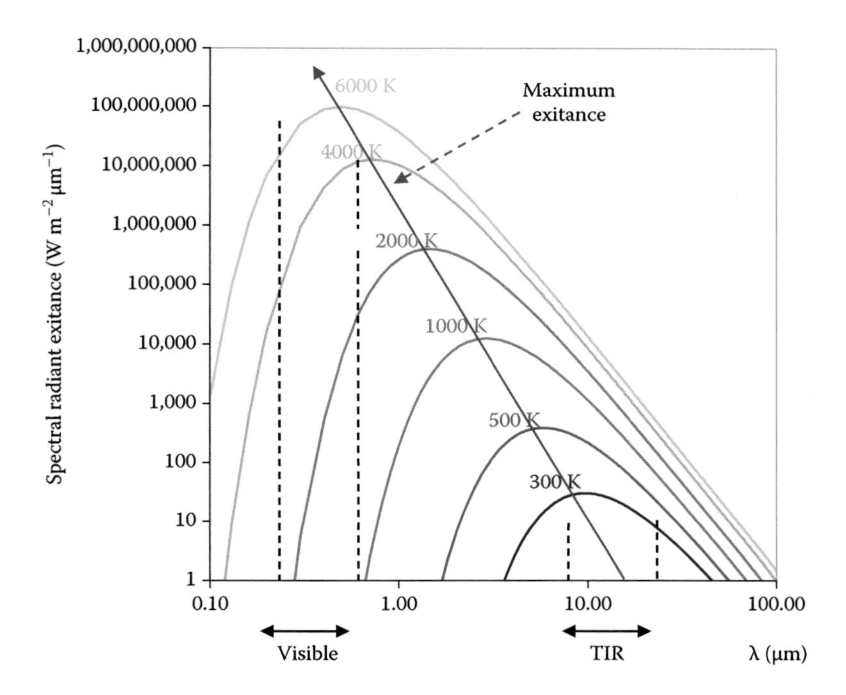


FIGURE 2.5 Blackbody spectral radiant exitance curves at various temperatures.

Figure 2.5, with increasing temperature, an object will radiate more energy and with higher exitance in shorter wavelengths.

The total radiant energy per unit surface area is a function of the object's temperature. The value can be obtained by integrating the spectral radiant exitance over all wavelengths. This is known as the Stefan–Boltzmann law:

$$M_n = \sigma T^4 \tag{2.9}$$

where

$\sigma$ is the Stefan–Boltzmann constant ($5.67 \times 10^{-8}$ W m$^{-2}$ K$^{-4}$)

$T$ is the temperature in Kelvin

As $M_n$ is the fourth power of $T$, small changes in temperature result in large variations in radiant exitance.

Kirchoff's law enables us to extend the foregoing relationships describing blackbody emission behavior to naturally emitting surfaces through an emissivity ($\varepsilon$) correction:

$$M = \varepsilon \sigma T^4 \tag{2.10}$$

A blackbody is a perfect emitter in that it absorbs and emits all the energy it receives. When an object does not absorb any of the incident energy, it is called a white body, completely reflecting all energy received (emissivity = 0). Gray bodies absorb and emit a fixed proportion of energy equally at all wavelengths. Most objects in nature have emissivity values that vary with wavelength and are referred to as *selective radiators*. Chapter 6 presents methods to convert the thermal radiation $M_\lambda$ measured by a sensor into the surface temperature $T$. The behavior of emissivity values with wavelength provides a mechanism for the discrimination of surface materials in the TIR.

The single maximum or wavelength of peak spectral radiant exitance of a blackbody may be derived from the first derivative of Planck's radiation law (Equation 2.8), described by Wien's displacement law:

$$\lambda_{max} = \frac{2898\ \mu\text{m K}}{T} \tag{2.11}$$

with the temperature ($T$) in Kelvin. Wien's displacement law is useful to determine the most sensitive band in the EM spectrum for the detection of a certain feature with high thermal contrast with their surroundings. For instance, a forest fire burning at 800–1000 K would be better detected between 3.6 and 2.98 μm (MIR region).

In summary, the amount and spectral distribution of energy radiated by an object vary according to (1) the temperature of the object and (2) the nature of the material, as depicted by its emissivity. From Equations 2.4 through 2.11, we can estimate the total spectral distribution and radiant exitance of an object by knowing its absolute temperature and emissivity. As with blackbodies, the energy emitted from a natural object is primarily a function of its temperature, but with important modifications dependent on its emissivity. Finally, knowing the temperature and emissivity of an object or surface allows us to determine the most suitable portion of the spectrum for optimal detection and discrimination.

In the following sections, we explore the behavior of EM radiation in more detail by focusing on the three regions of the spectrum where remote sensing measurements are made, namely, the solar spectrum (from VIS to SWIR), the TIR, and the MW portion of the spectrum. Our main interest from an environmental perspective is to better understand how the energy interacts with the main components of the Earth's surface: vegetation, water, soils, snow, and so on. Since these interactions do not occur in vacuum but are rather affected by atmospheric components, we will close this chapter by presenting the main effects of the atmosphere on the detected signal.

## 2.5 SPECTRAL SIGNATURES IN THE SOLAR SPECTRUM

### 2.5.1 INTRODUCTION

As previously mentioned, the Sun is the main source of EM energy. It is a gaseous body made up mostly of hydrogen. The internal temperature may reach 20 million K, but externally only the surface, called photosphere, is observable. The photosphere is at a temperature of ~6000 K. Using radiation laws we can calculate that most of the solar radiant exitance extends over a wavelength range from 0.3 to 3 μm, with a maximum output in the VIS region (0.4–0.7 μm) and peak exitance at 0.50 μm (Figure 2.5).

The actual solar radiation arriving at the top of Earth's atmosphere ($E_{0,\lambda}$) is a function of the Sun's temperature, the size of the Sun, and the distance between the Sun and Earth, and it can be computed as follows:

$$E_{0,\lambda} = M_{6000,\lambda} \frac{R^2}{D^2} \tag{2.12}$$

where

$R$ is the radius of the Sun ($6.96 \times 10^5$ km)
$D$ is the distance between the Sun and Earth

$D$ changes throughout the year, following the elliptical orbit of the Earth. The average values can be approximated to $149.6 \times 10^6$ km. For this reason, even though the Sun's radiation is always greater than the Earth's (as the Sun is much hotter), the actual solar radiation arriving on Earth is only relevant in the 0.4–2.5 μm band, which is consequently named the *solar spectrum* (Figure 2.6). For longer wavelengths, radiant exitance from the Earth or from hotter objects is higher than solar

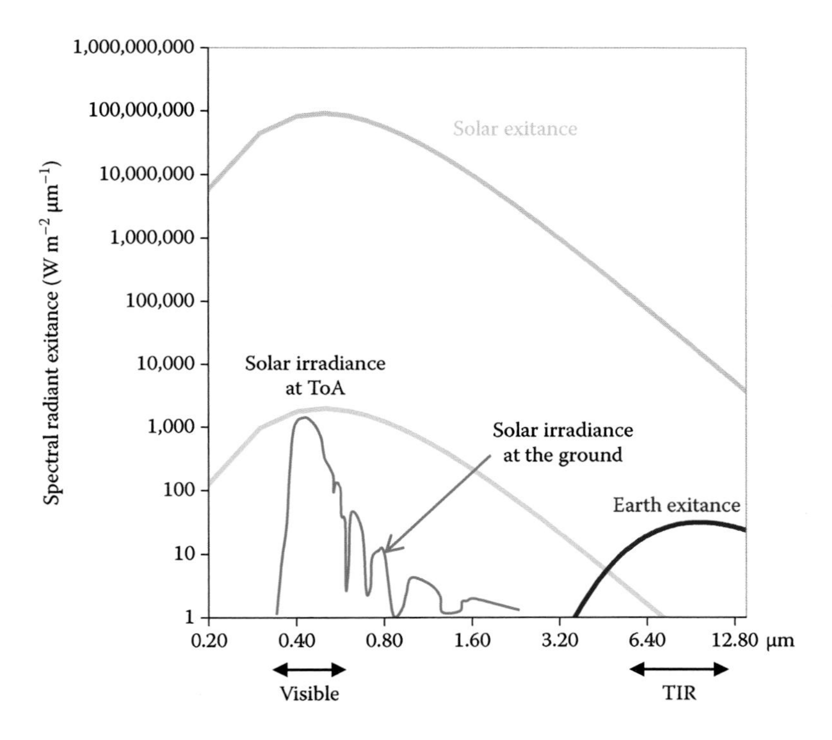


**FIGURE 2.6** Solar emittance and Sun irradiance at the top of the atmosphere (ToA) and at the ground versus Earth's emittance.

radiation. The Earth's surface has an internal temperature of 300 K with a maximum emittance in the 8–14 μm band (TIR) and a peak spectral exitance output at 10 μm.

The sum of total solar irradiance over all wavelengths is called the *solar constant* (=1370 W $m^{-2}$), but this amount varies largely throughout the Earth's surface as a function of latitude, day of the year, and time of the day. The actual solar irradiance reaching the ground is always lower than $E_o$ since the gases and aerosols of the atmosphere filter out different parts of the spectrum (Figure 2.6). Spectral regions where atmospheric transparency is high are called *atmospheric windows*. Most remote sensing missions include sensors sensitive to those wavelengths as they try to observe the ground. However, when they try to detect atmospheric components, such as ozone or water vapor, sensors on board are precisely sensitive to those regions where absorption of that target atmospheric component is more intense.

Solar radiation reaching the ground interacts with different land covers (soils, water, vegetation, asphalt, etc.). The incoming irradiance energy ($\phi_i$) will be either reflected ($\phi_r$), transmitted ($\phi_t$), or absorbed ($\phi_a$) by the cover elements (Figure 2.7), and therefore we can state

$$\phi_i = \phi_r + \phi_a + \phi_t \tag{2.13}$$

or expressed in terms of radiance (W $m^{-2}$ $sr^{-1}$)

$$L_i = L_r + L_a + L_t \tag{2.14}$$

We can also express these quantities in relative terms by dividing each of the radiances by the incident energy $L_i$:

$$\frac{L_i}{L_i} = \frac{L_r}{L_i} + \frac{L_a}{L_i} + \frac{L_t}{L_i} \tag{2.15}$$

$\phi_I$ Incident energy
$\phi_r$ Reflected energy
$\phi_a$ Absorbed energy
$\phi_t$ Transmitted energy

FIGURE 2.7 Interactions of electromagnetic energy with the Earth's covers.

which becomes

$$1 = \rho + \alpha + \tau \quad (2.16)$$

For any given surface, the magnitudes of the three components are not constant and will vary with wavelength. It is therefore more appropriate to express Equation 2.16 as

$$1 = \rho_\lambda + \alpha_\lambda + \tau_\lambda \quad (2.17)$$

The proportions of incident energy that are reflected, absorbed, and transmitted are a function of the unique characteristics of the surface, and these proportions vary with wavelength. In fact, the manner in which solar radiation interacts with the Earth's surface results in variations of the three components with wavelength, which can tell us much about the chemical and physical properties of the surface. For example, a leaf will appear green if its reflectance at green wavelengths is greater than its reflectance in the blue or red portions of the VIS spectrum. An object will appear blue if it reflects more energy in the blue region relative to the other wavelengths within the VIS spectral region. In the visible part of the spectrum, the variation in reflectance behavior of an object over the visible wavelengths results in what is called "color."

The reflectance behavior of an object over various wavelengths of the EM spectrum is commonly referred to as a *spectral reflectance signature* or just *spectral signature*. Figure 2.8 includes a series of spectral signatures for various covers. These materials have highly variable spectral signatures. The reflectance of snow is very high in the VIS region (blue, green, red), resulting in its "white" appearance (i.e., color theory tells us that high and equal amounts of blue, green, and red result in the color white). As we proceed to longer wavelengths, the reflectance of snow

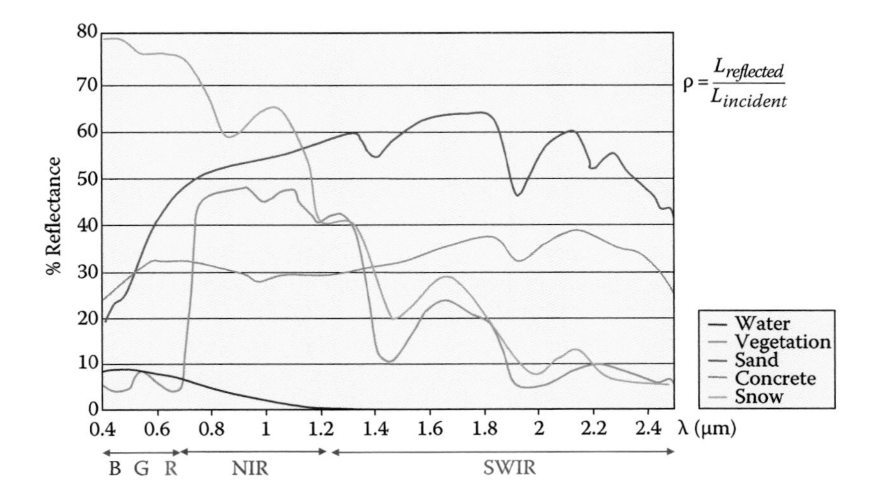


**FIGURE 2.8** Spectral reflectance signatures for representative Earth's surface materials.

decreases dramatically, and snow appears "dark grey" (near-zero reflectance) over the SWIR wavelengths. Water, on the other hand, has low reflectance in the visible wavelengths, which decreases toward the NIR and SWIR wavelengths. The presence of sediments, pollutants, phytoplankton, and other constituents, however, alters the spectral signature of water, which allows us to optically and quantitatively measure the amounts and turbidity of the constituents in water. Similarly, dust and pollutants will modify the spectral reflectance signature of snow such that the resulting changes in the remotely sensed spectral signatures can be used to assess the "age" of snow.

Vegetation has unique spectral reflectance signatures with low reflectance in the VIS, high reflectance in the NIR, and low reflectance in the SWIR portion of the spectrum. The spectral signatures of vegetation are modified by leaf type and morphology, leaf physiology, chlorophyll content, plant stress, and senescence. Soils, on the other hand, have spectral reflectance signatures that gradually increase with increasing wavelengths in a manner dependent on their iron, organic matter, water, mineral, and salt content. The spectral signature of a soil will also be modified by its structural and morphologic properties at the surface (e.g., roughness), as well as by the presence of plant litter and its stage of decomposition.

In summary, the relationship between the incident solar energy at the surface and the spectral composition of the remotely sensed reflected energy provides a wealth of information about the biogeochemical nature of the surface (leaf chemistry, soil mineralogy, water content) and the physical and structural characteristics of the surface (e.g., canopy height, leaf area, and soil roughness). The area observed by a sensor will most likely contain a variety of the surface materials (soil, vegetation, water, litter) in varying proportions and arrangements, and thus, remote sensing measurements often consist of mixed signals comprising multiple reflectance signatures. As we will comment later (see Section 7.1.5), there are various techniques to extract information from these spectral mixings.

Spectral signatures form the basis to discriminate objects from remote sensing measurements in the solar region of the EM spectrum. Unfortunately, these signatures are not constant for each cover, as the radiance flux detected by remote sensing depends not only on the intrinsic properties of the observed area but also on the external conditions of the measurement. The main factors affecting spectral signatures are the following (Figure 2.9):

(i) Atmospheric components, which affect both the absorption and the scattering of incoming and reflected radiation.
(ii) Land cover variations causing changes in chemical or physical composition, such as density, pigment contents, moisture, or roughness. They may be caused by vegetation or crop phenology, agricultural practices, grazing, etc.
(iii) Soil and geologic substrate, which are particularly important in open and sparse canopy covers, as the sensor will detect a stronger signal coming from the background.
(iv) Solar illumination conditions, which depend on the latitude, day of the year, and hour of the day.
(v) Terrain slope.
(vi) Aspect, both affecting the illumination conditions of a target cover.

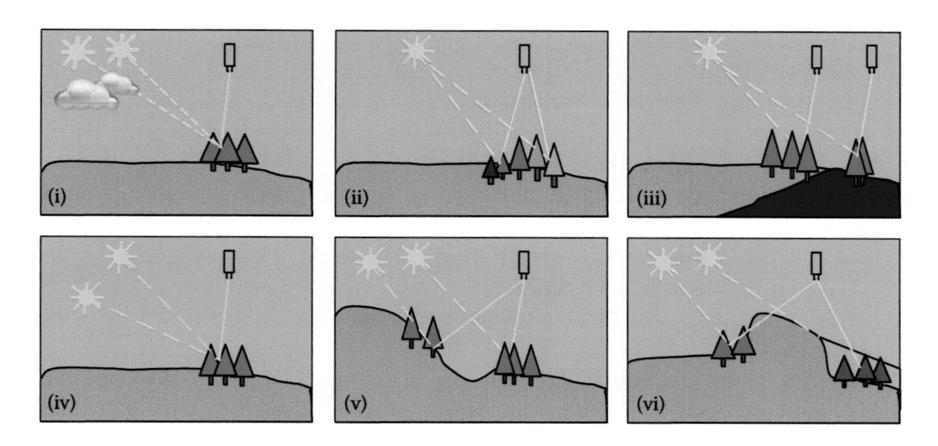


**FIGURE 2.9** Factors (see text) influencing spectral signatures.

Among these factors, it is particularly relevant to consider the impact of the geometric conditions of the observation, in particular, the incidence and viewing angles relative to the reflecting surface. This geometric relationship, along with the surface properties (primarily roughness), determines how the incoming radiation is scattered and the strength of the outgoing radiation. There are three types of surface scattering that may occur (Figure 2.10). The first type of scattering is known as

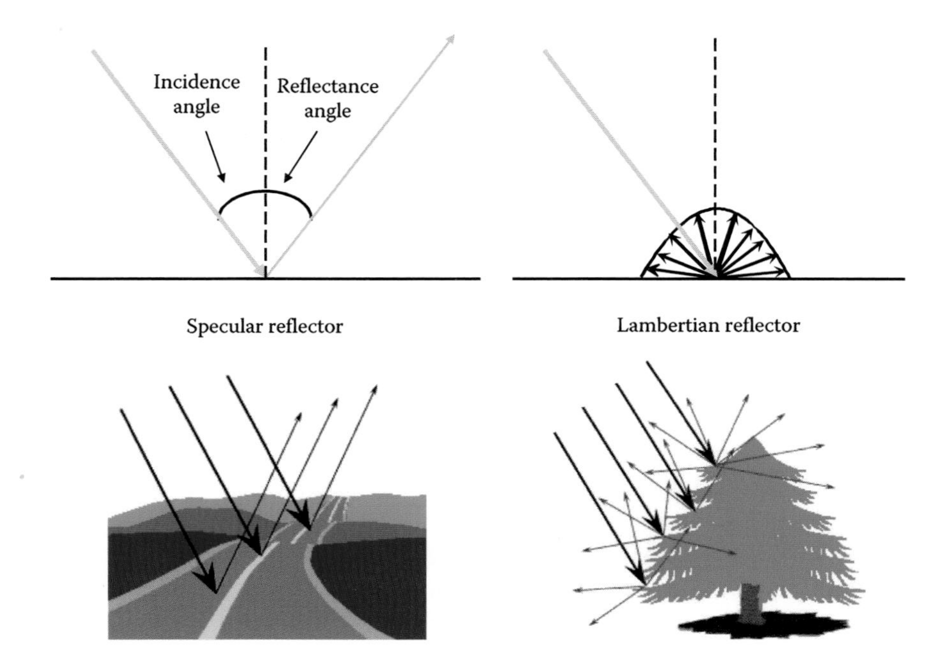


**FIGURE 2.10** Major types of surface reflection over a variety of surface roughness conditions.

*specular reflection*. In specular reflection, surface incident energy is reflected away from the Sun and at the same angle as the solar incident angle, and no energy is scattered in any other direction. In this case, the sensor can measure only the ground reflected energy at this particular viewing angle and will receive no energy at any other viewing directions. In the second type of surface scattering, the incident energy is reflected diffusely and equally, or isotropically, in all directions. When a surface is perfectly diffuse and exhibits the same reflected radiance for any angle of reflection, that is, independent of viewing angle to the surface normal, it is known as a Lambertian surface. Most surfaces exhibit the third type of scattering behavior, known as anisotropic reflectance, in which both diffuse and specular scattering occur.

These factors illustrate some of the challenges involved in the accurate assessment and characterization of land surface conditions with remote sensing data. Therefore, the spectral reflectance signatures presented in Figure 2.8 may be regarded just as reference signatures. In other words, land surface cover types will unlikely have the same observed reflectances as those referred in their expected spectral signatures. Rather, they show spectral variability caused by surface, illumination, and atmospheric variations, often making it difficult to unequivocally discriminate objects based solely on detected reflectance. Therefore, to obtain sound retrievals from remote sensing images, the interpreter needs to introduce different corrections aiming to remove the influences of those factors. As we will see in Chapter 6, different quantitative methods are available to eliminate or at least mitigate the impact of atmospheric, illumination, and terrain effects.

Despite this, the reference spectral reflectance signatures are very useful in understanding the images and can also be used to select the optimal bands or band combinations for discriminating certain surface properties, as well as to suggest technical requirements for future remote sensing missions.

In the following sections, we analyze in greater detail the spectral behavior and signatures of typical land surface components, namely, vegetation, soil, and water. We discuss the optical properties of these materials and relate their spectral signatures to key biochemical and mineralogical components.

### 2.5.2 Vegetation Reflectance

The spectral properties and characterization of vegetation canopies are one of the most important and challenging problems in remote sensing. The reflectance properties of a vegetation canopy are complex and a result of many biochemical and biophysical canopy attributes and the external factors that influence the signal detected at the sensor. A complete understanding of the reflectance behavior of a vegetation canopy includes the role of leaf biochemical, plant physiologic, and canopy structural and morphologic properties. Leaf biochemical constituents include pigments, lignins, and water. Differences in pigment concentrations are responsible for "color" changes, primarily in the visible portion of the spectrum, while leaf moisture content involves energy interactions in the SWIR portion of the spectrum. Plant physiologic condition (vigor, phenology, stress) involves nutrient, water, and light availability, which alters the pigments, lignins, and water-related biochemical interactions and affects plant structure. One must also consider the structural properties at the canopy level, which include the leaf area index (LAI), leaf angle distribution (LAD), fractional vegetation cover, plant height, crown diameter, leaf clumpiness, planting geometry, and associations with other species of shrubs, trees, and grasses.

Typically, many of these vegetation properties change simultaneously with plant canopy development, making it difficult to isolate the variation of specific vegetation biophysical components. For example, as a canopy develops, both fractional vegetation cover and LAI change simultaneously, along with pigment and water contents. Aside from the vegetation canopy itself, there is also the underlying canopy background consisting of soil, rock, litter, water, and snow, with optical properties that contribute a signal to the overlying vegetation canopy. External to the canopy are also the various influences that alter the signal at the sensor including Sun illumination and sensor view angles, landscape topography (slope, aspect), and atmospheric effects.

In this section, we introduce some of the most basic vegetation properties in the solar spectrum. More rigorous treatments involving laboratory-based studies of leaf reflectance, optical–geometric models of plant canopies, and numerical canopy radiative transfer models (RTM) can be found elsewhere (Asner 1998; Asner et al. 2000; Colwell 1974; Gates et al. 1965; Jacquemoud 1990; Jacquemoud et al. 1995; Knipling 1970; Liang 2004; Westman et al. 1988).

The reflectance properties of a leaf are mostly related to the levels of photosynthetic pigments and leaf water, as well as the leaf's structural characteristics. The overall spectral signature of a leaf or plant is further affected by leaf age, nutrient stress, and health (disease, vigor, etc.). A typical spectral reflectance signature of a green leaf obtained from laboratory measurements is shown in Figure 2.11. There are three main spectral domains influencing the optical properties of leaves, namely, the VIS, the NIR, and the SWIR regions.

The low reflectance in the VIS region (0.4–0.7 μm) is due to the absorbing effect of leaf pigments, mainly chlorophyll a, chlorophyll b, carotenoids, and xanthophylls,

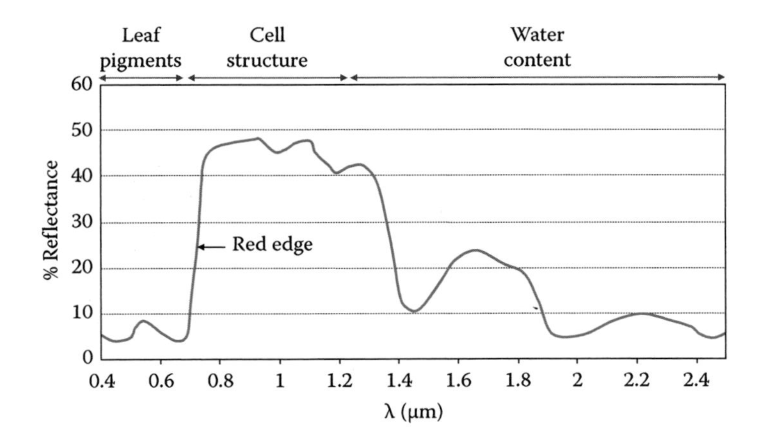


**FIGURE 2.11** Green leaf spectral reflectance.

with the chlorophyll pigments accounting for 60%–75% of the energy absorbed (Gates et al. 1965). All of these pigments absorb in the "blue" region of the EM spectrum centered on wavelengths ~0.45 μm, while chlorophyll also absorbs in the "red" portion of the spectrum, centered ~0.65 μm. Between the blue and red spectral regions, there is a spectral region of less intensely absorbed radiation, a "green reflectance peak" at 0.55 μm, which is responsible for the green appearance of healthy leaves. Because energy in the 0.4–0.7 μm range is absorbed by pigments to drive photosynthesis, the term *incident photosynthetically active radiation* (IPAR) is often used to describe the radiation in the visible part of the spectrum.

Other pigments also have an important effect on the spectral reflectance of leaves in the VIS spectrum. For example, the yellow to orange-red pigment carotene has a strong absorption in the 0.35–0.50 μm range and is responsible for the color of some flowers and fruits as well as leaves without chlorophyll. When leaves undergo senescence, chlorophyll levels and the associated absorptance decrease, causing a higher red reflectance, which combined with the green reflectance yields a yellowish color (green + red – yellow). The influence of the more persistent carotenoid pigments becomes more pronounced as the amount of chlorophyll in the leaves decreases during senescence. In some species, the red and blue pigment xanthophyll with strong absorption at 0.35–0.50 μm becomes prominent with leaf aging, resulting in many of the leaf colors in autumn (e.g., northern maples and Chilean Nothofagus).

Beyond the highly absorbing red region is the sharp "red edge" transition region at ~0.74–0.78 μm, in which leaf pigments and cellulose become transparent to NIR wavelengths. Leaves have very low absorptance (<10%) and high leaf reflectance that can reach 50%. Plant nutrient and mineral stress are known to cause shifts in the red edge. The region between 0.70 and 1.1 μm is called the NIR reflectance plateau, where reflectance is very high, except in two minor water-related absorption bands (0.96 and 1.1 μm), which depend on the internal cellular structure of the leaf.

Leaf structural properties strongly influence reflectance, particularly by the relative thickness of the mesophyll cell layer. The spongy mesophyll layer contains internal air cavities that scatter incident radiation (Figure 2.12). Leaf reflectance increases for more heterogeneous cell shapes and contents as well as with increases in the number of cell layers, intercellular spaces, and variations in cell size. The vegetation spectral reflectance curves are modified by the morphology of the leaf. Thus, needle leaves tend to exhibit greater absorptance across all wavelengths, while desert succulent plants will reflect more energy than other mesophytic species (Gates et al. 1965). As a result of NIR sensitivity to leaf structural properties, which varies across plant species, the NIR spectral region is very useful in plant biodiversity studies and in discriminating among plant species that are often not distinguishable in the VIS spectrum.

The NIR spectral domain also has a transition area between 1.1 and 1.3 μm, where reflectance decreases sharply from the NIR reflectance plateau to the low-reflecting SWIR domain (1.3–2.5 μm) (Figure 2.12). This region is characterized by strong absorption by leaf water. Leaf water strongly absorbs incident solar radiation in this range but is transparent to the shorter VIS or IPAR wavelengths. Reflectance of SWIR wavelengths generally increases as leaf liquid water content decreases; however, water absorbs radiation so strongly at 1.45 and 1.95 μm that these wavelengths

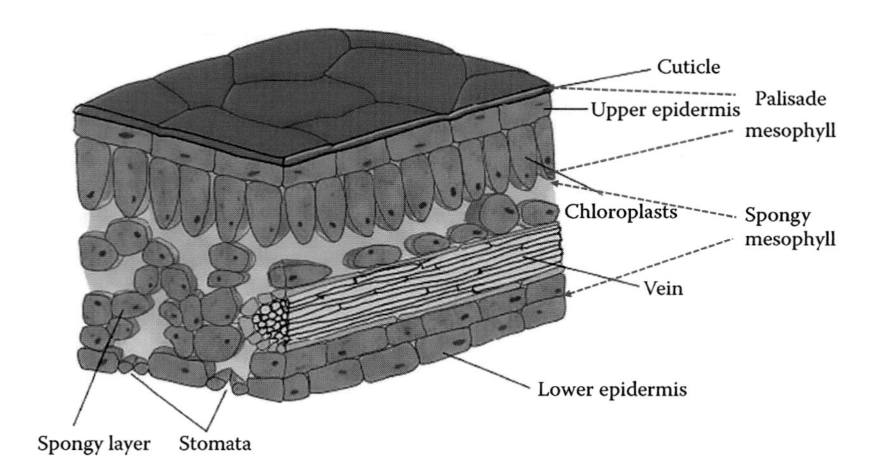


**FIGURE 2.12** Basic components and cellular structure of a leaf. (Adapted http://igbiologyy.blogspot.com.es/2013/01/46-leaf-structure.html, last access November 2015.)

cannot be used in land remote sensing because most of the solar radiation in these wavelengths is absorbed by the atmosphere before reaching the ground.

From laboratory measurements, we see dramatic differences between dry leaves and leaves infiltrated with water in this spectral region, especially at wavelengths near 1.45, 1.92, and 2.7 μm (Lusch 1989; Short 1982, p. 501; Yebra et al. 2013a; Figure 2.13), although these variations depend on the leaf type and plant species (Westman and Price 1988). A sensor placed above a canopy generally does not measure individual leaves but rather many leaves forming the vegetation canopy. The overall radiation transferred through the canopy also depends on the structural arrangement and quantity of leaves within a canopy. The canopy's structural or optical–geometric properties include canopy architecture, LAD, LAI, ground cover

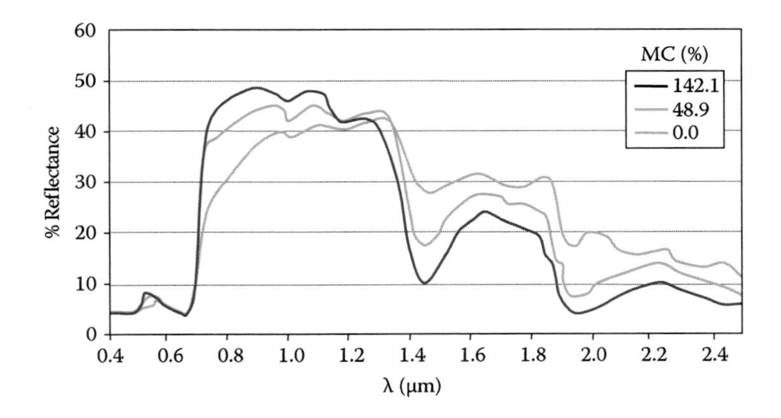


**FIGURE 2.13** Leaf spectral signatures as a function of moisture content (MC), estimated as water weight over leaf dry weight.

fraction, leaf "clumping," species composition, leaf morphology, leaf size and shape, and the underlying soil and litter.

In spite of this structural complexity, there are many common features in most vegetation spectra, such as the high contrast observed between the red band (~0.645 μm) and the NIR region (0.7–1.3 μm). In general, one can say that the greater the contrast between these two regions, the greater the amount and vigor of the vegetation. This theoretical spectral behavior of vegetation in the "red" and NIR forms the basis for the design and development of vegetation indices (see Section 7.1.3). Vegetation indices are constructed from combinations of these two bands (ratios, linear sums, etc.) when multispectral images are available. They are designed to isolate and enhance the vegetation signal in remotely sensed imagery, thereby facilitating the discrimination and extraction of useful vegetation information (Asrar et al. 1992; Gutman 1991; Huete et al. 1994; Huete et al. 1997).

It is implicit that any source of stress in the vegetation will cause a change in its spectral behavior. Senescent or stressed leaves tend to immediately reduce chlorophyll activity, resulting in less absorptance in the red band, with slight decreases in blue absorption (since carotenes persist and continue to absorb in the blue). The consequent increase in red reflectance and slight increase in blue reflectance alters the proportion of reflectance in the primary colors (blue, green, red), resulting in a change in leaf color. This is why leaves tend to show a yellowish color with senescence or stress. Stress may also reduce leaf reflectance in the NIR portion of the spectrum, due to the deterioration of the cellular structure of the leaf. The spectral curve often becomes flatter and less chromatic (Knipling 1970; Murtha 1978).

Such information is valuable in detecting damages produced by pollution, insects (Rock et al. 1986; Souza et al. 2005), or fires (Chuvieco 2009). The contrast between NIR and SWIR has also been extensively used to estimate leaf moisture content, provided an estimation of other controlling factors is available (Yebra et al. 2013a). In addition, it has been shown that certain factors in leaf stress are associated with a displacement in the red edge, that is, the slope change in the spectral curve between red and the NIR, toward shorter wavelengths. This phenomenon has been observed when the plants are affected by heavy metal contamination (Rock et al. 1986).

### 2.5.3 Soil Reflectance Properties

Much is known about soil properties through extensive laboratory and in situ field measurements. In contrast to vegetation, very little EM energy is transmitted through soils, and therefore their spectral signature is basically related to the most superficial conditions. The spectral composition of energy reflected and emitted by soil is mostly dependent on the biogeochemical (mineral and organic) constituents, optical–geometrical scattering properties (particle size, aspect, roughness), and moisture conditions of the immediate soil surface (Ben-Dor et al. 2008; Lusch 1989; Mulders 1987). For example, soils with high quartz content often reflect a large portion of incoming energy across the EM spectrum, and wet soils absorb most of the NIR and SWIR light they receive, while soils with high organic matter tend to absorb much of the incoming visible light.

Different soil spectral reflectance signatures result from the presence or absence, as well as the position and shape of specific absorption features, of a number of soil constituents. Absorption is due to various chemical/physical phenomena such as intermolecular vibrations and electronic processes in atoms. The VIS–NIR regions (0.4–1.1 μm) contain broad spectral absorption features such as strong absorption near 1 μm due to ferrous iron, and weaker absorptions at 0.7 and 0.87 μm are attributed to ferric iron. Strong Fe–O charge transfers in the blue and ultraviolet region result in fairly steep decreases in reflectance with shorter wavelengths. Iron is fairly ubiquitous, so most soils exhibit increasing reflectance with wavelength over the VIS to NIR portion of the spectrum (Mulders 1987). Soils have distinct spectral features in the SWIR region caused by atomic vibrational processes, which include two broad water absorption bands at 1.4 and 1.9 μm. Minerals with OH, $CO_3$ (calcite), and $SO_4$ (gypsum) exhibit absorption features in the 1.8–2.5 μm region, while layer silicates with OH absorb near 1.4 and 2.2 μm (Mulders 1987; Shepherd and Walsh 2002). Humus content also has a great influence on soil color, tending toward a low reflectance, especially around 0.7–0.75 μm (Curran et al. 1990).

Soils are mixtures of a number of inorganic and organic constituents, so it is not straightforward to evaluate the composition of soils from their spectral signatures (Ben-Dor et al. 2008). Many soil spectral signatures are fairly similar, making it difficult to distinguish them. As a result, only a limited number of soil spectral curves have been found to be distinguishable using remote sensing. Stoner and Baumgardner (1981) analyzed the spectral signatures of a great variety of soils (485 types), from 0.50 to 2.45 μm, and documented five unique soil spectral curve shapes primarily related to their relative contents of organic matter and iron and modulated by their textures. Other spectral libraries, such as the NASA Jet Propulsion Laboratory (JPL) Aster, include a wide range of soil spectral signatures (Figure 2.14). These, along with numerous

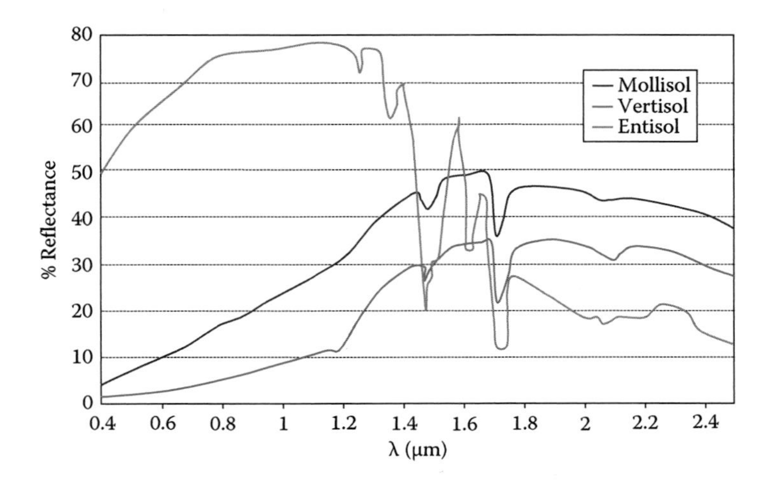


**FIGURE 2.14** Spectral reflectance for different soil types: mollisol (gray silt), vertisol (brown clay), and entisol (white gypsum). (From http://speclib.jpl.nasa.gov/, last access November 2015.)

laboratory and field studies, have shown that soil spectral signatures are largely controlled by the iron oxides, organic molecules, and water that coat soil particles.

Most soil surfaces scatter incident radiation anisotropically, which is a consequence of a soil's 3D structure. Scattering takes the form of diffuse and specular reflection and is sensitive to the geometric properties of soil components (particle size, aspect, roughness), the macro soil surface, sensor viewing angle, solar illumination angle, and the relative azimuthal positions of the Sun and sensor relative to the surface (Ben-Dor et al. 2008; Stoner and Baumgardner 1981). With the shortest wavelengths most affected, roughness and Sun–soil–sensor geometries alter a soil's spectral signature and the inferences of basic soil properties such as soil mineralogy. Remote sensing data taken under different Sun and viewing geometries are not necessarily comparable without correction for these angular effects (see Section 6.6.2.4).

Particle size distribution and surface height variation (roughness) are the most important factors influencing the directional reflectance of bare soils. They cause a decrease in reflectance with increasing size of "roughness elements," as coarse aggregates contain many interaggregate spaces and "light traps." Smooth, crusted, compacted, and structureless soils generally reflect more energy and are brighter (Lusch 1989). Clayey soils, despite having a finer particle size distribution, tend to be darker than sandy soils because clays aggregate and behave as larger, "rougher" surfaces.

Soil moisture has a strong influence on the amount and composition of reflected and emitted energy from a soil surface, and thus, information about soil moisture condition can be derived from measurements in all parts of the EM spectrum. In the shortwave region, the major effect of adsorbed water on soil reflectance is a pronounced decrease in reflected energy, making soils darker when moistened, particularly in the water absorption bands centered at 1.45 and 1.9 μm (Reginato et al. 1977). The decrease in reflectance is proportional to the thickness of the water film around the soil particles and can be related to the gravimetric water content as well as energy status of the adsorbed water.

In the shortwave portion of the spectrum, the SWIR region is considered most sensitive to surface moisture content. Water absorption in these two bands can be expressed as a ratio, relative to the other bands, or in linear combination and then related to soil water content for discrete soil textural classes.

At the landscape level, it is much more difficult to measure soil properties and extract soil information with spaceborne sensors. This is due to the extreme spatial variability of soil properties and because the soil surface is often masked by vegetation and plant litter. The discrimination and mapping of soil types and soil properties becomes a function of not only the properties of the surface materials but also sensor characteristics such as number of wavebands, bandwidths, spatial resolution, and instrument noise. The wealth of knowledge available from laboratory, field, and model studies, however, provides a strong foundation and starting point for the extraction of soil information at the more heterogeneous landscape level.

Direct measurements provide information on soil spectral signatures, moisture, and textural properties and can be interpreted using the extensive knowledge gained from laboratory and field studies. In arid and hyperarid regions, there has been much success in mapping soils with remote sensing because of the dominance of the soil

**FIGURE 2.15** RapidEye image of the Colorado River Delta, Mexico, showing the spectral contrast of different soils in arid areas. (Image courtesy of BlackBridge, color combination NIR, G, B.)

spectral signal and the minimal interference from vegetation (Figure 2.15). In more densely vegetated areas, the soil signal may become significant again with land conversion activities that result in exposed soil surfaces, such as in clear-cutting of forests, biomass burning, wind storms, agricultural and pasture expansions, and natural disasters (Adams et al. 1995).

Where the surface is strongly vegetated, soil properties must be inferred from measurements of the vegetated surface. Many studies have demonstrated the possibility of inferring valuable information about a soil from the spectral patterns and characteristics observed over the vegetation (Bruce and Hormsby 1987). This technique is known as *geobotany* and relies on the detection of anomalies in the vegetation cover to obtain information on soil and geologic characteristics. Examples of such anomalies include unique patterns of plant species distribution, abnormal plant growth, and density patterns, as well as alterations in plant pigments or a plant's phenological cycle.

The presence of leaf litter and other nonphotosynthetic vegetation (NPV) on the soil surface influences the resulting soil spectral signatures. Many studies have investigated the chemical and optical properties of soil organic matter dynamics as a function of plant source and aging. Stoner and Baumgardner (1981) showed three unique spectral signatures representative of various stages of litter decomposition in soils. A *fibric* curve had the most tissue morphology intact and thus had high reflectance properties. *Hemic* curves resulted from intermediate levels of decomposition, while *sapric* curves represented mostly decomposed litter and were very low reflecting. Ben-Dor et al. (2008) found that the slopes in the VIS–NIR spectral region and specific absorption features in the NIR–SWIR region were useful in assessing the properties of soil organic matter at several stages of biological decomposition.

### 2.5.4 Water in the Solar Spectrum

Water bodies and aquatic surfaces absorb or transmit most of the radiation they receive. In general, the longer the wavelength, the greater the absorption, so the spectral reflectance signature of water bodies shows decreasing reflectances with increasing wavelengths. The highest reflectance of clear water takes place in the blue band, and there is a gradual decrease nearing the NIR and SWIR regions, where reflectance values approach zero. For this reason, it is generally straightforward to spectrally recognize water bodies on remotely sensed images, as they appear very different from soil or vegetation.

Variation in water optical properties is more easily detectable in shorter wavelengths (blue and green) and is related to differences in water depth, chlorophyll content, dissolved particles, and surface roughness. For example, reflectance increases with sediment content across all the visible bands (Figure 2.16), although the amount of increase depends on sediment particle diameters (Bhargava and Mariam 1990). Reflectance also increases with increasing algal concentrations.

The depth of water directly influences the reflectance contribution from materials at the bottom layer of water. In shallow waters, the bottom layer has a stronger optical contribution and will generally increase the remotely sensed reflectance from the water body. The greater the water depth, the greater the absorptance, and bathymetric (water depth) mapping in coastal zones is based on these relationships. From studies conducted with Landsat thematic mapper (TM) images, the maximum water depth that could be determined was 6.4 m in the blue band, 3 m in the green band, and 2.1 m in the red band (Ji et al. 1992).

Water composition also influences reflectance properties, and several studies have demonstrated that water chlorophyll content can be mapped from spaceborne sensors. As the chlorophyll levels in water increase, reflectance decreases in the blue

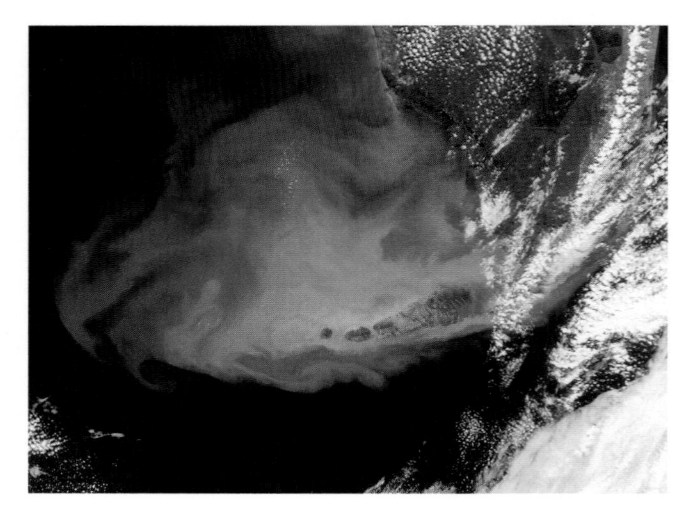

**FIGURE 2.16** MODIS-Aqua image of South Florida, showing different sediment concentrations in shallow waters. (From http://visibleearth.nasa.gov/.)

band. This has been used to map seaweed density (Robinson 1985) and to study eutrophication in reservoirs and lakes (Pulliainen et al. 2001).

Finally, water surface roughness promotes diffuse reflection and scattering, which result in a greater reflectance. With very calm waters, the surface behaves as a specular reflector (or mirror), with a highly anisotropic reflectance pattern and variable reflectance values depending on sensor view angles and direction (Figure 2.17). Ocean measuring sensors are designed to sense variations in very low reflecting waters while having orbits designed to avoid the strong specular signal, or Sun glint, which could easily damage the internal detectors.

Snow, on the other hand, provides a very different spectral signature to water. Snow has very high reflectance in the visible bands, lower reflectance in the NIR, and even lower reflectance in the SWIR region. Snow reflectance is most affected by the snow grain size, the depth and density of the snow layer, and the amount of impurities present (Dozier 1989). Reflectance is highest for fresh snow, lower for frozen snow, and lowest for dirty snow (Figure 2.18). Snow reflectance in the visible bands can decrease by up to 80% with increasing snow age and content of impurities (Hall 1988; Kelly and Hall 2008).

Snow is often difficult to distinguish from clouds in the VIS spectrum; however, differences are more evident in the SWIR because cloud ice crystals or drops are smaller than snow grains, which is why they absorb much less radiation in this portion of the spectrum. In addition, snow usually has a greater reflectance in the visible bands and a more homogeneous texture than clouds (Dozier 1989).

**FIGURE 2.17** Terra-MODIS image of Mexico and Central America acquired April 13, 2004, showing Sun glint effects. Notice the white colors of Lake Chapala (central section of the image) and the Pacific Ocean to the south. (From http://visibleearth.nasa.gov/.)

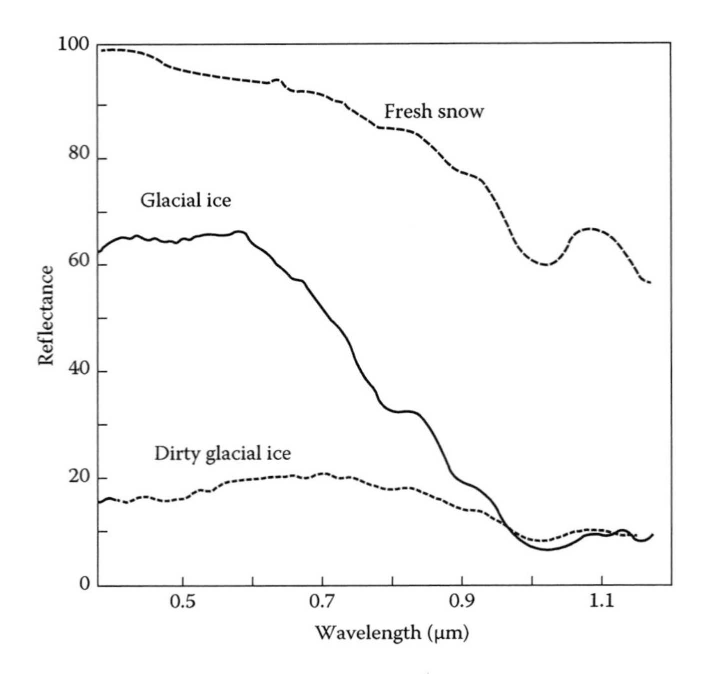


**FIGURE 2.18** Reflectance curves for different types of snow and ice. (Adapted from Hall, D.K. and Martinec, J., *Remote Sensing of Ice and Snow*, Chapman & Hall, London, U.K., 1985.)

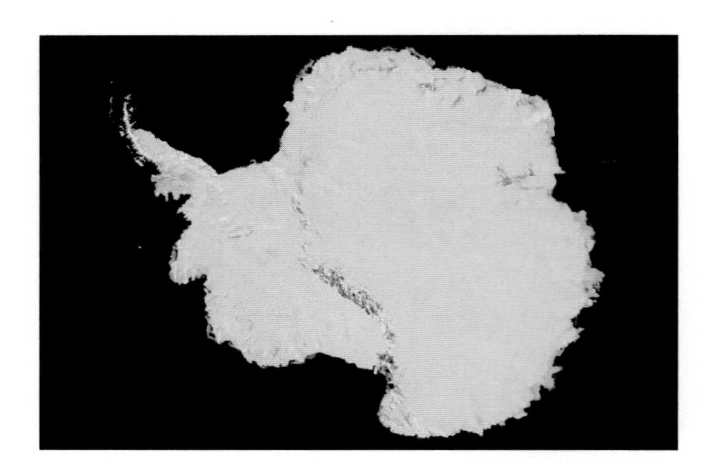

**FIGURE 2.19** Landsat image mosaic of Antarctica based on 1100 images from Landsat 7. (Image courtesy of USGS, NASA, National Science Foundation, and the British Antarctic Survey, http://visibleearth.nasa.gov/.)

The growing concern on climate warming and its impact on ice melting has increased the interest on using temporal series of satellite data to monitor glacier and ice cap trends. For instance, recent estimations of glacier extent and trends based on the analysis of satellite images have been able to provide more accurate estimations on the contribution of glacier loss to sea level rise (Gardner et al. 2013). Similar studies are being conducted for the ice caps of Greenland and Antarctica, using a combination of remote sensing techniques, including gravimetry, laser altimetry, and MW and optical observations (Figure 2.19).

## 2.6 THERMAL INFRARED DOMAIN

### 2.6.1 Characteristics of EM Radiation in the Thermal Infrared

According to Wien's (Equation 2.6) and Plank's laws (Equation 2.4), the spectral radiant exitance from the Earth is most pronounced in the spectral region between 8 and 14 μm, considering the Earth's temperature (about 300 K). This spectral band is named *thermal infrared* (TIR) because radiation in these wavelengths are closely linked to the Earth's temperature variations. As previously commented, the spectral signal received by the sensor in this band proceeds mainly from the Earth's emitted energy, as the solar reflection in this spectral region is very low. Therefore, observing in the TIR region is particularly useful to detect heat variations both from the ground (both ocean and land) and from the atmosphere (clouds, for instance).

The temperature of an object is closely linked to its capacity to absorb incoming solar radiation. Equation 2.16 states that the sum of reflectance, transmittance, and absorptance is 1. Considering that the transmittance is practically zero in the TIR, this equation can be simplified by stating that incident energy is either reflected or absorbed. Following Kirchoff's radiation laws, at thermal equilibrium the spectral emissivity of an object equals its absorptance. In other words, the greater the absorption, the greater the emission, which allows us to reformulate Equation 2.17 according to

$$1 = \rho + \varepsilon \tag{2.18}$$

which, in summary, means that surfaces with high reflectance (e.g., snow) are low emitters, while those with low reflectance are highly emissive, as is the case with water.

As already mentioned, emissivity is a relative measure that relates the radiant exitance of an object with that of a blackbody. High emissivity (close to 1) indicates that an object absorbs and emits a great proportion of the incident energy, while an emissivity considerably smaller than 1 refers to an object that absorbs and emits only a small proportion of the incident energy. In the case of land surfaces, water and dense vegetation have the greatest emissivity (0.99 and 0.98, respectively), while the lowest emissivity values correspond to sandy soils (0.90), snow (0.80), and metals (0.16). The moisture content modifies these values remarkably. Thus, in sandy soils, the moisture content increases of 8% imply a rise in emissivity from 0.90 to 0.94 (Mulders 1987). The emissivity values are calculated for an average temperature, in this case around 20°C (Curran 1985).

If we know the kinetic temperature of an object, the global radiant exitance can be estimated by measuring its emissivity, which is a critical parameter in the TIR. Two objects with the same temperature can produce different radiant exitances if having different emissivities. Emissivity values can be estimated if the land cover is known (see Section 6.7.4). Corrections can also be made based on empirical correlations of satellite radiances with temperatures taken in the field or from meteorological stations. Recent studies have shown that in sufficiently dense vegetation covers, there is no need to take the soil into account, which simplifies calculations remarkably.

In addition to emissivity, the thermal behavior of an object is related to other parameters. The most important are its thermal capacity, conductivity, diffusivity, thermal inertia, and heating index.

The thermal capacity ($C$) indicates the capacity of the cover to store heat. The thermal capacity per mass unit is known as *specific heat* ($c$). Water has the greatest thermal capacity, which is why it can store more heat than vegetation or soils. Thermal conductivity ($k$) measures the rate at which the heat in a land cover is transmitted. Urban areas are good thermal conductors, whereas vegetation and soils are not. For that reason, the diurnal temperature in zones with a vegetation cover is mainly a function of the temperature of the surface layer and not of lower layers.

The thermal diffusivity ($K$) is a measure of how well the temperature of a surface changes through heat conduction. In general, dry surfaces show low temperature changes in contrast to wet surfaces with higher conductivity.

Finally, thermal inertia ($P$) refers to the resistance of a material to temperature changes. This parameter is directly related to the thermal conductivity ($k$), the heat capacity ($C$), and the density of the material ($D$), according to the following formula:

$$P = \sqrt{DCK} \tag{2.19}$$

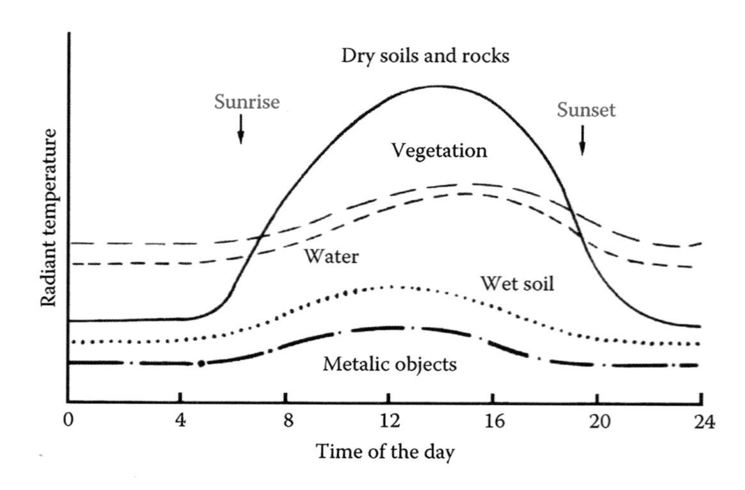


FIGURE 2.20 Thermal inertia of different land covers. (Adapted from Short, N.M. and Stuart, L.M., *The Heat Capacity Mapping Mission (HCMM) Anthology*, NASA, Washington, DC, 1982.)

Dry and sandy soils with low thermal inertia show severe contrasts between day and night temperatures. In contrast, humid and clayey soils are much more resistant to changes and will have lower daytime and higher nocturnal temperatures compared to dry soils (Figure 2.20). Thermal inertia is easily derived from periods of maximum and minimum sunlight, approximately at noon and midnight, respectively.

The heating index is a function of the intensity of the radiation and the absorptance of the object. The intensity is smaller in areas with forest, clouds, or buildings, which prevent direct radiation. Topography (slope or aspect) also directly affects radiation.

### 2.6.2 Thermal Properties of Vegetation

The thermal properties of vegetation are quite complex, since plants absorb a high amount of incident solar energy to drive photosynthesis. This energy is reemitted during the night to maintain the energy balance. In addition to this, vegetation has a high water content, and therefore will have a high thermal inertia. Another important factor controlling vegetation thermal patterns is evapotranspiration (ET), which regulates plant water and temperature. When plants transpire, the extra energy involved in releasing water vapor implies a reduction in sensible heat and a parallel decrease in plant surface temperature. For this reason, plants tend to be cooler during the day than their surroundings. On the other hand, the energy absorbed during the day is reemitted at night in the TIR region and the plant shows higher temperatures than its surrounding areas. Water stress or other physiological problems reduce this day–night thermal contrast in vegetation, which make it possible to detect plant health conditions. Although most of the studies on vegetation health are based on the solar spectrum, thermal images have also been extensively used to look at vegetation changes resulting from deforestation processes, desertification, or water stress (Curran 1985; Jackson et al. 1981; Moran et al. 1994; Nieto et al. 2011; Sandholt et al. 2002). Plant stress indices, such as the surface moisture index (SMI) (Nemani et al. 1993) and the water deficit index (WDI) (Moran et al. 1994), utilized thermal and vegetation index measurements of soil and vegetated surfaces to estimate the moisture content through the rate of evaporative water loss. When soil moisture is abundant, plants are able to transpire water at maximum rates for the given meteorological conditions. As the soil dries out, it becomes more difficult for plants to extract the necessary water to meet the atmosphere's evaporative demands and transpiration is reduced, resulting in higher leaf temperatures than those with ample water supply.

Several external factors may modify this thermal behavior. For example, ET rates are influenced by atmospheric humidity, wind speed, availability of light, air temperature, and soil moisture. Atmospheric humidity reduces ET, while the other factors generally increase ET (Allen et al. 1989). An important factor that affects vegetation emissivity is percent cover and density. Some authors have tried to estimate globally ET from satellite data, combining thermal and optical data. The former provide an estimation of energy balance, while the latter facilitate canopy cover and vegetation fraction (Li and Lyons 2002; Yebra et al. 2013b).

### 2.6.3 Soils in the Thermal Domain

Soil thermal behavior is largely a function of soil moisture, which modulates the heating and cooling of the soil through a partitioning of radiant energy into latent and sensible heat components. The difference in the amplitude of the diurnal variation in temperature across soil surfaces is a result of their differences in thermal inertia, which is primarily related to their moisture content and texture properties (Mulders 1987). Reginato et al. (1976) showed how daily maximum–minimum surface soil temperatures and maximum–minimum soil–air temperatures were inversely related to the water content (0–2 cm) in soils.

An indirect approach to estimate the water status of soil is to measure the TIR energy emitted by the overlying vegetation canopy. The greater the moisture content, the greater the soil's thermal inertia, because this increases its thermal capacity and conductivity. This makes wet soils appear cooler than dry soils during the day and warmer during the night.

Other parameters to consider include density, thermal capacity, and conductivity. Several laboratory studies have shown how quartz exhibits a density and thermal capacity similar to those of clay but has much higher values of conductivity. For this reason, its thermal inertia is greater, and it also has low values of emissivity (0.90). Soils with a high content of organic matter offer the lowest values of conductivity, which is why they tend to more clearly show temperature differences between day and night.

### 2.6.4 Thermal Signature of Water and Snow

From what has been mentioned, it is readily noted that water has a greater thermal inertia than other types of cover. This resistance to temperature changes is due to its high conductivity: the incident radiation is strongly absorbed and transmitted throughout the surface by convection movements, which explains the relative stability of water temperature. Thermal observation of oceans is very relevant to monitor a wide range of processes that affect both climatology and meteorology. Among them are the geographical variation of temperature throughout the oceans (Figure 2.21), the interannual variability of oceanic oscillations ("El Niño" being the most relevant), the meanders and eddies on major fronts and boundary events, the diurnal warming cycle, and the coral bleaching events (Merchant 2013). Estimating water temperatures is also quite useful for detecting fishing banks (Santos 2000) or plumes produced by power plants or other sources of water pollution.

In the case of snow, the temperature, size of the crystals, and liquid water content affect the measurement of radiating temperature. As we saw earlier, a snow cover has the lowest emissivity (or the greatest reflectance), which is why snow tends to register lower temperatures than the surroundings. The observation of snow in the TIR helps in differentiating the types of snow and monitoring snow cover, as well as estimating snow volume (Ferris and Congalton 1989; Kelly and Hall 2008).

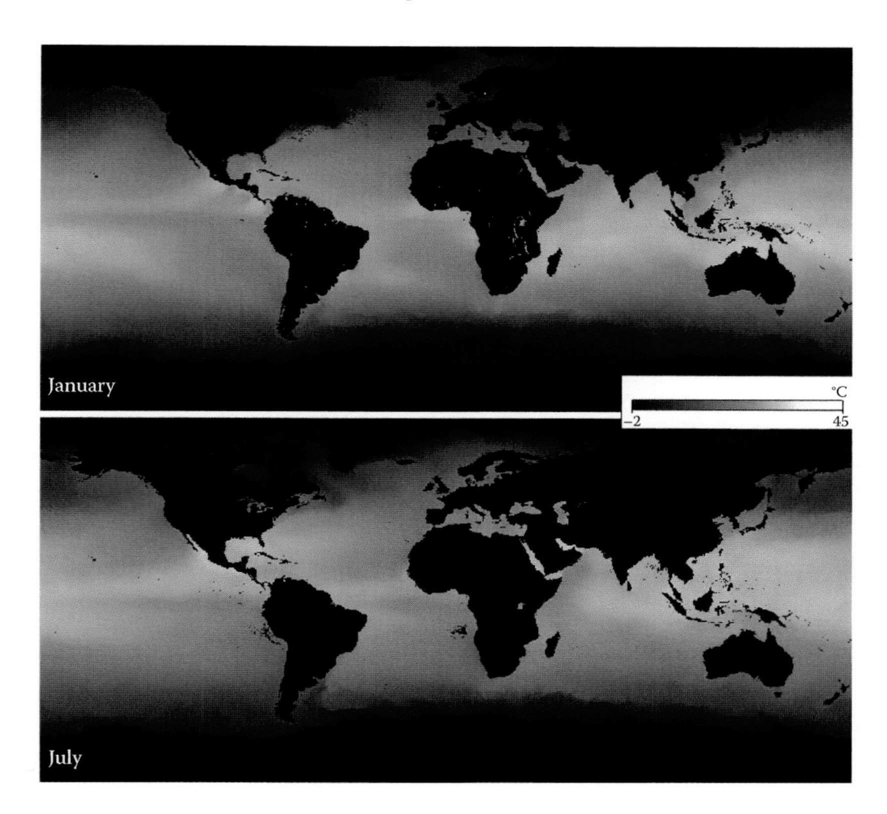


FIGURE 2.21 Global sea surface temperature estimated from 2014 MODIS Aqua data. Note thermal contrast between the two hemispheres in the two seasons. (Data extracted from http://neo.sci.gsfc.nasa.gov/, last access November 2015.)

## 2.7 MICROWAVE REGION

### 2.7.1 Characteristics of Electromagnetic Radiation in the Microwave Region

Despite its name "microwave" (MW), this spectral region comprises the longest wavelengths used in remote sensing. They also have other civilian and military applications (communications, tracking of moving objects, heating food, etc.), which have made this region quite popular in everyday use.

As a legacy of the military tradition of naming radar bands, MW bands are designated by a series of letters, referring to specific wavelengths or frequencies (Table 2.2).

From a remote sensing perspective, observations in the MW region are complementary and more complex than those performed in the optical region. The thematic applications of this spectral region are less extended than those using optical data, although they have remarkably grown over the last few years, thanks to the increasing availability of satellite missions equipped with MW sensors (ERS, Radarsat, Envisat, Sentinel-1, TOPEX, SMOS, TerraSAR, Cosmos, and ALOS, among them). The main

**TABLE 2.2**
**Bands Frequently Used in Microwave Remote Sensing**

| Name | Width (cm) | | Central Value | Frequency (GHz) | |
|---|---|---|---|---|---|
| Ka | 0.75 | 1.10 | | | |
| K | 1.10 | 1.67 | 1.0 | 10.9 | 36 |
| Ku | 1.67 | 2.40 | | | |
| X | 2.40 | 3.75 | 3.0 | 5.75 | 10.90 |
| C | 3.75 | 7.50 | 5.6 | 3.90 | 5.75 |
| S | 7.50 | 15.00 | 10.0 | 1.55 | 3.90 |
| L | 15.00 | 30.00 | 23.0 | 0.39 | 1.55 |
| P | 30.00 | 100.00 | 70.0 | >0.39 | |

advantage of the MW region is its high atmospheric transmissivity, which makes it possible to observe areas regardless of weather conditions. This is particularly useful in cloudy areas, such as equatorial or subpolar regions, which are particularly difficult to observe in the solar spectrum. MW wavelengths used in remote sensing are much larger than the usual size of atmospheric particles and therefore atmospheric effects in this region are almost negligible (particularly for $\lambda > 3$ cm, Figure 2.22).

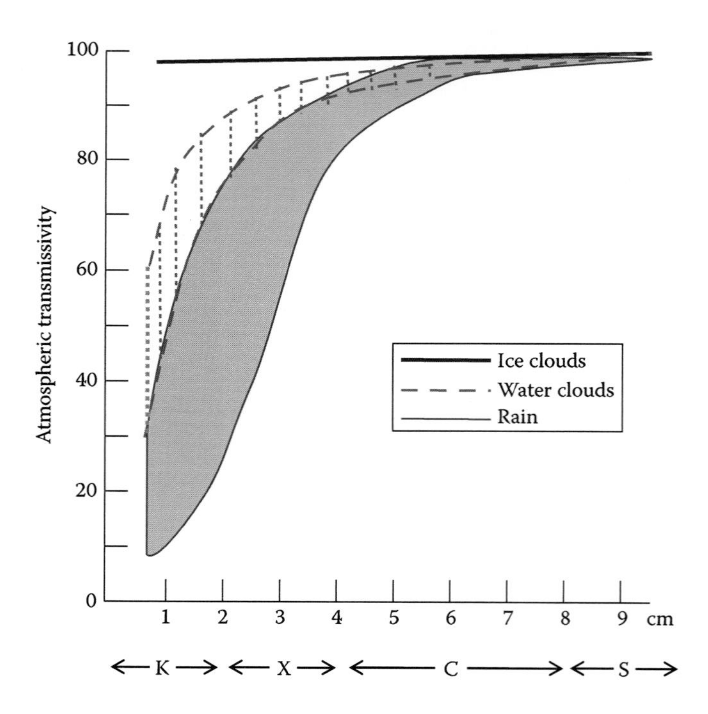


FIGURE 2.22 Atmospheric transmissivity in the microwave region. (Adapted from ESA, 1995, RADAR course notes.)

Remote observations in this spectral region have been mainly performed with active sensors (radar), which both emit MW energy and detect its return from the ground. However, there are also passive MW radiometers, which collect radiation from the surface, in a manner similar to optical sensors. We will review both types of sensors in Chapter 3, describing their technical characteristics and their impact on image interpretation, focusing here on the interaction of MW radiation with the main Earth covers. For better understanding this interaction, we need to examine first the radiation properties in this spectral band. In the case of passive sensors, the energy observed in the MW region is mostly derived from very cold targets, following Planck's and Wien's laws, and it has been used in mapping and monitoring snowcaps and ice caps. At these wavelengths, an adaptation of Plank's law calculated by Rayleigh–Jeans (Elachi 1987) is used:

$$M_\lambda = \varepsilon \frac{2\pi k T}{\lambda^4} \quad (2.20)$$

where

$M_\lambda$ is the spectral radiant exitance
$\varepsilon$ is the emissivity
$T$ is the temperature (K)
$k$ is the Boltzmann constant ($1.38 \times 10^{-23}$ W s$^2$ K$^{-1}$)
$\lambda$ is the wavelength

In this region, the values of $M_\lambda$ are very low, and there are frequent interferences between emitted, reflected, and transmitted signals. In this spectral region, different covers have very diverse emissivities, ranging from 0.41 for liquid water to close to 1 for ice. Since at Earth's temperatures the signal is very weak at these wavelengths, passive MW radiometers need to observe a wide area to obtain longer integration rates and better signal-to-noise ratios, and therefore the spatial resolution of those instruments is very low.

In the case of the radar sensor, the energy is generated by the sensor itself, which is also able to detect the returned energy. Radar measures both the time elapsed between the emission and reflection of the MW pulse and the intensity of the returned signal. The former is mainly used in remote sensing to estimate the distance between the sensor and the surface (see Radar altimeters in Chapter 3), while the latter can be used to extract relevant properties of observed covers. The intensity of returned energy (commonly named *backscatter pulse*) to the radar can be computed from

$$P_r = \frac{P_t G^2 \lambda^2 \sigma}{(4\pi)^3 r^4} \quad (2.21)$$

where

$P_r$ indicates the reflected backscattered power
$P_t$ is the power emitted by the radar
$G$ is the gain factor of the antenna
$r$ is the distance between the sensor and the surface cover
$\lambda$ is the wavelength
$\sigma$ is the backscatter coefficient

The σ values are the most complex to estimate, since they depend on a wide variety of factors: surface roughness, dielectric conditions, terrain characteristics (mainly slope and aspect) in relation to the emitted beam, wavelength, polarization, and incidence angle. In other words, as it is the case with other regions of the spectrum, land surface covers do not yield unique and constant backscatter coefficients. Different environmental factors modify the returned signal.

Surface roughness affects the intensity of the return signal. In simple terms, we can state that the higher the surface roughness, the greater the return signal. Depending on the nature of the substrate, the radar beam may be scattered upward in different directions (in the case of the soil) or internally (in the case of the vegetation), or it may be specularly reflected (in the case of the water). Depending on these three behaviors, the radar will register very different return signals (Figure 2.23).

Roughness is commonly defined as height variations within the target surface, but the impact of those variations on the incident beam depends on the wavelength and incidence angle of the emitted beam as well. At short wavelengths, a relatively flat surface can appear as rough, while at greater wavelengths it can behave as smooth. According to the Rayleigh criterion for roughness, a land surface is considered rough when

$$s_h \geq \frac{\lambda}{8} \sin \gamma \tag{2.22}$$

where

$s_h$ is the standard deviation of the height of the surface
$\lambda$ is the wavelength of observation
$\gamma$ is the radar incidence angle (between the radar beam and the normal to surface)

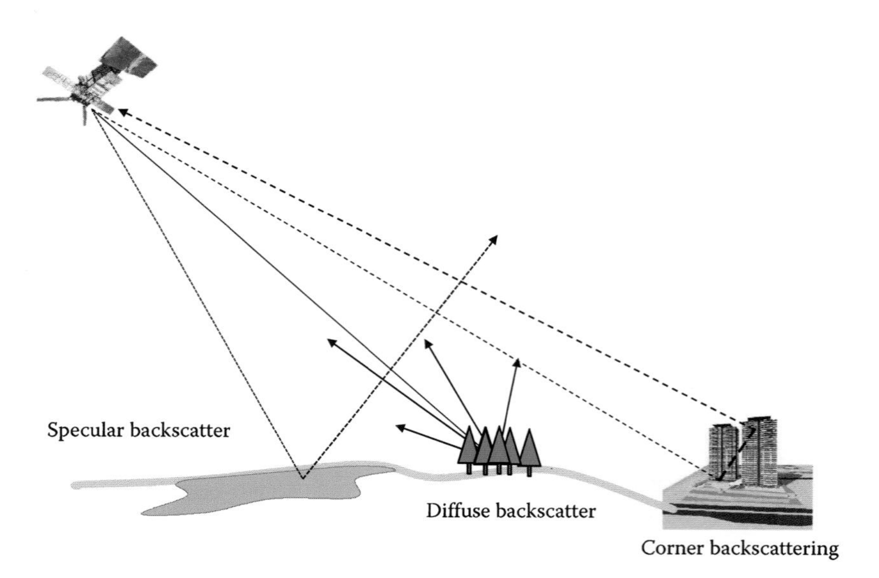


**FIGURE 2.23** Different types of radar backscatter (Adapted from Campbell, J.B. and Wyme, R.H., *Introduction to Remote Sensing*, The Guilford Press, New York, 1996.)

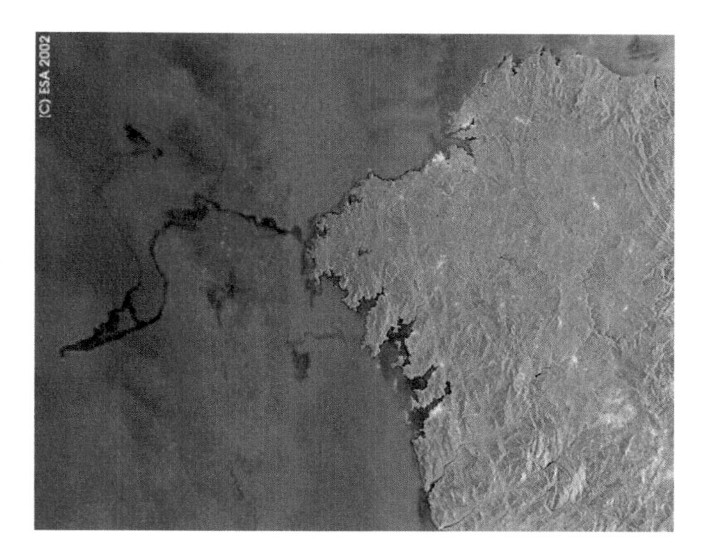


**FIGURE 2.24** Envisat radar image of the oil spill caused by the sinking of the Prestige tanker affecting the northwest of the Iberian Peninsula in 2002. (From ESA—Envisat mission.)

Briefly, this indicates that land covers with the same roughness will appear smoother at larger wavelengths and/or at greater incidence angles (more oblique observation).

The rougher the surface, the more diffuse the backscatter that will be derived from the target area, and the more intense the signal that will return to the radar antenna. However, if the surface cover is smooth, the backscatter is very low, since the reflection will be directed away from the radar antenna. This is the case in calm water bodies, which appear as dark tones in a radar image. Oil spills tend to make the water surface even smoother and, therefore, are clearly observable in radar images when wind speeds, and therefore ocean waves, are not very high (Figure 2.24). An extreme case of roughness is found in urban areas where building shapes frequently act as corner reflectors, which provide a high-intensity backscatter signal.

Along with surface roughness, the electrical characteristics of the surface play a critical role in determining the intensity of the backscattered signal. The ability of a surface to store and transmit electrons is termed the *dielectric constant*, and it is closely related to conductivity. In general terms, the drier the substance, the lower the dielectric constant, and the lower the backscatter. Dry soils have dielectric constant values between 4 and 7, while water has values as high as 80. Consequently, the presence of water in the soil or in vegetation can significantly alter the return pulse. Wet soils will return greater pulses than dry soils. Other good conductors are metallic surfaces, which give quite high values of backscatter. The dielectric constant also affects the emissivity in the MW region, which is a key parameter of radiant exitance (see Equation 2.20). Following Kirchoff's law, emissivity and reflectivity are complementary, and therefore the higher the dielectric values, the lower the emissivity. Consequently, wet surfaces tend to have lower emissivity.

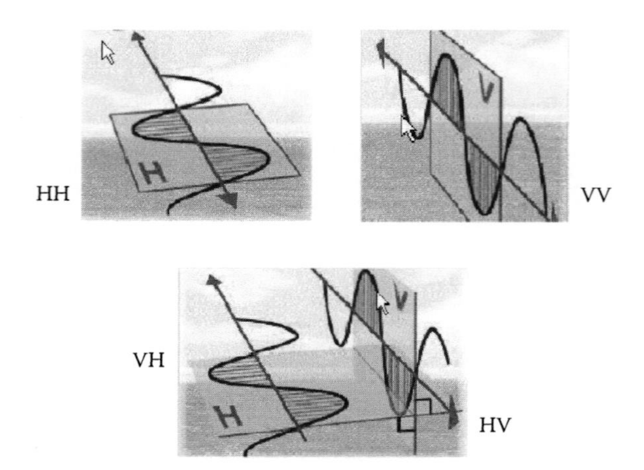


**FIGURE 2.25** Different polarization types of the radar pulses: H, horizontal; V, vertical. The first two are like polarization, and the last two cross-polarization.

In most remote sensors, the recorded energy comes from the most superficial layer observed by the detector. However, in the MW region, there is a great potential to obtain information from deeper layers of soil and vegetation. The penetration capacity depends on the wavelength (the longer the pulse, the higher penetration) and the dielectric constant (being higher for dry surfaces).

Finally, polarization refers to vibrational planes in which radar signals are transmitted and received (Figure 2.25). A polarized signal indicates that its vibration is restricted to a certain direction. The two most important forms are (1) like polarization and (2) cross-polarization; the former applies when both emitted and received signals have the same polarization (horizontal–horizontal or vertical–vertical), while in the latter the polarization changes from the emitted to the received beam (horizontal–vertical or vertical–horizontal). The polarization can alter the backscatter coefficient (σ) significantly in conjunction with the incidence angle, typically reducing σ as the incidence angle increases and when cross-polarization is used.

### 2.7.2 Characteristics of Vegetation in the Microwave Region

Vegetation MW signals are closely associated with variations in roughness and dielectric constant. The roughness is very dependent on the size, shape, direction, and number of leaves. The retrieval of vegetation parameters from a single radar band has proved to be very complex because of the wide variations of shapes and leaf conditions even within the same species. Attempts have been made in several forest inventory projects to separate, for example, coniferous from deciduous trees, using the L band (Deane 1980) as well as the S, X, and C bands. Many difficulties were encountered because of the great variability of backscatter coefficients in coniferous forests (Leckie 1990).

With respect to conductivity, the dielectric constant increases when the moisture content of vegetation is greater during the growing period of the plant. When the vegetation loses vigor or the contribution of the soil becomes greater, the dielectric constant diminishes drastically. For this reason, MW images have been used for estimating vegetation water content, although results are very dependent on species composition (Leblon et al. 2002).

In a pilot study to measure the moisture content of prairie grass covers, the best results were observed using shorter bands (X band) and at an incidence angle of 30° ($r^2 = 0.9$). For lower incidence angles (up to 0°), $r^2$ values close to 0.5 were obtained and a similar but worse correlation was observed in the C band (Gogineni et al. 1991).

The depth of penetration in vegetation canopies is dependent on the wavelength used as well as moisture conditions and the polarization. Canopy penetration depths are greater at long wavelengths (L band), where the substrate soil conditions influence the return signal and with like-polarization signals (HH or VV). There is decreased radar signal penetration into vegetation canopies with shorter wavelengths (the X band is more sensitive to the geometry and direction of the leaves) or with cross-polarization signals (HV or VH). According to some authors, the roughness of the vegetation can be estimated using the X band, the leaf biomass using the C band, and the wood biomass using the L band (Curran and Foody 1994).

### 2.7.3 Characteristics of Soil in the Microwave Region

From the previous paragraphs, it can be inferred that rough and wet soils have a greater backscattering coefficient and, therefore, will have bright tones in MW images. In dry soils, the penetration capability of radar increases up to several meters if the wavelength is sufficiently long. Urban areas, which are very rough, in long wavelengths appear with bright tones as well, which allow us to discriminate them from surrounding cultivated areas.

Both active and passive MW remote sensing can accurately measure surface soil moisture contents in the top 5 cm of the soil. The longer wavelengths (>5 cm) are best suited for soil moisture determinations. The theoretical basis for MW remote sensing measurements of soil moisture results from the large contrast between the dielectric properties of liquid water and dry soil. An increase in soil moisture content results in higher dielectric constant and therefore higher backscatter to the radar pulses and lower emissivity and radiant exitance to the MW radiometers. Therefore, radar instruments will generally receive a higher returned pulse from wet than dry soils, while passive radiometers will receive less radiation.

Global soil moisture estimations are currently based on different passive MW sensors on board several meteorological satellites (Figure 2.26). The challenge to retrieve this variable is to discriminate the impact of moisture from other factors affecting radiant exitance, such as vegetation cover or surface roughness. In addition, since MW exitance is very low, the sensors need to have a high radiometric accuracy to be sensitive enough to soil moisture variations (Wagner et al. 2013a). A recent project has cross-analyzed passive and active MW retrievals of soil moisture conditions to build a global temporal series that may be related to precipitation or other climate anomalies (Al-Yaari et al. 2014; Wagner et al. 2013b).

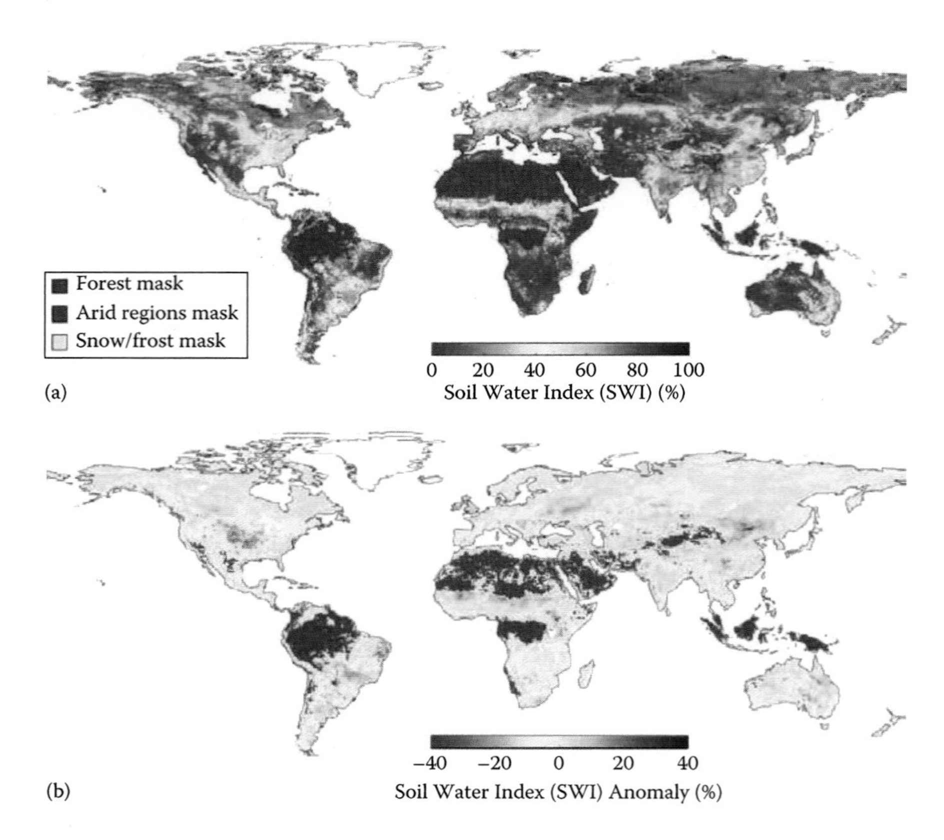


**FIGURE 2.26** Absolute soil moisture conditions (a) and anomalies (b) derived from the ASCAT sensor on board the METOP satellite. Period July–September 2012. (From Wagner, W. et al., The use of Earth observation satellites for soil moisture monitoring, WMO statement on the status of the global climate in 2012, World Meteorological Organization, Geneva, Switzerland, pp. 32–33.)

### 2.7.4 Water and Ice in the Microwave Region

The retrieval of water properties in the MW region is generally difficult since water surfaces tend to be very smooth and have very low backscatter signals. Wind or underwater relief may change the smooth character of water bodies, and therefore, wave patterns can be observed in radar images as long as the incidence angle is low. This application is very useful for detecting anomalies in the surface, for example, those caused by uncontrolled oil spills, as previously seen.

However, passive MW radiometers are very useful for monitoring sea ice, as they are more sensitive to cold temperatures than optical sensors. Different passive MW sensors have been launched to survey ice conditions, particularly in the Poles (Shepherd et al. 2012). A time series of these observations has detected a considerable reduction in Arctic sea ice, mainly at the end of the summer period.

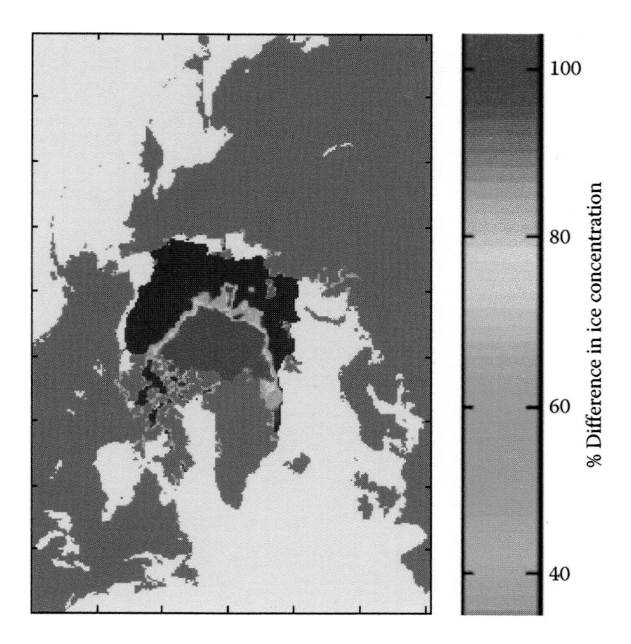


FIGURE 2.27 Comparison of summer sea ice in the Arctic based on passive microwave data: Blue area, ice covered in 1979 and open water in 2012. (Courtesy of Nansen Environmental and Remote Sensing Center, Bergen, Norway.)

In September 1979, the sea ice extent was estimated at 6.4 million km$^2$, while in September 2012 it was estimated to be 3.11 million km$^2$ (Figure 2.27). Passive MW sensors have also been used successfully to retrieve the snow water equivalent, while low-frequency radar has been used to discriminate between dry and wet snow (Kelly and Hall 2008).

On the other hand, MW nadir-looking radar can be used to estimate ocean wave heights, which are useful for monitoring ocean currents and oceanic relief. This has been the basis to monitor sea level trends. This information is very useful to predict global climate oscillations, such as the El Niño Southern Oscillation (ENSO), which implies an increase of ocean temperature and height in the South American Pacific coast and a parallel decrease in SE Asia. The average trends toward increasing sea level as a result of ice sheet and glacier melting are also well documented from radar altimeters (Figure 2.28; Ablain et al. 2015).

## 2.8 ATMOSPHERIC INTERACTIONS

We have assumed so far that the detection of EM radiation from the observed surface is dependent on the surface properties and observation conditions alone. However, we should also take into account the fact that this process does not happen in vacuum. Between the observed surface and the observing sensor, the atmosphere

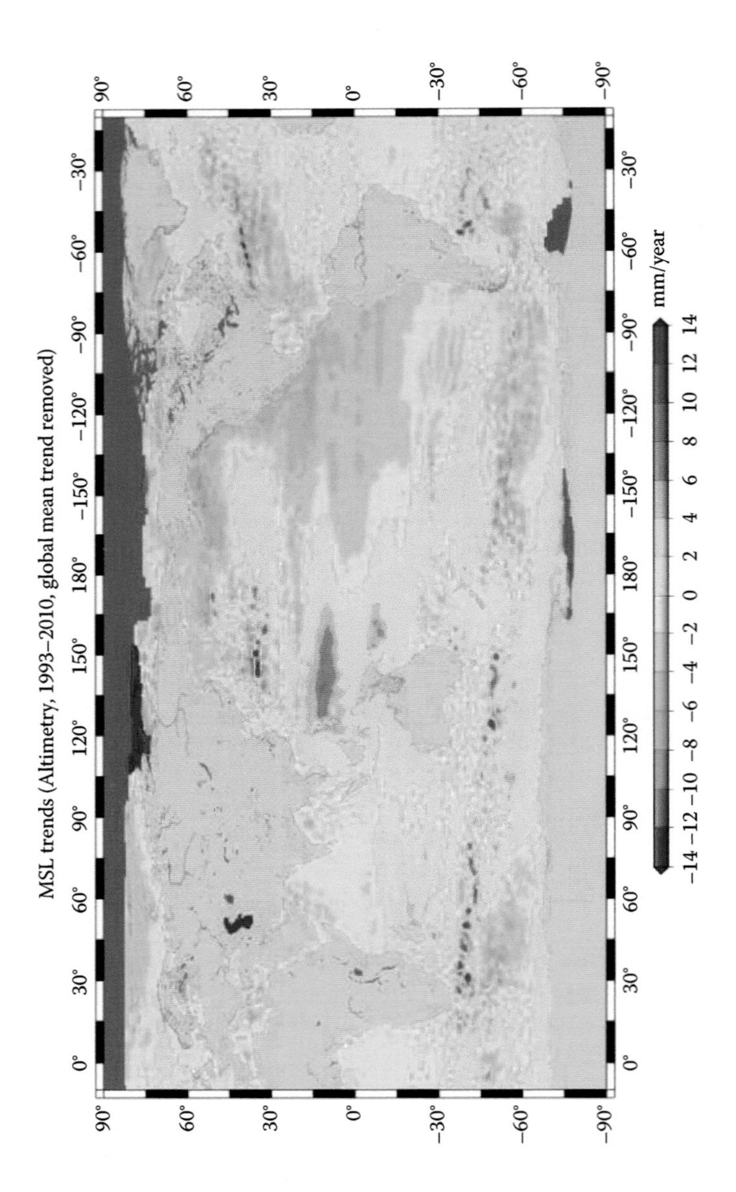


**FIGURE 2.28** Global mean sea level trends between 1993 and 2010 estimated from radar altimeters. (Courtesy of Anny Cazenave, Sea Level CCI project. Updated in Cazenave, A. and Cozannet, G.L., *Earth's Future*, 2, 15, 2014.)

interacts with both the incoming and outgoing radiation, thus distorting the signal that comes originally from the ground. This factor needs to be taken into account to accurately retrieve surface properties from satellite observations, as they observe the Earth from high elevations (commonly 600–900 km), and therefore the detected energy will most likely be affected by the different atmospheric layers.

The atmosphere is composed of many gases, including mainly nitrogen (N, 78%), oxygen ($O_2$, 21%), argon (Ar, 0.9%), water vapor ($H_2O$, 0.04%), carbon dioxide ($CO_2$, 0.033%), and ozone ($O_3$, 0.012%). In addition to these gases, there are also solid particles, called aerosols, that also interact with EM radiation. They are very diverse and include ocean salt, carbon, dust, and particulate matter produced by combustion processes.

The effects of these components can be grouped into three classes: (1) total or partial absorption of incoming or outgoing radiant energy; (2) reflection and scattering of incoming energy, which alters its path direction and intensity; and (3) emission, which adds new radiation to that emanating from the surface. These processes are wavelength dependent, and therefore they need to be tackled specifically for different spectral bands.

### 2.8.1 Atmospheric Absorption

The atmosphere behaves as a selective filter at different wavelengths, such that remote observations cannot be partially or totally conducted over certain spectral regions. The main causes for this absorption are the following atmospheric components:

- Atomic oxygen ($O_2$), which filters the ultraviolet radiation below 0.1 μm, as well as short ranges in the TIR and MW bands
- Ozone ($O_3$), which eliminates most of the ultraviolet radiation below 0.3 μm, as well as some narrow bands in the VIS spectrum
- Water vapor ($H_2O$), with one strong absorption around 6 μm and secondary absorption bands between 0.6 and 2 μm
- Carbon dioxide ($CO_2$), which absorbs in the TIR (15 μm), with important effects too in the MIR, between 2.5 and 4.5 μm

Since most remote sensors are designed to observe the Earth' surface, their spectral sensitivity is adjusted to focus on those spectral regions where atmospheric absorption is low (or alternatively atmospheric transmittance is high). As previously mentioned, these regions are known as atmospheric windows, the main ones being (1) VIS and NIR, located between 0.4 and 1.35 μm; (2) SWIR to MIR bands, from 1.5 to 1.8 μm, 2.0 to 2.4 μm, 2.9 to 4.2 μm, and 4.5 to 5.5 μm; (3) TIR, between 8 and 14 μm; and (4) MW, >1 cm, where the atmosphere is practically transparent (Figure 2.29).

Even though atmospheric transmittance at these wavelengths is generally high, it is not always 100%, as it depends on atmospheric conditions. The most obvious case is when the area is covered by thick clouds, which prevents surface observation in the solar and thermal bands. Solid particles (particulate matter or PM) derived from different sources have also an absorption effect, reducing visibility at long

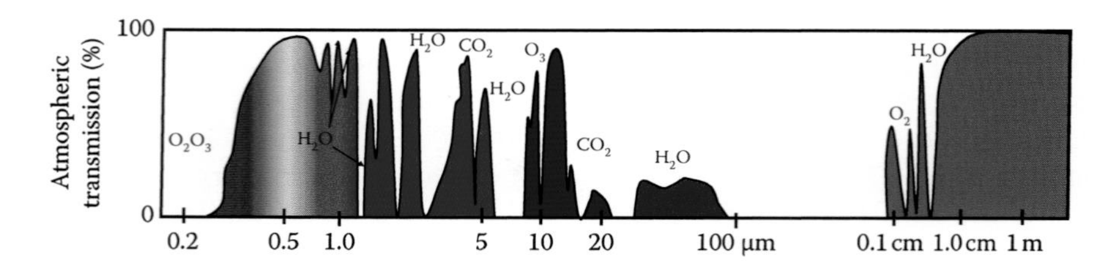


**FIGURE 2.29** Atmospheric windows suitable for Earth observations from space, along with the main gas responsible for absorption. (Adapted from http://earthobservatory.nasa.gov/.)

distances, as it is noticeable in a sunny day after recent rains, when the atmosphere gets much clearer as a result of PM precipitation.

Most terrestrial and oceanic sensors are adapted to atmospheric windows as they try to observe the ground. However, when the main goal of observation is actually the atmospheric components, remote sensors are designed to cover those regions where the target atmospheric component has a stronger absorption effect. More precisely, those sensors have generally observations bands with contrasting effects for that component: strong and weak absorption bands, since the estimation of their abundance is based on differential absorption retrieval.

This is the case, for instance, with greenhouse gases (GHGs), which are so relevant to better understand the Earth's climate. Several satellite missions have been developed to monitor carbon dioxide ($CO_2$) and methane ($CH_4$), two of the most relevant GHGs for climate change analysis (Dils et al. 2014). A dedicated sensor for these two gases, named SCIAMACHY, was launched in the Envisat's European Space Agency (ESA) satellite in 2002, followed by the TANSO-FTS sensor on board the Japanese mission GOSAT in 2009 (Figure 2.30).

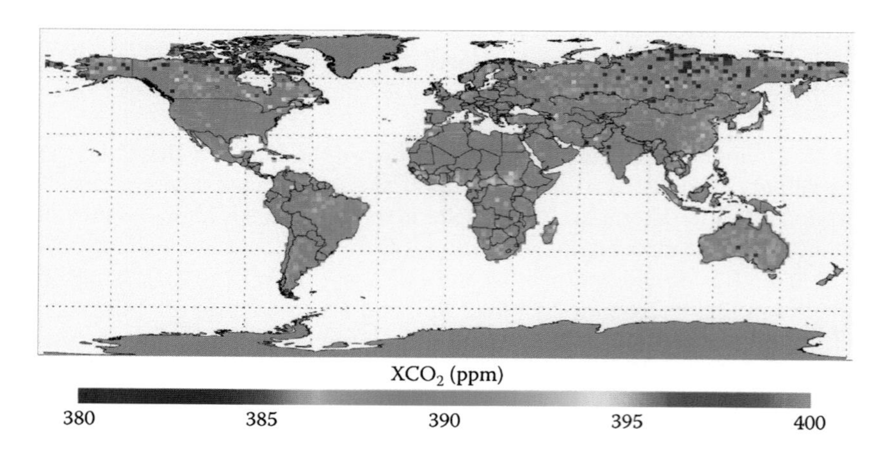


**FIGURE 2.30** Global $CO_2$ map retrieved from TANSO/GOSAT. (From Buchwitz et al., 2013, Fig. 3, http://www.esa-ghg-cci.org/index.php?q=webfm_send/148.)

### 2.8.2 Atmospheric Scattering

The scattering of EM radiation is caused by the reflection of incoming solar radiation by gases, aerosols, and water vapor present in the atmosphere. As a result, radiation detected by the sensor is a mixture of both surface and atmospheric reflected energy, and thus contains an undesirable noise that needs to be removed during the retrieval of the surface properties. At the surface, the direct irradiance is reduced, while diffuse radiance (coming from other objects) increases.

Aerosols originate from both natural and human causes. They can be oceanic, caused by the water movement, or continental, as in dust in suspension or particles emitted by combustion. Because these atmospheric particles are highly variable in time and space, correction of atmospheric scattering is rather problematic. However, this factor must be taken into account whenever a fine estimation of ground radiance is required: for instance, when satellite measures are compared with ground radiometric measurements, or when we try to detect changes between two images acquired several years apart. Normally, in situ atmospheric measurements are not available at satellite acquisition times, and therefore atmosphere correction often needs to rely on data extracted from the image itself. Some methods may involve changes in reflectance across several bands or from several observation angles (see Section 6.6).

Ideally, aerosol impacts need to be removed when the objective is the retrieval of ground properties. However, aerosols can also be the target of a remote sensing analysis, as they are very relevant for climate or human health, for instance. In this case, the retrieval of aerosol optical depth (AOD) is based on differential absorption (comparing reflectance measurements in wavelengths with high and low dispersion) or on differential observation angles (comparing reflectance measurements from nadir and off-nadir observations, analyzing different atmospheric thicknesses). Commonly, the satellite retrievals of AOD are compared with estimations from ground-based photometers, with very high correlations (de Leeuw et al. 2013).

Based on their origin and characteristics of aerosols, they vary widely in size, which causes different types of dispersion, since this process is highly dependent on the diameter of the scattering particles. The three main types of atmospheric dispersion are Rayleigh scattering, which is caused by particles smaller than the wavelength of the radiation; Mie scattering, when the particle is similar in size to the wavelength; and nonselective scattering, when the size of the particle is greater than the wavelength.

Rayleigh dispersion is highly dependent on the wavelength and mainly affects the shortest bands. It causes, for example, the blue color of the sky because blue is the lowest wavelength of the VIS spectrum. In aerial photography, the effect of Rayleigh scattering is very clear, as it leaves a bluish tone in photographs taken from high elevations.

Mie scattering is also dependent on the wavelength of incoming radiation, although to a smaller degree than Rayleigh scattering. Aerosols and atmospheric dust are the main causes of this type of dispersion, although it is also caused by forest fires or coastal mist.

Finally, the nonselective scattering affects various wavelengths equally. That is why clouds and fog tend to appear in gray tones, since they disperse the different bands of the VIS spectrum in a similar way.

### 2.8.3 Atmospheric Emission

There are also atmospheric emission influences that are important in the TIR. These must be corrected for in order to accurately retrieve the temperature of the surface from satellite acquired images. Similar to all materials with a temperature above absolute zero, the atmosphere emits radiant energy that will travel in the direction of and be detected by the sensor. Thus, as with atmospheric scattering, the atmospheric emission signal must be separated from the thermal exitance leaving the surface.

## 2.9 REVIEW QUESTIONS

1. According to Planck's radiation law, choose the correct sentence:
   a. Warm bodies emit more energy at longer wavelengths.
   b. Cold bodies emit more energy at longer wavelengths.
   c. Cold bodies emit more energy at shorter wavelengths.
   d. Warm bodies emit more energy at shorter wavelengths.
2. Which of these spectral regions is independent of Sun's energy?
   a. Thermal infrared
   b. Shortwave infrared
   c. Visible light
   d. Middle infrared
3. The solar spectrum is located between
   a. 0.4 and 0.7 μm
   b. 0.4 and 1.2 μm
   c. 0.4 and 2.5 μm
   d. None of the above
4. Which of the following parameters represents the energy detected by the satellite sensor?
   a. Irradiance
   b. Radiance
   c. Radiant intensity
   d. Radiant flux
5. Reflectance can be defined as the ratio between
   a. Reflected and emitted flux
   b. Reflected and scattered flux
   c. Reflected and incident flux
   d. Reflected and absorbed flux
6. Objects that have similar radiance from different observation angles are named
   a. Scatterers
   b. Lambertian
   c. Specular
   d. Amorphous
7. Low reflectance of healthy vegetation in the blue band is caused by
   a. Chlorophyll absorption
   b. $CO_2$ emission
   c. Thermal inertia
   d. Water absorption

8. When the plant is in a dark soil, as the leaf area index increases, it
   a. Increases reflectance in the visible spectrum
   b. Reduces reflectance in the NIR
   c. Increases reflectance in the NIR
   d. Increases reflectance in the SWIR
9. Which band is more sensitive to soil moisture discrimination?
   a. Visible
   b. SWIR
   c. TIR
   d. NIR
10. Which band is more sensitive for discriminating snow from clouds?
   a. Visible
   b. SWIR
   c. TIR
   d. NIR

# 3 Sensors and Remote Sensing Satellites

In this chapter, we describe the basic sensor systems that acquire remote sensing images, as well as the main Earth observation (EO) satellite missions. Understanding the basics of how images are acquired helps the interpretation process, while being familiar with satellite missions provides an overview of the information potentially available for different applications.

One of the most common ways to classify sensor systems is on the basis of the main mechanism they use to detect electromagnetic energy. In this regard, we could establish two broad groups of sensors: (1) passive sensors, which collect radiations derived from external sources; and (2) active sensors, in which the sensor system emits its own energy to the target and collects later the reflection of that flux to characterize the observed areas. A further distinction among passive sensors takes into account the method used to record the incoming energy. In the past, photographic cameras were the most common, but nowadays electronic systems are the most extended.

In addition to the characteristics of the sensor system, the user of remotely sensed data should also be aware of the different missions developed to observe different aspects of the Earth system. The beginnings of EO were characterized for general-purpose platforms, which provide information for a wide range of applications. As the prices of both sensor and satellite platforms are rapidly decreasing, it is more and more common nowadays to design missions for specific variables; therefore, they optimize the spatial, temporal, and spectral resolutions required for that target variable (clouds, $CO_2$, ocean color, etc.). We will review the most used EO satellite systems, including land, ocean, and atmospheric applications.

Before describing the different sensors and EO mission systems in more detail, we will review the aspects that affect the quality of a sensor system, in terms of the precision of its measurements (resolution). These aspects are relevant to select the most appropriate sensor for a particular application.

## 3.1 RESOLUTION OF A SENSOR SYSTEM

In general terms, we can define the resolution of a sensor system as its ability to discriminate between information (Estes and Simonett 1975). This concept includes several aspects that deserve detailed discussion. First, the resolution of a sensor refers to the system as a whole instead of its individual components. For example, improvement of the lens does not imply acquisition of higher photographic resolution if a more sensitive film is not used or if exposure conditions are not modified.

On the other hand, discrimination refers to the ability of the sensor to distinguish a specific object from other objects. This process may simply determine that the object *is there* (detection) or it may imply as well a precise delimitation of its shape (identification). Obviously, the latter task requires a higher spatial resolution than the former (Robin 1998). For instance, an object can be detected through its effect on the observed global radiance, such as an abrupt increment in temperature due to a volcanic eruption, but in order to identify and characterize the volcanic crater itself, the sensor should be able to resolve objects much smaller than the crater size.

The discrimination of information in remotely sensed images refers not only to the spatial detail but also to the number of spectral wavebands, their bandwidths, and spectral range covered, the temporal frequency of observation, and the signal-to-noise ratio or its ability to distinguish variations in the energy detected. One may also consider sensor potential to acquire information at different viewing angles or polarization channels. All of these informational dimensions are useful for discriminating target variables.

In summary, the concept of resolution embraces several aspects. The most common refers to the spatial resolution, but this feature is not necessary. The most important for image interpretation—the spectral, radiometric, or temporal resolutions—may be more relevant for specific applications.

### 3.1.1 Spatial Resolution

Spatial resolution identifies the smallest object that can be detected on an image. In a photographic system, spatial resolution identifies the minimum separation at which objects appear independent and isolated. It is measured in millimeters on the photograph or in meters on the ground, and it depends on the focal length of the camera and the height of the camera above the ground. In optical electronic sensors, the term "instantaneous field of view" (IFOV) is commonly used instead. The IFOV is defined as the angular section observed by the sensor, in radians, at a given moment in time. In practical terms, the most common definition of sensor spatial resolution is the size of the projected IFOV on the ground ($d$), computed as follows (Figure 3.1):

$$d = 2h \tan\left(\frac{\text{IFOV}}{2}\right) \tag{3.1}$$

where

$d$ is the distance on the ground per each information unit (pixel)
$h$ is the height of the observation

Since images are mostly acquired in digital form, the spatial resolution is commonly expressed as the length (in meters) of the minimum spatial units in the image, which are named pixels (from picture element). There are also more complex terms such as effective resolution element (ERE) or effective instantaneous field of view (EIFOV) that consider the detected signal as a composed modulation function (Townshend 1980). The total field of view (FOV) is the area observed by the satellite in a single image. It depends on the number of detectors and the optical system, and it is

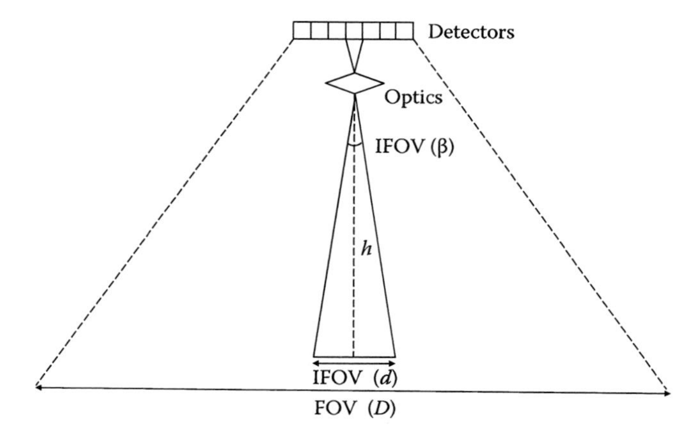


**FIGURE 3.1** Spatial resolution of a sensor system.

associated with the temporal frequency of the satellite overpass: the higher the FOV, the more frequent an image is acquired.

The spatial resolution of optical electronic sensors depends on orbital height, number of detectors, focal size, and system configuration. As for the antenna sensors, their resolution depends on the aperture angle, height of the platform, and the wavelength at which they work. The lower the height, the larger the aperture angle and the wavelength, and the more detailed the resolution.

Earth-observing sensors in operation cover a wide range of spatial resolutions (Figure 3.2). Sensors of higher spatial resolution now have 0.5–4.0 m pixel size; medium-resolution sensors have 20–50 m pixel size; sensors oriented toward global applications have a pixel size between 200 and 1000 m; and atmospheric oriented sensors typically have a very coarse spatial resolution (10–50 km).

Spatial resolution plays a major role in image interpretation because it affects the level of detail achieved. Obviously, the interpreter can only identify objects several times larger than the pixel size, although smaller features can be detected when enough radiometric contrast between the object and the background exists (for instance, a high-temperature fire can be detected even if its size is only a small proportion of the total pixel area).

The selection of the most convenient spatial resolution is closely related to the suitable scale for the particular problem under study. Additionally, spatial resolution affects the *purity* of a pixel signal. The smaller the size of the pixel, the smaller the probability that the pixel will be a mix of two or more cover types. A mixed pixel will likely have an average signal of the land covers present within the IFOV. As a result, that pixel may not resemble any of the mixing categories, which will make its identification difficult (Figure 3.3). However, an increase in spatial resolution does not always facilitate the discrimination of features since the internal heterogeneity within categories may increase as well (Cushnie 1987; Green 2000). Regarding visual analysis, the need to improve spatial resolution to facilitate a more accurate interpretation of the image is evident (Trolier and Philipson 1986).

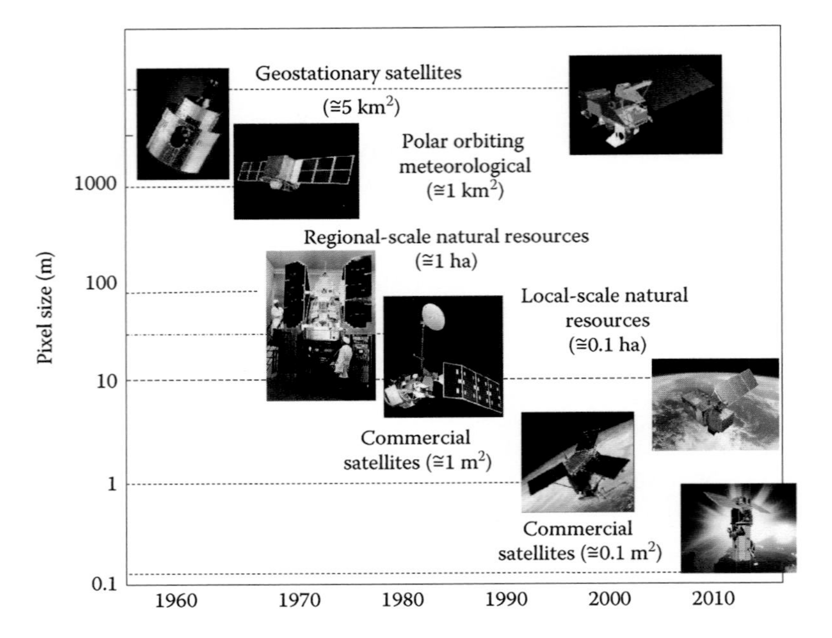


**FIGURE 3.2** Trends in sensor spatial resolution.

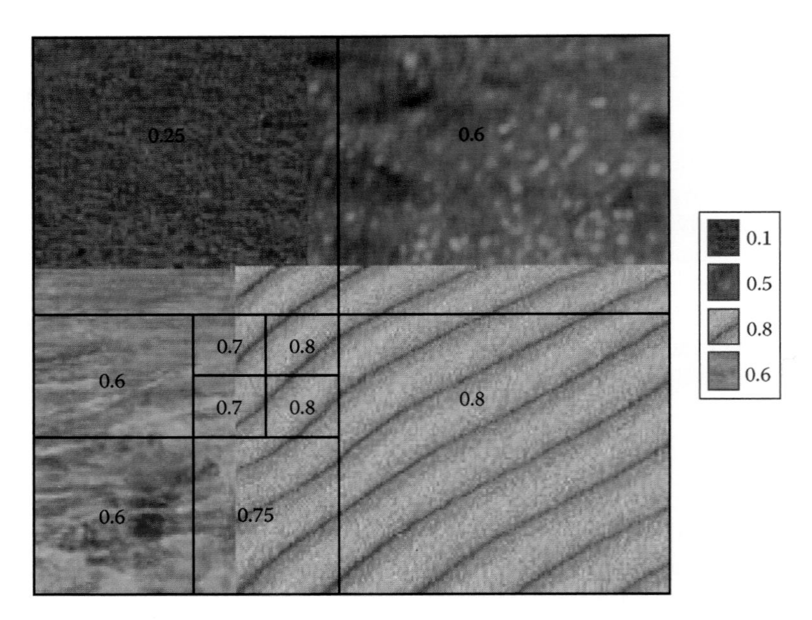


**FIGURE 3.3** Relationship between pixel size and discrimination power. As the spatial resolution increases (pixel size is smaller), the pixel is more likely to be homogeneous; therefore, the recorded signal will be more similar to the target cover. The upper right pixel has an average reflectance very similar to the blue category when, in fact, it does not have any area of this category.

### 3.1.2 Spectral Resolution

Spectral resolution refers to the number of bands provided by the sensor and their spectral bandwidths. Generally speaking, a sensor will provide better discrimination capacity as more bands are acquired. Ideally, those spectral bands should be narrow enough to identify specific absorption features that may be blurred otherwise (Figure 3.4).

Among EO satellite systems, the worst spectral resolution is found in the radar systems as commonly only a single band is acquired. Photographic systems acquire either panchromatic (one band) or color pictures (three bands). Digital sensors are generally multispectral (acquire a few bands, 4–7), although some of them have hyperspectral capabilities, acquiring dozens or cents of bands. The highest spectral resolution among the EO satellite sensors is the Hyperion, onboard the EO-1 satellite, which has 220 bands covering from 0.4 to 2.5 μm. NASA's Airborne Visible/Infrared Imaging Spectrometer (AVIRIS) has 224 bands with a similar spectral coverage but higher signal-to-noise ratio.

The selection of the number of bands, width, and spectral range measured by the sensor is related to the objectives that it is expected to achieve. While mining applications require multiple bands in the visible, near, and mid-infrared ranges, a meteorological satellite sensor may require just one or two bands in the visible range and several dozens in the middle infrared channels to estimate atmospheric gas composition.

### 3.1.3 Radiometric Resolution

Radiometric resolution denotes the sensitivity of the sensor, that is, its capacity to discriminate small variations in the recorded spectral radiance. In the case of

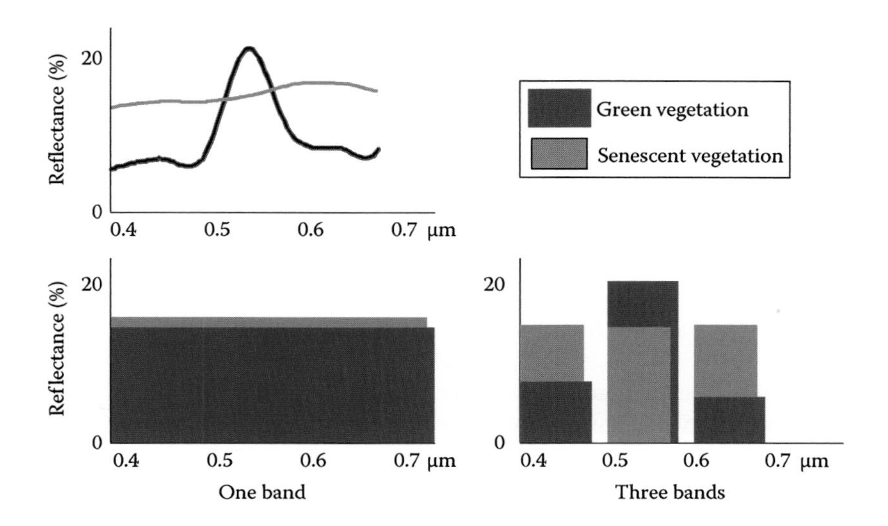


**FIGURE 3.4** Impact of bandwidth on discrimination of surface covers. A simple wide band does not discriminate between healthy and senescent vegetation, since the average value is similar, while this distinction is easily made from a multispectral observation.

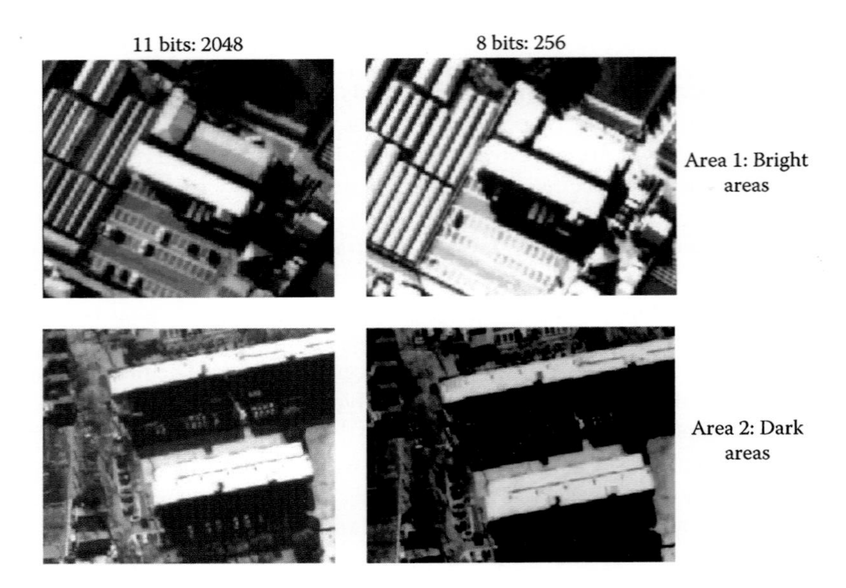


FIGURE 3.5 IKONOS images showing the impact of improving radiometric resolution. In the upper part of the figure, different roofs can only be discriminated when enough radiometric resolution is available, while in the lower part, the same applies to the darker areas. (Courtesy of Indra Espacio.)

photographic systems, the radiometric resolution of the sensor is given by the number of gray levels captured by the film. In optical electronic sensors, the image is acquired digitally, and the radiometric resolution is commonly expressed as the range of values used for coding the input radiance or, more precisely, as the number of bits used to store the input signal. An 8-bit sensor system may discriminate 256 different input radiances per pixel ($2^8 = 256$, therefore a range of 0–255). As more speed is available for data transmission and data storage is more accessible, the radiometric resolution of sensors is increasing. Formerly, most sensors provided 8-bit coding, while nowadays several have 11–16 bits per pixel (for instance, GeoEye, WorldView, MODIS, or Landsat 8 OLI).

Radiometric resolution is more critical in digital than in visual analysis. The number of gray levels that the human eye is able to discern does not exceed 64, and it cannot distinguish more than 200,000 colors. It seems redundant to have even 256 digital values per band (16.8 million values in a three-band color image). However, when the interpretation is digital, computers take advantage of all the available range, in which case high radiometric resolution is important to discriminate objects with similar spectral signatures, which would not be possible with less sensitive sensors (Figure 3.5).

### 3.1.4 Temporal Resolution

Temporal resolution refers to the observation frequency (revisiting period) provided by the sensor. This cycle is a function of the orbital characteristics of the satellite (height, speed, and declination), as well as the sensor FOV. Sensors with high temporal resolution have coarse spatial resolution as they will be able to observe a larger area in each image acquisition. Consequently, they also have wide FOV, which implies geometrical problems, as the same area may be observed from very different angles in a daily time series. We will later comment how this problem can be mitigated (see Section 6.7.3.4). If the sensor observes a smaller area (narrower FOV), it will need more orbits to repeat the same observed area; therefore, the acquisitions will have higher time gaps. To alleviate this problem, some sensors can make off-nadir observations, detecting adjacent areas to the orbit (pointing capacity typically goes from 0° to 20°). This is the case of cameras on board the SPOT or WorldView satellites. An alternative to nadir-looking sensors is to have several satellites with the same characteristics. This is the case of the RapidEye constellation (owned by a private company), which provides daily global acquisition frequency at 6.5 m resolution using five satellites.

The temporal resolution of EO sensor system varies according to the objectives set for the mission. Meteorological satellites observe very dynamic phenomena and, therefore, provide the highest temporal resolution: the geostationary satellites acquire images every 15 or 30 min, and the polar-orbiting satellites, such as the NOAA (United States), METOP (Europe), or Fengyun (China), provide world coverage every 12 h. Normally, higher-spatial-resolution sensors offer lower temporal resolutions, from 16 days (as the case of the Landsat series) to 28 days (for ERS radars). However, a growing tendency exists toward having a higher temporal resolution even for medium- or high-spatial-resolution satellites. For instance, the Sentinel-2 satellite of the European Space Agency (ESA) will provide 10–60 m resolution images for 13 spectral channels every 2–5 days when the two satellites of the constellation are active.

Before closing this section, it is important to consider that the effective temporal resolution of an EO system may be longer than the theoretical one, depending on the acquisition plans and the atmospheric conditions. If the satellite system is not programmed to register all possible images, the actual historical archive will have much less images than expected. Most privately owned satellites only acquire images on demand; therefore, the actual availability of historical images may have quite a long temporal gap. Finally, the cloud coverage is an important impediment of temporal resolution, since optical and thermal sensors cannot observe areas under clouds. In these areas, the historical archive even from systems with continuous acquisition programs may be quite limited.

### 3.1.5 Angular Resolution

This concept of resolution is very recent and refers to the sensor's capacity to make observations of the same area from different viewing angles (Diner et al. 1999). As an introductory concept, it is commonly assumed that terrestrial surfaces exhibit Lambertian reflection, which implies having similar reflectance, independently of the observation angle. This may lead to notable errors, especially in those surfaces with strong directional effects when observed from different angles. The variations in reflectance with viewing and illumination angles are commonly named

**FIGURE 3.6** Diagram of data acquisition by the MISR sensor. (From https://www-misr.jpl.nasa.gov/.)

bidirectional reflectance distribution function (BRDF; see Section 6.7.3.4). One way to model these effects is to observe the surface from different directions, thus facilitating a better characterization of the BRDF. Multiangular observations are also of great interest in estimating some atmospheric properties such as aerosol thickness or cloud height.

Several sensors with multiangular observation capabilities have been designed in the last decades, providing a better characterization of atmospheric components (particularly aerosol optical thickness). This is the case of ATSR 2 (Along-Track Scanning Radiometer), which was launched in 1995 onboard the European Earth Resources Satellite (ERS), which has a conic scanning mechanism providing almost simultaneous observations at nadir and 55° forward. The French mission Polarization and Directionality of the Earth's Reflectance (POLDER), installed in the Japanese satellites ADEOS-1 and 2, includes a spatial sampling at different observation and illumination angles. The Multi-Angle Imaging Spectroradiometer (MISR) sensor, launched onboard the Terra platform in 1999, provides nine observation angles almost simultaneously from the same zone and at different wavelengths (Figure 3.6).

### 3.1.6 Relationship between Different Resolution Types

It is important to emphasize that all aspects of resolution are closely interconnected. Acquiring high-spatial-resolution images usually implies reducing temporal resolution and/or spectral resolution. Considering limitations in data transmission or onboard recording, a sensor is able to optimize only one or two of the resolution types. Increasing any one of the five types of resolutions involves a considerable amount of additional data transmission and processing, for both the sensor and the ground segment.

Since the various types of resolutions have a great impact on data volumes, each remote sensing mission tries to emphasize a particular resolution that best fits its mission requirements. If the system is designed to detect dynamic phenomena, temporal coverage is critical, even at the expense of spatial resolution, as it is the case of meteorological satellites. However, if the mission is oriented for mineral exploration, the spectral and spatial details are more important, while the temporal coverage becomes less significant. For this reason, planning a general-purpose mission is very problematic because it is very difficult to accomplish the requirements of different end-user communities, and the compromises that are frequently made to meet budget constrains have historically resulted in remote sensing missions that serve experimental but not operational users, since a single mission cannot cover all needs.

As a summary of what has been discussed about the different types of resolutions, it should be emphasized that identifying the quality of a sensor based on its spatial resolution is very simplistic. High spatial resolution is required for certain applications (e.g., urban or agricultural mapping), but it might be secondary for other applications, such as fire detection or geological surveys. Assessment of natural hazards, for instance, requires high temporal resolution, so does weather forecasting. Other remote sensing applications must optimize spectral resolution, as in estimation of crop productivity or water demands. Consequently, the type of resolution to be prioritized is critical to selecting the most adequate sensor for a particular application.

## 3.2 PASSIVE SENSORS

Passive sensors are limited to measuring the electromagnetic radiation derived from external sources—energy either reflected from solar radiation or emitted by the Earth's surface. These sensors have been extensively used in remote sensing for the last few decades and include photographic cameras, electro-optical scanners, and MW radiometers.

### 3.2.1 Photographic Cameras

In the past, photographic cameras registered the incoming radiation of the target object using a photosensitive film. An optic system (lens, shutter) controlled the acquisition conditions and the time of film exposure. If the film grain and the optical system were fine enough, photographs provided a very high spatial resolution, even from long distances. For this reason, they were extensively used for both civilian and military surveillance.

Most aerial remote sensing used panchromatic film, in which the visible spectrum was captured on just one emulsion layer and images were displayed in gray tones. In color film, several photochemical layers were mixed, each of which was sensitive to one band of the visible spectrum (blue, green, or red). Infrared black-and-white films captured the incoming radiation between 0.7 and 0.9 μm in gray tones. Finally, color infrared film was sensitive to three bands in the green, red, and near-infrared (NIR) regions.

Photographic cameras were the most used systems to collect aerial remote sensing data, since they provide high-resolution images with a relatively simple technology. However, they had a more limited use on satellite platforms, because the recording capacity was constrained by the amount of film that a satellite could carry on board.

However, several photographic cameras were mounted in satellite missions, especially during the manned spacecraft missions of the 1960s. The first space photographs were acquired by the Mercury 4 mission, followed by the Gemini and Apollo programs. Initially, these photographs were taken by astronauts spontaneously, but later photographic missions were prepared for the observation of natural resources (see Figures 1.3 and 1.5).

The *Skylab* space station, active and with different crews during 1973, conducted several photographic experiments, within the Earth Resources Experiment Package (EREP). This program included a multispectral camera (S 190A), with six lenses sensitive to several wavelengths between 0.4 and 0.9 μm, and the Earth Terrain Camera (ETC), equipped with a long focal lens (457 mm), which provided high-resolution photography (15 m) (NASA 1974). These photographs were used for diverse thematic studies, including land cover, geological mapping, and vegetation and crop mapping (Links 1976; NASA 1974).

Following this experience, several camera systems were mounted on the Space Shuttle missions, the most common being the Hasselblad cameras with 70 mm film and diverse lenses (from 40 to 250 mm focal length). However, the most innovative cameras were the metric camera (RMK 30/23), the large-format camera (LFC), and the electronic still camera (ESC).

The metric camera RMK 30/23 was built by Zeiss to provide high-precision stereoscopic photography with a high level of coverage and a conventional format (23 × 23 cm). It was included on the first Spacelab mission of the ESA in 1983, onboard the Space Shuttle. This mission acquired around 1000 photographs with panchromatic and color infrared film over several parts of the world, covering an approximate region of 11 million km$^2$ (Konecny 1986). The focal length was 305 mm, which allowed an approximate scale of 1:820,000 at a height of 250 km, with an estimated spatial resolution between 20 and 30 m. The stereoscopic coverage was set at 60% overlap, although it was extended to 80% in rough terrain (see Figure 5.13).

The LFC was designed for precise mapping of the Earth's surface. Its main novelty was the extension of the area covered from the conventional 23 × 23 cm to 23 × 46 cm. The LFC was mounted on the Space Shuttle mission 41-G (1984), obtaining a wide collection of photographs using various panchromatic, color, and color infrared films with up to an 80% overlap. A 305 mm (12 in.) lens was used, which resulted in ground coverage of 225 × 450 km per photograph and resolution of 14–25 m, with a precision in height estimation of 30 m, at a 1:50,000 scale (Togliatti and Moriondo 1986). Photographs acquired by this camera proved especially valuable for natural resources assessment (Francis and Jones 1984; Lulla and Holland 1993). A good example of the information acquired by this camera is shown in Figure 3.7, which covers the urban region of Madrid in Central Spain. The fine spatial resolution of this camera can be seen in the clear display of urban structures such as street networks, main roads, and gardens.

FIGURE 3.7 LFC photograph over the city of Madrid. (Courtesy of R. Núñez de las Cuevas and F. Doyle.)

The ESC is a charge-coupled device that produces near-film-quality images in digital format. It was utilized on several missions since 1991. The current spectral response range was from 0.4 to 1.1 μm (Lulla and Dessinov 2000). The ESC was used in different missions of the International Space Station (ISS), providing a large collection of satellite photographs. Many other digital cameras have been used from the ISS. NASA created an image web server to facilitate the access to those photographs, as well as those from other photographic missions of the ISS and Space Shuttle (http://eol.jsc.nasa.gov/, last accessed in June 2015). Figure 3.8 shows an example of a photograph acquired from the ISS by a Nikon D3S camera with a focal

FIGURE 3.8 Example of an ISS photograph from South Africa. (From http://eol.jsc.nasa.gov/.)

length of 50 mm, showing the southwestern region of South Africa. The photograph was acquired on May 9, 2013, a few months before the death of Nelson Mandela.

Other interesting sources of photographic images were those acquired by the military surveillance missions, as they provided a historical perspective on land use–land cover transformation, particularly interested in those areas where aerial photography was not available. The White House announced in 1995 the decision to open part of the military satellite photographs archive to the general public. The released missions covered the period between 1960 and 1972 and provided a very valuable material for long-term change analysis. The archive includes photographs acquired by the Corona, Argon, and Lanyard satellites over more than 100 missions (Figure 3.9). They have very diverse formats, depending on the missions. Their spatial resolution varies from 2 to 150 m, although they are commonly between 2 and 10 m. The nominal scales vary between 1:275,000 and 1:4,250,000, and the area covered by the photographs ranges from 18 × 234 to 480 × 480 km (Day et al. 1998). Another set of historical photographs was declassified in 2002 covering the period up to 1980. These photographs are available at the U.S. National Archives, and some of them are digitized and can be obtained through a dedicated USGS image server (https://lta.cr.usgs.gov/declass_1, last accessed June 2015).

Russian satellites included several photographic sensors, some of high precision. One of the most outstanding was the multispectral camera MKF-6, installed on board the missions of the satellite Soyuz. Equipped with six lenses, the MKF-6 obtained photographs over six spectral bands, from blue to near infrared. Other cameras have also been widely used on board the RESURS-F satellite, such as the KFA-200 camera with a spatial resolution of 25–30 m and three spectral bands (green, red, and near infrared), the panchromatic KFA-1000 with a resolution of 8–10 m, and the multispectral camera MKF-4 with a 12–14 m spatial resolution (Kramer 2002). These photographs are valuable for natural resources studies (Strain and Engle 1993). The KVR-1000 camera was on board the Russian satellite Cosmos and provided a resolution of 2 m over an area of 160 × 160 km.

FIGURE 3.9 Corona photograph of the Santa Barbara coastline acquired in 1966. (From www.geog.ucsb.edu/~kclarke/Corona/gallery3.htm.)

### 3.2.2 Cross-Track Scanners

Digital sensors perform an onboard analog-to-digital conversion recording the incoming radiation in digital values that can be directly transmitted to the ground segment. The most commonly used in satellite missions have been the cross-track scanners. The incoming radiation is focused by a set of lenses and mirrors that rotate perpendicularly to the satellite track (Figure 3.10). The detected radiance is directed to a series of detectors that amplify the signal and convert it to a numerical value that can be stored on board or transmitted to a receiving antenna. The incoming radiation is commonly divided into several spectral bands, which are recorded by different detectors (most commonly Si photodiodes in the visible and near infrared and InGaAs in the SWIR). The information received on the ground is stored in magnetic media for further processing and distribution to end users.

The conversion from the analog radiance to digital values is constant in space and time, following a specific calibration procedure, although recalibration is often needed as the sensor degrades with time. In any case, digital values acquired by the sensor have a physical meaning, since they can be converted back to radiance and used for quantitative estimations of target variables.

The sensor samples the observed ground surface at systematic time intervals. Each of the instantaneous measurements becomes one digital value in the final image, which is the basis for the visual or digital interpretation of each pixel.

Multispectral scanner (MSS) systems were introduced in the early 1970s in EO satellite systems, such as the Landsat MSS, the Skylab S192, or the Nimbus Coastal Zone Color Scanner (CZCS). They presented several advantages over the then used analog cameras for EO satellites, as they were able to observe nonvisible bands

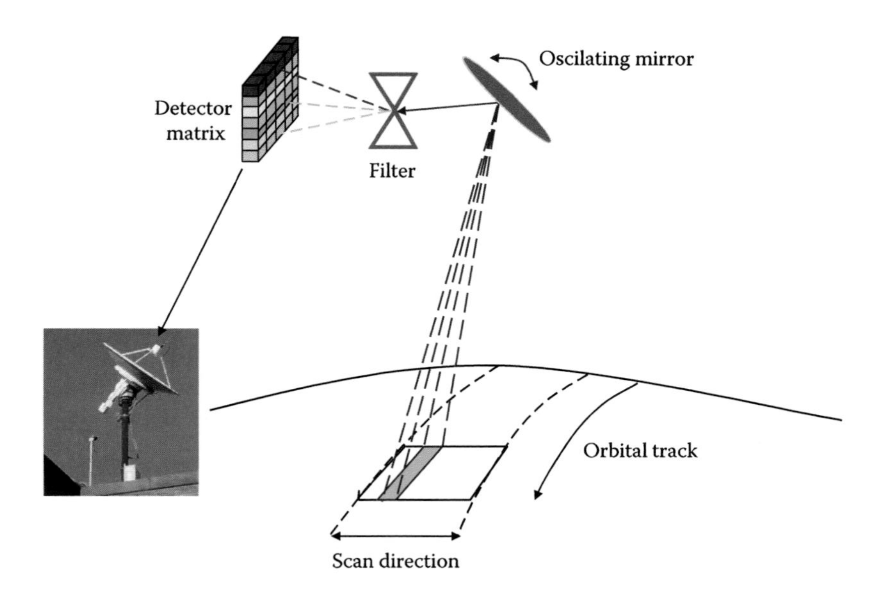


FIGURE 3.10 Scheme of the functioning of an across-track scanner.

(including the NIR, SWIR, and TIR wavelengths), and they did not need films to store the acquired images, therefore extending the useful life of the satellite. In addition, images were acquired in digital format; therefore, they were readily available for digital processing and more easily connected to geographical databases than analog photography.

Even though new detection techniques have been developed in the last decades, still across-track scanners are widely used in EO satellites. The MODIS sensor on board the Terra and Aqua satellites, the AVHRR on board the NOAA and METOP satellites, the SEVIRI instrument on board the Meteosat Second Generation, and many other atmospheric sensors used this technique. In addition, many historical missions have included across-track scanners, such as the Landsat satellites (MSS, Thematic Mapper [TM], and Enhanced Thematic Mapper [ETM]), the DMSP (Operational Linescan System [OLS]), HCMM (Heat Capacity Mapping Radiometer [HCMR]), and the Soviet program RESURS (MSU-E y MSU-SK), among others.

### 3.2.3 Along-Track (Push-Broom) Scanners

In the early 1980s, a new scanning technology was tested by the German space agency on board a Space Shuttle mission, which became operational in 1986 with the launch of the French satellite SPOT. Along-track scanners avoid the use of an oscillating mirror by detecting the whole FOV of the sensor system at once, using a linear array of detectors. The sensor explores each line simultaneously and creates the image along with the satellite orbital track (Figure 3.11). For this reason, these sensors are named push-broom or along-track scanners. Similar to the across-track

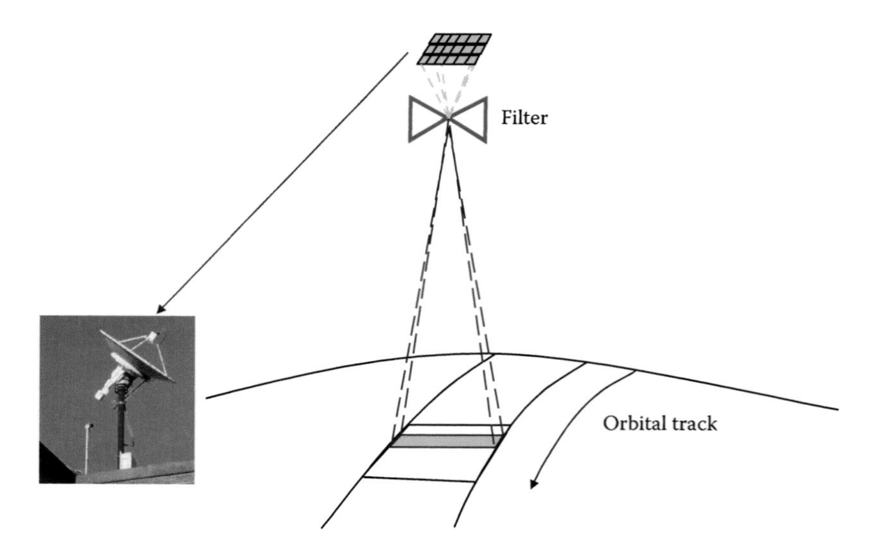


**FIGURE 3.11** Scheme of the functioning of an along-track scanner.

scanners, the incoming radiance is separated in different wavelengths and sent to solid detectors that record them in digital values.

This technology permits an improvement in the spatial resolution of the system over across-track scanners, as well as reduces the potential problems caused by the misalignment of the mirror oscillation and the satellite movement. Data are also acquired digitally and in different wavelengths beyond the visible spectrum.

Push-broom scanners may have more calibration problems than conventional scanners, since they have many more sensor arrays (3000 for SPOT-HRV versus 100 for Landsat TM). As their number increases, they require complex adjusting mechanisms to transform the received signal consistently; otherwise, the image would not be homogeneous across columns. For the same reason, push-broom scanners are more difficult to calibrate in the thermal infrared because they may be affected by heating of other detectors; therefore, most thermal scanners are still based on across-track systems.

Push-broom scanners are nowadays the most common digital cameras for EO satellites. The SPOT French family used several sensors on this kind (the *Haute Resolution Visible* [HRV]) and the VEGETATION. The Indian Remote Sensing (IRS) satellites have also used this technology since the late 1980s (PAN, LISS WIFS, and AWIFS sensors). NASA has also incorporated these sensors in the latest Landsat 8 mission (Operational Land Imager [OLI]), as well as in its precursor, the Advanced Land Imager (ALI). All high-spatial-resolution systems include push-broom scanners, such as the cameras on board the QuickBird, IKONOS, GeoEye, WorldView, or RapidEye systems.

### 3.2.4 Video Cameras

Video cameras were used in the first EO satellites programs, in both panchromatic and multispectral modes. The image was focused over a photoconductor, generating an electronic copy of the original image, and it is kept on this surface until a beam of electrons sweeps it again and is in equilibrium. The system was modified in the case of the Return Beam Vidicon (RBV) on board the first Landsat satellites. On these systems, the signal of the camera was derived from the unused portion of the beam of electrons that returned along the same trajectory as the incident ray and was then amplified electronically. This mechanism provided enhanced resolution over conventional video cameras, and it was included in some satellite missions: Apollo, Mariner, Nimbus, and Bhaskara-1 and 2, as well as the first three Landsat satellites.

The use of digital video cameras in local-scale studies is more recent. Videography then became defined as a new remote sensing technique that allows gathering of calibrated information with a great level of detail and low cost (especially if using conventional cameras) (Linden 2000).

### 3.2.5 Microwave Radiometers

Although the most common sensors in the MW region are active (radar), passive sensors have also been used at these wavelengths. They are commonly known as microwave radiometers. The radiometer is integrated by a directional element, a receptor that detects and amplifies the microwave signal, and a detector. These sensors operate in the spectral range corresponding to MW frequencies, usually between 6.8 and 90 GHz, which is why they are not affected by atmospheric situations or illumination conditions. The spatial resolution of MW radiometers is inversely proportional to the diameter of the aperture and directly proportional to the wavelength. Therefore, obtaining a fine spatial resolution would require an immense antenna, which is not feasible in satellite missions. For this reason, most MW radiometers have coarse spatial resolutions.

Microwave radiometers have a limited range of applications, since the radiant exitance of natural targets is hardly perceivable in this band, but they provide critical information for ice, rain, and snow monitoring. In fact, one of the first attempts to map Antarctica was based on the ESMR microwave radiometer on board the Nimbus 5 satellite. A thermal map of both polar caps was obtained from these images in 1977, revealing interesting information for the study of these regions hardly accessible through any other means (Hall and Martinec 1985). More recent studies for analyzing ice melting in oceans (Figure 3.12) are being conducted with data acquired by the Advanced Microwave Scanning Radiometer (AMSR-E) aboard the NASA satellite Aqua and the Japanese satellites ADEOS-II and Shizuku. Data of microwave radiometers are also very useful for estimating soil moisture conditions (Al-Yaari et al. 2014). The Tropical Rainfall Measuring Mission (TRMM) had a microwave radiometer designed to estimate rainfall over the ocean. This mission is continued by the Global Precipitation Mission (GPM: Smith et al. 2007), launched in 2014, which includes a core satellite with a precipitation radar and a multichannel passive microwave rain radiometer.

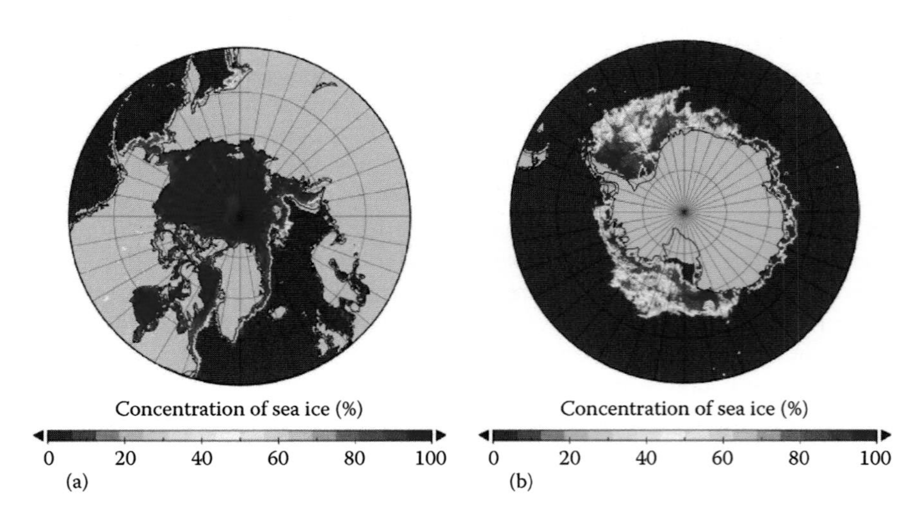


**FIGURE 3.12** Ice concentration map derived from the AMSR-E data—Arctic (a) and Antarctica (b)—from January 3, 2008. (Courtesy of Nansen Environmental and Remote Sensing Center, Bergen, Norway.)

## 3.3 ACTIVE SENSORS

### 3.3.1 Radar

Active systems have the capacity to generate energy pulses and collect them after the surface target reflects them back. Radar is the best-known active sensor system. It is an active MW system working in different spectral bands between 0.1 cm and 1 m. Each pixel in a radar image represents the backscattering coefficient of that area on the ground, and the stored value is greater as more intense signals are received. Working with microwaves, radar can acquire images under any atmospheric condition (Figure 3.13) and independently of solar illumination. Therefore, it has been greatly promoted in the last two decades (Henderson and Lewis 1998; Lewis et al. 1998). Currently, there are many radar satellite missions, acquiring images at different bands and with different polarization conditions. Band C is the most extended, being used by ERS-1 and 2, RADARSAT-1 and 2, ENVISAT ASAR, Sentinel-1, and Risat-1. Shorter wavelengths (band X) were used by the Shuttle Radar Topographic Mission (SRTM), TerraSAR-X, and COnstellation of small Satellites for the Mediterranean basin Observation (COSMO-SkyMed), while longer wavelengths (bands L or P) are used by JERS-1 and ALOS-PALSAR.

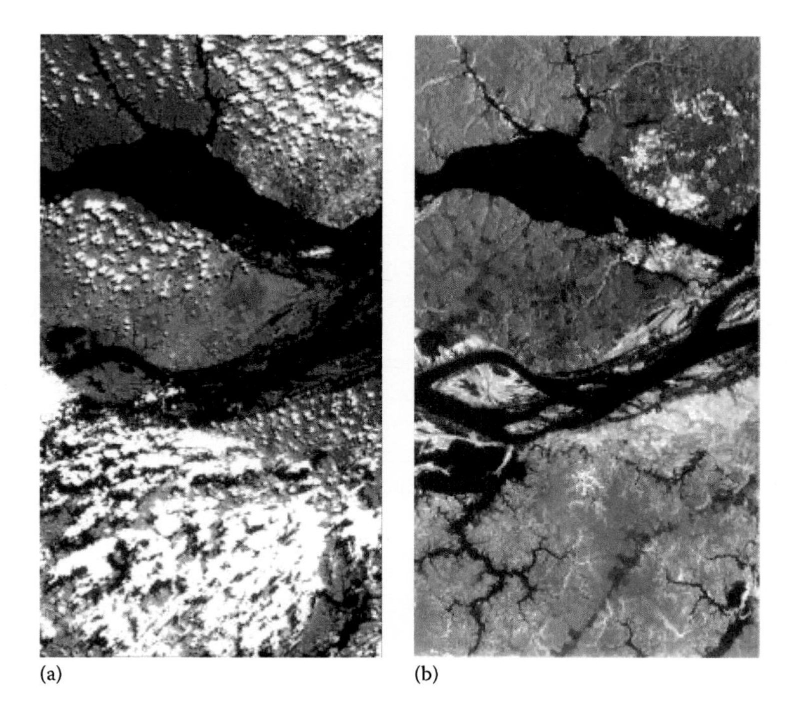

(a) (b)

FIGURE 3.13 Images acquired by the JERS-1 satellite in 1993 in the vicinity of Manaus (Brazil). (a) Visible image derived from the OPS sensor; (b) radar image of the same area and time. (Courtesy of NASDA, Arlington, VA.)

The first radar sensors were on board aerial platforms. The most widely used system was the side-looking airborne radar (SLAR), which was widely used in tropical regions (Allan 1983; Trevett 1986). These sensors do not provide good spatial resolution at long distances, since for a real aperture radar system, the minimum size of an object to be detected is directly proportional to the wavelength ($\lambda$) and the height of observation ($h$) and inversely related to the diameter of the antenna ($\phi$) following

$$R_s = \left( \frac{\lambda h}{\phi} \right) \quad (3.2)$$

Therefore, when $h$ becomes large, as in space platforms, obtaining an adequate resolution would imply using a very long antenna.

This problem can be avoided using an artificially synthesized virtual antenna. This is the concept behind the synthetic aperture radar (SAR) systems. SAR systems benefit from the Doppler effect, which is the change in radiation frequency resulting from the relative movement between the sensor and the observed surface. Space-borne radar emits continual pulses of microwave energy to the ground. As the satellite orbits, the emitted pulses are backscattered along the satellite path with a different frequency from what was originally transmitted. If the system is able to register from where the backscattered radiation was emitted, the discrimination of the observed surface would be as precise as when observing it with an antenna the size of the distance between the points of the initial and final phases of the satellite orbit. Following Equation 3.2, that long antenna provides high spatial resolution (Elachi 1987).

Both real and synthetic aperture radars have a side-looking observation, which results in a series of geometrical distortions that need to be properly addressed to retrieve significant information from the radar data. First, we should consider that the spatial resolution of a radar system is different in the parallel (range) and perpendicular (azimuth) directions to the satellite orbit, and therefore the pixel size changes in both directions.

The range resolution depends on the length of the pulses emitted (Figure 3.14). Two objects can be discriminated when their ground distance is greater than half the pulse wavelength (half because the beam has to travel back and forth). Additionally, the discrimination depends on the incidence angle ($\theta$). To increase the length of the pulse, it is necessary to decrease the frequency, but this will imply a higher noise. This paradox is solved by modulating the frequency of the emitted and received pulses. In summary, the range resolution is given by

$$r_{range} = \frac{c}{2B\sin\theta} \quad (3.3)$$

where

$c$ is the speed of the electromagnetic energy
$B$ is the spectral bandwidth
$\theta$ is the angle of incidence

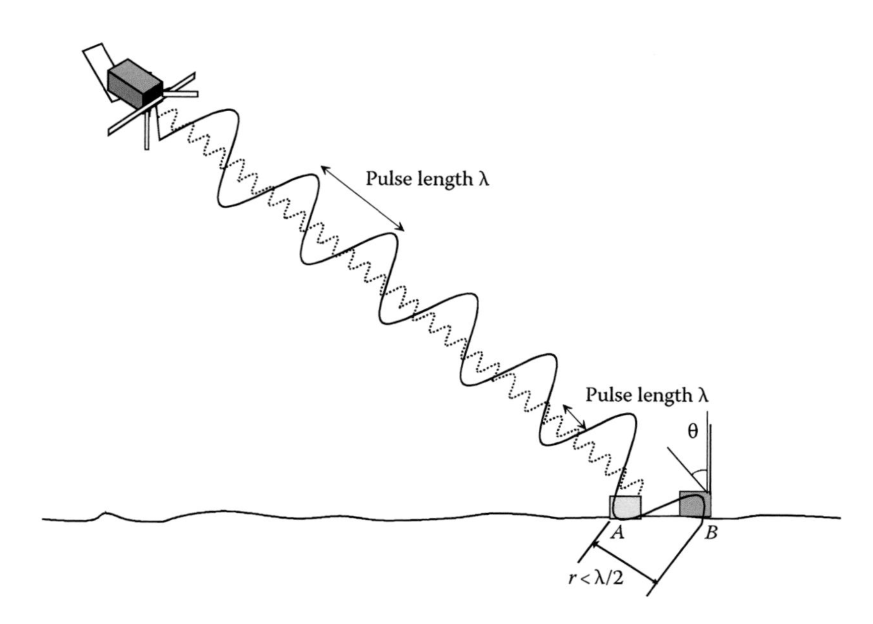


**FIGURE 3.14** Range resolution of a radar system.

As for the azimuth resolution, the minimum distance on the ground between two discernible objects on the image depends on the wavelength (λ), the distance between the antenna and the surface ($h$), and the antenna's length ($L_a$) (Figure 3.15):

$$r_{az} = \frac{h\lambda}{L_a} \tag{3.4}$$

Although in a synthetic aperture system the length of the antenna is incremented artificially, since it is generated from the two different signals registered in two different moments along the trajectory, the azimuth resolution is still dependent on the distance of the observed object. In consequence, since the distance between the object and the antenna changes from the nearest end to the most distant column of the image, it changes the effective spatial resolution across the image. Homogenization of the pixel size, therefore, requires postprocessing. A common method is slant-range correction, which takes into account the change in pixel size across the satellite trajectory:

$$Pr = \frac{c\tau_p}{(2\sin\theta)} \tag{3.5}$$

where

$\tau_p$ is the pulse length
$\theta$ is the incidence angle

The impact of the slant-range correction is seen in Figure 3.16.

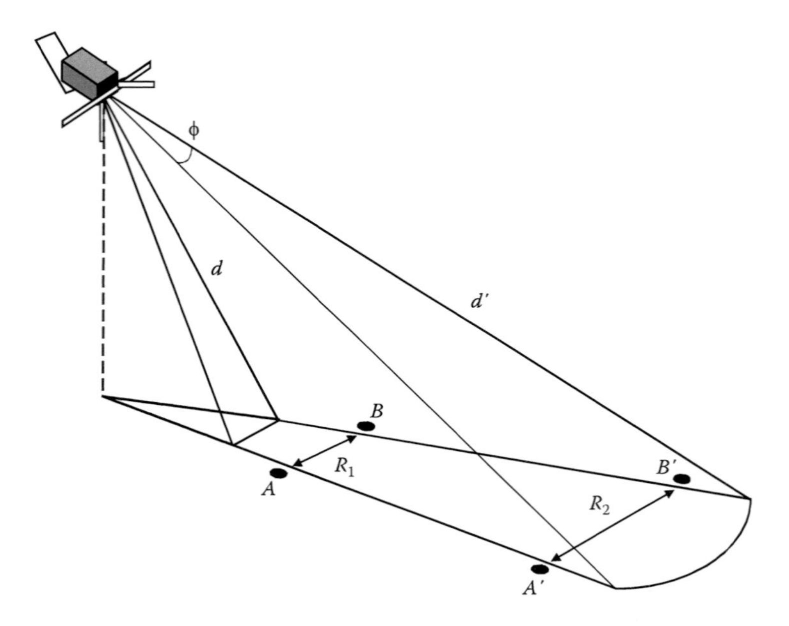


**FIGURE 3.15** Azimuth resolution of a radar system. (Adapted from Lillesand, T.M. and Kiefer, R.W., *Remote Sensing and Image Interpretation*, John Wiley & Sons, New York, 2000.)

**FIGURE 3.16** Slant-range correction of a SEASAT image near Geneva. (a) Raw image; (b) corrected image. (From http://earth.esa.int/applications/data_util/SARDOCS/_icons.)

As a result of the side-looking observation, radar data have several geometrical distortions, which are enhanced in the case of rough terrain (Figure 3.17). If the surface slope is smaller than the angle of incidence, the image tends to make distances shorter than in reality. Therefore, the slopes oriented toward the antenna look inclined in that direction. When the slope is steep enough, a *layover*

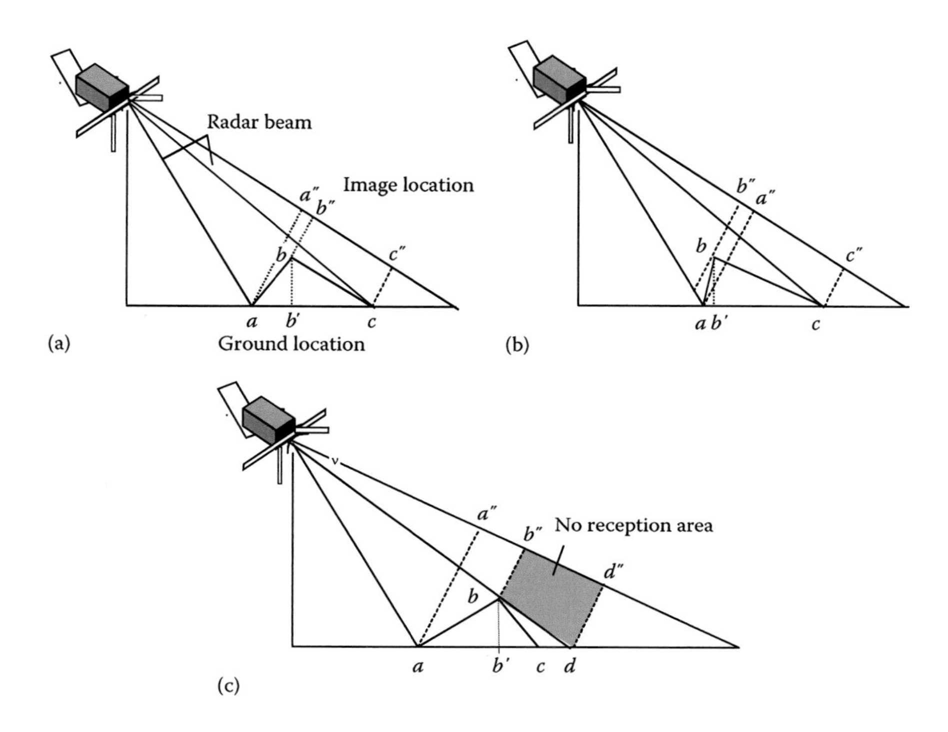


FIGURE 3.17 Terrain effects on a radar image: (a) shortening of radar-facing slopes, (b) layover, and (c) radar shadows.

effect may occur, that is, the inversion of the real position of objects throughout the slope, because the echoes reach the antenna in inverse order to their real location.

These factors can be corrected if detailed digital elevation models are available. Corrections are more complex when slopes facing away from the radar trajectory are not observed at all (radar shadows). That information may be retrieved if a satellite trajectory provides two or more observation angles, which frequently occurs with ascending and descending orbits of satellite radar missions. An example of two images acquired by the European satellite ERS-1 over a mountain area in Central Spain may help to understand better the impact of rough topography on radar images. The images were acquired 6 days apart, in the ascending and descending trajectory of the satellite (Figure 3.18).

In addition to geometrical problems, SAR images also present radiometric noise, known as *speckle*, which is caused by the random distribution of phase and amplitude, mainly related to surface roughness. The effect is very noticeable in raw images, which tend to offer a random white spot pattern, especially evident in areas of land cover transition.

In spite of these problems, SAR systems have proven very useful for different environmental applications: soil mapping, vegetation change as a result of forest fires or clear-cuts, forest biomass estimation, and vegetation regeneration after

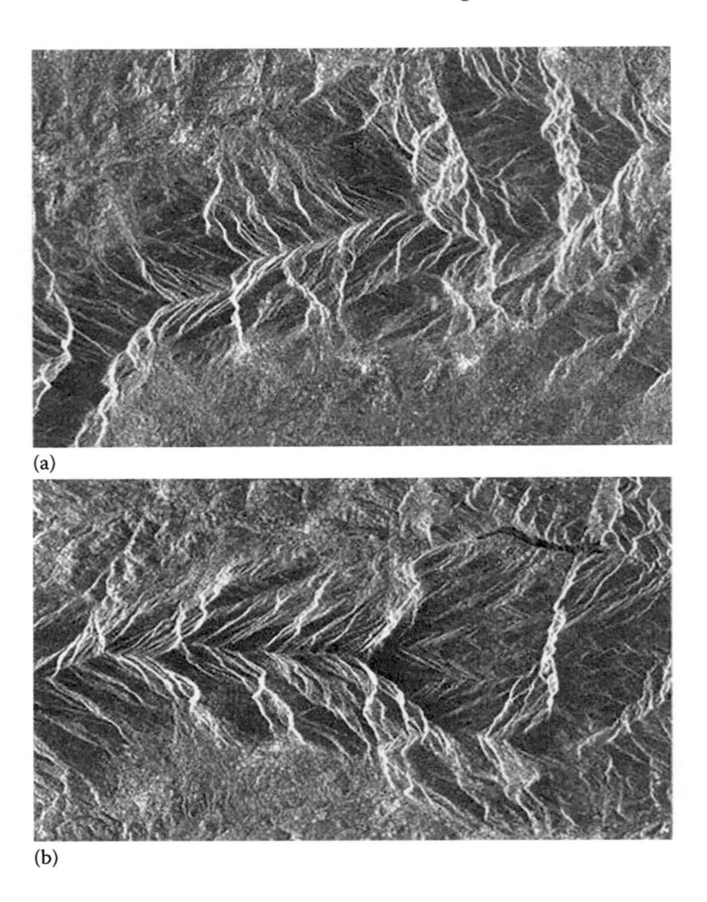


FIGURE 3.18 SAR images from the ERS-1 satellite over Central Spain (Sierra de Gredos). (a) Descending acquisition from April 23, 1993; (b) ascending acquisition from April 29, 1993.

catastrophic events (Goldshleger et al. 2010; Kasischke et al. 2007; Saatchi et al. 2007; Tanase et al. 2010a).

Technical developments during the last several years have increasingly favored the use of radar in natural resource management. The growing availability of multiangular, multifrequency, and multipolarization observations creates new possibilities for retrieval of additional information on vegetation, snow, and soil characteristics.

In addition to imaging radar systems, over the last decades radar altimeters have been widely employed. These systems provide vertical measurements of distances between the sensor and the surface, although data can only be precisely retrieved over water bodies when good measures of the Earth geoid are available. The basic altimeter equation is (Figure 3.19)

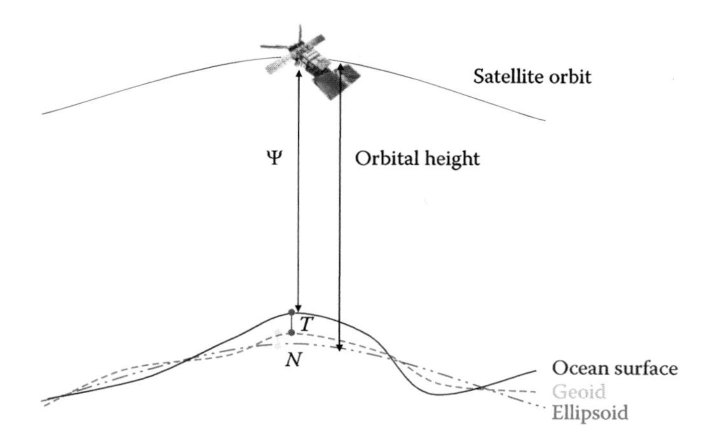


FIGURE 3.19 Principles of radar altimetry. (Adapted from Robin, M., La Télédétection, Nathan, Paris, France, 1998.)

$$r - Rs - \psi = N + T + Ar \quad (3.6)$$

where

$\psi$ is the radar altimeter measurement
$r - Rs$ is the elevation with respect to the reference ellipsoid
$N$ is the height of the geoid–ellipsoid at the point
$T$ is the sea level ellipsoid
$Ar$ is the measurement error

Radar altimeter measurements have been very useful to monitor ocean currents associated with El Niño events or other climatic phenomena. They are also very useful for monitoring water reservoirs, both of dams and natural lakes (Birkett 1998) as well as sea level rise (Figure 3.20), which is one of the main negative impacts of global warming.

In addition to the analysis of backscatter values and distances, the interferences of the radar signal have also been found very useful to obtain relevant information about terrain characteristics. Radar interferometry was developed during the late 1990s when the European ERS-2 satellite (launched in 1995) was used for acquisition in tandem with the existing ERS-1 (launched in 1991). The two satellites provided nearly simultaneous acquisitions, which were able to observe phase differences related to the surface conditions (Madsen and Zebker 1998; Quegan 1995). The principle of interferometry is simple, although the modeling of the signal is very complex. The goal is to measure the differences in phase from two backscatter signals that are produced by variations in terrain heights or object position. These phase differences can be deconvolved to estimate height variations in the range of the radar wavelength (for the ERS satellite, in band C = 5.7 cm, this implies an accuracy of a few centimeters). However, to obtain those precisions, a perfect adjustment of the two images should be obtained. For this reason, the acquisitions need to be almost

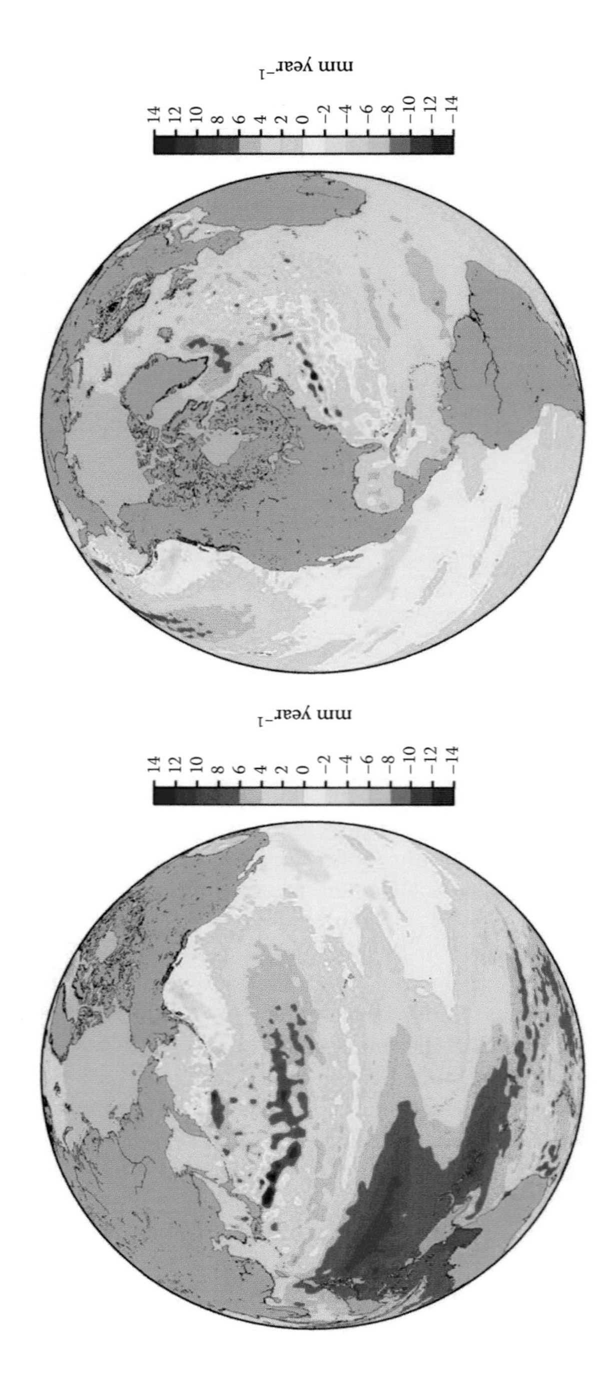


**FIGURE 3.20** Sea level trend patterns from multimission satellite altimetry (1993–2012). (Courtesy of Anny Cazenave, Sea Level CCI Project.)

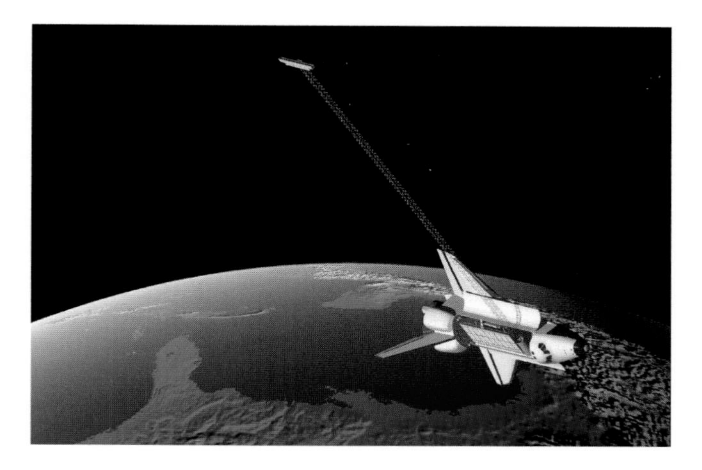

**FIGURE 3.21** Artistic representation of the interferometric Space Shuttle SRTM. (From http://www2.jpl.nasa.gov/srtm/multimed.htm.)

simultaneous, or at least without any perceptible change between the two observations. In semiarid regions, interferometric pairs have been obtained from images acquired even months apart (Massonet et al. 1993), while in more vegetated areas, minor changes imply losing the coherence.

Interferometric techniques were the basis to generate the first worldwide digital elevation model at medium resolution (30 m). The radar data were acquired by the SRTM, installed on board the STS-99 mission in February 2000 (Figure 3.21). The mission included two radar systems 60 m apart, separated by a very stable pole. The mission flew for 11 days with a 57° inclination orbit. It covered land surface between 60° north and 57° south, accumulating about 80% of the world emerged areas.

In addition to topographic cartography, interferometry has also been used very successfully for the detection and modeling of seismic movements, volcanoes, and glaciers (D'Errico et al. 1995; Massonet et al. 1993), achieving very precise adjustments with geologic and mathematical models. Recent studies have also shown good potential for interferometric measurements to retrieve forest parameters and degradation (Slatton et al. 2001; Tanase et al. 2010b).

### 3.3.2 Lidar

Lidar is the acronym of "light detection and ranging." Similar to radar, lidar is an active sensor system, emitting pulses to the ground, which are collected after being reflected from the target surface. However, there are several differences between the two sensors. First, they work at different wavelengths: radar with microwave energy, while lidar systems at the range from ultraviolet to NIR (Dubayah and Drake 2000; Fujii and Fukuchi 2005). This implies that lidar sensors are as affected by atmospheric perturbations as other sensors working on the solar spectrum. Radars observe mostly obliquely, while lidars mostly vertically (± low angle from nadir), and therefore they have different geometrical problems. Finally, the main interest of the radar signal for EO is the backscatter signal, while in lidar the time measurements. Radars also measure distances (radar altimeters), and lidars also measure return intensity, but these are not the most common uses of both.

Lidar systems use laser pulses (*light amplification by stimulated emission of radiation* [laser]), which are collimated, polarized, and have coherent radiation. This implies that laser emissions are very directional, the electrons are aligned, and the photons are in phase. These properties provide a very efficient source of radiation for many different applications. Spatial coherence makes it possible for a laser beam to stay narrow over long distances; therefore, a lidar is very useful for estimating concentrations, when the laser beam passes through a medium, and for measuring distances. These are the two main types of lidar sensors currently used in EO: differential absorption and ranging lidars.

Differential absorption lidars estimate medium properties based on how reflectance varies at two or more wavelengths crossing that medium. Frequently, green and NIR bands are used, as many materials present very different absorption coefficients in those wavelengths. The response time and intensity of those two returns provide an estimate of the properties of the absorption medium (for instance, water depth). Green light penetrates clean water, while infrared light is mostly absorbed. Therefore, the time delays between the two laser pulses will provide a good estimation of water components, particularly useful in coastal areas. Similar differential absorption principles have been applied to monitor atmospheric conditions, such as ozone or aerosol concentrations. An example of these measurements is the Cloud-Aerosol Lidar with Orthogonal Polarization (CALIOP) on board the Cloud-Aerosol Lidar and Infrared Pathfinder Satellite Observation (CALIPSO) satellite, which obtains aerosol and cloud height distribution from a vertical two-band lidar system (Figure 3.22).

On the other hand, ranging lidar systems base their measurements on a very precise estimation of the time that a laser pulse takes to move from the sensor to the ground and back to the sensor. From this time ($t$), the distance ($d$) can be easily computed from

$$d = \frac{ct}{2} \tag{3.7}$$

where $c$ is the speed of light ($3 \times 10^5$ km s$^{-1}$). If the sensor location is very precisely determined, the relative distances between the object and sensor can be converted to absolute coordinates ($X,Y,Z$); therefore, measurements of vertical properties can be obtained from lidar systems. The sensor coordinates are determined through an onboard global positioning system (GPS), which is commonly complemented with another ground GPS, and an inertial measurement unit (IMU), which measures variations in airplane speed and observation angles (pitch, roll, yaw).

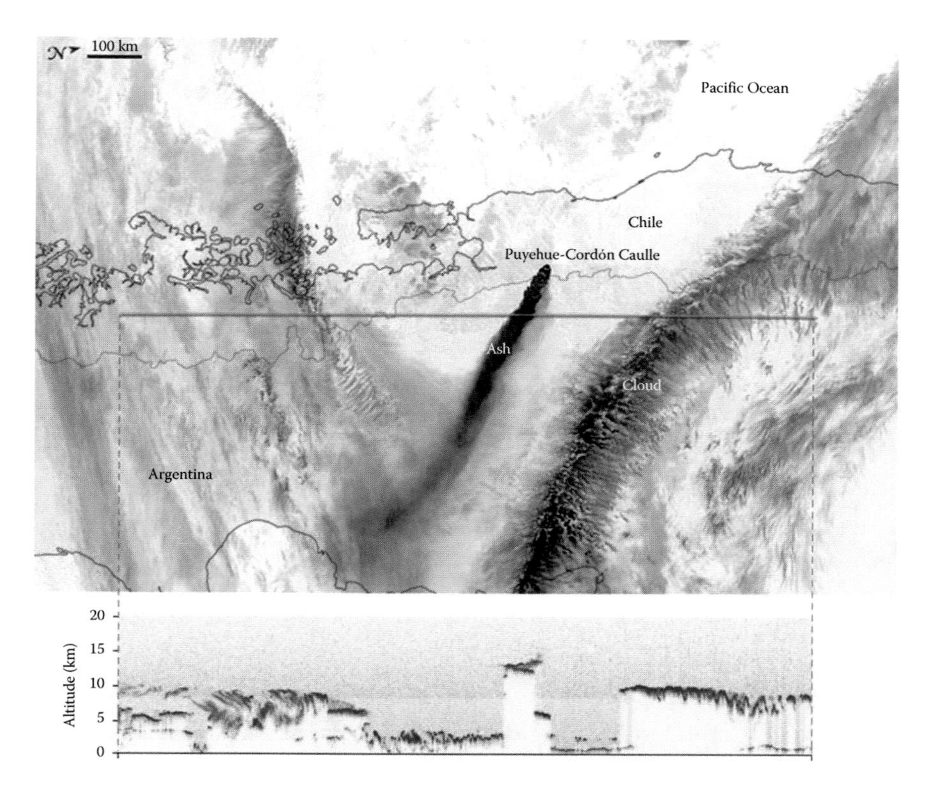


**FIGURE 3.22** CALIPSO 3D profile on the eruption of the Puyehue-Cordón Caulle volcano in Chile (June 2011). The upper part is a TIR nighttime image acquired by MODIS; the lower part includes the vertical profile acquired by CALIOP, where the volcanic plume can be clearly located at the center of the image. It reached almost 15 km, passing the lower limit of the stratosphere. (From http://earthobservatory.nasa.gov/.)

Within ranging lidars, a further distinction can be made between large- and small-footprint systems, based on the size of the ground projection of the laser pulse (Figure 3.23). Large-footprint systems cover areas of several meters in diameter (typically from 30 to 300 m), while small-footprint lidars refer to small observation areas (around 0.30 m or less). The large-footprint lidar systems normally collect the return signal at a very high frequency (1 ns), generating a waveform of pulses that can be used to model the vertical distribution of all elements within the lidar FOV. These systems are called full waveform. The small-footprint lidar systems typically collect a few discrete returns (the first and last one, and up to 2–3 in between), which may hit different parts of the cover (crown, branches, small trees, shrubs, and finally the ground). The large-footprint systems with full waveform record provide a more precise description of vertical diversity in the observed area than discrete pulse lidars but require a significantly higher amount of data to handle.

The size of the laser footprint on the ground can be easily computed from Figure 3.24.

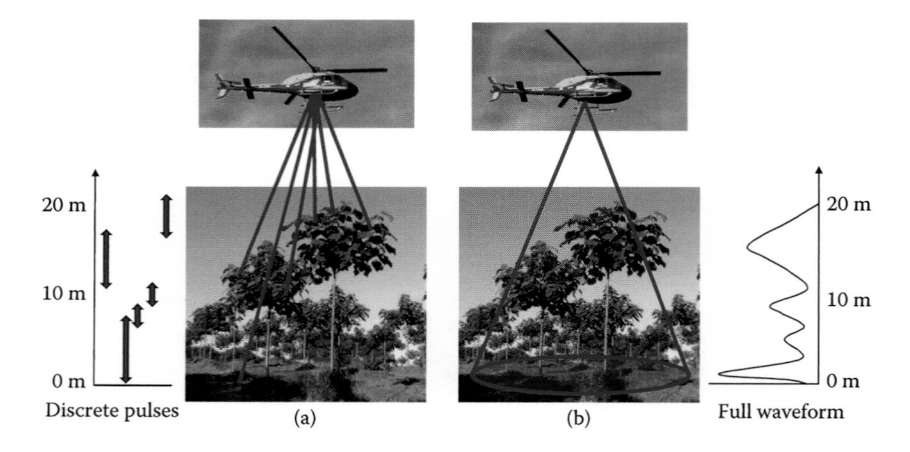


**FIGURE 3.23** Small- (a) and large-footprint (b) lidar systems. The small-footprint pulses may hit different sections of the target and record discrete pulses, while large-footprint systems cover the diverse heights of the observed area and record the full return waveforms.

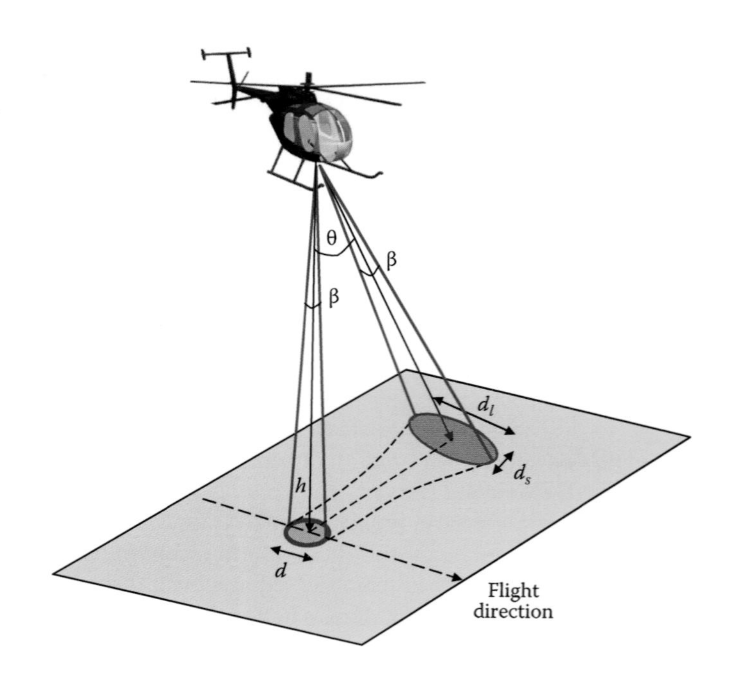


**FIGURE 3.24** Variations of lidar footprint size along and across the scanning line.

$$d = 2h \tan \frac{\beta}{2} \tag{3.8}$$

where

$d$ is footprint diameter
$h$ is flight elevation above ground
$\beta$ is the IFOV of the laser sensor (also named "laser divergence")

This formula assumes a vertical observation and flat terrain. When the observation is off-nadir, the footprint of an across-track laser scanner will be elliptical, and the size can be computed as follows (Sheng 2008; Figure 3.24):

$$d_l = 2h \frac{\sin \beta}{\cos 2\theta + \cos \beta} \tag{3.9}$$

$$d_s = 2h \frac{\sin \beta}{\cos \theta \left(1 + \cos \beta\right)} \tag{3.10}$$

where

$d_l$ and $d_s$ are the long and short axes of the elliptical laser footprint, respectively
$\theta$ is the instantaneous scanning angle.

The larger this angle, the higher the footprint. For a typical lidar scanning system, the footprint at 30° will be 54% larger than at nadir (Sheng 2008). Again these measurements refer to a flat terrain. In inclined terrain, the footprint size changes along with the slope gradient (see Sheng 2008).

Another classification of lidar systems is related to the observation patterns. The first lidars were basically profiling systems, sending single pulses along the platform trajectory. This implies nadir observation, monitoring areas underneath the platform. Most lidar systems now include scanning capabilities, which make surveying a wider area possible (Figure 3.25). Different scanning patterns are available in the current systems, which try to mitigate geometric errors while covering the whole target area (Beraldin et al. 2010).

Current lidar systems are mostly airborne, providing small footprints and discrete returns, but technology is evolving toward full waveform systems, even for small footprints. A few lidar systems have been mounted on satellite missions. We earlier referred to CALIOP, which was launched in 2006 and was developed for atmospheric applications. Besides this system, the GLAS lidar on board the ICESAT satellite was developed to monitor ice sheets (Shepherd et al. 2012), but it was also very useful for retrieving biophysical properties of vegetation with good results, particularly in flat terrain (García et al. 2012; Simard et al. 2011). This system includes a large footprint (70 m) and collects the full waveform. Future missions are being studied to extend the GLAS measurements and provide a worldwide coverage of lidar observations. ICESAT-2 and DESDYNI are on the decadal survey of NASA. The former will

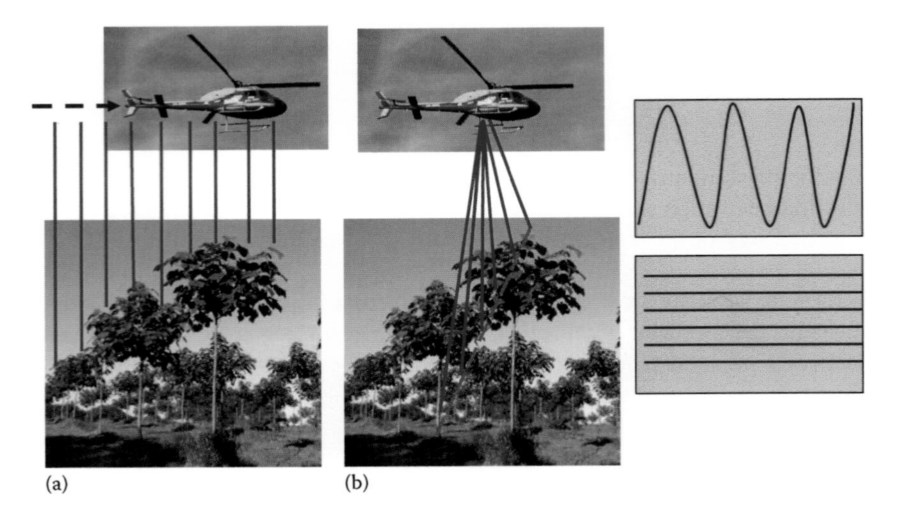


FIGURE 3.25 Profile (a) and scanning (b) lidar systems. Common scanning patterns are portrayed in (b), showing mirror scanning and multiple fiber optics.

have a laser altimeter, complementing the measurements of ICESAT-1, and the latter will host both an L-band SAR and a laser altimeter, aiming to better understand the crustal deformations and ice sheet trends.

Many lidar studies have been conducted in the last few years, almost exclusively from airborne systems, providing an extensive experience of the potential of this sensor for vegetation, crop, natural hazards, and urban mapping (Beraldin et al. 2010; Vosselman and Maas 2010). The estimations of vertical structure are very precise; therefore, a wide range of applications requiring vertical information can be derived. The most obvious application is the generation of digital elevation models (Vosselman 2000), with much higher accuracy than traditional photogrammetric methods, particularly in cities (Zhou et al. 2004; Figure 3.26) and floodplains

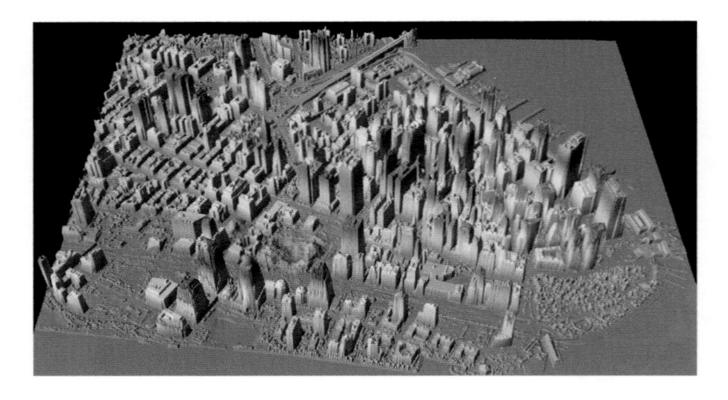

FIGURE 3.26 Three-dimensional view of Manhattan (New York) from a lidar survey acquired in September 2001. (Courtesy of NOAA/U.S. Army JPSD, Washington, DC.)

(Webster et al. 2006). Vegetation properties have also been derived from lidar data, with a wide range of parameters: biomass loads, basal area, crown size and shape, and leaf area index (García et al. 2010; Lefsky et al. 1999; Morsdorf et al. 2004; Naesset and Okland 2002; Riaño et al. 2004a,b, 2007a). Ground lidars are also becoming very useful to measure field parameters, such as leaf area index, heights, and carbon stocks (García et al. 2011a). Procedures to retrieve information from lidar data will be reviewed later (see Section 7.1.5).

## 3.4 SATELLITE REMOTE SENSING MISSIONS

### 3.4.1 Satellite Orbits

Section 3.4 reviews the main satellite remote sensing missions. This section of the book is particularly sensitive to updates because the status of the missions changes quite significantly even in short periods of time. For this reason, we suggest that the reader visits the different missions' web pages to get the most recent information available.

A first review of the EO satellite missions distinguishes between geostationary and polar-orbiting satellites. The former are located at very high equatorial orbits (36,000 km over the Earth) and always observe the same area (Figure 3.27). They have a wide FOV to register the whole visible hemisphere of the planet in a single image. A network of geostationary satellites from different longitudes ensures coverage of the whole Earth. Meteosat is located on the 0° meridian (to acquire images of Europe and Africa), and two Geostationary Operational Environmental Satellites (GOES) are located on the 70° and 140° W meridians, focused toward the east and west coasts of America. Similar satellites operate in Asia. Geostationary satellites provide the best temporal resolution possible, which ranges between 15 and 30 min, but at coarse spatial resolutions caused by the very high orbit.

To complement the observation of geostationary orbits, other lower-altitude EO missions have been launched. The most common orbits for EO satellite missions are near polar (Figure 3.28), either elliptic or circular, with different orbital heights

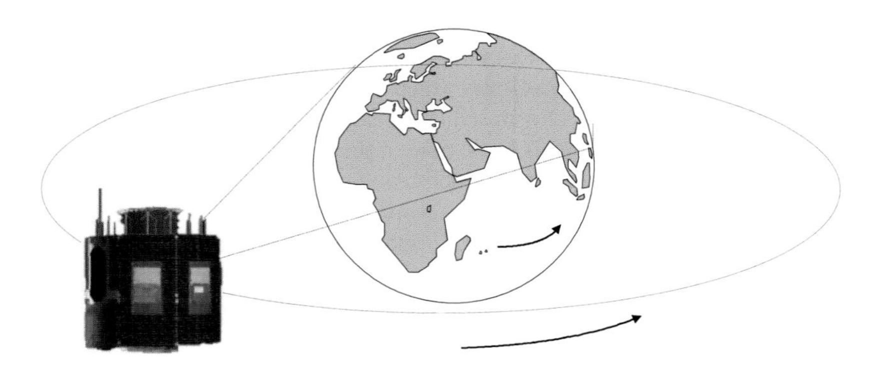

FIGURE 3.27 Scheme of a geosynchronous orbit.

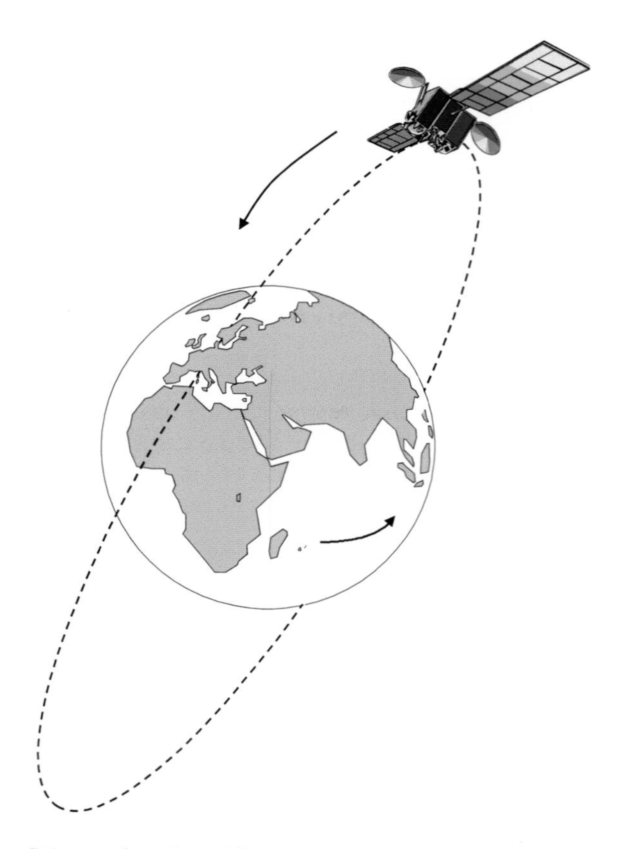

**FIGURE 3.28** Scheme of a polar orbit.

(typically between 700 and 900 km). Since the orbital plane is approximately perpendicular to the equatorial plane, as the Earth rotates these satellites can observe different areas and cover a whole view of the planet in a certain number of hours or days, depending on the height and the sensor FOV. One particular polar orbit is named *Sun-synchronous,* where it is possible to acquire images at the same solar time, therefore facilitating multitemporal comparisons (Figure 3.28). Some satellite systems have near-equatorial orbits, which are more suited to improve temporal resolution in tropical regions. Obviously, those satellites will not be able to observe areas outside the tropical belt (this is the case of the TRMM). The specific observed latitudes will depend on the equatorial inclination of the orbit.

The selection of the satellite orbit is determined by the characteristics of the mission. If the platform is designed to obtain global and very frequent observations, geostationary sensors or polar sensors with a large FOV are preferable. When higher spatial resolution is required, polar sensors with a narrow FOV are most commonly used.

After this summary review of orbits, we next analyze the main EO missions, with special emphasis on the technical characteristics of the sensor payload.

### 3.4.2 The Landsat Program

The success of the space photographic missions during the 1960s led NASA to develop the first program oriented toward monitoring Earth resources. This effort led to the launch of the first Landsat satellite, which was originally named Earth Resources Technology Satellite (ERTS), on July 23, 1972. After the second launch in 1975, the series was renamed Landsat and extended in 1978, 1982, and 1984 with the subsequent launches of Landsat 3, 4, and 5, respectively. Landsat 6 was lost shortly after the launch in 1993, while Landsat 7 was launched in 1999 and is still working, although it has a serious technical problem in the main sensor. Landsat 5 was working until 2012, when it was officially decommissioned. Landsat 8 was developed within the program termed Landsat Data Continuity Mission (LDCC: Irons et al. 2012). Finally launched in February 2013, Landsat 8 started operations 3 months later. In April 2015, the U.S. Congress approved the development of Landsat 9, expected to launch in 2023. The Landsat program has been the most important EO mission ever designed.

The first three Landsat satellites had a Sun-synchronous polar orbit at 917 km, with a revisiting period of 18 days and a 10:30 a.m. equatorial crossing time. For the fourth and fifth Landsat satellites, their instrument configuration and orbital characteristics were modified. The orbital height was reduced to 705 km, and the revisiting observation was improved to 16 days in temperate latitudes. Overpass time remained similar to the previous Landsat, crossing the Equator around 10:10 a.m. The next two satellites of the mission, Landsat 6 and 7, were changed in their appearance again (Figure 3.29), although they maintained the orbital characteristics of their two predecessors.

The first three Landsat satellites included the MSS sensor and three video cameras (RBV). The cameras did not work properly in the first two satellites; hence, the MSS was really the most useful sensor of the mission until the launch of Landsat 4. The MSS had an 11.56° FOV and could observe an area of 185 × 185 km at 57 × 79 m nominal pixel resolution. It covered four bands of the spectrum (green, red, and two in the near infrared, numbered 4, 5, 6, and 7, respectively), designed to imitate true and false color infrared photography. Landsat 3 included an updated version of the MSS with a thermal band. Data were digitally acquired and recorded on board or transmitted in real time to a set of acquisition antennas. Bands 4–6 were coded in 7 bits (between 0 and 127) and band 7 was in 6 bits (0 and 63). Table 3.1 includes the technical specifications of the different sensors on board the Landsat program.

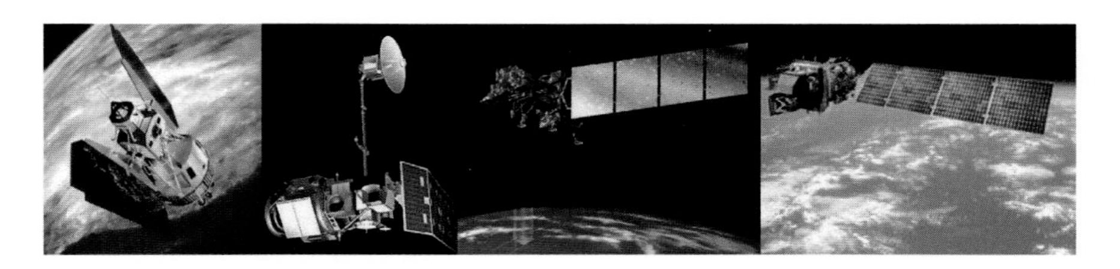

**FIGURE 3.29** Artist view of the different Landsat missions. From left to right: Landsat 1–3, Landsat 4 and 5, Landsat 6 and 7, and Landsat 8.

**TABLE 3.1**
**Landsat Sensor Characteristics**

| MSS[a] (μm) | | RBV (μm) | | TM[b] (μm) | | ETM+[c] (μm) | | OLI/TIR[d] (μm) | |
|---|---|---|---|---|---|---|---|---|---|
| 4 | 0.5–0.6 | 1[e] | 0.475–0.575 | 1 | 0.45–0.52 | 1 | 0.45–0.52 | 1 | 0.43–0.45 |
| 5 | 0.6–0.7 | 2[e] | 0.580–0.680 | 2 | 0.52–0.60 | 2 | 0.52–0.60 | 2 | 0.45–0.51 |
| 6 | 0.7–0.8 | 3[e] | 0.690–0.830 | 3 | 0.63–0.69 | 3 | 0.63–0.69 | 3 | 0.53–0.59 |
| 7 | 0.8–1.1 | 1[f] | 0.505–0.750 | 4 | 0.76–0.90 | 4 | 0.76–0.90 | 4 | 0.64–0.67 |
| 8[f] | 10.4–12.6 | | | 5 | 1.55–1.75 | 5 | 1.55–1.75 | 5 | 0.85–0.88 |
| | | | | 6 | 10.40–12.50 | 6 | 10.40–12.50 | 6 | 1.57–1.65 |
| | | | | 7 | 2.08–2.35 | 7 | 2.08–2.35 | 7 | 2.11–2.19 |
| | | | | | | 8 | 0.52–0.90 | 8 | 0.50–0.68 |
| | | | | | | | | 9 | 1.36–1.38 |
| | | | | | | | | 10 | 10.60–11.19 |
| | | | | | | | | 11 | 11.50–12.51 |
| | | | | **Spatial Resolution** | | | | | |
| 4–7 | 79 m | 1–3 | 80 m | 1–5 and 7 | 30 m | 1–5 and 7 | 30 m | 1–7 and 9 | 30 m |
| | | | | | | 6 | 120 m/60 m$^3$ | 8 | 15 m |
| 8 | 240 m | 1 | 40 m | 6 | 120 m | 8 | 15 m | 10–11 | 100 m |

[a] Only Landsat 1–3.
[b] Only Landsat 4 and 5.
[c] Only Landsat 7.
[d] Only Landsat 8.
[e] Only Landsat 1 and 2.
[f] Only Landsat 3.

The launch of Landsat 4 and 5 gave a boost to the program, by incorporating another across-track scanner with better spatial, spectral, and radiometric resolutions than the MSS. It was named Thematic Mapper (TM), which has been extensively used in the last three decades for a wide range of environmental studies (Wulder et al. 2012). Landsat 7 included an improved version of the TM, named ETM+, with similar characteristics as TM, but with an additional panchromatic band with a 15 m spatial resolution and improved resolution of the thermal band to 60 m (Table 3.1). The TM/ETM+ sensors included two new bands in the shortwave infrared (SWIR, 1.6–2.2 μm) as well as one in the thermal infrared, not contained previously in the MSS. The former were very valuable for estimating water content of soils and plants and improving atmospheric correction. Unfortunately, the ETM+ encountered a serious technical problem in May 2003 when the scan line corrector failed. This is a critical hardware component, which results in incomplete scan lines, with missing strips that become wider as the scan moves out from the center of the image (Figure 3.30). These images are commonly referred to as SLC-off images (scan line corrector).

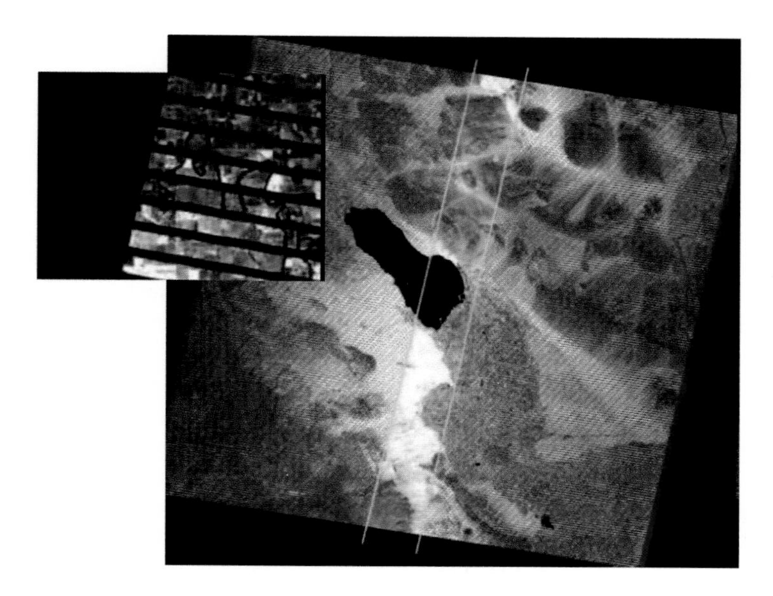

FIGURE 3.30 Impact of the malfunctioning of SLC. Errors increase from the center to the image edges. (From http://landsat.usgs.gov/products_slcoffbackground.php.)

Landsat 8 includes two sensors on board: OLI and TIR. The former is a successor of ETM+, although it includes the push-broom technology and two additional bands, while TIR is still an across-track scanner and contains two bands in the thermal infrared at lower spatial resolution than the other bands (100 vs 30 m for the multispectral and 15 m for the panchromatic). OLI has a high signal-to-noise ratio and has a radiometric resolution of 12 bits. The sensor incorporates two new bands over the TM/ETM: a deep-blue band (#1) for better discrimination of coastal waters and a cirrus cloud band (#9) to improve atmospheric correction.

Landsat images were recorded on board during the first three missions and directly transmitted to the ground receiving antennas in the last five. The network of receiving antennas practically covers the whole Earth's surface, as images can be transmitted to different communications satellites. Data access policy has undergone changes during the lifetime of the Landsat mission. During the 1990s, the management of the Landsat program was moved to the commercial sector through the EOSAT company, which operated Landsat 4 and 5. During those years, images were expensive and applications were broad but mostly focused on research and development. As a result of the Land Remote Sensing Policy Act of 1992, later in the 1990s, the U.S. government decided to return Landsat operations to the USGS (Williamson 2001), which reduced prices and extended operational applications.

The Landsat historical archive was made freely downloadable in October 2008 (Landsat 7) and in January 2009 (all Landsats), leading to a 60-fold increase in data downloads, which surpassed six million only 2 years later (Rocchio 2011). The immense benefits of this data access policy have been observed in the extension of operational uses of Landsat images and the potentials of multitemporal change detection to monitor Earth changes in the last 30 years (see, for instance, the atlas of our changing environment: http://na.unep.net/atlas/, last accessed in June 2015). In addition, other space agencies have also opened their historical archives for extending the opportunities of analyzing trends in critical environmental variables.

### 3.4.3 SPOT Satellites

After Landsat, the SPOT program has been the most extended and fruitful for land applications. The acronym stands for Systeme Pour l'Observation de la Terre (System for Earth Observation) and it was developed by the French *Centre National de Etudes Spatiales* (CNES), in collaboration with Belgium and Sweden. Now the program is run by a private company, SPOT Image, based in Toulouse, France. The first SPOT satellite was launched in 1986, followed by other platforms in 1990, 1993, 1998, 2002, 2012, and 2014. The last two satellites were developed by EADs Astrium and are still operated by SPOT Image, which is a subsidiary of Astrium.

The first SPOT satellite had a Sun-synchronous orbit at 822 km height, with a 98° inclination and an orbital period of 101 min. The complete coverage duration cycle was 26 days. The last two, SPOT-6 and 7, have an altitude of 694 km.

Among the most interesting innovations in this satellite was the use of two pushbroom scanners named initially HRV (*haute resolution visible*, high resolution visible). These two sensors offered two modes of image acquisition: panchromatic and multiband (green, red, and NIR in the first three satellites), with a spatial resolution of 10 and 20 m, respectively (Table 3.2). The area covered by each image was 60×60 km. The high-resolution sensor installed on SPOT-4 and 5 provided an additional band in the SWIR and reduced the pixel size from 20 to 10 m in the multiband

**TABLE 3.2**
**Sensors Onboard SPOT Satellites**

| HRV[a] (μm) | | HRVIR[b] (μm) | | HRG[c] (μm) | | HR[d] | | VEGETATION[b,c] (μm) | |
|---|---|---|---|---|---|---|---|---|---|
| 1 | 0.50–0.59 | 1 | 0.50–0.59 | 1 | 0.50–0.59 | 1 | 0.455–0.525 | 1 | 0.43–0.47 |
| 2 | 0.61–0.68 | 2 | 0.61–0.68 | 2 | 0.61–0.68 | 2 | 0.530–0.590 | 2 | 0.61–0.68 |
| 3 | 0.79–0.89 | 3 | 0.79–0.89 | 3 | 0.78–0.89 | 3 | 0.625–0.695 | 3 | 0.78–0.89 |
| P | 0.51–0.73 | 4 | 1.58–1.75 | 4 | 1.58–1.75 | 4 | 0.760–0.890 | 4 | 1.58–1.75 |
| | | P | 0.51–0.73 | P | 0.48–0.71 | P | 0.45–0.74 | | |
| | | | | **Spatial Resolution** | | | | | |
| 1–3 | 20 m | 1–3 | 20 m | 1–3 | 10 m | 1–4 | 6 m | 1–4 | 1000 m |
| | | | | 4 | 20 m | | | | |
| P | 10 m | P | 10 m | P | 2.5–5 m | P | 1.5 m | | |

[a] Only in SPOT-1 to 3.
[b] Only in SPOT-4.
[c] Only in SPOT-5.
[d] Only in SPOT-6 and SPOT-7.

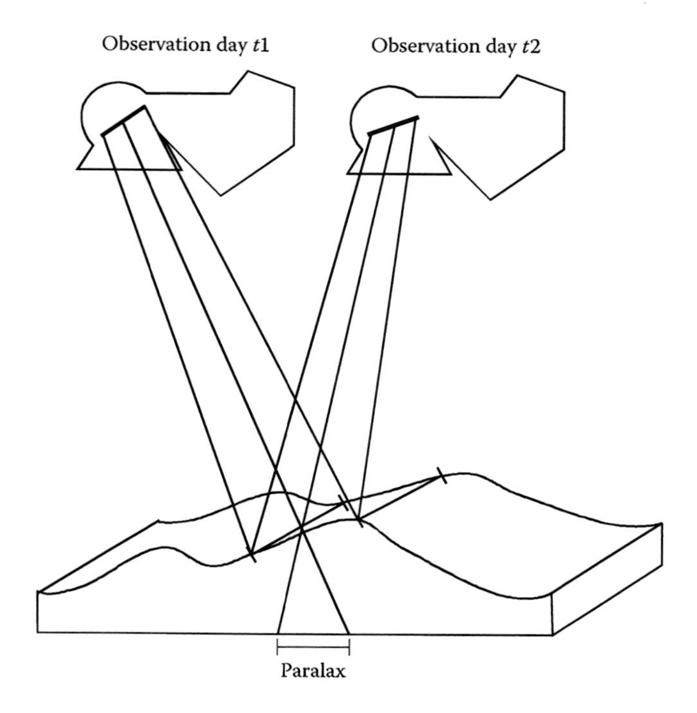


**FIGURE 3.31** Stereoscopic acquisitions of the SPOT-HRV sensor.

mode and from 10 to 2.5 in the panchromatic one. Finally, the last two sensors include a higher spatial resolution instrument with 6 m in the multiband mode (from B to NIR, SWIR has been removed) and 1.5 in the panchromatic.

These high-resolution cameras have the capacity for off-nadir acquisitions by using a pointable mirror (Figure 3.31). This makes it possible to improve the nominal observation cycle (from 26 days to 2–3 days depending on the latitude) and to take stereoscopic images, which can be used to generate digital terrain models.

In addition to the HRV sensor, the first SPOT satellite included other instruments, such as Doppler Orbitography and Radiopositioning Integrated by Satellite (DORIS), which provided a very precise position of the satellite, and Polar Ozone and Aerosol Measurement (POAM) in SPOT-3, which measured the content of ozone, aerosols, nitrogen, oxygen, and water vapor, with nine channels between the ultraviolet and near infrared. SPOT-4 and SPOT-5 also carried a coarse-resolution sensor named VEGETATION, intended for global assessment of crops and natural vegetation. This instrument had a spatial resolution of 1 km, four spectral bands (matching the HRV-HRG instruments), and a coverage area of 2250 km, with daily acquisition over the whole planet. The SPOT VEGETATION images have been very valuable for global analysis of land cover and vegetation trends (Liu et al. 2002; Silva et al. 2005; Tansey et al. 2004a). The successor of SPOT VEGETATION sensor is the PROBA-V, developed by the Belgian Science Policy Office and operated by ESA. It was launched in 2013 and includes four bands at 333 m resolution (G, R, and NIR) and 666 m (SWIR).

### 3.4.4 Other Medium-Resolution Optical Sensors

Other EO satellite missions oriented toward monitoring natural resources have been developed by different space agencies, from Japan, India, China, Brazil, Argentina, Korea, or Taiwan, as well as private companies (Surrey Technologies). A brief review of the most relevant follows.

The Japanese space agency has developed several EO projects in the last 30 years. The first one, named the Marine Observation Satellite (MOS), was launched in February 1987 and 1990 and remained active until 1996. It was first designed for oceanic investigations, although it had interesting features for terrestrial applications. The payload included the Multispectral Electronic Self-Scanning Radiometer (MESSR), with 50 m resolution and four spectral bands, two in the visible and two in the NIR; the Visible and Thermal Infrared Radiometer (VTIR), with one visible band and three in the TIR (spatial resolution 900 and 2700 m, respectively); and a microwave radiometer (Tsuchiya et al. 1987). The continuation was the JERS-1, launched in 1992, with mainly a radar focus, although it also had an optical radiometer, the OPS, with eight bands in the VIS, NIR, and SWIR and 18 m spatial resolution. Band 8 provided stereoscopic coverage over band 3 (0.76–0.86 μm). The ADEOS series (1996 and 2002) and the Advanced Land Observing System (ALOS) satellite 2006 were the last EO satellites launched by the Japanese space agency. However, the agency cooperates with other international missions and develops atmospheric observation satellites, such as GOSAT (to monitor greenhouse gases) and GPM (for rainfall monitoring).

The Indian Space Research Organization (ISRO) has been very active in satellite observation program development since the 1970s with the Bhaskara series. Among the most recent missions, the most outstanding are the four IRS satellites launched in 1988, 1994, 1995, and 1997. The main objective of this mission was to improve the knowledge of natural resources in this Asian country, but its images have been used in many other countries. The IRS had a push-broom scanner system, with two sensors called Linear Imaging Self-Scanning (LISS). LISS-I had a resolution of 72.5 m when only one camera was active and 36.25 m when both cameras were active (LISS-II). The area covered by each image was 148 km for LISS-I and 74 km for LISS-II, with a cycle of 22 days. The sensor LISS provided information on four bands of the spectrum. The two last satellites of the series (IRS-1C and 1D) included two additional sensors: a high-resolution panchromatic camera (5.8 m) and a sensor for regional observation, the WIFS, with a resolution of 188 m and 810 km coverage (improved later with the AWIFS [60 m], and the same coverage: Figure 3.32). A similar configuration had the Resourcesat-1 (IRS-P6) and two satellites, launched in 2003 and 2011, respectively, which included the three sensors of previous IRS satellites.

Additional satellites of ISRO are Cartosat-1, 2, 2A, and 2B, launched in 2005, 2007, 2008, and 2010, which carried two stereo cameras with 2.5 m spatial resolution, covering an area of 30 km. In addition to these platforms, ISRO manages an oceanographic satellite, known as IRS-P4 (Oceansat) launched in 1999, with an eight-channel sensor and a microwave radiometer, radar, and geostationary satellites.

The Russian space agency has undergone several transformations in the last few decades, weakening considerably its role in EOs. The most interesting programs were the meteorological satellites METEOR-3 and GOMS, and the natural resources

**FIGURE 3.32** AWIFS image of a large wildland fire occurring in Central Spain in August 2005.

satellites RESURS and OKEAN. The METEOR incorporated a series of television cameras of medium low resolution (0.7–35 km), as well as equipment designed by other countries to obtain atmospheric profiles (aerosols and ozone layer). GOMS was a geostationary satellite similar to Meteosat first generation (three bands and similar spatial resolution) located at longitude 76°50′. RESURS took images at intermediate resolution between Landsat MSS and NOAA-AVHRR, which are very useful for regional to continental studies. It has two multispectral across-track scanners, the MSU-E with higher spatial resolution and MSU-SK with medium resolution (140 m in the optical channels and 550 in the thermal), providing information in the green, red, NIR, and TIR bands.

Other countries have joined in the last two decades the list of space agencies involved in EO satellite development. In Asia, South Korea launched in 1999 the Kompsat-1, equipped with a panchromatic camera (6 m) and other oceanographic sensors. The program was continued by Kompsat-2 in 2006, Kompsat-3 in 2012, and Kompsat-5 in 2013. This latter equipped with a SAR instrument. Taiwan launched its first EO satellite, Formosat-1, in 1999 with six bands and 800 m resolution. This mission has been extended with four additional satellites, two of which have high-spatial-resolution capabilities.

In Latin America, the most active countries are Brazil and Argentina. Brazil has launched a series of EO satellites in partnership with China, named CBERS. The first two satellites were launched in 1999 and 2003. They had a medium-resolution camera system (20 m) in four bands and a regional coverage scanner, WFI, with two bands and 260 m resolution. The most recent CBERS-4 was launched in 2014 and includes four instruments, with spatial resolutions ranging from 5 to 866 m. The Argentinean EO program developed the SAC-C in 1998, which included a medium-resolution sensor, with 350 and 175 m and five spectral bands, as well as other sensors for the analysis of geomagnetic fields, ground positioning systems, and space radiation of low orbit. The SAC-D mission (Aquarius) launched in 2011 is a cooperative international mission between the Argentinean Space Agency and NASA and includes several scientific instruments.

The ESA will launch in 2015 the Sentinel-2A series of satellites, as part of the Global Monitoring for Environment and Security (GMES) initiative of the ESA/European Commission (later named Copernicus). This mission is planned as a continuation of the Landsat and SPOT series, incorporating a medium-resolution sensor named Multispectral Instrument (MSI), with 13 spectral bands and spatial resolution between 10 and 20 m (60 m for channels used for atmospheric correction), with a very wide swath of 290 km, thus providing a 5-day coverage at the Equator when the two satellites of the series are launched.

Finally, it is worth mentioning the Disaster Monitoring Constellation (DMC), which was developed by a British company (Surrey Satellite Technology Ltd.) for different countries. The constellation was designed to take advantage of low-cost technology that made it possible to operate an EO satellite with standard configurations at a reduced budget. The company has developed 43 satellites for different agencies and applications, including global positioning. The most relevant for EO images were the Alsat-1 (operated by Algeria), Bilsat-1 (Turkey), Nigeriasat-1 (Nigeria), Beijing-1 (China), Deimos-1 (Spain), and the DMC series, providing Landsat-like resolution at a high temporal resolution.

### 3.4.5 High-Spatial-Resolution Satellites

In the late 1990s, a new generation of high-spatial-resolution satellites (between 0.5 and 4 m) was developed by different private consortiums with the goal of opening new markets for satellite images. Urban applications and precision farming have extensively benefited from these new high-quality images, although other applications have also been opened to EO satellites from these high-quality data, from national security to natural hazard assessment (Baker 2001).

Three private U.S. consortiums took the leadership in developing high-resolution commercial satellites: Space Imaging, Orbimage, and Earthwatch Inc. Space Imaging (later named DigitalGlobe) successfully launched the IKONOS-2 in September 1999 after a failed launch in 1998. The sensor on board this satellite provided a $1 \times 1$ m resolution in the panchromatic channel and $4 \times 4$ m in the multispectral channels (blue, green, red, and NIR: Figure 3.33). The area covered in each image was 11 km and swaths up to $11 \times 1000$ km could be acquired. Images were coded with 11-bit radiometric resolution, which offers a high sensitivity for land cover discrimination. Data were commercialized through a net of commercial distributors, but the reception of images was centralized. The satellite was deactivated in April 2015.

Orbimage (later renamed GeoEye) was the pioneer company to own commercial satellites with the launch of Orbview-1 in 1995, mainly oriented toward meteorological forecast. It also owned Orbview-2, which included the oceanographic sensor SeaWiFS, developed in collaboration with NASA. SeaWiFS was a radiometer with eight high-resolution spectral bands between 0.4 and 0.88 μm, mainly oriented to monitor various ocean phenomena. It was active until 2010. The company also launched Orbview-3 in 2003, with a panchromatic camera at 1 m resolution and a multispectral with 4 m resolution, operating in four spectral bands. It was active until 2007. Recent developments of GeoEye included the GeoEye-1 and GeoEye-2

FIGURE 3.33 Image acquired by the IKONOS camera on 2008 over irrigated areas in Spain. (Courtesy of European Space Imaging/DigitalGlobe, Munich, Germany.)

satellites. The former was launched in 2008 and provides 41 cm panchromatic and 1.65 multispectral images. The system has a very fine location accuracy (>3 m). The image swath is 15.2 km. GeoEye-2 was initially developed with even better resolution, but as a result of the fusion between GeoEye and DigitalGlobe in 2013, it was decided to be held in storage until needed.

The third commercial owner of satellite missions was DigitalGlobe, which is now the heir of the previous two companies after fusion with Space Imaging first and later with GeoEye. It was responsible for the operation of the QuickBird satellite, launched on October 18, 2001, and deactivated in 2015. This satellite incorporated two cameras: one panchromatic, with a 61 cm spatial resolution, and another multispectral, with 2.5 m resolution. It had a 17 km area of coverage and could generate images of up to 17 × 225 km. It provided 11 bits radiometric resolution and could take oblique observations to improve temporal frequency up to 3 days.

DigitalGlobe launched another high-resolution satellite named WorldView-1 in 2007, with improved geolocation capabilities. It provided a single panchromatic channel at 0.5 m spatial resolution with 11-bit radiometric sensitivity and up to 2 days revisiting time. WorldView-2 was launched in 2009. It includes a panchromatic sensor with a 0.46 m pixel size and a multispectral sensor with eight bands (from blue to NIR) and 1.84 m of spatial resolution (Figure 3.34). In 2014, the company launched WorldView-3 with a panchromatic band of 0.31 m, 8 multispectral bands at 1.24 m, 8 SWIR bands at 3.7 m, and 12 additional bands at 30 m resolution. The revisiting time may be better than 2 days in oblique acquisitions.

FIGURE 3.34 Natural color image acquired by Worldview-2 camera in 2014 over the Vatican City. (Courtesy of European Space Imaging/DigitalGlobe, Munich, Germany.)

FIGURE 3.35 Natural color image acquired in 2014 by the SkySat-1 over the Shenzhen port, China. (Courtesy of Skybox Imaging, Inc., Mountain View, CA, Provided by European Space Imaging, Munich, Germany.)

In 2009, a new provider of high-spatial-resolution images was created, Skybox Imaging, which developed two satellites named SkySat-1 and SkySat-2. These satellites obtained images and videos. In the image mode, they have a 0.9 resolution in the panchromatic band and 2 m in the B, G, R, and NIR bands for 8 km swaths (Figure 3.35). The company was purchased by Google in 2014, which plans to build a full constellation of high-spatial-resolution sensors for very frequent revisiting access.

Outside the United States, other private consortiums are leading the development and operation of high-spatial-resolution systems. A good example is the German company BlackBridge, which currently owns the RapidEye constellation. This program includes five small satellites, providing world daily coverage (using off-nadir acquisitions) at 6.5 m resolution in five spectral bands (B, G, R, R-NIR, NIR) with a dynamic range of 12 bits (Figure 3.36). Another example is ImageSat International,

FIGURE 3.36 RapidEye natural color image of the city of Madrid, Spain, acquired in 2011. (Courtesy of BlackBridge, Lethbridge, Alberta, Canada.)

established in 1997, which involves the participation of several countries and is headed by an Israeli consortium that has its technical headquarters in Chipre. In 2000, ImageSat launched its first satellite of the EROS series, which had a panchromatic camera with a 2.1 m resolution, a swath of 15 km, 11 bits radiometry, and offers stereoscopic coverage. In 2006 the EROS-2 was launched, providing 0.7 m spatial resolution with a single band, 10 bits of sensitivity, and a swath of 7 km. These satellites were designed primarily for intelligence, homeland security, and national defense uses.

### 3.4.6 Geostationary Meteorological Satellites

Geostationary satellite images are well known because they are commonly used in TV weather forecasting. These satellites are coordinated by the World Meteorological Organization (WMO) through the Global Atmospheric Research Program (GARP). The first geostationary satellite, Applications Technology Satellite (ATS), was launched in 1966 by NASA. In operational form, this satellite became the series SMS, known as Geostationary Operational Environmental Satellite (GOES) since 1975. Other geostationary satellites are the European Meteosat, the Indian INSAT, the Russian Elektro-L, the Chinese FY-2D, and the Japanese MTSAT. Geosynchronous satellites orbit the Earth at 36,000 km in the Equatorial plane, therefore observing the same area. A scanning system provides coverage of the full observable Earth disc at a certain longitude (Figure 3.37).

Currently, there are two operational GOES satellites: one located at 75° west (eastern coast of the United States) and the other at 135° west (between Hawaii and California). Both satellites include different sensors for atmospheric observation. The most important is named Imager, a multispectral scanning sensor with five channels (visible, two in the MIR, and two in the TIR), with a spatial resolution of 1 km for the visible, 8 km for the long MIR, and 4 km for the rest of the bands (Figure 3.38). The Earth disc full scanning takes 26 min. The GOES satellites also include an atmospheric profiler, called Sounder, with 19 spectral bands.

FIGURE 3.37 Meteosat-10 image captured on April 22, 2015, with the SEVIRI instrument. (From www.esa.int.)

FIGURE 3.38 GOES-12 image of the Hurricane Katrina acquired on August 29, 2005. (From http://apod.nasa.gov/apod/ap050829.html.)

Meteosat is the European counterpart of the GOES platform. It was first launched in 1977, and nine more have been orbited since then, providing critical weather information for Europe and Africa. Currently, three satellites are working in parallel: one at 0° longitude (the primary one) and another two in nearby orbits (for backup and to improve temporal observation) and another one at 57° east. All are orbiting at 36,000 km. These satellites are built by the ESA and operated by the EUMETSAT consortium.

The first generation of Meteosat satellites had an MSS with three spectral bands—VIS, MIR, and TIR—with 2.5×2.5 km spatial resolution for the former and 5×5 km for the latter two. In 2002, a new series of Meteosat satellites (named Meteosat Second Generation, MSG) was launched, incorporating notable improvements over the previous ones. The MSG included two observation instruments: the Geostationary Earth Radiation Budget (GERB) instrument, a visible-infrared radiometer for Earth radiation budget studies, and the Spinning Enhanced Visible and InfraRed Imager (SEVIRI), which observes the Earth in 12 spectral channels every 15 min, with a spatial resolution up to 1 km in the visible band and 3 km in the other channels. A third generation of Meteosat is currently being developed. It will include new sensors and improve the presently operational. The new scanning instrument will provide 16 channels at higher spatial resolution (from 1 to 2 km), at a higher frequency observation of 10 min. This satellite series is expected to be launched in 2019.

In addition to atmospheric products, information acquired by geostationary satellites is very useful for oceanographic (chlorophyll concentration, surface temperature) and land applications (fire detections, fire risk, land temperature, snow cover, vegetation trends, etc.) (Calle et al. 2006; Nieto et al. 2010, 2011; Trishchenko 2006).

### 3.4.7 Polar-Orbiting Meteorological Satellites

The main advantage of geostationary satellites is the very frequent observation they provide, which is ideal for atmospheric monitoring. However, since the geosynchronous orbit is very far, the images have coarse resolution. In addition, they only observe a certain area of the Earth, always the same. For these two reasons, a complementary series of polar-orbiting meteorological satellites have been developed by national weather services and space agencies.

The first and most relevant so far has been the Television Infrared Observation Satellite (TIROS), first launched during the 1960s, and renamed NOAA after the deployment of the sixth satellite in 1979. The NOAA designation comes from the U.S. weather service (the National Oceanic and Atmospheric Administration), which is the agency in charge of the satellites' management.

The orbital height of the NOAA satellites has varied from 833 to 870 km, along with the development of the series (the last one launched in 2009, and it was named NOAA-19).

The NOAA series carries out several weather-oriented sensors, the main one being the AVHRR. This system is an across-track MSS providing 1.1 km nadir resolution at nadir in five spectral bands (R, NIR, MIR, and two in the TIR). The FOV is 55.4° on both sides of the satellite track, covering a swath of 3000 km. The observation cycle of AVHRR is 12 h, but this frequency is improved to every 6 h since two satellites have operated simultaneously, one crossing the equator at 7:30 and 19:30 GMT, and the other one at 14:30 and 2:30 GMT. These hours change throughout the sensor's life as a result of orbit degradation (Privette et al. 1995).

The AVHRR has provided one of the longer time series of satellite data, from 1979 to the present. Its coarse spatial resolution is balanced by its good spectral and very good temporal resolution; therefore, it has been successfully used for many different terrestrial and oceanic applications (Chuvieco et al. 2008a; Gitelson et al. 1998; Goetz et al. 2006; Loveland et al. 2000; Peters et al. 2002; Riaño et al. 2007b; Soja et al. 2004; Figure 3.39). The sensor has important radiometric and geometrical problems (Cracknell 1997), especially because of the large-view zenith angle and orbital degradation, but it is still an important source of temporal trends for many dynamic variables at global scales.

AVHRR images are available in three different formats: the highest resolution is known as Local Area Coverage (LAC), when recorded onboard, or High Resolution Picture Transmission (HRPT), when data are sent in real time to the receiving stations. Since the recording system on NOAA was quite reduced, only a coarser-resolution product (4×4 km), named Global Area Coverage (GAC), was usually stored on board.

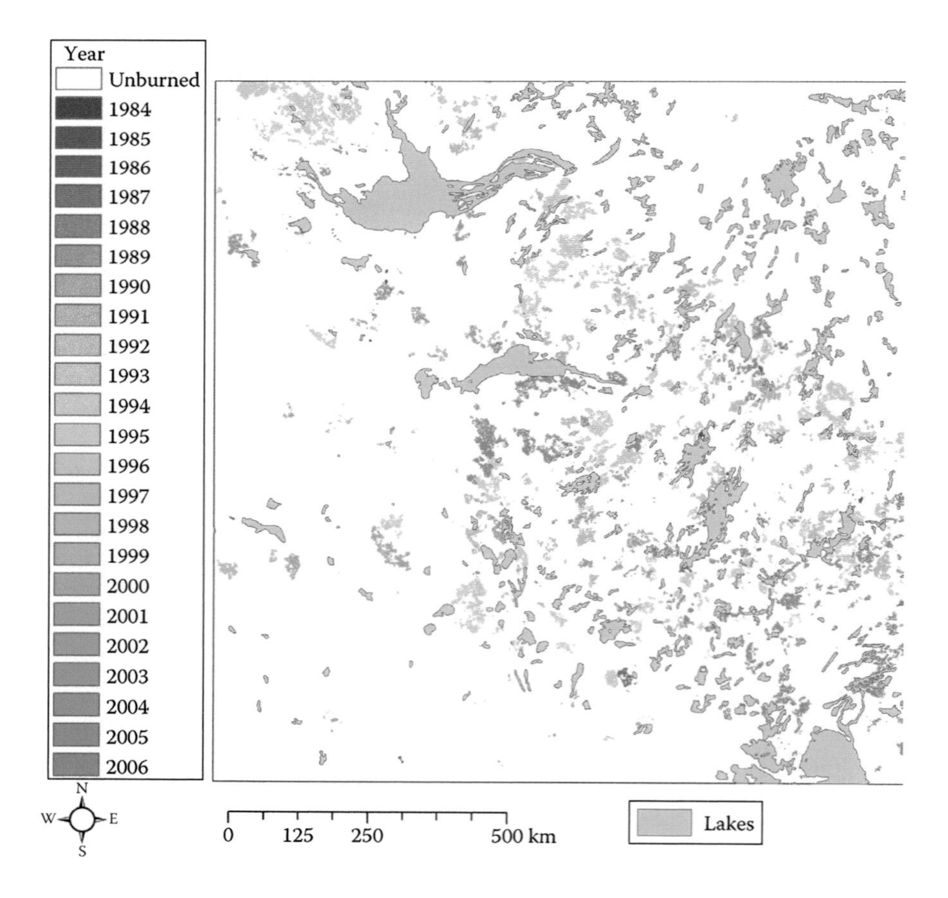


**FIGURE 3.39** Multitemporal series of burned areas in Northern Canada extracted from AVHRR images. (After Chuvieco, E. et al., *Remote Sens. Environ.*, 112, 2381, 2008a.)

From GAC data, historical time series of AVHRR have been released for the analysis of global temporal trends. The most used was the Pathfinder AVHRR Land data sets (PAL) at 8 km resolution, including the raw bands from 1981 to 1996, and the Global Inventory Modeling and Mapping Studies (GIMMS) data set, which includes only a spectral vegetation index (the NDVI; see Chapter 7), also at 8 km spatial resolution for a 25-year period spanning 1981 to 2006. More recently, the PAL database has been improved with a new time series named Land Long-Term Data Record (LTDR) (http://ltdr.nascom.nasa.gov/cgi-bin/ltdr/ltdrPage.cgi, last accessed in June 2015), which includes global coverages of both AVHRR and MODIS images at 0.05° resolution from 1981 to 2014.

In addition to the AVHRR, the NOAA satellite series includes other meteorological sensors such as the Solar Backscatter Ultraviolet Spectral Radiometer (SBUV/2) to measure the vertical concentrations of ozone in the atmosphere; the Earth Radiation Budget Experiment (ERBE), with a resolution of 200–250 km at nadir, dedicated to the analysis of global radiation; and the TIROS Operational Vertical Sounder (TOVS), which measures a vertical profile of atmospheric temperatures, humidity, and ozone contents. The TOVS includes three sensors: the High Resolution Infrared Radiation Sounder (HIRS/3), an MSS with 19 channels in the infrared (3.76–14.95 μm) and one in the visible (0.69 μm), with a spatial resolution of 17 km at nadir; the Microwave Sounding Unit (MSU), a microwave radiometer with 4 channels close to 55 GHz and a spatial resolution of 109 km at nadir; and the Stratosphere Sounding Unit (SSU), an infrared radiometer of modular precision with 3 channels in 15 μm and a spatial resolution at nadir of 147 km. The MSU and SSU sensors have been substituted after NOAA-15 by the AMSU-A and B; the first is a microwave radiometer with 15 channels that provide more precise sounding of temperature and atmospheric moisture. The AMSU-B is a microwave radiometer with five channels especially adapted to measuring water vapor.

Although NOAA has been the longest civilian satellite series used for meteorological and global environmental analysis, it is also worth mentioning the Defense Meteorological Satellite Program (DMSP) series, which were initially limited to military applications but were released in the late 1990s for civilian uses as well. The DMSP series were initially designed for facilitating global diurnal information of cloud coverage. Initially, they included only video cameras, but they were improved in the late 1970s with passive microwave radiometers. The last DMSP satellite (F-19) was launched in 2014.

For terrestrial applications, the most interesting sensors on board the DMSP are the Special Sensor Microwave Imaging (SSM/I), a microwave radiometer that completes the measurements on soil moisture and rainfall taken by the Nimbus in the 1970s, and the Operational Linescan System (OLS), which is an MSS that provides diurnal and nocturnal images in two bands, visible (0.5–0.9 μm) and TIR (10.5–12.6 μm). The spatial resolution can be up to 0.56 km, although only a degraded version of 2.7 km is usually recorded. This sensor has a very high radiometric sensitivity at visible wavelengths, which makes it possible to detect light sources at night (urban areas, oil wells, or fires), obtaining spectacular images of urban patterns at global scale (Figure 3.40).

**FIGURE 3.40** World mosaic of DMSP images showing urban lights. (From http://visibleearth.nasa.gov/view.php?id=55167.)

In the near future, both the NOAA and DMSP satellites will be replaced by the Joint Polar Satellite System (JPSS), which will incorporate advanced sensors to make more detailed observations over the Earth system. The preparatory mission for this new constellation of satellites was launched in 2011 and named Suomi after a U.S. meteorologist, while the next two are expected for 2017 and 2021. JPSS series include different atmospheric sensors for temperature and moisture, clouds and earth radiation budget, ozone concentration and aerosols retrieval. The main sensor on board is the Visible/Infrared Imager/Radiometer Suite (VIIRS), with 21 bands and a nadir spatial resolution of 375 m, which can obtain sea and land temperature and aerosol optical depth; monitor vegetation and crop trends, snow, and ice melt; detect active fires and volcanoes; and many other applications. In addition, this satellite incorporates the Advanced Technology Microwave Sounder (ATMS), a cross-track scanner with 22 channels, aiming to retrieve atmospheric profiles; the Cross-track Infrared Sounder (CrIS), a Fourier transform spectrometer with 1305 spectral channels, oriented to produce 3D temperature, pressure, and moisture profiles; the Ozone Mapping Profiler Suite (OMPS), an ozone-profile detector; and the Clouds and the Earth's Radiant Energy System (CERES), a three-channel radiometer, measuring both solar-reflected and Earth-emitted radiation.

Other countries have also developed polar meteorological satellites. The most widely known are the Russian METEOR, active since 1991, with two bands in the visible and in the TIR; and the Chinese Feng Yun, launched for the first time in 1988. In the last versions, this satellite includes 10 channels (VIS, NIR, SWIR, and TIR), with a spatial resolution of approximately 1 km. More recently, ESA and EUMETSAT have developed and launched the polar-orbiting European platform named METOP-A. The first satellite of this constellation is named and it was launched in October 2006 and the second one in 2012. These satellites will replace the morning NOAA satellites in the near future. The payload of METOP is very similar to that of NOAA, with AVHRR-3, HIRS, AMSU, and MHS. The only sensors not included in the NOAA series are the Infrared Interferometer, for temperature sounding; an advanced scatterometer, for retrieval of oceanic winds; and Global Ozone Monitoring Experiment (GOME), for measuring ozone concentrations.

The ESA's Copernicus constellation will include additional missions that either have a meteorological scope or will include coarse-resolution sensors that could serve the needs of climate researchers. This is the case of Sentinel-3 and 5. The former will include different sensors for measuring ocean color and temperature, with updated instruments over the AATSR and MERIS sensors onboard the ENVISAT satellite. Sentinel-5 will include sensors for meteorological observation to prepare the second generation of METOP satellites.

### 3.4.8 Terra–Aqua

The Terra and Aqua platforms are part of the NASA Earth-Observing System (EOS) program (Parkinson et al. 2006), which includes several other platforms for terrestrial observations. Terra was successfully launched in December 1999, and it is located in a near-polar, Sun-synchronous orbit, crossing the equator around 10:30 a.m. and p.m. The orbital height is 705 km, with a period of 98.88 min and a repeat cycle of 16 days. It carries five sensors (MODIS, CERES, MISR, MOPITT, and ASTER) designed for global observations of critical land, oceans, and atmospheric variables. Aqua was launched in May 2002, and it has similar orbital characteristics as Terra, but with a lag period of 3 h and, therefore, crosses the equator at 1:30 and 13:30. This satellite carries six sensors (AIRS, AMSR-E, CERES, HSB, and MODIS), with an instrument configuration oriented toward oceanographic studies.

The main sensor on board the two platforms is MODIS. It is the most advanced MSS ever used for remote sensing studies and acquires worldwide daily data from 36 different spectral bands, at different resolutions, from 250 to 1000 m. The first two bands have finer spatial resolution (250 m) and include the red and NIR spectral wavelengths to monitor vegetation activity. Another five bands are acquired at 500 m resolution, covering the visible and SWIR, and are mainly intended for retrieval of vegetation, snow, and soil properties. All the other bands have 1000 m resolution and cover additional wavelengths in the visible, MIR, and TIR. The swath covered by MODIS is 2300 km, and observation frequency is daily (twice when the two satellites are accounted for). The MODIS program was built around a wide range of scientists working together to obtain both calibrated radiances and final products. These products are grouped in three domains: atmosphere products, including aerosol, total precipitable water, clouds, and atmospheric profiles; land products, including surface reflectance, surface temperature, land cover, vegetation indices, active fires, incoming radiation, evapotranspiration, plant productivity, albedo, burned areas, snow cover, and sea ice; and ocean products, which includes sea temperature, chlorophyll-a concentration, particulate carbon, fluorescence, and incoming photosynthetic radiation. Images are free and can be accessed through the data distribution network of NASA, named Distributed Active Archive Center (DAAC). Figure 3.41 shows an example of a global albedo image derived from MODIS three reflectance channels (B, G, R). Snow and desert areas are shown in more intense values, while vegetated areas present more reflectance in the green region, as expected.

On board the Terra and Aqua missions are also other sensors worth mentioning. One of those included in Terra is the Advanced Space-Borne Thermal Emission

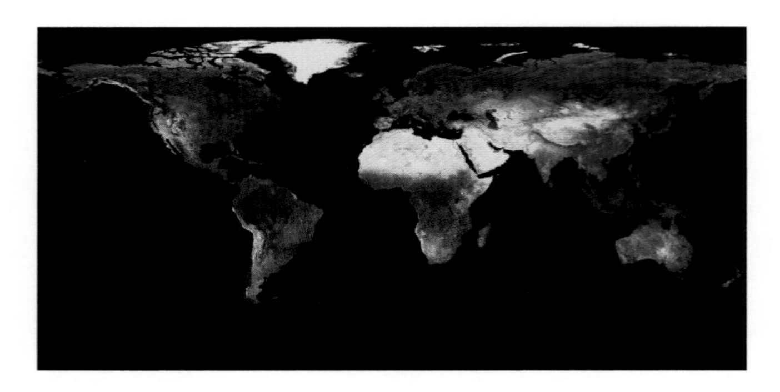

**FIGURE 3.41** Spectral gap-filled snow-free albedo in the MODIS blue, green, and red channels showing the percentage of solar radiation reflected by the land surface. Images represent July12, 2010 (range 0–0.25 albedo). (Courtesy of Crystal Schaaf and Qingsong Sun, University of Massachusetts Boston, Boston, MA.)

and Reflection Radiometer (ASTER), developed by the Japanese Ministry of Economy. ASTER had 15 channels: 4 with 15 m resolution (G, R, NIR1, NIR2), the last two bands duplicated for stereoscopic view; 6 with 30 m resolution in the SWIR; and 5 other TIR bands with 90 m. Its main objectives were the measurement of cloud and soil properties, vegetation and crop monitoring, land temperatures, and generation of digital elevation models. Since images were acquired simultaneously with other sensors, ASTER has been used for validation of coarse-resolution sensors such as MODIS (Schroeder et al. 2008). From the stereoscopic bands, a freely available digital elevation model with 15 m resolution was created from photogrammetric methods (Tachikawa et al. 2011). ASTER was not operated continuously, but only for a few minutes of each orbit because of energy constraints; therefore, it does not provide a systematic coverage, as Landsat sensors do. Figure 3.42 shows a beautiful example of an ASTER image for the vicinity of Arequipa, the second largest city of Peru.

The Multi-Angle Imaging Spectroradiometer (MISR) is one of the first sensors providing a multiangular observation of the whole planet. It acquires images at nine different angles (at nadir plus at 26.1°, 45.6°, 60°, and 70.5°, forward and backward) and in four spectral bands (B, G, R, and NIR). This configuration facilitates the analysis of atmospheric optical thickness of the atmosphere (Figure 3.43), cloud type and height, leaf area index, and fraction of absorbed photosynthetically active radiation. Each image covers 360 km at a variable spatial resolution, which depends on the observation angle, although the most common are 275, 550, and 1100 m. MISR flies on the Terra platform.

The Measurements of Pollution in the Troposphere (MOPITT) instrument was primarily designed to measure carbon monoxide and methane concentrations in the atmosphere. The sensor uses a correlation spectroscopy technique, which estimates the concentration of a gas by spectral selection of emitted or absorbed radiance, using a sample of the same gas as a filter. The sensor has a resolution of 22 km with a swath width of 640 km. It was included in the Terra platform, as part of a cooperative agreement with Canada.

FIGURE 3.42 Three-dimensional ASTER image of the Arequipa region (Peru). The Misti volcano at more than 5800 m is shown prominently in the image. (From http://visibleearth.nasa.gov/.)

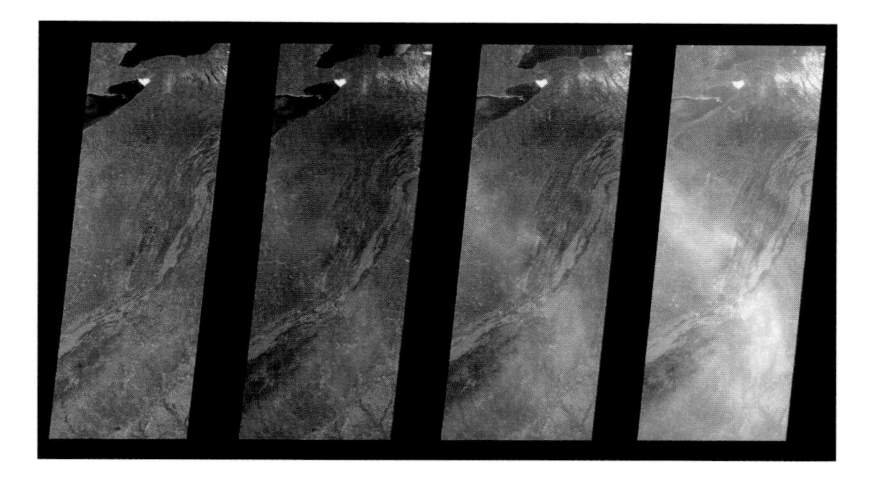

FIGURE 3.43 Images acquired by the MISR multiangular mission over the Appalachian range. Increasing atmospheric haze is observed in the most oblique viewing angles (to the right). (From http://visibleearth.nasa.gov/.)

The Atmospheric Infrared Sounder (AIRS) is a high-spectral-resolution grating spectrometer containing 2378 infrared channels (from 3.7 to 15.4 μm) for obtaining atmospheric temperature profiles. AIRS also has four visible/NIR channels (from 0.4 to 1 μm) for characterizing cloud and surface properties and obtaining higher spatial resolution than the infrared measurements. It is included on the Aqua platform.

The AMSR-E is a Japanese instrument on board the Aqua platform. It includes a 12-channel microwave radiometer designed to monitor a broad range of hydrologic variables, including precipitation, cloud water, water vapor, sea surface winds, sea surface temperature, sea ice, snow, and soil moisture.

The Advanced Microwave Sounding Unit-A (AMSU-A) is a 15-channel microwave sounder designed to obtain temperature profiles in the upper atmosphere and to provide a cloud-filtering capability for the AIRS infrared measurements, for increased accuracy in troposphere temperature profiles. It is on board the Aqua satellite as well as the latest in the NOAA Polar-Orbiting Environmental Satellite (POES) series (after NOAA-15).

The Humidity Sounder for Brazil (HSB) provided atmospheric water vapor profile measurements until February 2003 when it lost operation. It included a four-channel microwave sounder designed to obtain atmospheric humidity profiles under cloudy conditions and to detect heavy precipitation under clouds. It was included on the Aqua platform.

As previously commented, the next generation of U.S. polar-orbiting weather satellites will incorporate some of the sensors tested in the EOS program, such as VIIRS (the legacy of MODIS) and CERES.

### 3.4.9 Radar Missions

Even though the SEASAT was the first civilian satellite incorporating a radar sensor in 1978, the first long-term mission based on SAR instruments was the European Remote Sensing Satellite (ERS), launched in 1991 by the ESA, and continued in 1995 by a similar satellite (ERS-2). The ERS focused on ocean and ice monitoring. These satellites had an orbital height of 780 km and included a synthetic aperture imaging radar, radar altimeter, an across-track scanner named Along-Track Scanning Radiometer (ATSR), and other instruments to measure ozone (GOME), wind speed, and atmospheric profiles.

The most innovative sensor was the Active Microwave Instrument (AMI) and included a radar imager and a wind scaterometer. The first instrument could be used in image mode working in band C (5.3 GHz), vertical polarization with a spatial resolution of 26 m in range and between 6 and 30 m in azimuth, and a coverage area of 102 km. The average angle of incidence was 23°. This equipment could also work in "wave" mode, directed toward the study of oceanic waves. The second MW instrument on board the ERS-1 and 2 was the radar altimeter, working in the K band (13.8 GHz). The system was capable of calculating the height of ocean waves with a 10 cm precision, which was extensively used for monitoring sea level alteration and oceanic current movements. The ERS mission ended in 2011.

In 2002, ESA launched an ambitious satellite mission named ENVISAT. In addition to continuing with ERS observations, this new satellite intended to improve the role of Europe in global observation missions, with special emphasis on atmospheric chemistry and oceanographic studies. ENVISAT followed a Sun-synchronous orbit at 800 km, with a 35-day repeat cycle. It carried a wide variety of scientific instruments. It included an advanced version of radar instruments previously flown on the ERS, which was named Advanced Synthetic Aperture Radar (ASAR), also working in the C band with 30 m spatial resolution, but with up to five polarization modes and different swath modes. The radar altimeter on board the ENVISAT worked in band R, but with higher precision than the previous versions on board the ERS. The ENVISAT satellite also included an improved version of ATSR, a hyperspectral sensor named Medium Resolution Imaging Spectrometer (MERIS), with 15 channels and 300 m pixel size, and several other sensors for measuring aerosols, trace gases, and atmospheric constituents. The mission ended suddenly in 2012 as a result of unexpected communication problems.

The ESA continued radar observations after the launch of the Sentinel-1A satellite in 2014, which is part of the Copernicus Operational Services being developed by the agency. Sentinel-1 orbits at 693 km and includes an advanced SAR, able to detect 80 km swaths at a resolution of up to 5 m spatial resolution (Figure 3.44). A second satellite with the same instruments is expected to be launched in 2016.

In addition to ESA, the Japanese and Canadian space agencies have been very active in developing radar missions. The JERS-1 was launched by the Japan Aerospace Exploration Agency (JAXA) in 1992 and incorporated an L-band SAR, with HH polarization, 35° angle of observation, spatial resolution of 20 m, and swath coverage of 75 km. Unfortunately, it stopped functioning in 1998. The successor of this Japanese mission was the ALOS, launched in January 2006. ALOS had three sensors: the Panchromatic Remote Sensing Instrument for Stereo Mapping (PRISM), which has three sets of optical systems to measure land elevation; the Advanced Visible and Near-Infrared Radiometer Type 2 (AVNIR-2), for land cover monitoring; and the Phased Array Type L-band Synthetic Aperture Radar (PALSAR). This advanced SAR instrument was able to work at two different imaging modes, with resolutions between 7 and 44 m in the fine mode (up to 70 km swaths) and 100 m in the scan mode (up to 350 km swaths), with different polarizations. The ALOS satellite was lost in 2011, replaced in 2014 by the ALOS-2 satellite, which included a more sophisticated L-band Radar system capable of acquiring up to 3 m resolution

FIGURE 3.44 Sentinel-1A multitemporal radar image of the Chinese city of Tianjin. (From www.esa.int/spaceinimages/Images/2015/06/Tianjin_China.)

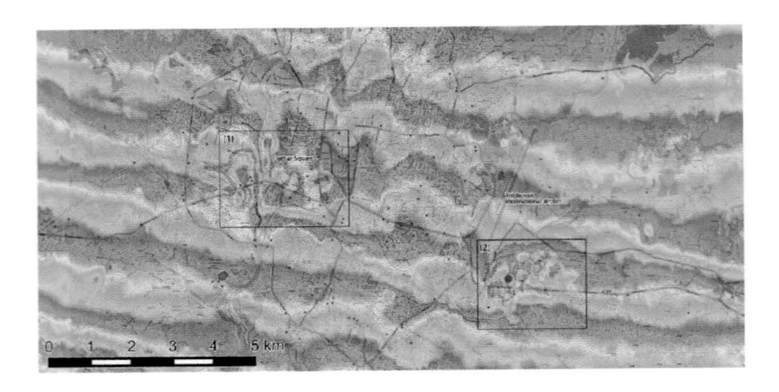


FIGURE 3.45 PALSAR-2 interferogram around Katmandu processed by JAXA. (From PALSAR-2 data: JAXA, www.eorc.jaxa.jp/ALOS-2/en/img_up/dis_pal2_npl-eq_20150502.htm; Map data: Copyright OpenStreetMap contributors.)

images in the ultrafine mode, keeping the scan mode of the previous PALSAR. A striking figure of the local displacements produced by the Nepal earthquake of 2015 is obtained from the interferogram portrayed in Figure 3.45, which was based on the analysis of two PALSAR-2 images acquired before and after the quake.

The Canadian RADARSAT program had also been very successful with almost continuous series in the last 20 years. The first satellite was launched in November 1995. The satellite was placed in a 798 km height polar orbit, with an inclination of 98.6° over the equatorial plane. It incorporates a highly advanced radar that can operate in several modes, with several spatial resolutions, areas of coverage, and incidence angles. It used the same frequency as the ERS satellites (band C, 5.3 GHz) and horizontal polarization (HH). The spatial resolution could vary between 11 and 100 m and the area covered between 50 and 500 km. For the standard mode of operation, the temporal frequency was 24 days. As for the angle of incidence, it could vary its operative mode between 20° and 50°, reaching up to 10° and 60°, in experimental mode. A successor of this mission was the RADARSAT-2 launched in 2007, which improved the SAR instrument characteristics with a finer mode spatial resolution that could reach 3 m. It included multipolarization (HH, HV, VV, and VH) and more flexibility in the incidence angles than RADARSAT-1.

The Italian Space Agency and Ministry of Research built and successfully launched the COSMO-SkyMed, formed by four identical medium-sized satellites equipped with SAR instruments and launched between 2007 and 2010. The SARs work in band X with two polarizations and provide different observation modes and spatial resolutions, being the finest 1 m for 10 × 10 km areas.

Finally, some radar missions have also been promoted by private companies, such as TerraSAR-X, a joint venture from the German space agency and EADS Astrium. This system launched in 2007 includes an X-band SAR with different image modes that can provide up to 1 m resolution for small areas. Different polarization modes are provided. The system was established in tandem with another mission named TanDEM-X (launched in 2010) to derive a new global high-resolution digital terrain model using interferometric techniques. The model was finally released in

2014, at 12 × 12 m spatial resolution, with an estimated vertical accuracy of 2–4 m, and it is commercially available as WorldDEM™ by Airbus.

### 3.4.10 Programs with Hyperspectral Sensors

Hyperspectral instruments (more properly named imaging spectrometers) provide very high spectral resolution as they can collect simultaneous images in numerous spectral bands. Until recently, these instruments were limited to airborne observation, since the data flow and energy consumption were very high for reliable spaceborne operation.

The Airborne Visible/Infrared Imaging Spectrometer (AVIRIS) has been the most extensively used imaging spectrometer to date because of its sound technical characteristics. This sensor has been flown in four different airplanes at different elevations, and it is managed by the NASA Jet Propulsion Laboratory. The sensor has 224 continuous channels, covering from 0.4 to 2.5 µm, and 12-bit radiometry. This spectral resolution provides continuous spectra or surface features at different spatial resolutions from 5 to 20 m, depending on the height of the flight. The range of studies performed with these sensors is enormous in vegetation studies (Bo Cai and Goetz 1995; Riaño et al. 2002; Roberts et al. 1993, 1997; Serrano et al. 2000) and soil analysis (Kruse et al. 1993a; Okin et al. 2001; Palacios-Orueta and Ustin 1996, 1998; Smith et al. 1990). The AVIRIS data allow the extraction of several parameters of the structure of forest cover, the age of tree plantations, the importance of soils, or the effect of shadows (Ustin and Trabucco 2000). Figure 3.46 shows an AVIRIS image of a forest fire in Southern California. The same phenomenon is observed in several wavelengths, showing different types of information. While in the short bands (0.5 µm), the smoke is the most evident feature, in the NIR (1 µm), the recently burned area clearly contrasts with unburned vegetation, and the SWIR (1.5–2 µm) shows the active fire front. In this band, the dispersion of smoke is smaller than in the shorter wavelengths, which is the reason the fire can be seen more clearly.

In addition to AVIRIS, there are other airborne hyperspectral sensors worth mentioning. The Digital Airborne Imaging Spectrometer (DAIS), managed by the

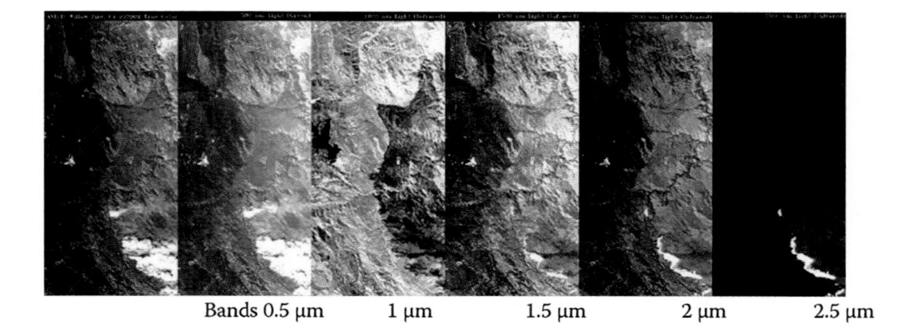


**FIGURE 3.46** AVIRIS image of the San Bernardino forest fire of 1999. The bands selected for display are (from left to right) 0.5, 1, 1.5, 2, and 2.5 µm. (From visibleearth.nasa.gov.)

German DLR, is capable of registering 79 channels, most of them in the VIS–NIR–SWIR, but also some in the MIR and TIR; also included are the Canadian Compact Airborne Spectrographic Imager (CASI), with 288 bands between 0.43 and 0.87 μm, and the Australian HyMap, with 125 bands between 0.4 and 2.5 μm.

Imaging spectrometers have been widely used by atmospheric remote sensing scientists, as well as by extraterrestrial observation. However, the first sensor with proper hyperspectral capacity was the Hyperion, launched in 2000 on board the EO-1 satellite. This sensor has 220 channels between 400 and 2500 nm, a 30 m spatial resolution and a coverage area of 7.5 × 180 km. Even though it did not have a good signal-to-noise ratio, it was very useful to test different potential applications of satellite imaging spectrometers (Goodenough et al. 2003; Roberts et al. 2003). In addition to this sensor, the EO-1 mission included additional sensors to test the use of new systems in future remote sensing missions. In fact, the ALI was a precursor to the Landsat 8 OLI, being a push-broom sensor with similar spectral bands but higher spatial resolution than ETM+ (10 m in the panchromatic band), as well as better radiometric sensitivity.

Hyperspectral images have also been built by ESA, particularly with the Compact High Resolution Imaging Spectrometer (CHRIS) on PROBA-1, launched in 2001, as a technology demonstration mission. The CHRIS instrument can be programmed for different acquisition modes, from 18 to 61 spectral bands with widths from 1.3 to 11.3 nm, and different spatial resolutions. MERIS, on board the ENVISAT satellite, could also be considered an imaging spectrometer, although it had only 15 bands, as they were continuous and fairly narrow.

Future hyperspectral missions for Earth resources observation are being planned. Among the most developed is the HyspIRI satellite, managed by NASA-JPL, which will incorporate two imaging spectrometers, one focused on the solar spectrum (400–2,500 nm) with 10 nm contiguous bands and 60 m spatial resolution, and another one in the middle and thermal infrared (3,000–12,000 nm). Launch is not yet fixed, but it may occur in the early 2020s. The Japanese agency is also developing a hyperspectral instrument named Hyperspectral Imager Suite (HISUI), to be included in ALOS-3 (expected to be launched in 2017), which will include 57 bands in the VIS–NIR range and 128 additional bands in the SWIR at 30 m spatial resolution.

## 3.5 REVIEW QUESTIONS

1. Which one of the following resolutions is more relevant for urban studies?
   a. Spatial
   b. Spectral
   c. Temporal
   d. Radiometric
2. Which one of these sensors has higher spatial resolution?
   a. Landsat 8 OLI
   b. GeoEye
   c. SPOT-HRG
   d. Terra MODIS

3. The radiometric resolution of a sensor is commonly measured as
   a. The number of spectral bands
   b. The pixel size
   c. The bits per pixel
   d. The orbiting cycle
4. Which one of the following geometrical errors is exclusive of a radar image?
   a. Scan angle
   b. Panoramic distortion
   c. Fore shorting
   d. Earth's curvature
5. What is the main innovation of the MISR sensor?
   a. Multiple polarization
   b. Multiple angles
   c. Multiple bands
   d. None of the above
6. Which one of these sensors was not on board the Terra platform?
   a. SAR
   b. MODIS
   c. MOPPITT
   d. MISR
7. Which one of these sensors acquires stereoscopic images?
   a. Landsat TM
   b. ASTER
   c. SAR
   d. Terra MODIS
8. Which one of the following sensors has hyperspectral capacity?
   a. MODIS
   b. Hyperion
   c. VEGETATION
   d. SAR
9. Which of the following missions includes a radar sensor?
   a. Terra
   b. Sentinel-1
   c. EO-1
   d. SPOT
10. What is the main observation advantage of a geostationary satellite?
    a. Temporal resolution
    b. Spatial resolution
    c. Angular resolution
    d. Spectral resolution

# 4 Basis for Analyzing EO Satellite Images

As mentioned in Section 1.1, this book has an environmental scope, and it is intended to improve the readers' understanding of how remote sensing (RS) techniques are used to extract relevant information from the Earth system. Therefore, we will place more emphasis on the interpretation of the images rather than on the acquisition process, which has been reviewed in previous chapters. Before addressing visual or digital interpretation methods, this chapter reviews a series of introductory concepts that may help the reader to better manage a project dealing with RS data.

In the organization of an Earth observation (EO) project, the interpreter should make decisions mainly based on two considerations: (1) the objectives of the project and (2) the means available to accomplish them (Figure 4.1). The project objectives include end-user requirements as well, which affect the number and precision of the categories to be discriminated, the scale of the output results, the deadline of the project, and budget considerations. On the basis of those requirements, the interpreter should select the most suitable sensor, the number and dates of the images, the processing steps, etc. The appropriate decisions are always a balance between what is ideal and what is feasible, with time, cost, and labor limitations taken into account.

## 4.1 CONSTRAINTS IN USING REMOTE SENSING DATA

### 4.1.1 What Can Be Estimated from the EO Images

Three types of variables can be retrieved from RS imagery, depending on whether they are retrieved directly or indirectly from sensor measurements (Jensen 2000). Primary variables are those directly associated with the physical quantity measured by the sensor system: radiance, when working on detecting incoming radiation flux, or time, when measuring the temporal gap between sending and receiving a certain flux (this is the case of lidar or radar altimeters). Secondary variables are those associated with the interactions between radiation and ground covers; they are factors that explain how much a particular cover radiates energy at different wavelengths. For instance, water or chlorophyll content impacts the outgoing spectral radiance from a vegetated area, absorbing incoming solar energy and thus reducing outgoing radiation (reflectance). There are many other variables that are closely linked to the way a cover reflects or emits energy, such as leaf area, leaf angle distribution, or dry matter for plants; mineral content, roughness, or water content for soils; or turbidity, salinity, and acidification for water. We analyzed the impact of those variables in Chapter 2. Finally, a third group of variables are those not directly affecting observed radiation but related to the secondary variables. To retrieve these variables, a certain level of abstraction is required. For example, we can determine when the

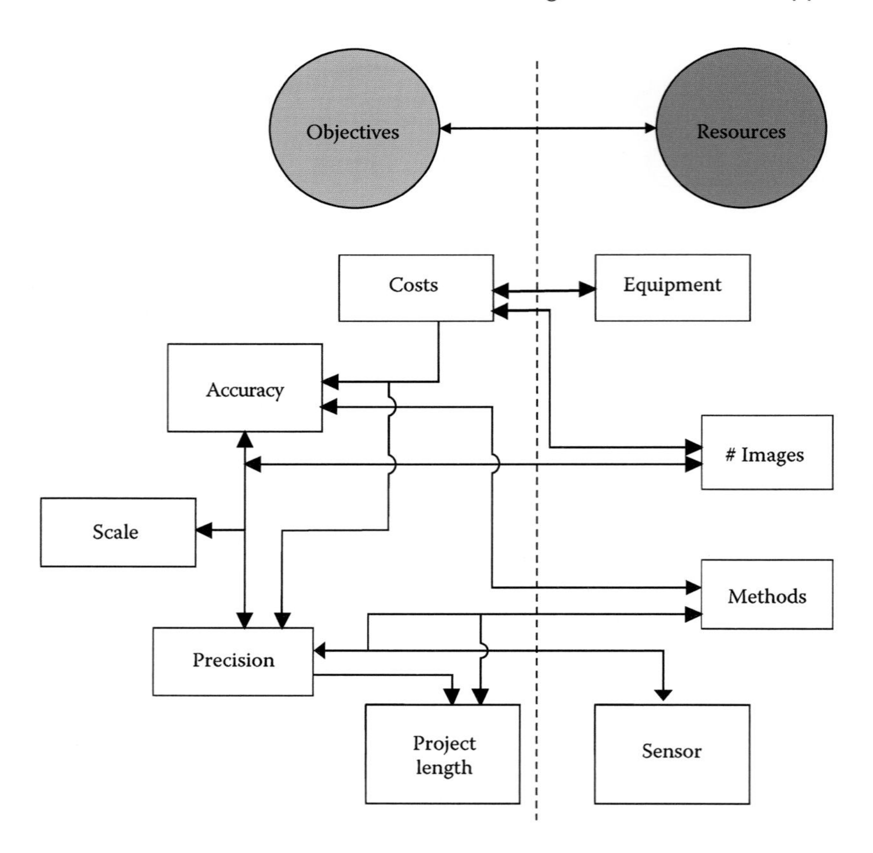


**FIGURE 4.1** The effective incorporation of remote sensing into a project requires a series of decisions to meet the proposed objectives with available resources.

plant is stressed from the temperature, water, or chlorophyll content, but the stress itself is not directly measured. In the same way, we can identify the land cover of a particular area by its reflectance or texture in several spectral wavelengths, but land cover per se is not directly measured from satellite images.

Any interpreter should keep in mind the advantages and limitations of RS data for any particular application. Obviously, remote sensing is not a panacea for all requirements of environmental information. Unfulfilled overoptimistic expectations often lead to the opposite position. When an unrealistic target is carried out, poor results can lead to the opposite conclusion that RS techniques are useless. To avoid both extremes, any project should start with a critical attitude toward the advantage and limitations of current sensors and whether they have the potential to properly satisfy end-user requirements. The range of applications of remote sensing is constantly growing, thanks to the innovations in sensor performance and processing methods. However, when proposing a new application, the EO analyst should clearly state whether that proposal is based on previous successful results or rather is a research area with potential for failure: overselling is even worse than ignorance.

With the current state of EO development, some applications can be considered quite mature, while others are still in the research or development realm. Land cover and precision farming, active fire detection, weather forecast, mineral exploration, ozone monitoring, ocean color or temperature, or snow or flood monitoring can be considered very reliable. Other fields are still in experimental phases, such as the retrieval of vegetation biomass, biomass consumed by fires, snow volume inventories, or earthquake prediction. Finally, there are other projects currently unfeasible, either because of the level of precision required (cattle or wildlife inventories, demographic estimations) or insufficient spatial or temporal resolution (traffic control). In any case, a literature review on what has been done by previous researchers may avoid unrealistic expectations and unnecessary expenses.

### 4.1.2 Costs of Data Acquisition

A couple of decades ago, a common question when presenting a certain analysis based on EO satellite data to a non-RS specialist was: how expensive are those images? Certainly, the common image of satellite technology was to be unaffordable. Nowadays, the growing availability of satellite information makes this perception more unusual, but still many scientists or managers interested in using satellite images consider this technique quite expensive.

First of all we should clarify that accessing accurate and updated information is always expensive. Any expert in developing corporative geographic information system (GIS) knows well that a high percentage of implementing and maintaining the GIS will be the cost of information. Traditional sources of spatial data were aerial photographs, field work, and meteorological or hydrological sensors. Aerial photographs were relatively cheap, but covered small areas, and were only adequate when acquired at the period when the survey was required. In case of requiring a specific aerial survey, prices went much higher. On top of this cost, analog photographs were interpreted manually and commonly need manual digitizing for creating digital databases.

Satellite images are acquired at the date needed, since there are now many sensors that can be used to observe almost anywhere in the world at different spatial resolutions. These images are acquired in digital format and therefore their connection with GIS is straightforward. Finally, the costs of high-spatial-resolution satellite imagery are decreasing as more satellite missions are developed, particularly in the high-spatial-resolution range, which is still the most expensive (Figure 4.2). Medium-resolution sensors (such as Landsat OLI, IRS LISS, Sentinel-2 MSI, etc.) are now freely accessible and therefore new approaches oriented toward large territories and long time series are now affordable. Coarse-resolution sensors, including those used for atmospheric and oceanographic purposes, are most commonly freely accessible as well.

In any case, even though the investment in satellite data could be high, one should always consider the benefits of having accurate and updated geographic information. If decisions are made with poor or outdated information, the outcomes may be far more costly than the required EO images. This is especially true in the case of natural disasters, where the lack of timely data can accentuate the catastrophic

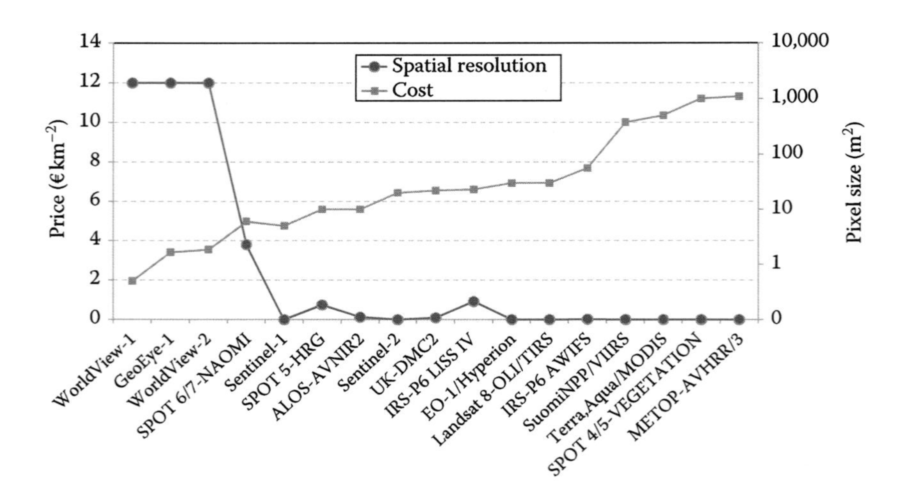


**FIGURE 4.2** Relations between cost and resolution for some environment-oriented sensors.

consequences of these events. Mitigation measures may be too late and inefficient, increasing the loss of lives and damage to infrastructures (Baker et al. 2001; Linnerooth-Bayer et al. 2005; Nourbakhsh et al. 2006).

### 4.1.3 End-User Requirements

The use of remote sensing has been generalized in many applications, but for others it is still scarce or nonexistent. This may be either because the available data do not meet the real requirements of that user community or because the RS community has not properly understood user needs.

The beginning of EO satellite missions was characterized by few and general-purpose systems, which tried to maximize the benefits of the acquired information. This implied the selection of the spatial, spectral, and temporal resolutions that would serve a wide user community, as the investments to develop and launch satellite missions were very high.

Nowadays, the decrease in cost of technical components and weights makes an EO satellite mission affordable, even for medium-size countries and private corporations. This also implies that the needs of specific user communities can be met, optimizing the technical requirements. This has been the case of the atmospheric community, which has been able to develop specific missions for retrieving the concentration of certain gases, such as $CO_2$, CO, $CH_4$, ozone, water vapor and clouds, aerosols, etc. A similar conclusion can be drawn of oceanographic systems, but this is not yet the case of land applications community, which still lack specific missions that could serve the needs of a certain service. For instance, if satellite systems are intended to replace terrestrial observation networks for fire detection, they should provide very frequent observation (≤30 min) of high-temperature sources for small areas (<50 m$^2$). There is not yet a single satellite mission that accomplishes all these requirements, and it is unlikely that a mission will be designed for this purpose in the near future. Consequently, satellite observation would not be an alternative for fire managers in those countries with good ground fire prevention infrastructures. However, it is very useful for countries that do not have field observations, since even one or two daily detections would significantly improve their current information. For those end users, current satellite missions may already be considered operational for detection of fires.

Something similar can be said about the estimation of vegetation biomass, which would need dedicated radar or lidar missions, which are still in the process of development.

## 4.2 TYPES OF INTERPRETATION

The interpretation of digital values acquired by the sensor can be approached in different ways, leading to several procedures to extract information from remote sensing data. Considering the final goal of the interpretation process and the target variables, four types of interpretation approaches can be distinguished (Figure 4.3): thematic classification, generation of biophysical variables, change detection, and spatial patterns.

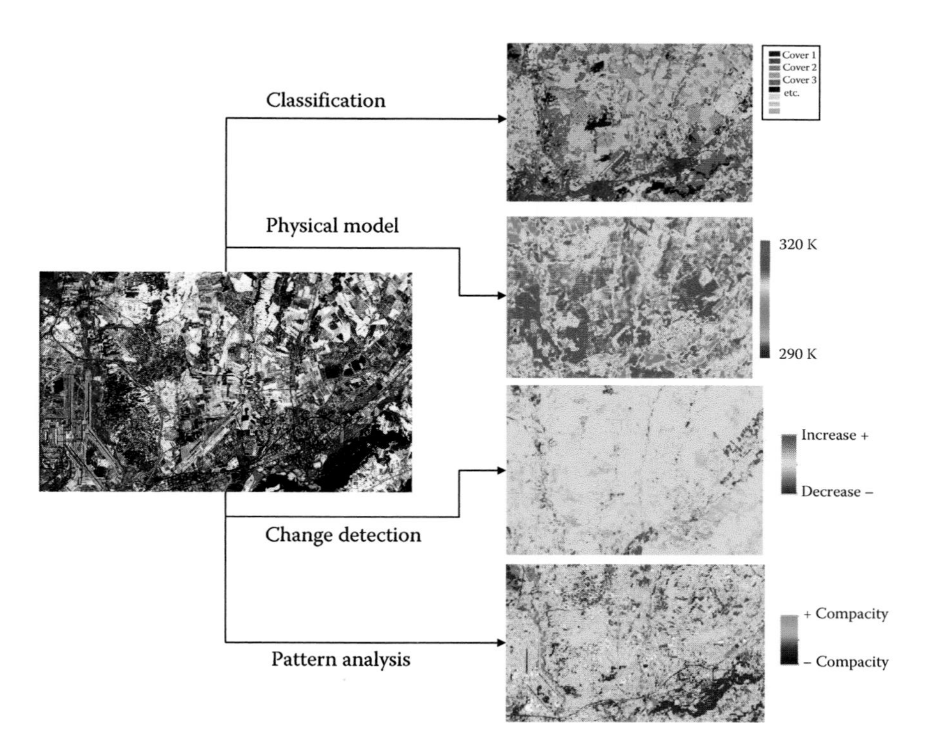


FIGURE 4.3 Different approaches to image interpretation.

### 4.2.1 Thematic Classification

This analysis involves assigning each element of an image to a certain category (land cover, crop type, vegetation species, etc.) through either visual or digital interpretation. This approach aims to label each pixel with the most appropriate thematic class.

This has been the most common application of image interpretation, as a direct inheritance of classic photointerpretation. Conceptually, this approach implies converting the original interval scale of the raw image to a categorical one. Visual analysis is appropriate for this goal, since the human interpreter can integrate a wide variety of criteria to differentiate neighbor surfaces (see Chapter 5). Digital procedures can also be used for the same goal by classifying the image based on different statistical techniques (Tso and Mather 2001) that aim to compare every element of the image with a set of reference patterns (see Section 7.2).

### 4.2.2 Generation of Biophysical Variables

A second goal of remote sensing interpretation aims to generate the spatial distribution of a biophysical variable that is derived from the sensor measurements through an empirical or physical model (Section 7.1). In this case, the goal is not to obtain categories but to convert raw digital values to calibrated energy measurements, and ultimately to a biophysical variable (temperature, ice thickness, moisture, etc.). The quantitative nature of the sensor measurements is fully exploited in this case, as those measurements have a physical meaning: they are not just an expression of how brighter some areas are versus others, but are in fact a quantitative measurement of a radiometric signal. For instance, acquiring a thermal image is similar to measuring temperatures at regular intervals, the distance between measurements being the pixel size.

Even in the case of meteorological satellites of low spatial resolution, spatial sampling provided by satellite images is much denser than any observation network of ground sensors, and therefore they provide a more spatially representative measurement of the radiometric variation of the observed area (and those variables linked to radiometric variability). For example, to create a temperature map of the sea surface, thermal samples taken by buoys or ships' transects can be used to interpolate a continuous surface. With satellite observations, measurements of temperature can be obtained twice a day every kilometer from the MODIS or AVHRR sensors and every 15–30 min at coarser resolutions from geostationary meteorological satellites. In both cases, the spatial coverage would be much denser than ground observation. This also applies to other oceanographic variables, such as ocean height, chlorophyll content, or water turbidity.

Obviously, this interpretation approach is only applicable when digital and calibrated information is available, that is, when there is a consistent relationship between the data of the image and the given physical variable.

### 4.2.3 Change Detection

In this case, the interpretation is oriented toward analyzing what has changed between two or more images acquired with a certain time gap. Satellites observe areas in a periodic way, depending on sensor field of view (FOV) and orbital characteristics.

Therefore, EO satellite is an ideal tool to monitor dynamic phenomena. This monitoring potential is invaluable in many environmental processes—urban growth, deforestation, wetland drying, floods, etc.—as well as to monitor agricultural yields or pasture condition. From this point of view, the goal of the interpretation is to identify image changes between two or more dates. Change detection can be performed over classified images or over biophysical parameters (see Section 7.3).

### 4.2.4 Spatial Patterns

Finally, satellite imagery can also be considered as a representation of the landscape mosaic. From this standpoint, the image can be used to measure spatial properties of objects within the observed area: size, connectivity, shape, etc. (Frohn 1998). The spatial structure can also be used to improve the discrimination of thematic categories, measuring their spatial diversity or homogeneity (see Section 7.4).

We will further comment on these four types of interpretation in Chapters 6 and 7. From now on, the interpreter should be aware that EO is not only valuable for thematic classification but also has many other potentials that extend its possibilities for retrieving interesting environmental variables and fuel many operational services requiring updated and accurate information.

## 4.3 ORGANIZATION OF REMOTE SENSING PROJECT

Assuming that a certain project can be carried out and is attainable through remote sensing techniques, some aspects have to be defined to better organize the project workflow. For instance, suppose that a fishing company tries to use satellite images to better locate certain fish species, providing the interpreter with the thermal and chlorophyll range of water this species is more likely to live in. The company demands from an EO consultant company the required decisions to meet its information needs. Several aspects have to be looked into: which sensor is most suitable for the analysis, how many images are required, what dates are preferred, which methods will be applied, and how much and when fieldwork and ground truth are required. These decisions must be made on the basis of the project requirements and objectives, the current potentials of different sensor systems, and the most suitable interpretation techniques. A review of these aspects would help improve the management of the project.

### 4.3.1 Description of Objectives

A clear definition of project objectives is crucial for making proper decisions on the role of EO images for any particular task. The project requirements should guide the decisions regarding which remote sensing data are most suitable for any particular project: sensor, number of images, dates, and so on, as well as the appropriate methodology.

A broad range of remote sensing projects can be established based on whether the projects are defined by the end user or proposed by the remote sensing interpreter. The former implies that an institution, either a government body or a private company, states its requirements and the remote sensing specialist has to figure out how to meet them in a cost-effective and timely way. This would be the case of the fishing company, where the interpreter does not have any role in defining, but rather in answering properly to, the requirements.

Whenever the interpreter is forwarding a proposal to either a research-funding organization or a potential EO end user, the interpreter is responsible for defining the objectives and the methods to answer them. As a general rule, any proposal of any EO project should be based on a clear and well-defined set of objectives that are compatible with the remote sensing capabilities. In the case of a research project, a set of hypotheses need to be made explicit, as science advances by hypotheses that are verified or discarded. Often, we tend to forget the second part and assume that a project failed when basic assumptions were not verified. However, it is worth remembering that a project is just as scientifically sound when the initial hypotheses are confirmed as when they are refuted. What matters is following a sound procedure for assessing the assumptions. Generally speaking, a research project always entails risks and uncertainties associated with the exploration itself, but this is what investigation is all about. Proposing an application project with the same amount of risks would not be appropriate since it would create false expectations to the client that may not be met. Therefore, the remote sensing scientist should clearly identify whether the project is research-oriented or operational-oriented and identify, as honestly as possible, the potential failures and expected successes of the project.

Confusion between operational and research applications can diminish the use of remote sensing in institutions that otherwise would be open to this technique. In conclusion, it is important to use realistic objectives, based on previous experience or sound references, and adapted to the spatial, spectral, and temporal resolutions of the sensor to be used. The definition of objectives should make reference not only to the generic purpose of the work (e.g., inventory of crops) but to the project specifications, such as target scale, precision, accuracy, time length, and budget limits. In the case of projects that are defined by the end user, these specifications are part of user requirements that the remote sensing interpreter should properly answer.

### 4.3.2 Scale and Resolution

The word *scale* has two different meanings, which are almost contradictory. From a mapping point of view, *scale* refers to the level of reduction of reality on a particular map: that is, how many times items represented in the map are smaller than in reality. Therefore, scale is a fraction that relates measures in the map with measures in reality. Following this concept, a large scale (for instance, 1:5,000) implies a low reduction of reality, while a small scale (1:1,000,000) implies a large reduction of reality.

In everyday language, however, large scale commonly refers to the amount of surface covered by a particular study. A large-scale inventory of redwoods, for instance, would imply in this sense that the study will cover a large area (e.g., the whole western United States). However, since the larger the areas covered, the smaller the level of detail, typically, a large-scale study is done at a small mapping scale, and vice versa, a very small surface study is typically done at a very detailed (large) mapping scale.

This conceptual precision helps us to define the relationship between scale and image resolution. Large mapping scales require high-spatial-resolution images, while small-scale studies are commonly undertaken by coarse-spatial-resolution sensors.

Another aspect to consider in this regard is the relation between scale and the level of precision, meaning the level of desegregation that is intended. Projects at large scale generally have more detailed legends, since it does not make much sense to include many categories when the level of cartographic generalization is large. The more categories in a map, the more polygons, and the more difficult it is to represent them at smaller scales.

Similarly, project scale is associated with the smallest unit of information that can be included in a thematic map, usually known as the *minimum mapping unit* (MMU). Several authors have recommended that the MMU should not be smaller than 4 $mm^2$, measured on the scale of the map (Anderson et al. 1976). This rule provides some insights into defining which sensor is more appropriate for a particular scale (Robin 1998; Figure 4.4). A land cover map at the 1:50,000 scale using the AVHRR sensor would imply an MMU of ~4 $cm^2$ (10 times bigger than the recommended MMU). Similarly, it would not be reasonable to undertake a small-scale study (e.g., 1:1,000,000) using high-resolution GeoEye images, because the results would have to be tremendously degraded afterward, and therefore the quality of the input data would be lost in the generalization process. Following these rules, we can consider an appropriate scale for the range of AVHRR or MODIS between 1:2,000,000 and 1:1,000,000, the range of Landsat TM/OLI between 1:100,000 and 1:50,000, and the range of GeoEye–WorldView and other high-spatial-resolution sensors between 1:25,000 and 1:10,000.

The MMU should not be confused with the smallest-size object observed on an image. Obviously, the MMU has to be larger than this; otherwise, the features will not be recognized. In this sense, it is convenient to recall the differences between

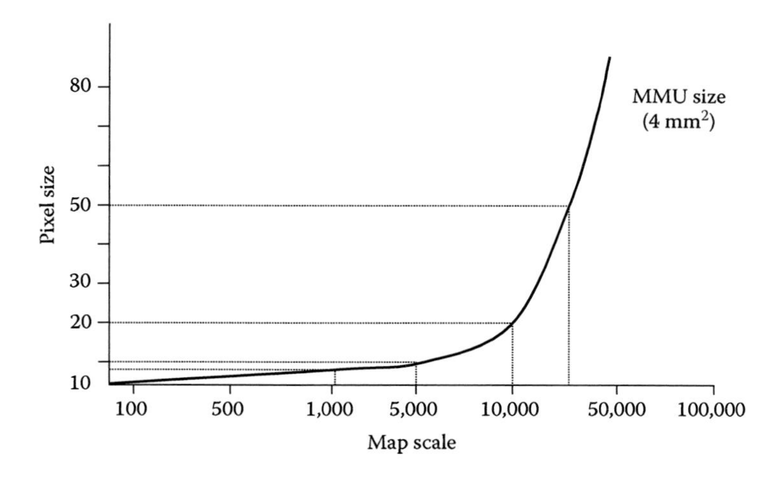


**FIGURE 4.4** Relationship between resolution and the MMU size. (After Robin, M., *Télédétection*, Nathan, Paris, France, 1998.)

object identification and object detection. For example, a river can be detected even if its width is smaller than the spatial resolution of an image because of the high contrast between the low reflectance (high absorption) of water pixels and the adjacent drier upland pixels. Consequently, even for narrow rivers, their presence can be detected on medium-resolution images, but often not their characteristics (water turbidity, for instance), which will be perceptible only from higher-resolution images (Figure 4.5). In conclusion, the spatial resolution of an image will determine whether specific features can be detected or interpreted.

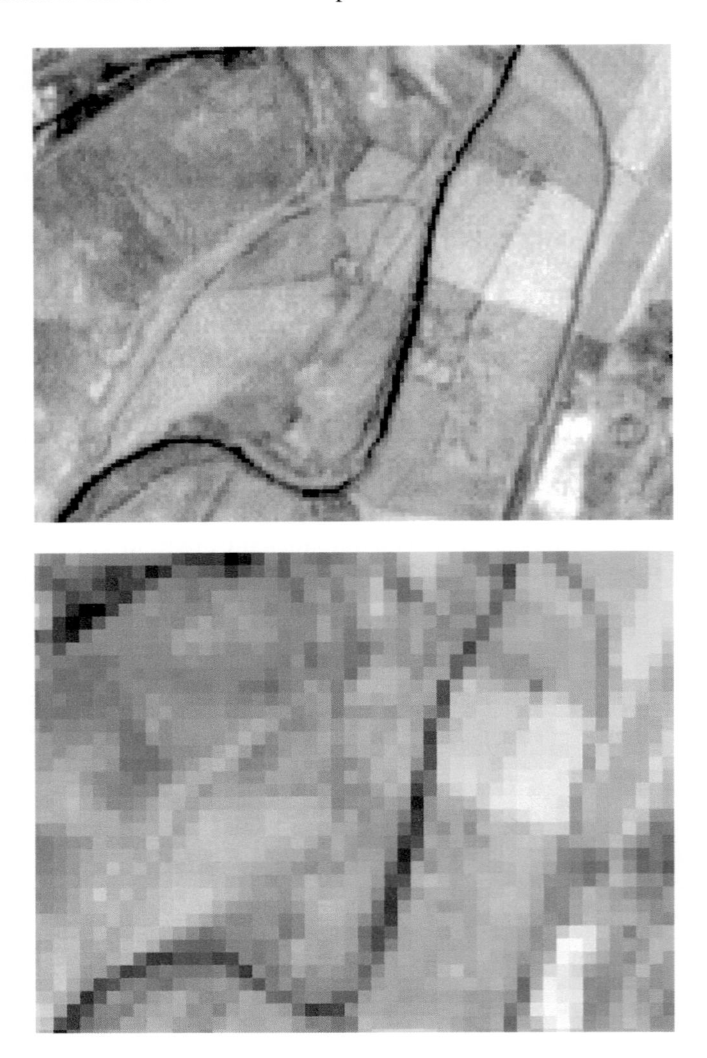

**FIGURE 4.5** Two images of the Henares River in Central Spain. The 15 m resolution of the top image provides an identification of the river, while the 180 m resolution of the bottom image only detects it.

With the increasing use of satellite images in different applications, one can find numerous examples of mapping projects at different scales. These include "global"-scale imagery from the NOAA-AVHRR, SPOT-VGT, Envisat-MERIS, or Terra/Aqua-MODIS sensors that cover the entire planet at resolutions ranging from 8 × 8 km to 0.3 × 0.3 km (Alonso-Canas and Chuvieco 2015; Bartholome and Belward 2005; Giglio et al. 2010; Riaño et al. 2007b; Tansey et al. 2004b) to regional and continental inventories at scales of 1:100,000 to 1:50,000 based on the Landsat TM/ETM+, SPOT-HRV/HRG, and ASTER sensors (Boschetti et al. 2015; Goward 2001; Woodcock et al. 2001). Finally, high-spatial-resolution sensors provide sound information for large scales, 1:10,000 and higher (Baker et al. 2001; Green 2000).

### 4.3.3 Classification Typology

A thematic map legend identifies the categories that the map includes, which should be adapted to the project requirements. As mentioned earlier, the legend is very much dependent on the project scale, but it is also related to the complexity of the target territory. Consequently, it has become common practice in land cover and land use mapping projects to use specific legends that are applicable to the needs of certain study areas while being unsuitable to other areas. The lack of a common criterion for building thematic legends makes it difficult to extrapolate mapping efforts to other study areas and inhibits the generalization of classes for systematic mapping of larger areas and territories.

In situations in which a legend is not already established for a project, that is, the categories to be discriminated are not fixed a priori, the interpreter must be careful to establish classes that are consistent among them and with the data and tools used in the analysis. It is particularly important that the legend capture all the possible diversity in the study area, that the classes do not overlap, and that all surface features are included in one class. In other words, these classes should be mutually exclusive (the same area cannot be placed in two classes) and exhaustive (at least one class for each area).

Many of the published thematic classifications do not always follow these two basic logical rules, especially when several criteria for legend definition are mixed. For example, a category named "shrub on north-facing slopes" implies a mixture of two different terms, land cover and slope aspect, which creates problems when classifying. In this example, shrubs occurring over other aspects will not fit the class, and therefore the typology would not be exhaustive. Similarly, the classes "regrowth pine" and "pine on limestone" are not mutually exclusive since one may also encounter "regenerating pine on limestone rocks."

In an effort to establish more robust legends, several classification schemes have been proposed that incorporate a hierarchical character, allowing one to obtain different levels of detail as a function of the available data. An example of this scheme was designed by the U.S. Geological Survey in 1976 for using remote sensing imagery in land cover mapping (Anderson et al. 1976). This scheme included four hierarchical levels, each of which was designed to be obtained through different remote sensing imagery, such as satellite images at the most general level and large-scale aerial photography, in combination with fieldwork, for the more detailed classes. Moreover, this classification was flexible and could be applied to diverse areas, guaranteeing the extension and integration of different environmental inventories. One weakness of the classification, in our opinion, was the inclusion of land use and land cover in the same scheme of classes, which leads to ambiguity in the classification process. For example, an area could be assigned to several categories according to their designation as land use or land cover: quarries (land use) and bare rock (land cover) or grazing fields (land use) and grassland (land cover). Logically, remote sensing systems take into account only the land cover since the signal detected by the sensor corresponds to the objects that appear on the terrestrial surface and not to their human uses. In some instances, land use could be deduced by the interpreter: when the land cover is "crops," one can easily infer an agricultural land use. However, there are other cases where this is not straightforward, as in a shrub cover class from which one cannot deduce whether the land use is as a natural reserve or is dedicated to recreation or cattle grazing.

In spite of this problem, the work of Anderson and collaborators has been fruitful and useful for similar attempts. Such was the case of the class legend used for the CORINE-Land Cover program (Table 4.1), which was designed specifically for satellite images. This project was promoted by the European Commission for the European Environmental Agency (EEA) in the late 1980s and intended to monitor land cover in Europe at 1:100,000 scale from satellite images (Büttner et al. 2000a; European Commision 1993). The project was one of the first terrestrial-based, operational applications of satellite remote sensing, with the aim of generating a systematic inventory of the land cover of the European territory. The first inventory was completed at the beginning of the 1990s, the second inventory was based on images from 2000, and the third one was from images of 2006. The CORINE Land Cover legend employs a hierarchical approach in such a way that individual countries can develop common schemes according to their own needs while maintaining consistencies among them to ensure that the results are comparable. After the second Corine land cover inventory, the program also included some technical guidelines for monitoring changes between reference periods (Büttner et al. 2000b).

Another interesting effort at unifying classification legends is the Land Cover Classification System (LCCS), developed by the FAO (Di Gregorio 2000). The proposed scheme tries to harmonize the different classification schemes by using a hierarchical approach based on dichotomous and modular properties.

The level of complexity of a legend is associated with potential errors in the interpretation phase. It is very likely that a subdivision of a land cover class into more detailed categories will involve dealing with discriminating similar spectral categories and therefore increasing the likelihood of misclassification. If the last division is impossible to undertake using the images available or the proposed methods of analysis (interpretation), the user will have to either shift toward a more generalized legend or use more refined input imagery (higher spectral and spatial resolution) or advanced processing methods.

**TABLE 4.1**
**CORINE Land Cover Project Class Legend**

| First Level | Second Level | Third Level |
|---|---|---|
| 1. Built-up area | 1.1 Urban fabric | 1.1.1 Continuous urban fabric surfaces |
| | 1.2 Industrial, commercial | 1.1.2 Discontinuous urban fabric |
| | 1.3 Mine, dump | 1.2.1 Industrial or commercial units and transport units |
| | 1.4 Artificial nonagricultural | 1.2.2 Road and rail networks and associated land |
| | | 1.2.3 Port areas |
| | | 1.2.4 Airports |
| | | 1.3.1 Mineral extraction sites and construction sites |
| | | 1.3.2 Dump sites |
| | | 1.3.3 Construction sites |
| | | 1.4.1 Green urban areas and vegetated areas |
| | | 1.4.2 Sport and leisure facilities |
| 2. Agricultural area | 2.1 Arable land | 2.1.1 Nonirrigated arable land |
| | | 2.1.2 Permanently irrigated land |
| | | 2.1.3 Rice fields |
| | 2.2 Permanent crops | 2.2.1 Vineyards |
| | | 2.2.2 Fruit trees and berry plantations |
| | | 2.2.3 Olive groves |
| | 2.3 Pastures | 2.3.1 Pastures |
| | 2.4 Heterogeneous | 2.4.1 Annual crops associated with permanent crops |
| | | 2.4.2 Complex cultivation patterns |
| | | 2.4.3 Land principally occupied by agriculture, with significant areas of natural vegetation |
| | | 2.4.4 Agroforestry areas |
| 3. Forest and natural area | 3.1 Forests | 3.1.1 Broad-leaved forest |
| | | 3.1.2 Coniferous forest |
| | | 3.1.3 Mixed forest |
| | 3.2 Shrub and/or herbaceous | 3.2.1 Natural grassland |
| | | 3.2.2 Moors and heathland |
| | | 3.2.3 Sclerophyllous vegetation |
| | | 3.2.4 Transitional woodland—shrub |
| | 3.3 Open spaces with little or no vegetation | 3.3.1 Beaches, dunes, and sand plains |
| | | 3.3.2 Bare rock |
| | | 3.3.3 Sparsely vegetated areas |
| | | 3.3.4 Burnt areas |
| | | 3.3.5 Glaciers and perpetual snow |

(*Continued*)

**TABLE 4.1** *(Continued)*
**CORINE Land Cover Project Class Legend**

| First Level | Second Level | Third Level |
|---|---|---|
| 4. Wetland, salt | 4.1 Inland wetlands | 4.1.1 Inland marshes |
| | | 4.1.2 Peat bogs |
| | 4.2 Coastal wetlands | 4.2.1 Salt marshes |
| | | 4.2.2 Salines |
| | | 4.2.3 Intertidal flats |
| 5. Water bodies | 5.1 Inland waters | 5.1.1 Water courses |
| | | 5.1.2 Water bodies |
| | 5.2 Marina waters | 5.2.1 Coastal lagoons |
| | | 5.2.2 Estuaries |
| | | 5.2.3 Sea and ocean |

*Source:* Büttner, G. et al., *Corine Land Cover Technical Guide: Addendum 2000*, European Environmental Agency, Copenhagen, Denmark, 2000a.

### 4.3.4 Selection of Imagery

Once the legend and the scale of work are determined, the selection of the imagery to better achieve the desired goals follows. Global inventories and maps should rely on coarse- and moderate-spatial-resolution sensors (MODIS, AVHRR, VEGETATION, MERIS, etc.), while those projects oriented at larger scales require sensors that offer more detailed spatial resolution (TM/ETM/OLI, HRV, ASTER, IKONOS, QuickBird, etc.).

However, spatial resolution is not the only criterion to be taken into account. Spectral and temporal resolutions may be more critical than spatial details. For example, if the project is intended to estimate crop yields, the input sensor should provide very frequent updates of crop conditions, and therefore the use of high-temporal-resolution sensors such as MODIS or VEGETATION will be preferred, even over those with a higher spatial resolution. On other occasions, the spectral dimension will be the dominant criterion for sensor selection. For instance, when evapotranspiration rates need to be estimated, the input sensor should provide TIR radiances as well as VIS and NIR radiances for estimating ground emissivity.

Once the input sensor is selected, the remote sensing specialist needs to define the most convenient period for data acquisition, depending on the seasonality of the phenomenon under study. Logically, the most suitable period for image acquisition would be when the phenomenon of interest is best discriminated, as opposed to other times when it cannot be easily distinguished from similar covers. For instance, in geomorphologic studies, a winter date may be desirable since shadow effects are greater, rendering the detection of topographic features more evident. On the other hand, an inventory of irrigated lands in the Mediterranean region is best accomplished during the summer time when there is maximum contrast with the rainfed crops.

Sometimes, to ensure discriminability of certain land covers, several dates are required. For example, one may wish to follow the growth cycle of a certain crop over the growing season to discriminate it from another type, or study vegetation changes between two critical moments. In these cases, it is convenient to create a phenology calendar for the various vegetation cover types to select optimal times for their discrimination.

If one can acquire only a single image for vegetation cover mapping, one may select either a single date that maximizes the separation among all the cover types present or a time period that coincides with the least amount of cloud cover. Generally, summer months, when the average solar radiation is high, topographic shadows less evident, and vegetation is green, are the most preferred times, particularly right after the rainy season when the atmosphere is more transparent.

### 4.3.5 Image Formats and Media

The format and media of the input image depends on the type of processing to be applied. For visual analysis, paper or photographic products were very common during the past 20 years, but nowadays most visual analysis is done by digitizing on the screen. For older images, the most common media were negative film, positive film, and hardcopy paper. As most historical archives are being digitized, today almost all old photographic products are available in digital format. As in the case of digital images, all those products are commonly available through dedicated web pages, where image files are stored. Although some users with slow Internet connection may prefer magnetic or optical devices, most images are nowadays transmitted through Internet file transfer protocols. These transmissions are quicker than ordinary mail services and generally cheaper.

### 4.3.6 Selection of Interpretation Method: Visual or Digital Processing?

As was reviewed in Chapter 3, most satellite images are originally acquired in digital format, and therefore, computer image processing seems the most obvious option for image interpretation. However, digital values can be transformed into brightness or color tones, making it possible to apply visual interpretation techniques as well. Both digital and visual interpretation have advantages and drawbacks. To choose which is more convenient for a particular application, several aspects should be considered: (1) economic and human resources available, (2) timeliness and accuracy required, (3) type and continuity of the tasks, and (4) homogeneity of the target surface.

In spite of the large reductions in costs of hardware and software for image processing, visual interpretation is still cheaper than computer processing, especially for small study sites. On the other hand, visual interpretation provides a wide range of interpretation criteria that are difficult to translate into computer code, and therefore it may provide better results than computer interpretation when the areas are highly heterogeneous and the objects are large enough to be recognized at the resolution of the image. A clear example is urban environments, which are very difficult to map from computer processing (because of the great mixture of different materials) and are relatively simple to interpret visually (see Chapter 5). In other categories, the relation is the opposite, since the human eye is not able to detect tonal variations that are numerically distinguishable (marshland, vegetation transitions, and different crops). Some studies have compared both techniques for land cover mapping, finding advantages and disadvantages of visual versus digital interpretation (Chuvieco and Martínez Vega 1990; European Commision 1993).

Computer image processing, in principle, provides faster and cheaper processes for repetitive procedures, which are common in large study areas or multitemporal analysis. It provides enough flexibility to access a wide range of operations, from image corrections to the creation of synthetic bands and different classification strategies. Additionally, computer interpretation is more objective than visual analysis because the process is independent of the analyst and can be replicated. The disadvantage of computer interpretation is the higher initial investment required compared to visual analysis, considering both hardware and software costs as well as the investment in training and education.

Visual analysis applies similar identification methods such as classic photointerpretation (tone, texture, size, location, forms, etc.) and does not require sophisticated training. Obviously, visual analysis greatly benefits from the interpreter's experience and visual and mental sharpness, but in general, it can easily be performed by an environmental scientist having a good knowledge of the application field (geomorphology, ecology, land cover, etc.). Digital processing, on the other hand, requires training in fields that may not be included in the standard curriculum of an environmental scientist, for example, statistics, computer systems, or programming languages. However, statistics and computer sciences are only a means of improving interpretation and not the objective itself.

In summary, digital processing requires greater economic investment and training than visual analysis. However, it provides greater versatility, speed, and objectivity. In addition, it provides good accuracy and the ability to discriminate homogeneous areas or subtle changes in reflectance that are not observable visually, as in the case of spatial distribution of biophysical variables (temperature, chlorophyll in water, etc.), and simplifies the identification of temporal changes.

Actually, both methods should be regarded as complementary. Computer-assisted interpretation performs operations that are too complex, expensive, or inaccessible for visual analysis (Büttner et al. 2000b). On the other hand, visual analysis is a good alternative in order to update existing maps, identify heterogeneous classes, or classify areas at medium scales. It aids digital classification by isolating sectors of potential confusion on the image or stratifying sectors for further processing. Digital processing needs a high initial investment, but the costs decrease as the area of the analyzed surface increases. In contrast, the costs of visual analysis are linear with small starting investment costs.

Assuming that digital processing equipment is available, image interpretation can be achieved with both visual and digital criteria. The former may be used to extract complex forms or texture information, while the latter helps in image color enhancing, geolocation, or retrieval of biophysical properties. A hybrid approach might make use of computer processing for preprocessing (geographic location, color compositing, or image enhancement, for instance) and then visual analysis to extract features of interest (new roads, clouds, burn scars, etc.) by on-screen digitizing.

## 4.4 INTERPRETATION PHASE

The applications of satellite remote sensing are numerous and diverse, making it difficult to establish a general framework applicable to all of them. However, we believe it is worth describing a general workflow that may help use remotely sensed imagery for a particular application with the necessary adaptations. We will review this process using as an example the case of a fishing company trying to locate a certain species breeding area, once it knows the ideal range of thermal and chlorophyll conditions.

The proposed scheme is included in Figure 4.6 and involves the following phases:

1. A clear and well-defined set of objectives is needed, indicating the limitations of the study area (heterogeneity), time, or scale required for the application (duration of project, maximum cost, thematic categories to be discriminated) and the resources available. In the case of the fishing industry, these requirements should be set by the company following its end needs.
2. Fieldwork is used in several stages of the interpretation phase. The first stage involves the general characterization of the study site, its environmental characteristics, and associated human activity, which may greatly help the selection of images (dates of the season, selecting the period when it is easier to discriminate the target phenomenon). During this phase, measurements with field radiometers can be taken (Figure 4.7) for a better spectral characterization of the area as well as to select the most adequate sensor and bands to discriminate the target variable. This work may also help identify ancillary data that may be needed for later interpretation and analysis of the image. Obviously for oceanographic applications, these analyses are

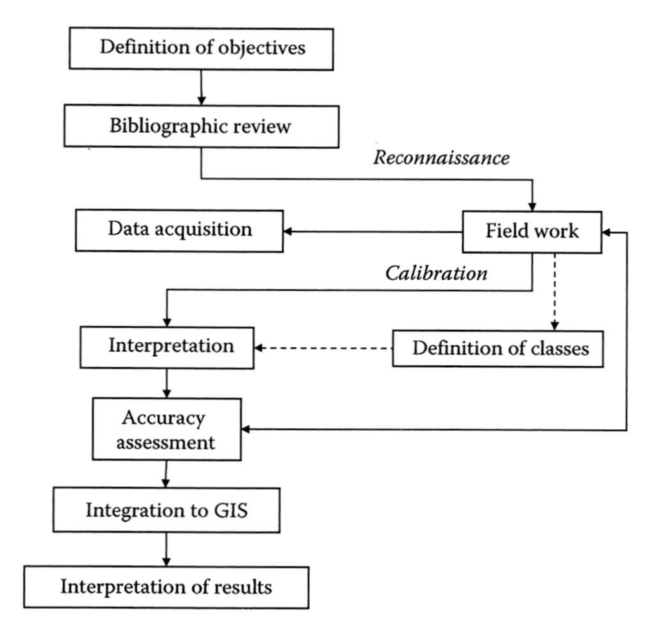


**FIGURE 4.6** Generalized procedure for the interpretation of remote sensing imagery.

FIGURE 4.7 Spectroradiometric measurements in the field are useful in selecting the most useful sensor and spectral bands for a particular application. (Courtesy of M. Pilar Martin, SpecLab, CSIC, Madrid, Spain.)

more limited. However, chlorophyll or thermal sensors are also adapted to surveying ships or buoys.

3. Selection of the desired satellite data including the sensors, the dates of image acquisition, number of images, and formats. For the fishing problem, we obviously need to rely on sensors that detect images in the blue band, the most sensitive to chlorophyll concentration, and the thermal bands: MODIS include both regions, MERIS only the former, ATSR or AVHRR the later.
4. When the classification or discrimination categories (legend) are not known, the user must establish a legend or set of classes suitable to the project, based on the characteristics of the area and the potential offered by the sensor. In case of quantitative variables, the requirements should also establish a certain level of accuracy and precision to be met.
5. Once the images are acquired and processed, field data are very useful for adjusting models of interpretation or validating the results obtained through the interpretation process. Normally, this phase is conducted nearly simultaneously with satellite acquisitions, in order to ensure that the conditions observed by the sensor are the same as those measured in the field. This is particularly important for retrieval of biophysical properties (leaf chlorophyll, leaf area index, soil moisture, for instance) that may greatly change throughout the season. Finally, fieldwork is also important for validation purposes, to measure the degree of correspondence with sampled values of the target variable. In the case of oceanographic analysis, this step is based on observation networks already established.
6. The interpretation phase, either through visual analysis or computer-based processing. In this phase, the interpreter aims to classify the target categories over the whole area of interest. Whenever biophysical variables are targeted

(as is the case of water chlorophyll or temperature), the estimations are based on physical models that retrieve those variables from radiometric measurements.

7. Once the results are obtained, they are commonly integrated with other relevant variables within a GIS (see Chapter 9 for details). Working with digital images makes this process much simpler than when using analog photographs, as digitizing of cover limits is not required.
8. Finally, the results are given to the end user. They can be used as a final product or integrated with other critical information to be used in the decision-making process.

## 4.5 PRESENTATION OF STUDY AREAS

This section is dedicated to presenting some of the images that will be used to illustrate the methods of visual and computer-based image interpretation. From a pedagogical point of view, we intend to apply the different interpretation procedures reviewed in this book to a few study cases, so the student can better understand the implications of those procedures by comparing them in the same scenarios.

Since it is difficult to find a single site that is suitable for all techniques covered in this book, we decided to select instead a large study region from which specific sites will be extracted in the different sections of the following chapters. The region we selected covers an area of more than 1 million km$^2$ in southwest United States and northwest Mexico (Figure 4.8). The elevation varies from sea level at the Colorado River delta to 3000 m in the White Mountains and San Francisco Peaks region near Flagstaff, Arizona. The MODIS image covering this area contains a wide range of landscape features and vegetation types, including a high-plateau shrub desert, a conifer-forested

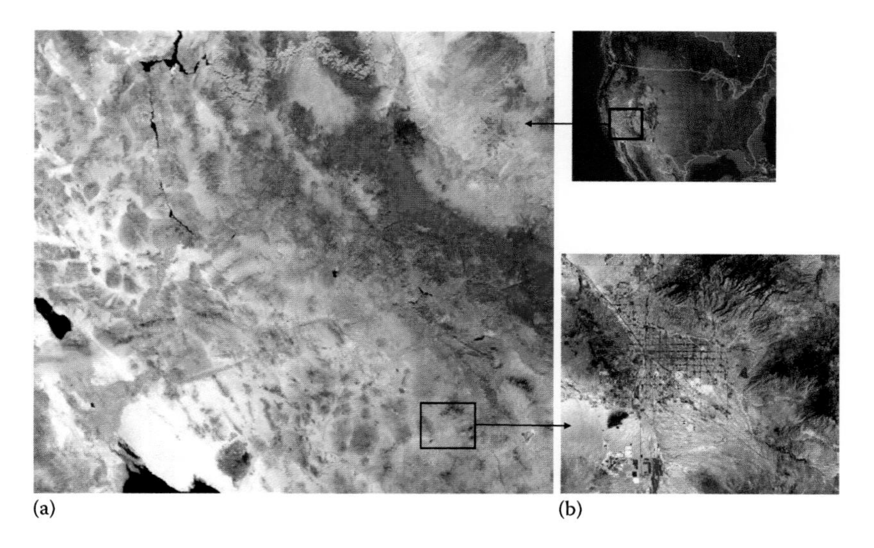


FIGURE 4.8 Study areas used for image interpretation: (a) MODIS image of southwest United States acquired on February 19, 2001, and (b) Landsat 8 image of the Tucson area acquired on May 14, 2014.

mountain zone with snow cover (the image was acquired in February 19, 2001), geologic formations such as the Grand Canyon and volcanic fields, major urban areas such as Phoenix and Tucson, two lakes (Lake Powell and Lake Mead), extensive agriculture cropland with different management practices to the southwest of the image, and wetland and riparian areas. Remarkable features can be observed in the images, such as the Colorado river delta, the Imperial Valley with its vast irrigated agricultural areas, the Californian and Nevada deserts, the Grand Canyon, the Petrified Forest National Park, and Saguaro State Parks, as well as interesting human-made features, such as the Hoover Dam. This image will be used mainly in Chapters 5 through 7.

A small area within this broad image acquired by the Landsat 8 on May 13, 2014, has been selected to show examples of image processing methods applied to finer-resolution data. The L-8 image includes spectral bands acquired by the OLI and TIRS sensors and covers the Tucson city limits, along with the Santa Catalina Mountains that hold Mt. Lemmon (2791 m) and the Coronado National Forest. In addition to the recent Landsat 8 acquisition, we will also present examples of another image of the same area acquired by the Landsat 5 TM sensor in 1986. These images will also be used in Chapters 6 and 7 to show examples of classification and multitemporal analysis. Landsat TM/ETM+/OLI images have provided the most valuable source of environmental information for the past 25 years, and therefore, the student may easily apply similar concepts to his or her own area of interest.

As an example of the relevance of high-resolution images for visual interpretation, Chapter 5 incorporates an example of a RapidEye image acquired over the region of Madrid, Spain (Figure 4.9).

Finally, in the change detection section, we will present some examples of image processing techniques using as study case two Landsat images acquired by the

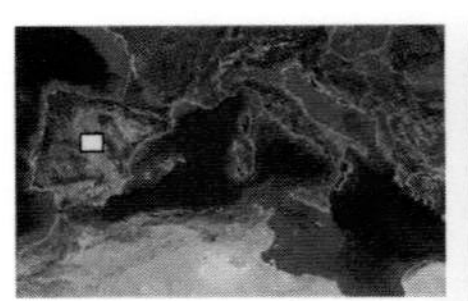

**FIGURE 4.9** RapidEye image from the city of Madrid, Spain, acquired on October 2011. (Image courtesy of BlackBridge©, Berlin, Germany, color combination NIR, R, G.)

**FIGURE 4.10** Multitemporal Landsat-TM images acquired over the state of Acre, Brazil, in 1990 and 2010.

Thematic Mapper sensor in 1990 and 2010 (Figure 4.10). The area is located in the state of Acre, on the Brazilian Amazonia, and shows severe changes between the two reference dates due mostly to severe deforestation processes.

## 4.6 REVIEW QUESTIONS

1. Which factor is associated with map scale?
   a. Spatial resolution
   b. Spectral resolution
   c. Temporal resolution
   d. Radiometric resolution
2. When the scale is larger, the legend of a land cover project should be
   a. More general
   b. More detailed
   c. The same at all scales
   d. None of the above
3. Which one of these scale–sensor combinations is particularly unsuitable?
   a. IKONOS: 1:5,000
   b. Landsat OLI: 1:10,000
   c. SPOT-HRV: 1:100,000
   d. NOAA-AVHRR: 1:2,000,000
4. Which of these characteristics favors visual versus digital image interpretation?
   a. Geometric accuracy
   b. Use of multiple interpretation criteria
   c. Inventory of categories
   d. Repetitive processes
5. Which of these applications should be easier to carry out with visual interpretation than with digital interpretation?
   a. Crop mapping
   b. Chlorophyll content estimation
   c. Monitoring seasonal trends
   d. Updating urban features

# 5 Visual Interpretation

## 5.1 CHARACTERISTICS OF PHOTOGRAPHIC IMAGES

Until a few years ago, visual interpretations were based mostly on photographic products. Nowadays, however, the vast majority of products are digital, either originally acquired as such or digitized afterwards to facilitate delivery. Digital products include auxiliary information on acquisition conditions (metadata), which is required to carry out the processing, including acquisition date and time, sensor calibration coefficients, sensor spectral wavelengths, and cartographic information (in case the image was previously georeferenced). Similarly, the photographic products included auxiliary information for the interpreter, which was commonly printed on the borders of the image on the same paper or film. An example is shown in Figure 5.1, which shows one of the first Landsat images processed by NASA and distributed by the EROS Data Center (U.S. Geological Survey). In the image margins, the following information was included: date of acquisition (February 2, 1975); approximate center coordinates of the image (37°21′ north and 8°26′ west); coordinates at nadir (point of intersection between the terrestrial surface and a perpendicular line from the acquisition center); sensor and band used (multispectral scanner, band 7); Sun's elevation angle (26°), measured between the horizontal surface and the direction of the solar rays; Sun's azimuth angle (143°), measured clockwise from the geographic north; processing and corrections, including geometric corrections applied, scale, projection, compression, etc.; agency and project (National Aeronautics and Space Administration ERTS); scene identification code, which includes the number of the satellite; days since the launch of the satellite; and the time of acquisition (hour, minutes, and seconds). The image also included a "grayscale" to assess the quality of the reproduction. Similar information was included in other photographic products generated by other space agencies.

## 5.2 FEATURE IDENTIFICATION

One way to become familiar with an image is to compare it with conventional maps. This exercise can be performed in several phases. First, we can compare the image with a topographic map of the area at the same scale and begin to locate some of the most important features, such as water bodies and dams. A second inspection of the image may then enable the identification of human-made features, such as cities, roadways, and networks of communication. Topographic features, such as mountains, can be identified in the third stage since they are represented more abstractly on conventional maps and are, therefore, more difficult to delineate. Vegetation and ephemeral features on the image (e.g., fires or pollutants) can be identified in the later stages since they rarely appear on a conventional map. This exercise serves as a preamble to a discussion of the principal criteria used in visual interpretation of images. For instance, it will be useful for the reader to compare the MODIS image of our study

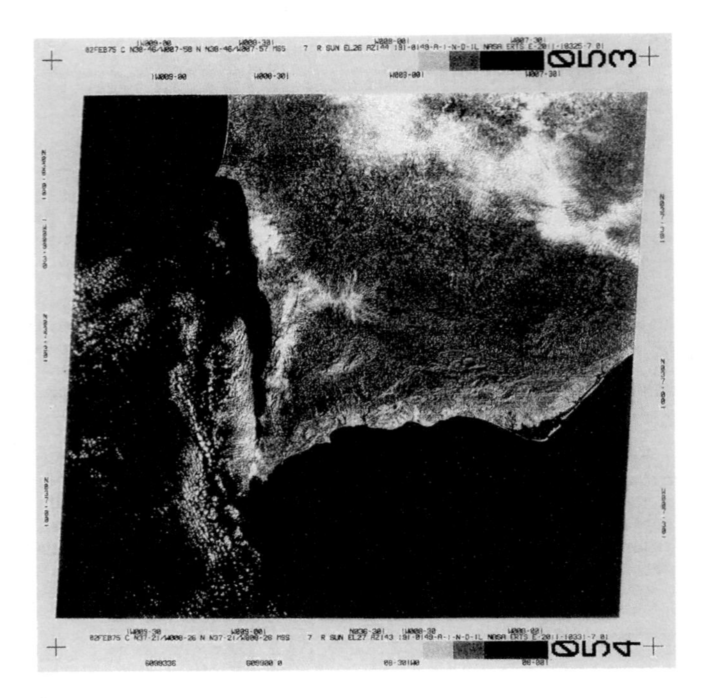

**FIGURE 5.1** Landsat image of the Portuguese coast in the original NASA format.

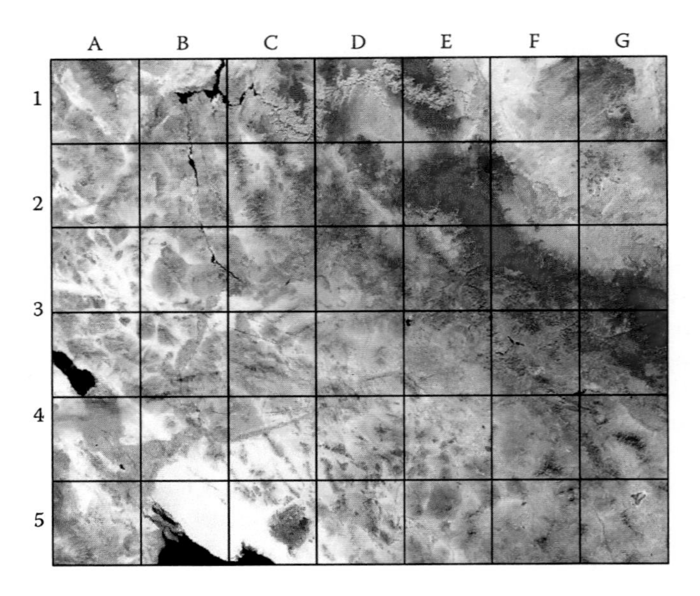


**FIGURE 5.2** MODIS image of the study region: A grid reference has been overlaid to help identify outstanding features.

region with an atlas of the southwestern region of the United States, to identify some of the outstanding features. For instance, the reader may locate, using the reference grid of Figure 5.2, the following geographic elements: Salton Sea, Lake Mead, Grand Canyon, Colorado River delta, Cerro del Pinacate (México), Gila River, San Pedro River, San Francisco Mountains, Roosevelt Lake, and the city of Phoenix.

## 5.3 CRITERIA FOR VISUAL INTERPRETATION

As mentioned in Chapter 4, one of the main advantages of visual over digital interpretation is the capability of the former to incorporate complex criteria to detect and identify features in an image. While digital image processing is mostly based on the spectral radiances of the different image pixels, visual interpretation of images uses other criteria, such as texture, arrangement, location, or spatial patterns, which are very difficult to define in digital terms. Therefore, visual analysis makes it possible to discriminate categories that have similar spectral properties but different spatial features. For example, some irrigated crops might present similar spectral reflectance in summer time as lawns in urban parks, because they have similar herbaceous vegetation components. However, they have very different thematic meanings and should be differentiated in any land cover map. With digital interpretation, their discrimination may be complex because they will probably have similar radiances. However, a simple visual criterion, such as geographic context, enables a visual interpreter to distinguish them quite accurately.

Many of the visual interpretation criteria that are applied to satellite images are common to traditional aerial photointerpretation, such as tone, texture, structure, shadows, or context, while other criteria are more exclusive to satellite images, such as those associated with multispectral and multitemporal observations, which are very restricted in traditional aerial photography. A third group of criteria would be those that are used routinely in aerial photointerpretation but are rarely accessible to satellite images, such as stereoscopic coverage, which is vital for geomorphology studies. As reviewed in Chapter 3, some satellite sensors (ASTER, SPOT-HRV, Cartosat, Space Shuttle or ISS photographs, etc.) provide stereoscopic coverage, but this feature is found more commonly in aerial photography.

In comparison, some criteria for interpreting aerial photography are similar to those for visual analysis of satellite images, but they have some uncommon ones too. Additionally, it is important to emphasize that satellite images have very different scales from those used in aerial photographs (typically from 1:18,000 to 1:30,000), and therefore the recognition of objects and the geometry of acquisition are very different. However, the growing use of high-spatial-resolution images (GeoEye, WorldView, SkySat, etc.) makes this distinction nowadays less relevant.

Criteria for visual interpretation of satellite images can be grouped in a hierarchical way, according to their degree of complexity (Figure 5.3). Brightness and color are the most elemental criteria, since they characterize spectrally the cover under consideration. Shape, size, and texture are related to the spatial properties of the object. Shadow and location describe the relation of the target with the surrounding objects. Finally, the temporal dimension refers to the seasonal evolution of the land cover. The following sections discuss these interpretation criteria in greater detail.

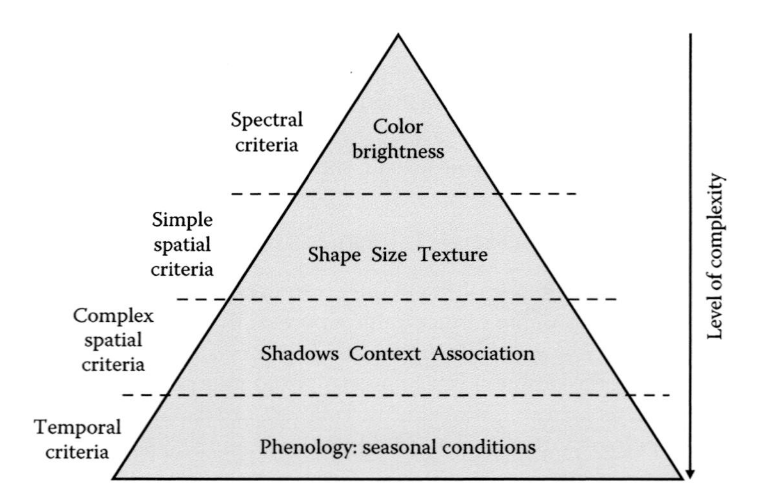


**FIGURE 5.3** Hierarchical organization of visual interpretation criteria. (After European Commission, *Corine Land Cover: Guide Technique*, Office for Official Publications of the European Union, Luxembourg, Europe, 1993.)

### 5.3.1 Brightness

The brightness of a particular cover refers to the intensity of radiance received by a particular pixel for a given spectral band. In a photographic product, pixels with dark tones indicate areas where the sensor detected low radiances, while lighter zones represent higher intensities. Therefore, brightness can be associated with the spectral reflectance of any particular area and band. As a general remark, we should remember that the number of gray levels one can visually distinguish is limited by our eye perception and the sensitivity of the medium (paper or computer monitor). A visual interpretation always implies a reduction in radiometric sensitivity from the raw image. The human eye is restricted to discriminating 64 or fewer gray levels, and therefore the full radiometric resolution of the image (from 256 to 2048 different digital values) is not exploited in visual interpretation.

As seen in Chapter 2, reflectance of most covers changes at different wavelengths, and therefore their brightness in different spectral bands will also be different. For instance, since vegetation pigments absorb incoming visible light, densely vegetated surfaces will show dark tones in the visible (VIS) bands, especially in the B and R regions. In contrast, those surfaces will have light tones in the near-infrared (NIR) region, since leaf pigments do not absorb light in this spectral region. Consequently, to discriminate between vegetated areas, a contrast of brightness tones between the R and NIR bands will be a clear clue.

To detect a particular category of interest, the visual interpreter should be familiar with the spectral behavior of that cover. In this way, the analyst can select which spectral regions should be more appropriate to detect that feature and fully understand the reasons for the tonal differences observed in different spectral regions. For instance, drainage patterns and morphology should be more detectable in the NIR,

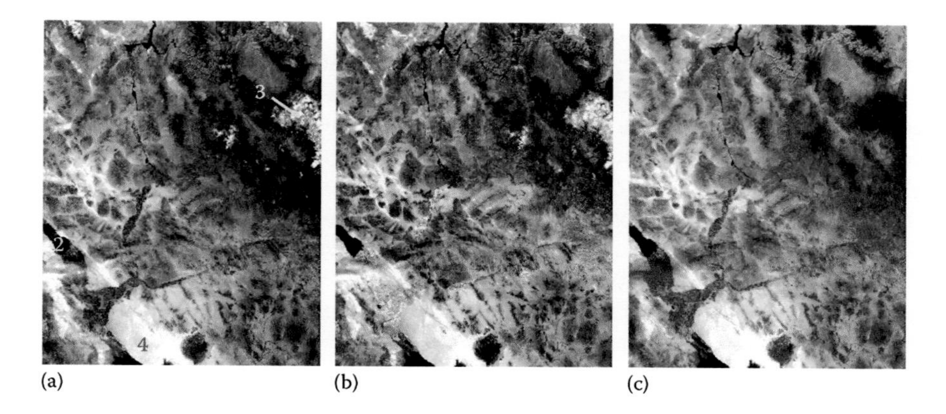


**FIGURE 5.4** Variation in brightness of different spectral bands of MODIS for our sample region: image was acquired on February 19, 2001. (a) R band (B1); (b) NIR band (B2); and (c) SWIR (B7). Green vegetation (1), water (2), snow (3), and bare soils (4).

urban layout in the G–R space, water turbidity in the B band, soil–water boundary in the NIR, and vegetation in the R–NIR bands.

An example of the spectral diversity of brightness may be observed in Figure 5.4, which includes three bands of a MODIS image of our study region acquired in the winter season of 2001. The image contains a wide diversity of land covers, and so the spectral differences can be observed by comparing brightness tones in the different spectral channels. The R band (620–670 nm) shows green vegetation (1) and water (2) with low brightness tones, while snow (3) and bare soils (4) are shown in light to very light gray tones. The NIR band (841–876 nm) shows lighter tones for green vegetation (see, e.g., the large crop area in the Imperial Valley: 1), while snow and bare soils are still shown in bright tones and water presents dark tones (strong absorption) in this band too. Finally, the shortwave infrared (SWIR) band (2105–2155 nm) presents a clear contrast in snow covers from other bands, since at longer wavelengths the snow absorbs, which therefore has darker tones than in the NIR or VIS bands. Green vegetation also absorbs in this band, as well as water and dark soils, and therefore they are shown in dark grays. Only bare soils have bright tones in this spectral band.

### 5.3.2 Color

It is well known that the human eye is more sensitive to chromatic variations than brightness variations. When mixing three spectral bands in a color composite, more information is available for detection and identification of features in the images. Therefore, color becomes a critical element in visual interpretation of images.

Our perception of color is a result of the selective reflectance of objects observed at different wavelengths. Surfaces with high reflectance in short wavelengths (blue) and low in the remaining wavelengths appear blue, while red objects absorb the short wavelengths and reflect the long visible ones. Our eyes only perceive wavelengths between 400 and 700 nm, separating the energy received into three color components on our own spectral sensitivities. These three components are known as "primary colors" (red, green, and blue; RGB), from which any other color can be derived.

The colors we perceive with our eyes can be created by assigning the primary colors (RGB) to the spectral bands red, green, and blue, respectively. This natural combination can be modified by the interpreter by assigning RGB to three spectral bands in any convenient order. Obviously, this new color composite will not correspond to the colors we normally perceive (what we refer to as "natural colors"); the resulting colors will be unusual to our eyes, and therefore we name them "false color" composites. False color composites are not perceived by our eyes, but they are usually more interesting to discriminate different ground features and therefore are very common in EO visual analysis.

A color composite can be obtained by combining three spectral bands using two different processes: *additive* and *subtractive*. In the additive process, a given color is obtained by adding light of different wavelengths. The more reflectance a pixel has, the more light (the brighter) that color will have in the composite. The three basic color guns are blue, green, and red. In the additive process, any other color is obtained by summing up light from two or more basic colors. For instance, cyan is the sum of blue and green, magenta is the sum of blue and red, and yellow is the sum of green and red. The three color guns together yield white color (Figure 5.5a). The additive process is the one used in electronic display systems (computer monitors, tablets, screen projectors), and it is the most commonly used for digital image processing systems.

The subtractive method is used in mechanical reproduction of color, typically in printed media (journals, books, etc.). The printed colors are obtained by mixing three basic inks: yellow, magenta, and cyan. In contrast to the additive process, the more reflectance a pixel has, the less ink it will have, since it needs to be brighter in the final composite. In other words, the more the ink, the darker the color, and the less reflectance in that particular band. As mentioned earlier, the subtractive process uses three elementary inks, and the other colors are obtained by mixing them (Figure 5.5b).

Among the multiple color combinations used in visual analysis, the most outstanding is known as "color infrared." This combination is obtained by shifting the visible bands toward longer wavelengths and applying RGB to the NIR, R, and G,

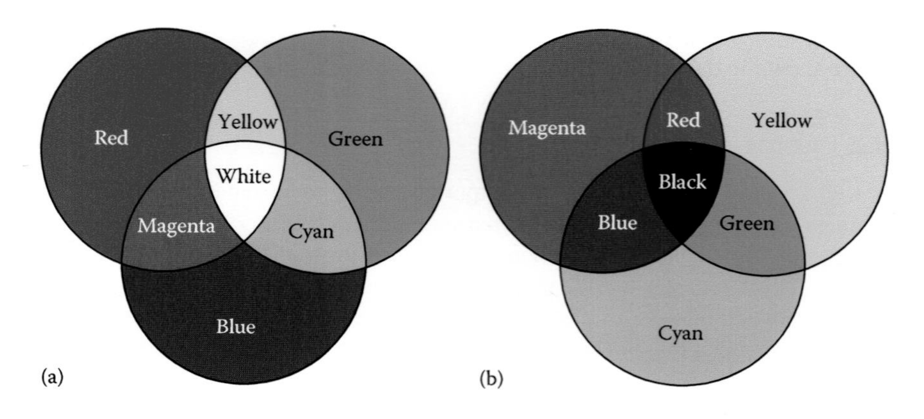


**FIGURE 5.5** Processes in color formation: (a) additive and (b) subtractive.

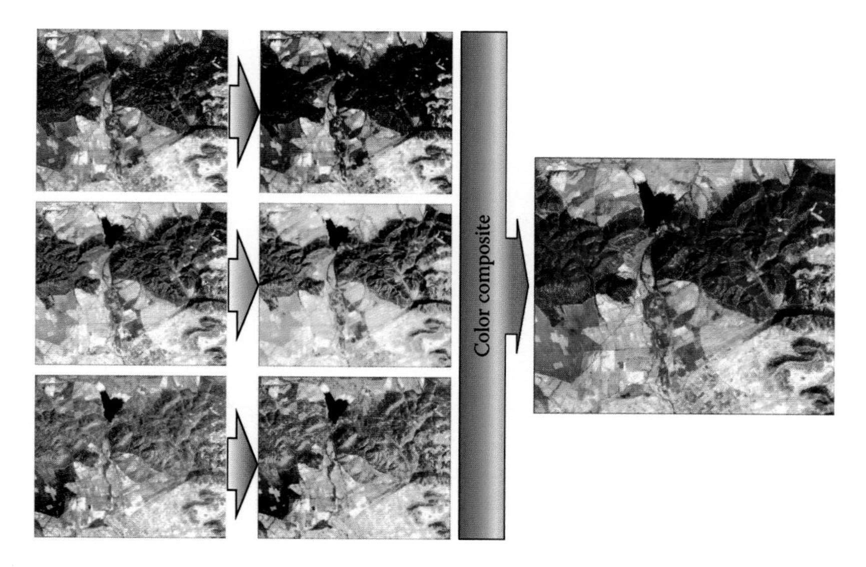


FIGURE 5.6 Process to generate a false color composite: example of color infrared.

respectively (Figure 5.6). It has been widely used because most satellite sensors include these three spectral bands. In order to facilitate color interpretation, it is convenient to include a simple color key to familiarize the interpreter with these types of composites. The most usual colors correspond to the following covers (Tindal 1978):

- *Red-magenta* represents vigorous vegetation, such as irrigated crops, mountain meadows, or deciduous forests in summer images and herbaceous crops of nonirrigated land in spring images. The detailed study of the intensity and saturation of red in these composites makes it possible to identify several vegetation covers, as well as to estimate their growth cycles and vigor.
- *Pink* represents less densely vegetated areas and/or vegetation in early stages of growth and suburban areas close to large cities when they have gardens and scattered trees.
- *White* represents sparsely vegetated area, clouds, sand, saline deposits, quarries, bare soils, and snow.
- *Dark blue to black* represents surfaces totally or partially covered by water, rivers, lakes, and dams. In volcanic zones, black tones might represent lava flows or soils with low vegetation cover.
- *Gray to metallic blue* represents cities and bare rocks.
- *Brown* represents shrub land, variable according to the density and ground material.
- *Beige* represents zones of transition, dry meadows frequently associated with sparse scrub.

When Landsat TM images started to become available at the beginning of the 1980s, the number of new color combinations was greatly expanded. The inclusion of the blue band made it possible to obtain a natural color composite from this sensor, while the addition of SWIR remarkably extended the possibilities of combining multiband information. Several authors have tried to establish an objective indicator to aid in the selection of the most suitable color composite by maximizing nonredundant information. For doing so, various statistical techniques have been used, such as principal component analysis, maximum variation index, and maximum variation ellipsoid (see Section 6.5.2).

For visual analysis, some studies have been performed to identify the color composite images that offer the best land cover discrimination. This approach is based on selecting interpreters with different levels of experience to quantify the number of covers that can be adequately discriminated in diverse color composites (Trolier and Philipson 1986). For Landsat TM-ETM+ and now OLI, the most highly graded combinations included a band from the VIS spectrum, one from NIR, and another from the SWIR region. The usual order is to apply red to NIR, green to SWIR or R, and blue to R or G (see Figure 5.7). These results are very similar to the ones obtained by other authors through statistical methods (Hord 1986).

Following these results, the European CORINE land cover project selected the color composite NIR, SWIR, and R for Landsat TM/ETM images as the basis for visual

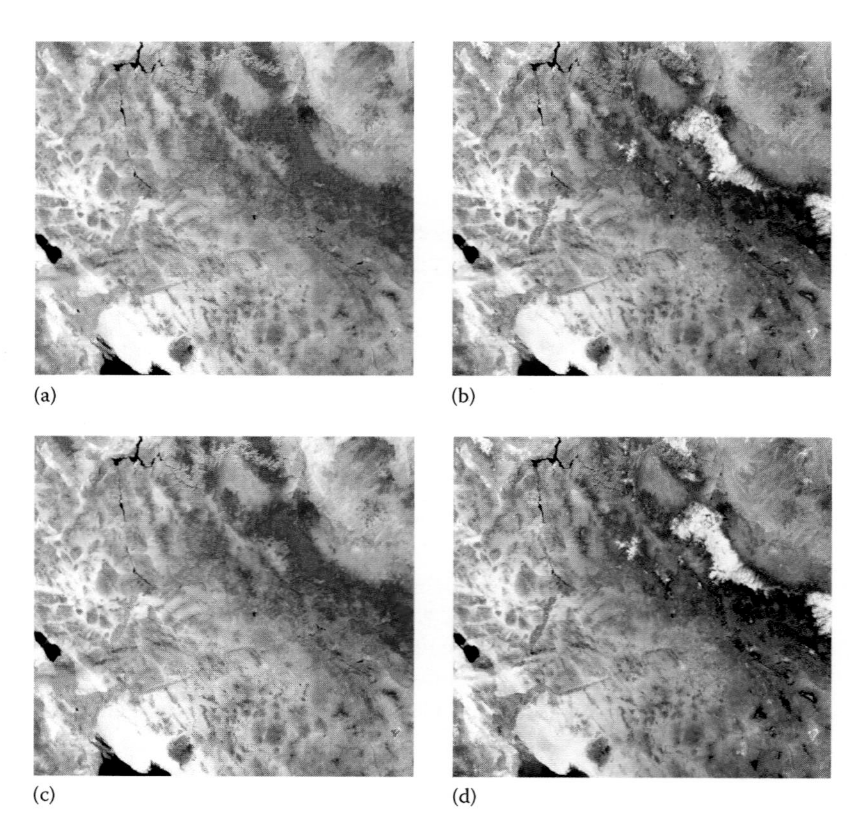


FIGURE 5.7 Different color composites for the MODIS February image: (a) MODIS 261 (NIR, SWIR, B), (b) MODIS 214 (NIR, R, G), (c) MODIS 275 (NIR, SWIR2, SWIR1), and (d) MODIS 143 (R, G, B).

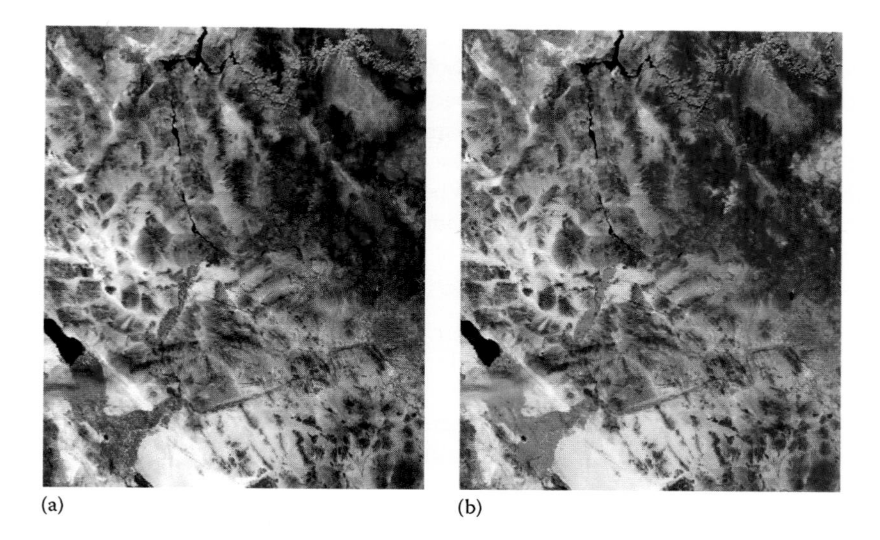


**FIGURE 5.8** Same color composite but different color gun assignments for the MODIS February image: (a) MODIS (NIR, SWIR, R) and (b) MODIS (SWIR, NIR, R).

image interpretation of land covers. For particular objectives, other combinations may be more useful. For instance, again with TM images, the combination SWIR, NIR, and B proved to be efficient in detecting burned areas (Koutsias et al. 2000), SWIR, NIR, and R helped in discriminating irrigated crops or flooded zones, and SWIR/R/B is useful in oceanographic applications. In spite of these interesting composites, the traditional false color composite (NIR, R, G) is still widely used (Figure 5.8a).

The interpretation of color in a three-band composite depends on the three channels selected for the composition as well as the order of the three primary colors assigned to that combination. When new bands are included, changes in color are associated with different spectral information. When different composites use the same spectral bands but in different orders, the chromaticity becomes different but the spectral information remains the same. For example, if we modify the aforementioned conventional infrared composite (NIR, SWIR, R) and instead assign SWIR, NIR, and R, healthy vegetation will appear in green tones instead of red ones (Figure 5.8b). This change does not mean that we can talk of a natural color, as two of the bands used (NIR and SWIR) cannot be observed by our eyes. However, since vegetation is expected to be green to ordinary people, this color composite has been commonly used by different mapping agencies. Even if they have the same spectral information, technically speaking, the potential for interpreting the image may be lower in this latter case because it is widely recognized that the human eye is more sensitive to red than to green variations.

### 5.3.3 Texture

Texture refers to the spatial heterogeneity of a given cover, that is, the spatial contrast between the elements of which it is made. Visually, it appears as the roughness or smoothness of the tones in the image. The more similar the tones in the image, the more homogeneous the cover and the smoother the texture. On the other hand, if the cover has a greater variation of reflectance values in the vicinity, the heterogeneity will be greater as well, and the area will appear with a rough texture.

Texture depends on the relation between the size of the objects in a particular cover and the resolution of the sensor. When an object occupies a surface <1 $mm^2$ in the image, it cannot be identified individually but can be done through the spatial variability that it produces. On the basis of size of the objects that make up a cover, three types of texture are possible:

1. *Rough texture*: When the objects are between 0.25 and 1 $mm^2$ on the scale of the image
2. *Medium texture*: When the objects are between 0.04 and 0.25 $mm^2$
3. *Fine texture*: When objects are smaller than 0.04 $mm^2$

In addition to the size of the objects, the texture also depends on other factors, such as the sensor observation angle (a more oblique observation increases the roughness), the illumination conditions (the lower the solar angle, the greater the shadow effect), and the wavelength (a smooth surface at long wavelengths can appear rough at short wavelengths). In the VIS and NIR parts of the spectrum, roughness is mainly modified by changes in solar illumination. For instance, a forested area can present different textures during the year depending on the Sun's elevation angle and its seasonal changes.

In any case, the texture criterion is very important in discriminating certain covers that show similar spectral behavior. For example, citrus trees in irrigated areas and some deciduous trees in forested areas may present very similar reflectance. In this case, texture can facilitate the distinction. Fruit trees show a regular plantation with a low spatial variability in cover, while deciduous forests develop in a much more heterogeneous manner. Consequently, fruit trees show a finer texture, which enables the interpreter to separate them from natural forest species (Gordon and Philipson 1986).

Figure 5.9 shows several types of textures in the northern part of the city of Tucson, Arizona, based on a Landsat-8 OLI image. The area is located toward the south of MODIS image. The urban area includes a great diversity of covers (roofs, asphalt, vegetation, etc.) that most commonly are smaller than the pixel size, and therefore the grain of those areas is coarse. In contrast, irrigated areas and grasslands tend to present a smooth surface because they have little variations in reflectance within a pixel size. Forested areas may appear rougher or smoother, depending on whether they are natural or plantation forest and whether there are many shadows.

With higher-resolution images, a different perspective of texture can be obtained (Figure 5.10), since small objects appear rougher than in lower-resolution images. Different textures within an urban area are noticeable in an IKONOS image, while in an OLI image those texture distinctions would be blurred.

FIGURE 5.9 Different textures in a Landsat 8 OLI image of northern Tucson, Arizona. Image was acquired on May 2014: (1) smooth, irrigated crops; (2) medium–low, shrubs and grasslands; (3) medium–high, forested areas; and (4) rough, built-up areas.

FIGURE 5.10 Textures on high-resolution images: (1) smooth, grass; (2) medium, concrete roof; (3) medium–high, gardened single-family housing; and (4) rough, trees. Fragment of the RapidEye image from the city of Madrid, Spain. (Image courtesy of BlackBridge, Berlin, Germany.)

### 5.3.4 Spatial Context

The relative location of an object includes the vicinity of the object, which is taken into account for detection or interpretation. In other words, the context criterion considers the surrounding elements of a particular cover for its interpretation.

The spatial context is a good indicator of a thematic category in many cases. For instance, green vegetation within an urban area is most likely an indicator of an urban park. This park may be composed of deciduous trees with tone, color, and texture similar to that of a deciduous forest. However, the meaning of an urban park is very different from that of a natural forest, and therefore it is logical to discriminate between them. In this case, since spectral reflectance and texture may be similar, the only criterion left would be the geographic context: those trees inside a built-up area should be considered urban parks (see for instance #1 in Figure 5.11), while those located in an area of natural vegetation will be assigned to forest (#2 in Figure 5.11). This association could also be applied to discriminate riparian vegetation from other deciduous species, since their distribution is associated with river streams.

Context is more easily applied to visual than digital interpretation, as the human eye system is very powerful to understand relations and spatial connections. However, in Chapter 7, some possibilities for contextual analysis will be considered.

### 5.3.5 Shape and Size

The shape of an object greatly helps to identify it, since its contour allows its association with familiar patterns. As in other visual criteria, shape is more easily

**FIGURE 5.11** Spatial context and shapes in high-resolution images: (1) urban park, (2) natural forest, (3) golf courses, and (4) hippodrome. Fragment of the RapidEye image from the city of Madrid, Spain. (Image courtesy of BlackBridge, Berlin, Germany.)

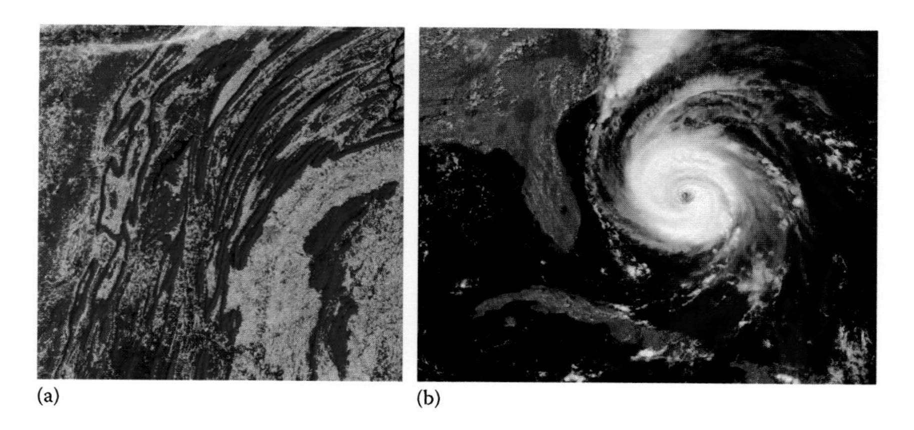


FIGURE 5.12 Large shapes visible in medium- and coarse-resolution satellite images. (a) Appalachian relief in a Landsat MSS image and (b) tropical cyclone in an AVHRR image.

recognizable in high-resolution images, while in medium- or coarse-resolution images it is restricted to large features. Different urban objects can be determined from shape such as sport complexes, race or golf courses (Figure 5.11), manufacturing complexes, airports, storage, and transport networks. It also helps to discriminate between highways and railroads (which tend to have wider curves) or between rivers and artificial channels (which are more geometrically regular). As far as large structures are concerned, shape can help in the identification of regional geologic structures such as volcanoes, domes, or lineaments, as well as some meteorological phenomena (Figure 5.12).

In the case of uncertainty and confusion generated from shape, size complements the identification of objects. For example, the shapes of a tennis court and a soccer field are similar but their sizes are not (and in many cases neither the surfaces: lawn versus clay). The use of the shape criterion is determined by the spatial resolution of the image. Images with higher resolution can identify single objects (e.g., swimming pools), while the ones with lower resolution only indicate ranges of size for different surfaces (e.g., dams, flooded areas). Consequently, the shape of an object might not be identifiable but that of a group of objects will be, and in the end this will help identify a cover type. For example, a pine plantation will appear with geometric borders, while a natural forest will have an irregular shape.

### 5.3.6 Shadows

Variations of illumination conditions may create severe problems in discriminating particular surfaces, and therefore shadows are considered as added noise in image interpretation (see Chapter 6 for shadow removal techniques). However, they can be useful in object recognition, since shadows give an estimate of the height and depth of objects. In addition, shadows enhance the interpretation of geomorphologic features and texture of the image, especially in forested zones.

Object height can be precisely determined if the Sun's angle at the time of image acquisition is available. A simple trigonometric equation applies here (Figure 5.13):

$$h = \frac{l_s}{\tan \theta} \tag{5.1}$$

where
$h$ is the height of the building
$l_s$ is the length of its shadow
$\theta$ is the solar zenith angle

This method is adequate for isolated buildings. In dense urban zones, this method can be more difficult to apply, since shadows of different buildings may overlap and it will be difficult to determine the full length of a single building.

An example of the relevance of shadows for building height estimation is included in Figure 5.14, which shows a fragment of the RapidEye image acquired over the city of Madrid. The image shows the northern section of the city, dominated by four skyscrapers (~250 m tall). The actual shape of the buildings is not easily observable, but the long shadow evidences its position. In the surroundings, the contrast between apartment buildings and one-story, single-family buildings can also be easily detected.

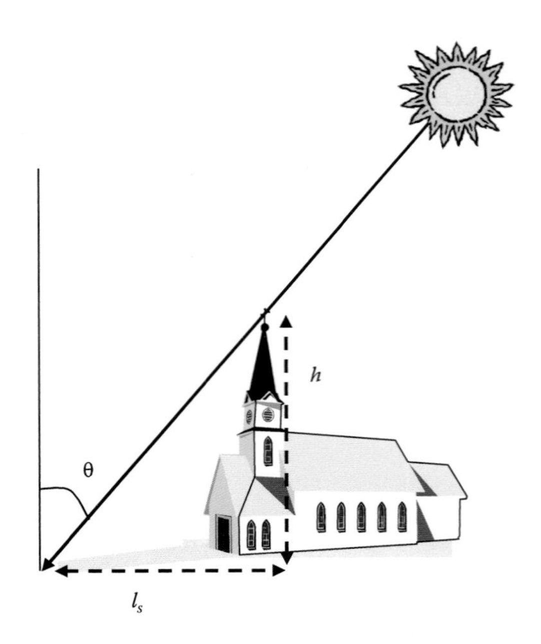


**FIGURE 5.13** Estimation of a building height from its shadow length.

**FIGURE 5.14** Shadows help to identify different heights in urban areas: (1) skyscrapers, (2) apartment buildings, and (3) single-family houses. Fragment of the RapidEye image from the city of Madrid, Spain. (Image courtesy of BlackBridge, Berlin, Germany.)

### 5.3.7 Spatial Pattern

This concept refers to the spatial arrangement of objects that belong to a particular cover. For instance, an olive orchard can be distinguished by its regular plantation patterns from a sparsely forested area. This criterion has been rather limited with satellite images because of the low resolution of most sensors, but it can be effective with high-spatial-resolution images.

However, even in medium-resolution images some features can be discriminated by their spatial pattern, such as forest plantations (that may show a regular sequence of trees) or golf courses that in the NIR region show a peculiar alternation of sectors of high reflectances (the golf lanes and the greens) and medium reflectances (the original surrounding vegetation). A similar principle can be used to identify geological structures or marine features (Figure 5.15).

### 5.3.8 Stereoscopic View

Stereoscopic vision is fundamental to geomorphologic interpretation because it provides a 3D view of the areas observed. Most satellite sensors do not provide stereoscopic coverage; however, SPOT-HRV, Terra-Aster, IRS-Cartosat, and some missions on board the Space Shuttle have this capability. For other satellite images, stereoscopic vision can be obtained only in overlapping zones between orbits, restricting this criterion to a narrow portion of the image.

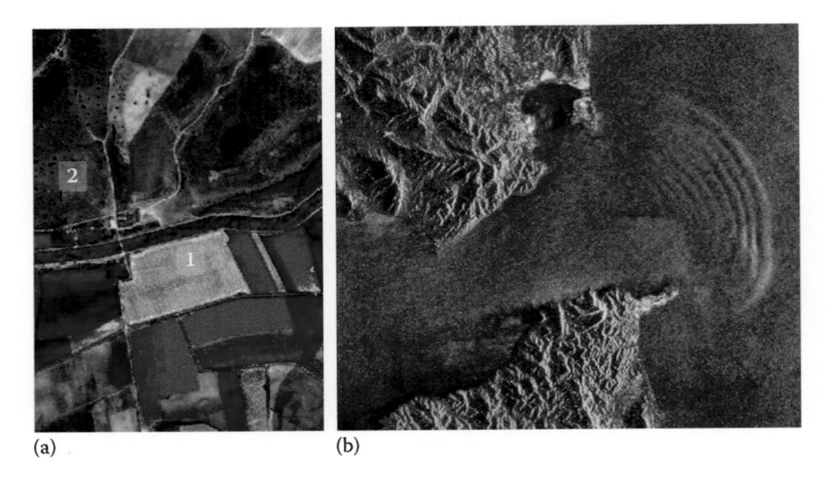


FIGURE 5.15 Spatial pattern facilitates visual discrimination. (a) QuickBird image of an agricultural region in Salamanca (Spain) showing contrast between crops (linearly arranged, 1) and natural vegetation (scattered trees, 2). (b) SAR image of the Gibraltar Strait, showing ocean waveforms.

### 5.3.9 Period of Acquisition

Most investigators agree that systematic observation of the Earth is one of the main advantages of satellite remote sensing. The characteristics of the satellite's orbit allow periodic acquisitions of images under similar conditions of observation, facilitating any study that requires a temporal dimension.

This multitemporal analysis has been carried out from two main perspectives:

1. The main goal is the determination of changes between two or more dates of reference. The examples of application studies are very diverse, from analysis of flooding or volcanic eruptions to monitoring of deforestation or land erosion processes. Commonly, the images are well separated in time. For instance, when the process being monitored is slow (urban growth, desertification, etc.), a time frame of several years is very usual. For this reason, we can term this approach *multiannual*. The images are commonly acquired around the same time of the year to reduce the impact of seasonal variations in discriminating the permanent change in the images.
2. The second perspective deals with changes in shorter periods, and it is frequently associated with seasonal trends such as crop growth or snow-cover variations. In these cases, the approach is termed *multiseasonal* and is typically based on many images, sometimes even dozens or hundreds if the required time resolution is very high. In this case, the images are acquired at all seasons, as the common goal is precisely to characterize seasonal trends in the target areas (Figure 5.16).

Both multitemporal approaches require previous georeferencing of images or at least a common rectification among them, so that the spatial information is common and

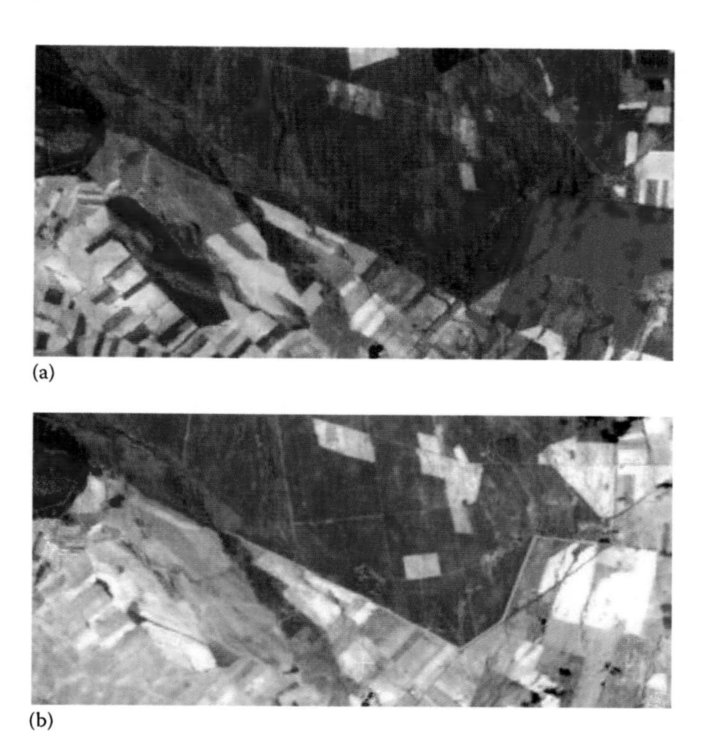

FIGURE 5.16 Seasonal changes in Landsat TM images over an agricultural area in Central Spain. (a) Spring image (May), and (b) summer image (July).

both refer to the same ground sites. The multitemporal capability in remote sensing is a very important identification criterion in visual analysis, both in multiannual (Bruzzone and Fernández-Prieto 2000; Hermosilla et al. 2015) and multiseasonal studies (Alcaraz-Segura et al. 2010; Goetz et al. 2005).

Some multitemporal studies have been based on color composition techniques to enhance changes between two reference dates. In a study to detect patches of infected forest in Canada, primary colors were applied to the NIR and R bands of the two reference dates (Hall et al. 1984; Sader and Winne 1992).

## 5.4 ELEMENTS OF VISUAL ANALYSIS

Having reviewed the main criteria for visual interpretation, we will now review a series of elements that must be taken into account when working with satellite images.

### 5.4.1 Geometric Characteristics of a Satellite Image

Although satellite images have fewer geometric errors than aerial photographs, because of greater stability and higher flight altitude of the platform, they are not free from geometric errors, and therefore they cannot be directly overlaid on standard

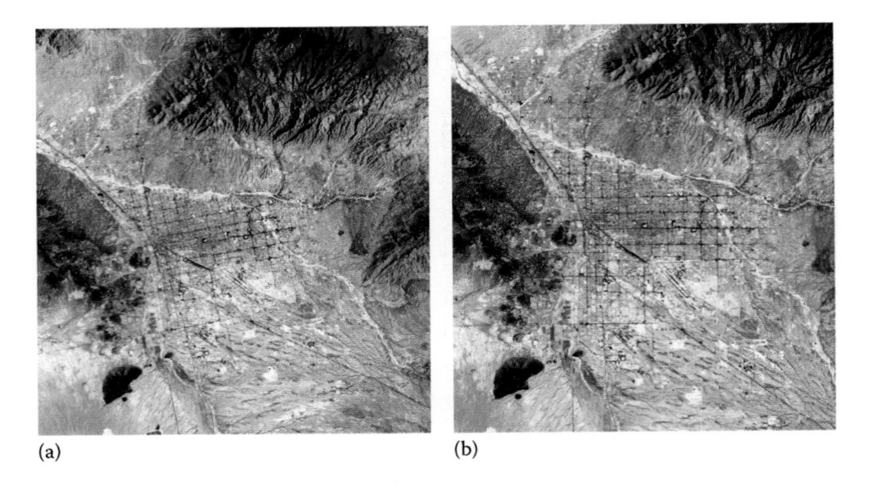


**FIGURE 5.17** Effect of geometric distortions on image interpretation. (a) Tucson TM image acquired June 9, 1989; (b) Tucson ETM+ image acquired on September 3, 2000.

maps. These geometric differences between map and image position are caused by several factors: (1) Earth's shape and rotation, (2) terrain characteristics, (3) sensor's field of view, and (4) cartographic projection. In Chapter 6, procedures to correct such geometric distortions are discussed.

Figure 5.17 shows the effect of geometric correction on an image of the Tucson metropolitan area. The two images were acquired by a similar sensor (TM and ETM+) with the same spatial resolution and orbital characteristics, but the second one was corrected to a Universal Transverse Mercator (UTM) projection. The direction of the main streets of the city, leading true north–south in the corrected image and NW to SE in the uncorrected one, is noticeable. This orientation of the raw image is caused by the combination of satellite orbit and Earth's movement, which results in the original image columns to not follow an exact north–south direction.

Once the geometric deformations of an image are solved using different correction methods (see Section 6.6), satellite images can be used to update features on the map at different scales, depending on the spatial resolution of the input sensor (Martínez Vega 1996).

### 5.4.2 Effect of Spatial Resolution in Visual Analysis

As previously mentioned, spatial resolution refers to the size of the minimum discernible unit on the image. This concept, in visual analysis, is related to the pixel size as well as the scale at which the image is represented. Therefore, spatial resolution has an important impact on the interpretation of the scene and its minimum discriminating unit. As we have seen, elements smaller than the pixel size will not be discriminated in the image; this is essential for selecting the most adequate sensor to meet the desired objectives. Higher resolution implies better definition of the

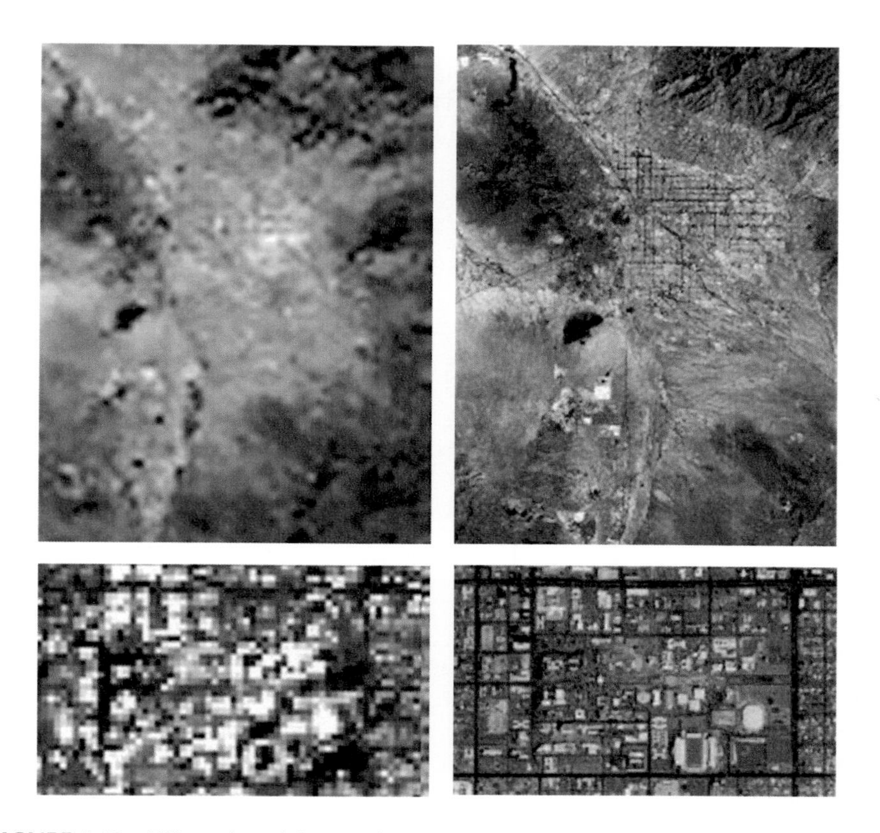

FIGURE 5.18 Effect of spatial resolution on image visual interpretation. Top, comparison between (left) MODIS and (right) ETM+ images of the Tucson area; bottom, comparison between (left) ETM+ and (right) IKONOS of the University of Arizona's campus.

pixel because it will more likely represent a single cover type. On the other hand, if the pixel size is large, the signal frequently represents several cover types, making its interpretation more complicated.

The effect of spatial resolution on image interpretation is shown in Figure 5.18, which shows three images of the Tucson area at very different resolutions: MODIS (500 m), ETM+ (30 m), and IKONOS (1 m). The general distribution of gray levels is similar across all sensors, but the details for recognizing single objects are very different. MODIS provides good regional to continental coverages, while ETM+ is suitable for local to regional coverages, and IKONOS is adequate for detailed local areas.

### 5.4.3 Effect of Spectral Resolution in Visual Analysis

As we have pointed out in several sections of this book, multispectral observations are the basis of identification and detection of surface features and covers in satellite images. The possibility of observing surface features in several bands of the spectrum extends remarkably our capacity to recognize them. The spectral signature

A. Urban areas

| Brightness | B3 | B4 | B1 | B2 | B6 | B7 |
|---|---|---|---|---|---|---|
| Very light | | | | | | |
| Light | | | | | | |
| Dark | | | | | | |
| Very dark | | | | | | |

B. Water

| Brightness | B3 | B4 | B1 | B2 | B6 | B7 |
|---|---|---|---|---|---|---|
| Very light | | | | | | |
| Light | | | | | | |
| Dark | | | | | | |
| Very dark | | | | | | |

C. Bare soils

| Brightness | B3 | B4 | B1 | B2 | B6 | B7 |
|---|---|---|---|---|---|---|
| Very light | | | | | | |
| Light | | | | | | |
| Dark | | | | | | |
| Very dark | | | | | | |

D. Irrigated crops

| Brightness | B3 | B4 | B1 | B2 | B6 | B7 |
|---|---|---|---|---|---|---|
| Very light | | | | | | |
| Light | | | | | | |
| Dark | | | | | | |
| Very dark | | | | | | |

E. Shrubs—Grasslands

| Brightness | B3 | B4 | B1 | B2 | B6 | B7 |
|---|---|---|---|---|---|---|
| Very light | | | | | | |
| Light | | | | | | |
| Dark | | | | | | |
| Very dark | | | | | | |

F. Snow

| Brightness | B3 | B4 | B1 | B2 | B6 | B7 |
|---|---|---|---|---|---|---|
| Very light | | | | | | |
| Light | | | | | | |
| Dark | | | | | | |
| Very dark | | | | | | |

**FIGURE 5.19** Template for visual analysis of MODIS spectral signatures.

provides the physical basis for land cover recognition (see Chapter 2). This spectral signature is modified by different factors, such as atmospheric conditions, plant seasonal cycle, topographic position, or soil substratum. However, the use of spectral signatures remains important as reference points.

The spectral signature of a particular cover can be approached by analyzing differences in gray levels over different sensor bands. In fact, brightness levels display visually these digital values recorded by the sensor. Within this context, we propose a single exercise for qualitative recognition of spectral signatures, based on visual analysis of MODIS spectral bands.* The reader should complete the tables in Figure 5.19 after observing how gray tones of different covers change in MODIS spectral bands of the study region shown in Figure 5.20. Consulting a land cover map of the region will be useful for the exercise, but not fully necessary, since the reader can rely on previous sections of this book for finding representative areas of the different covers. The reader should keep in mind that the image was acquired during

* Adapted from Short (1982).

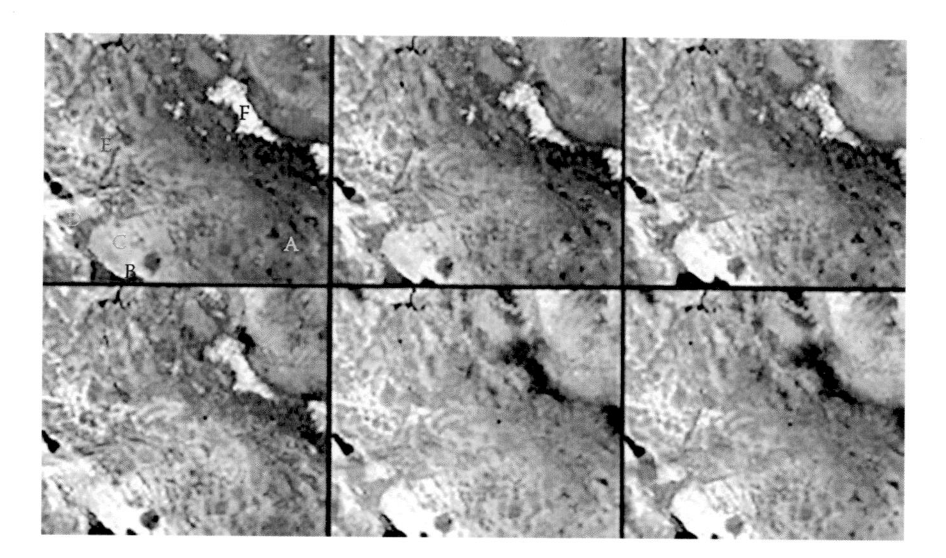


**FIGURE 5.20** MODIS spectral bands of the study region (images acquired on February 19, 2001). From left to right and top to bottom: B3, 459–479 nm; B4, 545–565 nm; B1, 620–670 nm; B2, 841–876 nm; B6, 1628–1652 nm; and B7, 2105–2155 nm. Letters refer to Figure 5.19.

the winter season; as gray levels are influenced by vegetation phenology, they change throughout the seasons.

### 5.4.4 Color Composites

Once the brightness levels for the different land covers have been identified, it becomes simple to infer the color they will present in different color composites, keeping in mind what we previously learned regarding the additive process of color compositing. For instance, if using a false color composite (B2, B4, and B3 for MODIS applied to RGB, respectively, see Figure 5.8a), vegetation will be shown in red tones, since the brightness of vegetation in B2 is medium to high, while it is dark to very dark for B4 and B3. Similarly, snow will be white, since brightness levels are high in B2, B4, and B3, and water will be black for the opposite reason.

Following the same line of reasoning, the reader can answer the questions regarding the color in which bare soils or urban areas will appear, which color composite yields vegetation in green tones, and what colors the shrubs and grasslands will be in that particular combination.

The reader can also observe again Figure 5.7, which includes several color composites for the same MODIS image used here for the exercise. The color composites stand out with much stronger contrast than the gray-level images, and the reader can clearly distinguish between the different land cover types. Using auxiliary cartography, different features can be identified, such as dams and reservoirs, main mountain areas, rivers, roads, and cities. When elevation data are available, some relationships between vegetation types and topography can be derived; for instance, it is interesting to note that higher elevations in the arid regions of the U.S. Southwest are the only zones where tree-covered areas can be found. The reader can confirm this relation in the vicinity of the city of Tucson, located in the southern portion of the image.

### 5.4.5 Multitemporal Approaches

The analysis of temporal change is very relevant for discriminating land covers, as well as for better understanding their seasonal trends. Figure 5.21 includes three MODIS images acquired at different times of the year: winter, spring, and the beginning of fall. This seasonal depiction enables us to discriminate the snow cover (winter image, white; spring and fall in dark), the permanence of irrigated crops (all dates dark), and the variations of turbidity in the Colorado River delta (note that in the May image, the sediments make the water appear very light in the red band). The bottom part of the figure shows the areas around Hoover Dam and part of the Grand Canyon. The variations of accumulated water (flooded areas) are notable, as well as the different shadowing effects caused by the variations in Sun's illumination on the canyon fringes.

The second approach to multitemporal analysis focuses on change detection between two or more distant dates in order to study the temporal dynamics of an area such as urban expansion, agricultural changes, deforestation, and land degradation. This comparison makes it possible to evaluate changes caused by ephemeral phenomena, such as floods or volcanic eruptions, as well as to monitor the temporal evolution of these effects (see Section 7.3 for further details).

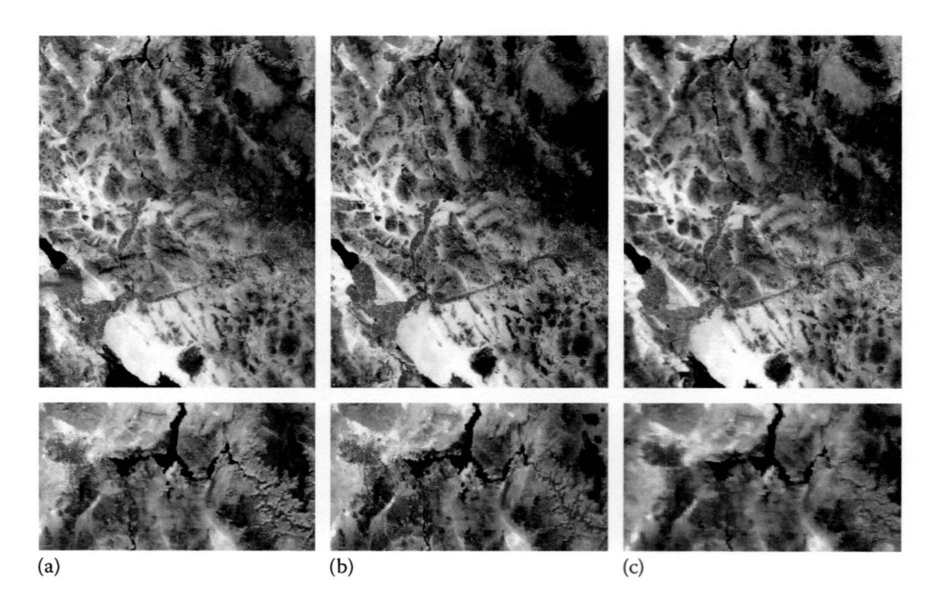


**FIGURE 5.21** MODIS images of the study region acquired at different times of year. (a) February 19; (b) May 21; and (c) September 25. The bottom window shows the status of the Hoover Dam on the same dates.

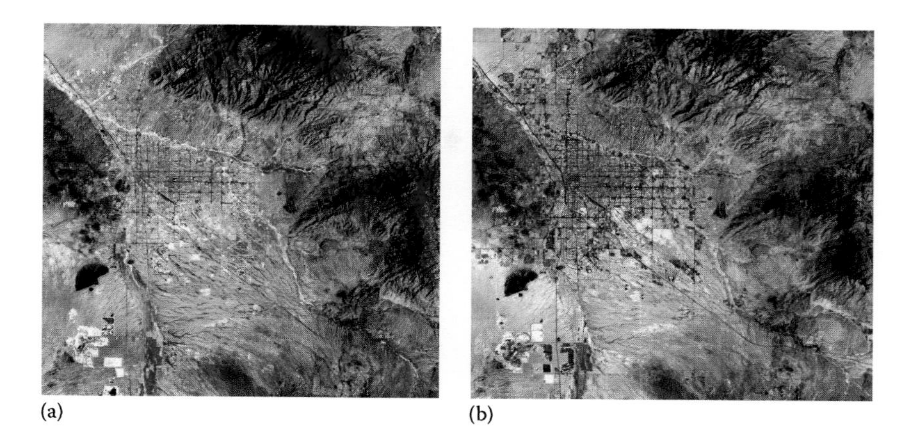


**FIGURE 5.22** Multitemporal change of the Tucson city observed through the analysis of Landsat images: (a) TM acquired in 1985, and (b) OLI image acquired in 2014.

Figure 5.22 illustrates this type of multitemporal analysis within the city of Tucson, using two Landsat images acquired exactly the same day (May 14, with 29 years of difference). Changes in both urban areas, mining and agricultural zones are noticeable. According to recent statistics, Tucson has almost doubled its population and size in the last 40 years, going from a little more than 260,000 people in 1970 to more than 525,000 people in 2010. The images partially show this urban growth, since the increase in urbanized area is prominent, both to the south and north of the original area. Many more examples of these trends are found in the *Atlas of Our Changing Environment* compiled by the UNEP a few years ago and are continuously being updated.

Additional examples of visual interpretation could be associated with geological interpretation (Woldai 1983), forest mapping (Morain and Klankamsorn 1978; Sadar et al. 1982), landscape stratification (Astaras 1984; Hilwig 1980), land cover mapping (Chuvieco and Martínez Vega 1990; Harnapp 1978), and analysis of urban patterns (Adeniyi 1987; Olorunfemi 1987).

## 5.5 REVIEW QUESTIONS

1. When an area shows dark tones in both the red and SWIR bands, then it may be
   a. Vegetation
   b. Snow
   c. Soil
   d. Vegetation or water
2. Which one of these criteria is not spatial?
   a. Brightness
   b. Context
   c. Shape
   d. Association

3. What should be the main criteria to discriminate an irrigated crop from an urban garden?
   a. Color
   b. Texture
   c. Spatial context
   d. Shadows
4. What is the dominant color of vegetation in the standard false color composite?
   a. Green
   b. Red
   c. Yellow
   d. White
5. Which of the following color assignments would display vegetation in green (RGB)?
   a. NIR, SWIR, R
   b. SWIR, R, NIR
   c. NIR, R, SWIR
   d. R, NIR, SWIR
6. What is the main criterion to discriminate a deciduous forest from irrigated fruit trees in a summer image?
   a. Texture
   b. Color
   c. Tone
   d. Context
7. Which of the following color compositions have the greatest unique information?
   a. NIR, R, B
   b. R, G, B
   c. SWIR, NIR, R
   d. NIR, R, G
8. To estimate the height of objects in a visual image we should know
   a. Sensor calibration conditions
   b. Solar angles
   c. Spectral characteristics
   d. Solar angles and spatial resolution
9. The relevance of shapes in visual analysis depends on
   a. Spatial resolution
   b. Spectral resolution
   c. Texture
   d. Color composite
10. Multitemporal comparisons by visual analysis require
    a. Geometrical matching
    b. Same spatial resolution
    c. Same spectral resolution
    d. None of the above

# 6 Digital Image Processing (I)

## *Enhancements and Corrections*

### 6.1 STRUCTURE OF A DIGITAL IMAGE

Remote sensing images were traditionally distributed in two different media: photographic products and computer files. Nowadays, both currently acquired and historical images (such as those acquired by spy satellites or airborne cameras) are distributed in digital format, and therefore digital image interpretation has become the standard for Earth observation (EO) information retrieval. Digital images are accessible through user-friendly web image services, which provide a first visualization to the image contents before it is downloaded.

A digital image is a numeric translation of the original radiances received by the sensor, forming a 2D matrix of numbers (actually a 3D matrix if we include the spectral dimension). Those values represent the optical properties of the area sampled within the sensor coverage. Radiances of the observed area are sampled every $x$ meters, $x$ being the projected area on the ground of the sensor instantaneous field of view (IFOV) (Figure 6.1). Radiance received at the sensor is converted to a discrete numerical value for each of the spectral bands to which the sensor is sensitive.

The instantaneously observed area is named pixel, a contraction of picture element, which is the minimum unit of information in a digital image. They can be easily seen in the image when it is amplified (Figure 6.2). The value defining the brightness of a pixel is denoted as a *digital number* (DN; Lillesand and Kiefer 2000), *pixel values* (Hord 1986; Mather 1998), *brightness values* (Jensen 1996), *gray levels* (Schowengerdt 2007), and *digital counts* (Richards 1993). In this book, we will use the term *DL* (digital level) to emphasize that they are digital values with a clear pictorial content easily converted to gray or color levels. There are as many *DL*s in an image as the radiometric resolution of the sensor. As previously seen (Section 3.1.3), the coding range of *DL* can be identified with the radiometric resolution of the system sensor and is expressed in the number of bits per pixel required to store that *DL*.

The actual *DL*s of any image can be manipulated through a variety of methods, which commonly result in new arrays of transformed numbers. In this sense, it is important to distinguish between the *DL* of a pixel, as stored in the computer disk, and the brightness levels (BLs) or values displayed in the monitor or printer. Those BLs are useful to facilitate visual image interpretations, while the original *DL*s are physically based and are related to the actual radiances received at a sensor.

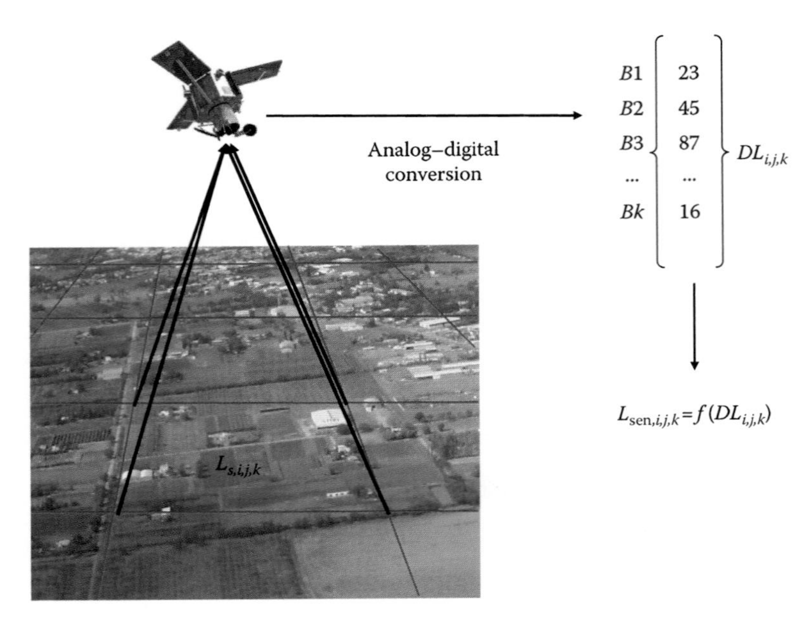


**FIGURE 6.1** Digital image acquisition process. Soil radiance ($L_s$) at a certain coordinate ($i,j$) and wavelength ($k$) is converted to a list of digital levels (DLs) one for each band. Later, those values can be converted back to radiances.

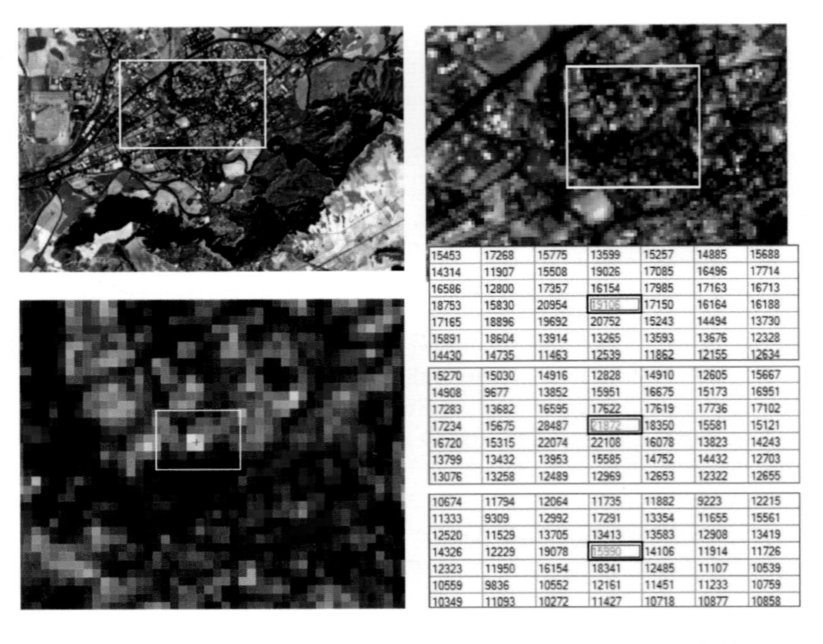

| 15453 | 17268 | 15775 | 13599 | 15257 | 14885 | 15688 |
|---|---|---|---|---|---|---|
| 14314 | 11907 | 15508 | 19026 | 17085 | 16496 | 17714 |
| 16586 | 12800 | 17357 | 16154 | 17985 | 17163 | 16713 |
| 18753 | 15830 | 20954 | 19106 | 17150 | 16164 | 16188 |
| 17165 | 18896 | 19692 | 20752 | 15243 | 14494 | 13730 |
| 15891 | 18604 | 13914 | 13265 | 13593 | 13676 | 12328 |
| 14430 | 14735 | 11463 | 12539 | 11862 | 12155 | 12634 |

| 15270 | 15030 | 14916 | 12828 | 14910 | 12605 | 15667 |
|---|---|---|---|---|---|---|
| 14908 | 9677 | 13852 | 15951 | 16675 | 15173 | 16951 |
| 17283 | 13682 | 16595 | 17622 | 17619 | 17736 | 17102 |
| 17234 | 15675 | 28487 | 21872 | 18350 | 15581 | 15121 |
| 16720 | 15315 | 22074 | 22108 | 16078 | 13823 | 14243 |
| 13799 | 13432 | 13953 | 15585 | 14752 | 14432 | 12703 |
| 13076 | 13258 | 12489 | 12969 | 12653 | 12322 | 12655 |

| 10674 | 11794 | 12064 | 11735 | 11882 | 9223 | 12215 |
|---|---|---|---|---|---|---|
| 11333 | 9309 | 12992 | 17291 | 13354 | 11655 | 15561 |
| 12520 | 11529 | 13705 | 13413 | 13583 | 12908 | 13419 |
| 14326 | 12229 | 19078 | 15990 | 14106 | 11914 | 11726 |
| 12323 | 11950 | 16154 | 18341 | 12485 | 11107 | 10539 |
| 10559 | 9836 | 10552 | 12161 | 11451 | 11233 | 10759 |
| 10349 | 11093 | 10272 | 11427 | 10718 | 10877 | 10858 |

**FIGURE 6.2** Digital image consisting of picture elements (pixels) each coded by a numeric value ($DL$). Inner boxes represent the areas that are magnified. $DL$ values for each of the three color bands are displayed.

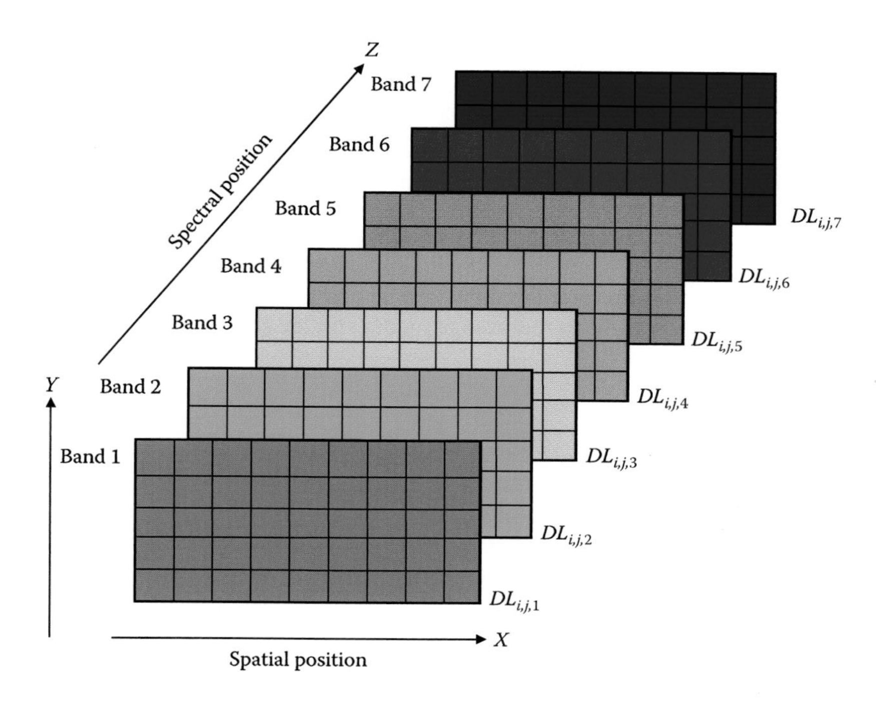


**FIGURE 6.3** Structure and organization of numeric data within a digital image.

Thus *DL*s can be used to retrieve physical parameters (e.g., reflectances or temperatures) from remotely sensed imagery and serve as a reference for digital analysis and interpretation (see Section 6.7).

In summary, the minimum spatial unit of measurement in a digital image is known as a pixel, defined by a *DL* that represents the detected average radiance within the ground area of the IFOV.

The structure and organization of the data within a digital image can be illustrated as a 3D numerical matrix (Figure 6.3). The first two axes correspond to the geographic or spatial range of coordinates in the image, while the third axis describes the spectral dimension or how the area is sampled through different spectral filters and band passes. The spatial matrix consists of *rows* or *lines*, numbered in a north–south orientation, and *columns* that indicate the west–east orientation of the image. In this image matrix, the origin of the image starts at the upper left corner (line 1, column 1) rather than the lower left corner as in the case with Cartesian axes. This is a result of the image acquisition sequence, which generally proceeds from north to south in many polar orbiting satellite paths. The third dimension corresponds to the spectral properties of the sensor and consists of as many layers as there are available bands on the sensor; for example, 36 for MODIS, 9 for Landsat 8 OLI, and 5 for AVHRR.

A thorough understanding of the 3D structure of digital images and data matrix formats readily enables one to perform digital image transformations and analysis. For example, one can perform many statistical operations commonly applied to other numerical matrices, such as computing the mean and standard deviation, building histograms, generating new matrices by multidimensional combinations across bands (e.g., principal components), filtering, feature extraction, classification, image registration, and many others. In the sections that follow, we first discuss the structure of digital images and then the presentation of a wide range of image processing techniques to correct and extract relevant information from the original data.

## 6.2 MEDIA AND DATA ORGANIZATION

The first step in digital image analysis is to read and organize the raw image data. Data access depends mainly on the storage media and file formats.

### 6.2.1 Data Storage

Several years ago, most satellite images were archived and distributed on computer-compatible tapes (CCTs), an economical and standard data storage medium for large volumes of information. The main problem with CCTs was their sequential access of data, which slowed down data reading, the high cost of tape readers, and their low storage density, compared with recent technologies. Today, images are commonly distributed through the Internet protocols (file transfer protocol [FTP] or Secure SHell [SSH]), particularly when a high-speed network connection is available. This is the quickest way to obtain the images, and therefore it is the most common when real-time access to the data is critical (such as in natural hazards applications). When working on large areas with medium-resolution images (such as Landsat 8 or Sentinel 2), or with long-term time series of images, the storing demands for a particular project may reach up to several terabytes (1 Tb = $10^{12}$ bytes). Transferring this amount of data may take very long when internet connections are slow, and therefore sending magnetic discs by courier services may still be a good alternative.

### 6.2.2 Image File Formats

To access the image contents for starting the digital processing, the first requirement is to understand how the images have been recorded. This implies knowing the logical organization of the data and how it was coded. The *DL*s of an image are recorded in binary numbers. Most sensors used to code the raw radiances in 8 bits per pixel. For 8 bits, each *DL* is defined by a byte with a range of 256 gray levels ($2^8$) that vary from 0 (darkest) to 255 (brightest). Now, 11 or 12 bits per pixel is increasingly becoming more common (IKONOS, MODIS, or Landsat 8). This implies that a wider range is available to storage *DL*s and therefore the sensitivity to discriminate ground reflectances will be larger.

Digital images are organized in a series of *files*, each of which is divided into *records*. The Committee on Earth Observation Systems (CEOS) has recommended format conventions for remotely sensed digital images and the media in which they are stored. The media in CEOS format contain a *volume directory file*, a *leader file*, and a *data file*. The data file is organized into a *header record* and *N* data records corresponding to the number of image products or bands. The *header file* contains information on the organization of *DL*s, the acquisition conditions (type of sensor, location of the scene, day of year, elevation and solar azimuth angle, etc.), and the type of processing, and data corrections applied at the receiving station. Now, the header file is commonly named "metadata," and it follows geographic documentation standards, such as the ISO 19115 and the ISO 19121, currently the most accepted schema to describe geographic information and services with vector and gridded data, respectively. Metadata include all the necessary information to manage the contents of the images.

The organization and format of files to archive a particular image follow the description found in the image header or metadata. There are file formats generally accepted as standards for transferring geographic information, such as the GeoTIFF (a version of the well-known TIF (tagged image file) format that includes geographical coordinates), the Hierarchical Data Format (HDF), EOS, and Network Common Data Form (NetCDF) formats. HDF is an open format used for handling NASA's Earth Observing System data products, particularly those from Terra and Aqua missions. It was originally developed in 1988 by the National Center for Supercomputing Applications (NCSA) at the University of Illinois at Urbana–Champaign, to handle large mathematical arrays in diverse computing environments. HDF is a self-describing and portable, platform-independent data format for sharing science data, as it can store many different kinds of data objects, including multidimensional arrays, metadata, raster images, color palettes, and tables in a single file (Folk et al. 1999). All the HDF source codes are available in the public domain. The NetCDF is also a self-describing, portable, scalable format that is currently widely used by climate modelers (Zender 2008).

## 6.3 DIGITAL IMAGE PROCESSING SYSTEMS

Processing and manipulation of digital images are supported by computer systems and image processing software (IPS). A brief review of some the basic characteristics and components of image analysis equipment is summarized in the following, although it is somewhat difficult to keep up to date with the new advancements in image processing hardware and software available, given the market dynamism. The development of operating systems, for both personal computers (Windows or IOS) and workstations (e.g., Linux), has also aided in integrating the main differences between these two systems. Even tablets or cell phones now have the potential to host image processing programs.

To carry out the main functions of digital image analysis, the interpreter needs a series of hardware and software components, including internal and external memory, processing units, display devices, and output peripherals (Figure 6.4). Today's personal computer systems have internal memories of at least 8 Gb, processing speeds exceeding 3 GHz, and disk capacities of over 1 Tb.

There is a variety of IPS available, both in the commercial and public domain, that offers the most common and basic image analysis routines. These software packages are available in UNIX or Linux, Windows (XP, Vista, W-7-10). Apple IOS, and Android operating systems. For the most part, the user and interpreter will only need to write software code for very specific tasks, since most common routines are available in both commercial and public domain software. Most commercial systems

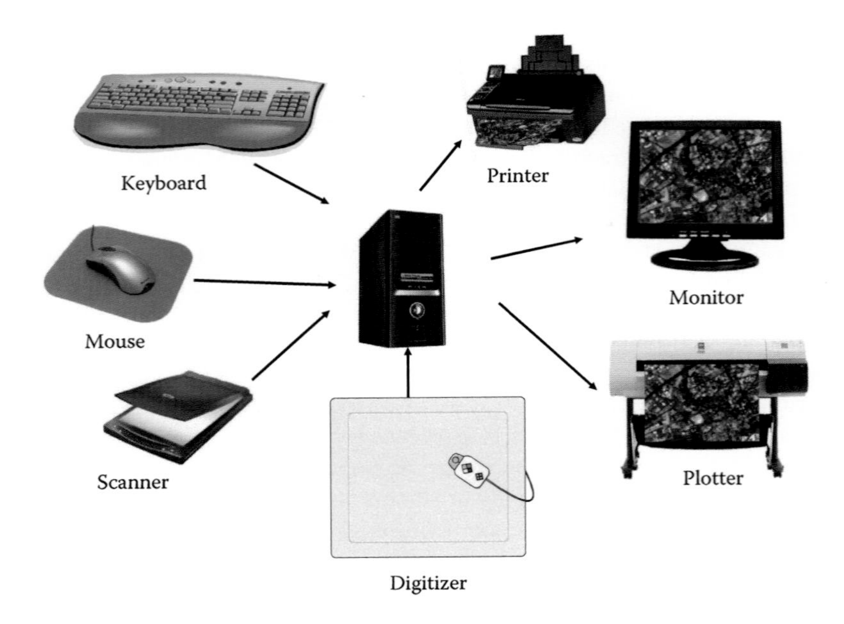


FIGURE 6.4 Components of a digital image processing system.

offer programming tools to help the interpreters in writing their own code, by using either graphic tools or program libraries that can be accessed through standard programming languages (e.g., IDL [interface definition language], Python).

IPS systems nowadays are hardware independent and user friendly and have more open environments, distributed access, and closer links between image processing and geographic information systems (GIS). Among commercial software packages, the most commonly used include ERDAS, PCI Geomatics, and ENVI. There are very powerful open source IPS, such as GRASS, originally written by the U.S. Army Construction Engineering Research Laboratories (USA-CERL) and now in the public domain as part of the Open Source Geospatial Foundation; SPRING, developed by the Brazilian National Space Research Institute; MultiSpec, designed by Purdue University; and ILWIS originally developed by the ITC in the Netherlands. In addition, several libraries have been created for specific IPS needs, such as VICAR, a collection of Unix and Windows platform image processing programs developed at the Jet Propulsion Laboratory for the analysis of data from the NASA family of airborne and spaceborne imaging spectrometers, and BEAM, developed by Brockman consulting for processing datasets from the European Space Agency missions. Updated links to these resources can be easily searched with standard Internet browsers.

In general, there are no strict criteria in selecting between one IPS or another. In many cases, the best selection will vary with the intended application. A user interested in geologic applications may be more interested in the visualization capabilities of the system or the transformation and image enhancement capabilities. On the other hand, a user focusing on land cover mapping may be more concerned with available classification algorithms in a software package. In all cases, it is important to consider some general characteristics:

1. Processing speeds, based on programming language and the structure of algorithms used. The time used on common tasks is also important (filters and classification).
2. User friendliness: most of the new equipment facilitates the handling capacities with a system of interactive menus and instructions aimed at helping the user. It is important to consider this aspect to obtain results in less time.
3. File export/import capabilities and type of formats accepted for reading and writing, paying special attention to the most standardized.
4. Programming capabilities, macro languages, libraries, or available programming languages.

## 6.4 GENERAL FILE OPERATIONS

In this section, we discuss the operations employed in IPS to access and maintain image files. To illustrate various image processing techniques, we utilize the same MODIS and Landsat 8 images used in Chapter 5. The selection of a few images from two sensors enables the reader to observe, compare, and understand the transformations created by various image processing techniques. However, to extend the illustration of some processes, we will also use examples of images from other areas when appropriate.

The basic digital image processing techniques that we have classified as general utility include those that are applied for a variety of purposes, and serve as an intermediate step to more specific operations. We include in this section procedures related to file management, display utilities, statistics, and image histograms.

### 6.4.1 File Management

Most IPS contain an image processing toolbox or a series of modules that include general-purpose utilities and functions for reading and displaying images, copying and deleting files, spatial transformations, image enhancements, image registration, image displays, and on-the-screen digitizing.

IPS commonly starts by reading and displaying image files. The software should easily allow one to import the raw data and access the metadata, which should include data formats, image size, cartographic characteristics, acquisition conditions, sensor calibration, and any other ancillary information necessary to read and process the image data.

Several display utilities help in the visual inspection of the selected image. For instance, observing the entire image is useful to display a reduced resolution version of the image, in order to locate sectors of interest. A reduction factor is obtained through systematically sampling the image (every $n$th line and column for viewing the image, with $n$ being the reduction factor). Once the target area is precisely located, it can be read at full resolution.

Once the data have been read, the IPS should import header information to properly document the image for further processing. Metadata fields should at least include image dimensions (number of lines and columns), number of bands, and number of bits per pixel of the image. If the image is georeferenced, cartographic projection details, as well as the boundary coordinates (NW and SE corners) and pixel size in the *x*- and *y*-directions, are required to handle pixel position properly. Other relevant information includes basic statistical properties (mean, variance, histograms, etc.), sensor radiometric calibration, and image acquisition conditions (e.g., Sun elevation and azimuth angles, cloud cover, atmospheric haze). In some IPS, these fields are attached to the image, normally in the first bytes of the file, while in other cases they are stored as auxiliary files with the same name as the original image but with a different extension. The former option facilitates the compactness of the information, since all data referring to that image are recorded in the same file (Figure 6.5), while the latter alternative facilitates accessing the image by other programs since the image file will contain only *DL*s and not the metadata.

A common utility in IPS is the ability to display several images simultaneously, which helps visual comparison between different dates or bands. For example, Figure 6.6 displays four bands of the L-8 image. The ability to observe simultaneously several bands facilitates understanding image spectral variations. For instance, in the L-8 image, band 2 (located in the blue spectral band) has overall dark tones and strong contrasts between mining districts and surrounding vegetated zones, with

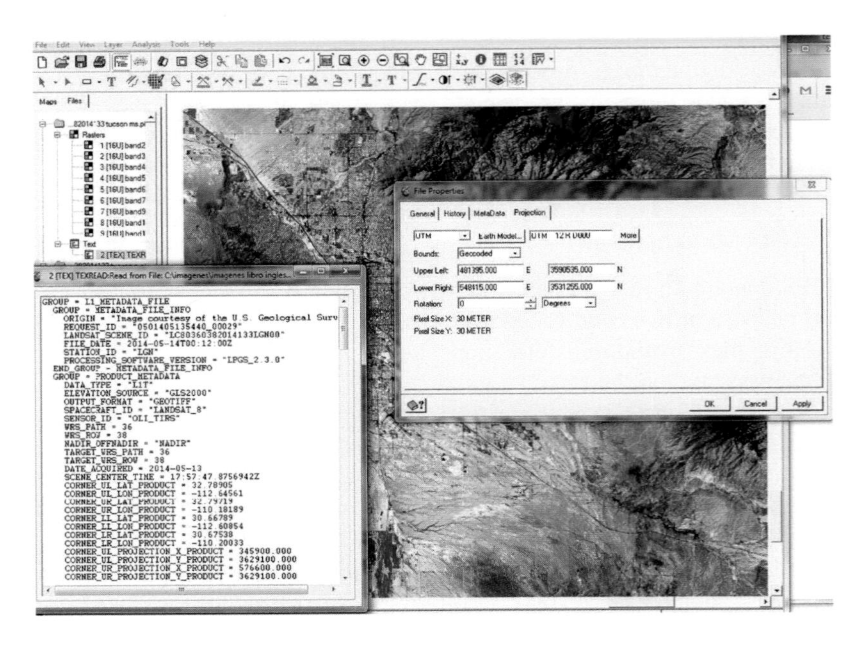


**FIGURE 6.5** Image documentation stored in PCI Geomatics showing cartographic information from the Tucson OLI image. The standard USGS metadata format for Landsat 8 images is also displayed.

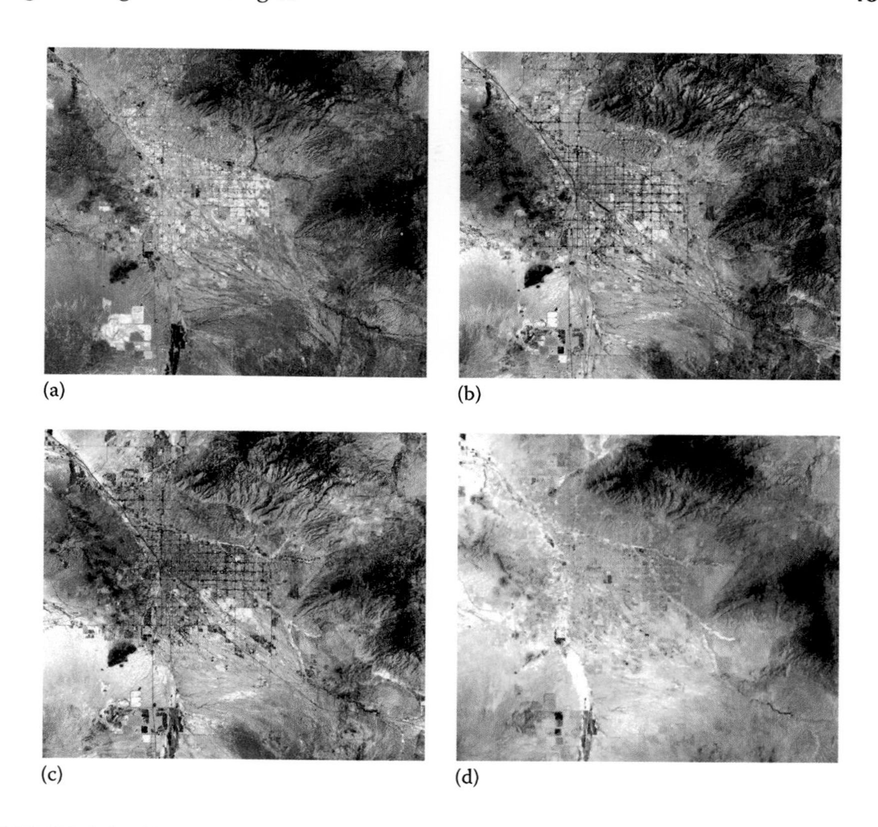


**FIGURE 6.6** Four spectral bands of the L-8 image: (a) band 2 (blue), (b) band 5 (NIR), (c) band 6 (SWIR), and (d) band 10 (TIR).

dark tones for irrigated crops. The NIR image (band 5) has very light tones over the irrigated agriculture areas, while the SWIR (band 6) shows more contrast between dry and green vegetation and emphasizes absorbed surfaces such as asphalt and desert scrubs. Finally, the thermal band shows higher temperature for built-up areas and dry streams and lower temperature for the vegetated upper elevations of the Santa Catalina and Mica mountains.

A useful utility for certain applications is to delineate and isolate sections of the image, defined by regular or even irregular polygons, in order to apply specific processing techniques and analysis. This is known as *masking*, and it is commonly linked to external datasets stored in a GIS. For instance, one may want to obtain crop estimations for specific agricultural districts, or evapotranspiration estimations for certain water basins. To mask out an image, we should first define the boundaries of the mask (commonly by the enclosing polygon), either by importing them from an external file or by digitizing them using a digitizer or electronic mouse. Then, the masking polygon is used to extract pixel values that lie inside the polygon, assigning a *DL* of 0 to those outside. Figure 6.7 illustrates this process for selecting an irrigated agricultural section the MODIS image.

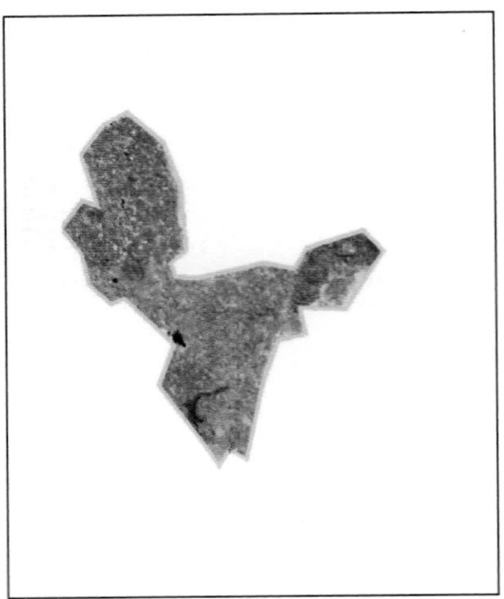

FIGURE 6.7 Example of a mask applied to a sector of the MODIS image.

### 6.4.2 Display Utilities

The ability to display an image is an essential property of any IPS. Some of the most common visualization tools and utilities include the following:

1. *Analog-to-digital conversion of DLs stored in the image, which allows their visual representation on the monitor.* The technical process applied can be seen in other publications (Muller 1988a; Richards 1993, Appendix F). Here, it is sufficient to note that it involves a conversion from a digital value to a BL for its visual representation. The old cathode ray tubes to display digital images have now been changed to a diversity of electronic systems: light-emitting diode (LED) display, electroluminescent display (ELD), plasma display panel (PDP), liquid crystal display (LCD), thin-film transistor (TFT) display, and organic light-emitting diode (OLED) display. To keep a fixed image on the screen, the data are stored temporarily in the refresh buffer that facilitates high-speed scanning (25–30 times per second), enough to be unnoticed by the human eye. Data are stored in three refresh buffers, corresponding to the three primary colors (red, green, and blue [RGB]), providing a composite color of the three bands. Commonly, refresh memory used to be a part of the video or graphic card, an external unit with high-speed access memory. Nowadays, video memory is being included in the motherboard of the CPU.
2. *Extraction of coordinates.* With the aid of a digitizer or electronic mouse, the interpreter can locate the coordinates or the *DL*s of points of interest (Figure 6.8). This is a very common task in image processing, for instance, to perform a geometric correction of the image or to characterize spectral values of different land covers. Usually the cursor is represented by a cross or an arrow on the screen.

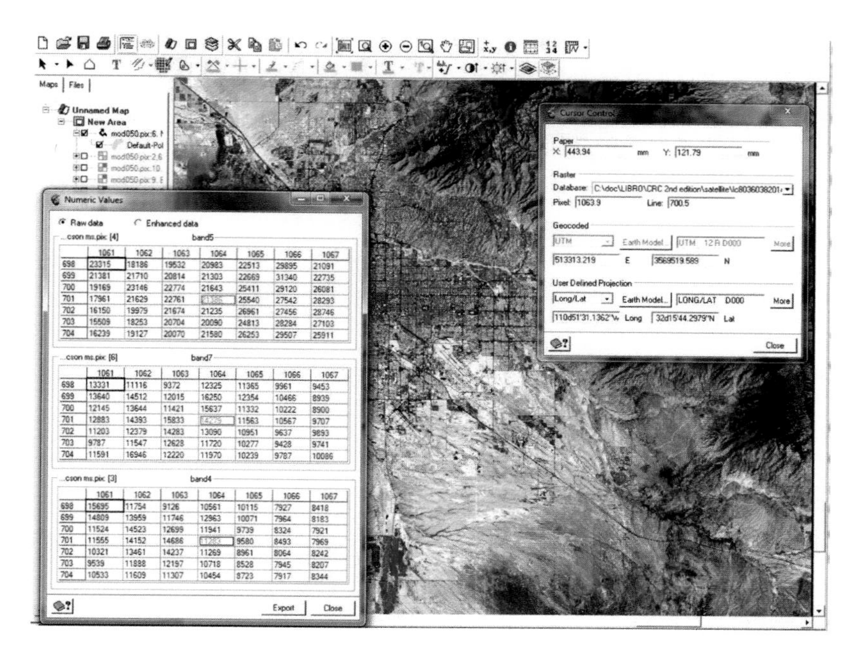


**FIGURE 6.8** An image processing software (IPS) provides a wide range of functions to obtain information from pixels being located by the mouse or other pointing device. In this case, file, cartographic, and geographic coordinates plus *DL*s at the cursor and neighbors pixels are shown.

3. *Digitizing areas on the image, again using a digitizer or mouse*. The features to delimit can be training areas for classification, linear elements (roads, rivers) to overlay on classified images, or areas to be masked out (as previously commented). Similarly, text and symbols (headers, orientation arrows, UTM grids, etc.) can be added for cartographic purposes.
4. *Changes of scale* (*zoom in and out*). This operation is utilized to reduce or enlarge the image and observe an area in smaller or greater detail, respectively. This process takes place in the refresh buffer, occurring almost instantaneously, making it very interactive to the user. In addition to increasing the size of the image, the user can scroll a zoomed image and move through it at a more detailed scale, which greatly simplifies locating points of interest. A more detailed image enlargement can be obtained using an interpolated zoom. Here, instead of duplicating the original *DL*s, a new matrix with a size equivalent to the magnification level is created. This new matrix is based on interpolation techniques from the original *DL*s. The output image is sharper than one enlarged by duplicating *DL*s because it maintains a transition between neighboring pixels (Figure 6.9).
5. *Creating image files*. It is often convenient to keep a copy of the image displayed on the monitor. This operation can be done through a variety of

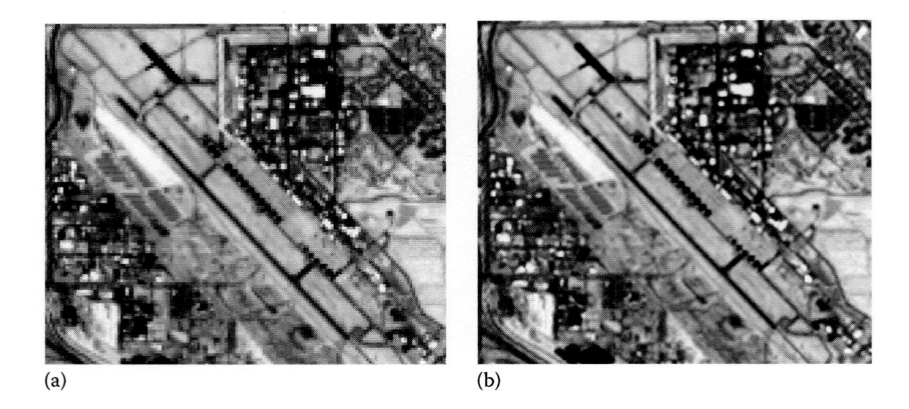


FIGURE 6.9 Comparison of a zoom obtained (a) by duplication and (b) by interpolation of original pixels in the OLI image. The area surrounding Tucson airport is shown.

auxiliary programs that capture the information stored of the refresh buffer and save it to a new graphic file. This capability is particularly useful for instructional purposes, as the interpreter can display a set of images and/or associated graphic elements without the need to recreate the process. In this case, the resolution of that graphic composition would depend on screen resolution. An alternative is to save a batch file that stores the operations used to build that particular visual representation, which can later be invoked from the same IPS software.

### 6.4.3 Image Statistics and Histograms

Like with any numeric variable, digital images can be described using basic statistical parameters, such as the mean and standard deviation. The former accounts for the central tendency of the image *DL*s, while the latter informs about their variability. The mean is computed as

$$\overline{DL_k} = \frac{\sum_{i=1,n} DL_{i,k}}{n_k} \tag{6.1}$$

where
$k$ is the target band
$n$ is the number of image pixels

The standard deviation ($SD_k$) is computed as

$$SD_k = \sqrt{\frac{\sum_{i=1,n} (DL_{i,k} - \overline{DL_k})^2}{(n_k - 1)}} \tag{6.2}$$

**TABLE 6.1**
**Basic Statistical Properties for the Tucson OLI Image**

| | Mean | Min | Max | *SD* |
|---|---|---|---|---|
| Band 2 | 12,118 | 7,958 | 61,376 | 1,424 |
| Band 3 | 12,632 | 6,875 | 62,528 | 1,783 |
| Band 4 | 14,166 | 6,200 | 55,117 | 2,450 |
| Band 5 | 18,230 | 5,315 | 65,536 | 2,535 |
| Band 6 | 19,737 | 5,031 | 63,997 | 3,143 |
| Band 7 | 16,762 | 5,211 | 60,888 | 2,928 |

Both parameters for the OLI image of Tucson are included in Table 6.1, along with the minimum and maximum *DL*s. Thermal and atmospheric correction bands have not been included in the table. The range of *DL*s goes from 0 to 65,536, following the 16-bit codification range of the L-8 OLI sensor. Band 2 (blue) has the highest minimum *DL* and the lowest mean, while band 6 (SWIR 1) has the highest mean and range (maximum–minimum) values, followed by band 6 and 7 (SWIR 2), which have the highest variability, and the visible bands (band 2–4) with the lowest value. Therefore, the SWIR and NIR bands offer a higher variability and therefore the most information content, and land covers should be more distinguishable in those bands than in the shorter ones (B2–B4).

In addition to the average and standard deviation values, it is also useful to compute the histogram for each band, which describes graphically the distribution of *DL*s for the image. In order to facilitate the representation of the histogram, absolute frequencies are usually turned into relative frequencies, according to a simple formula:

$$RF(DL_i) = \frac{F(DL_i)}{\sum_{t=1,m} F(DL_i)} \tag{6.3}$$

The relative frequency (*RF*) of a given *DL* (e.g., the value 8123) is calculated as the quotient between the number of pixels with that *DL* value and the total number of pixels in the image. This ratio serves to scale the histogram from the highest relative frequency.

The histogram provides a first evaluation of the data distribution of each band. The center of the histogram indicates the dominant brightness of each band (mean), while its width is related to band variability (*SD*). The presence of relative peaks might represent certain land covers in case they have distinct absorption to other features and high frequency. Band histograms for the Tucson image (Figure 6.10) show that only a small portion of the *DL*'s range (0–65,536) is actually being used in the image. Bands 5, 6, and 7 have the widest histograms, while bands 2 and 3 have narrower ones.

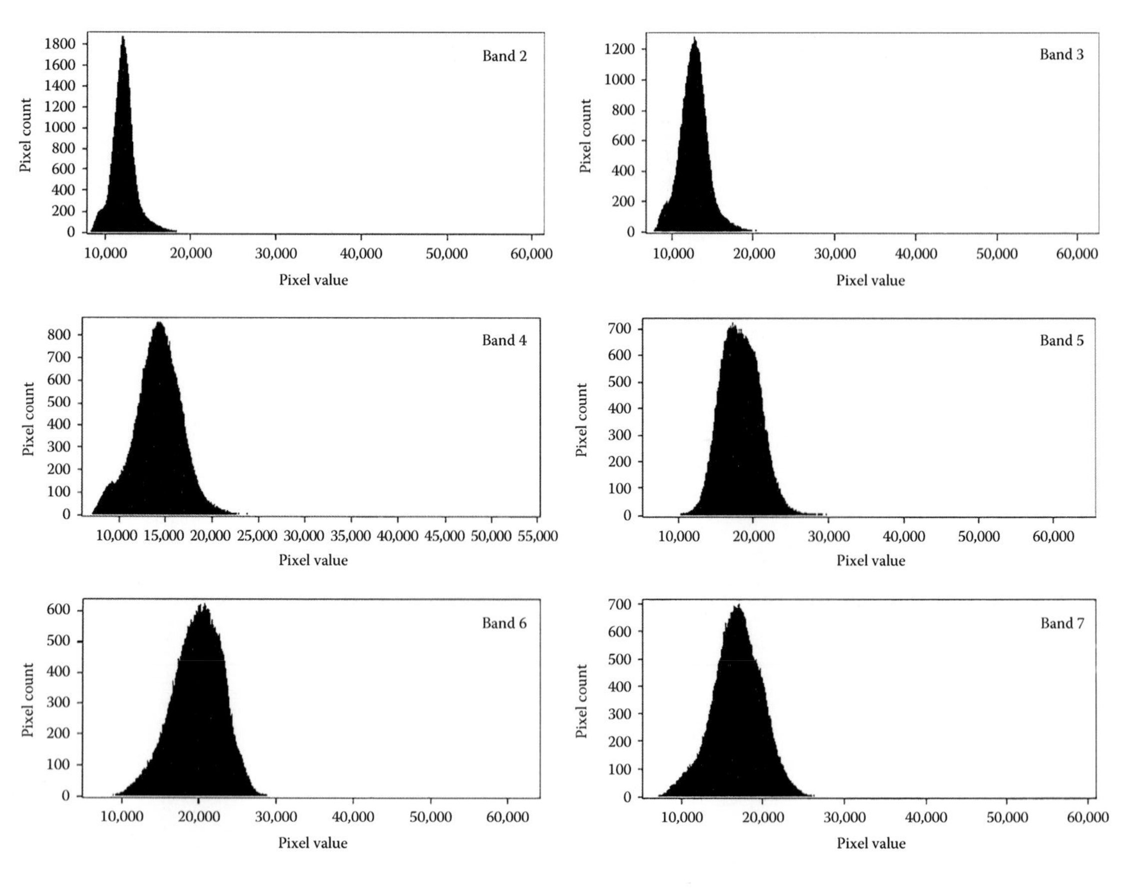


**FIGURE 6.10** Histograms of six OLI optical bands of the Tucson image.

In addition to analyzing each band separately, it is also interesting to look at the relationship between bands to determine the degree of original information they provide. This analysis can be done graphically, through bivariate scatter graphs that show the *DL* values of two bands simultaneously. When the distribution of points tends to form a line (or a narrow ellipse), the two bands are highly correlated and contain similar information. If this is the case, it may be sufficient to use only one of the bands (the one with higher variance) without losing much of the original information. This often occurs with the visible bands or the bands located in the SWIR. If the distribution of points is more scattered, as it is commonly the case with the NIR and R bands, the correlation between these bands is lower, since both offer unique and contrasting information. In this case, the interpreter should include them together for further analysis. As an example of this graphic analysis, the scatter graphs of the B and R bands and R and NIR bands over the Tucson image are shown in Figure 6.11. The first two bands are in the visible region and are highly correlated, as most of the land covers have similar reflectances in both bands. In contrast, the red and NIR bands show much less similarity, which is due to the greater spectral contrasts for different land covers (high reflectance in NIR and low in red).

The relationship between two bands can also be measured numerically using the Pearson's linear correlation coefficient ($r$). The $r$ statistic ranges between −1 and +1, meaning perfect negative and positive correlation, respectively. The Pearson correlation is computed as the ratio of the covariance of two bands, $k$ and $m$, and the product of their standard deviations:

$$r_{k,m} = \frac{\left(\sum_{i=1,n} (DL_{m,i} - \overline{DL}_m)(DL_{k,i} - \overline{DL}_k)\right)/n}{s_k s_m} \tag{6.4}$$

For the Tucson image, the correlation between visible bands 2 (B) and 3 (G) is very high ($r = 0.97$), as expected, while bands 4 (R) and 5 (NIR) are less correlated ($r = 0.83$) although they still show similar trends, particularly because of the importance of bare soils in the area (which have similar reflectances in the NIR and visible bands).

Another technique to achieve a better understanding of the radiometric information content of an image is through the construction of spatial or spectral profiles. The former are also named *transects*, and they are created from extracting the *DL*s that intersect a line digitized by the user. These profiles are useful to better understand the spatial variation of the *DL*s in different spectral bands. In the example of Figure 6.12, a small transect was digitized from the MODIS image. It crosses the irrigated areas of the Imperial Valley, the wetlands of the Colorado River delta, and the Sea of Cortez. The three bands used are the NIR (MODIS band 2), the SWIR (MODIS band 6), and the red (MODIS band 1). The NIR band shows high reflectance for crops and other vegetated areas, while the SWIR and red bands show

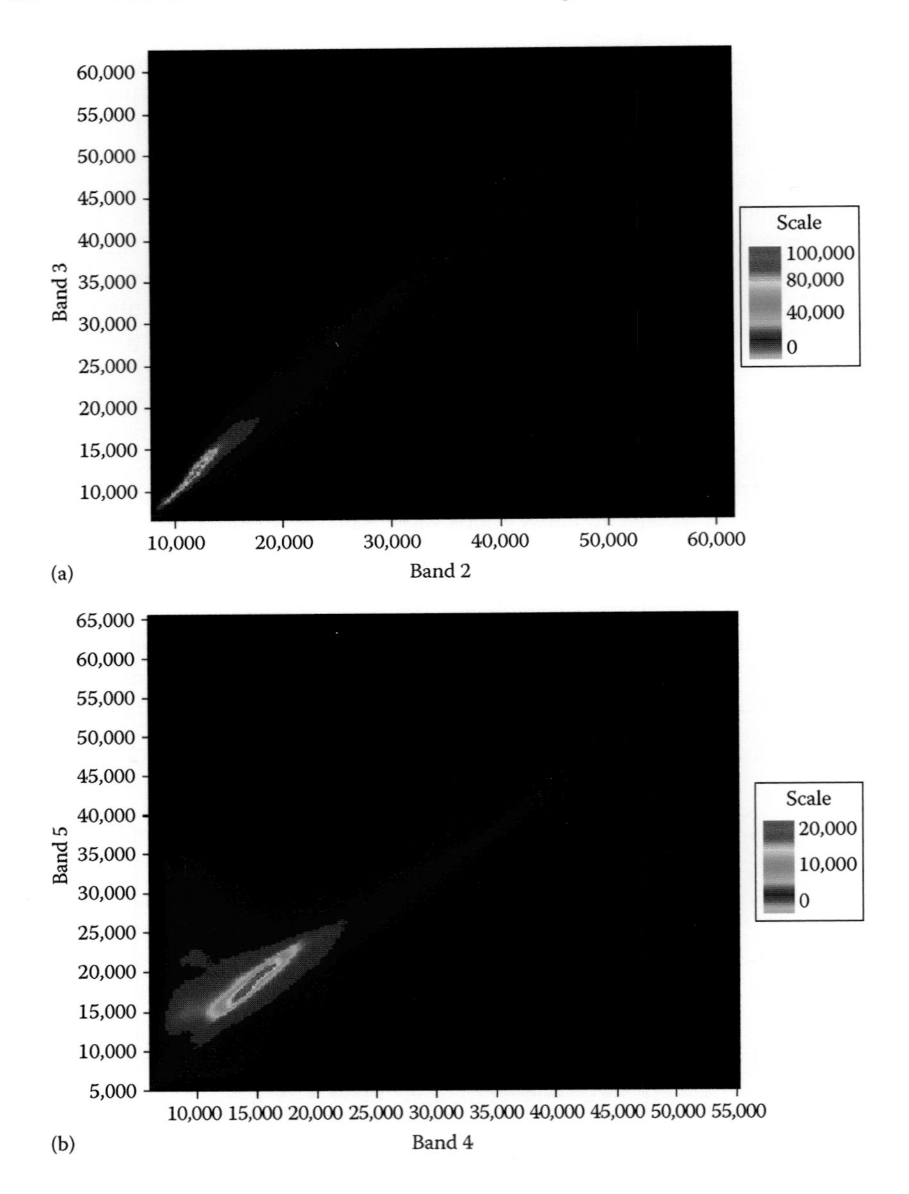


**FIGURE 6.11** Scatter graphs of bands 2–3 (a) and 4–5 (b) of the OLI image.

similar trends in this area although with lower red values. Water bodies result in lower reflectance for all bands, which is especially clear at the end of the transect (Sea of Cortez). In between the irrigated crops and the sea, different mineral deposits offer differential absorption features, with much higher absorption for the SWIR band than the NIR or red bands around 60 km from the origin.

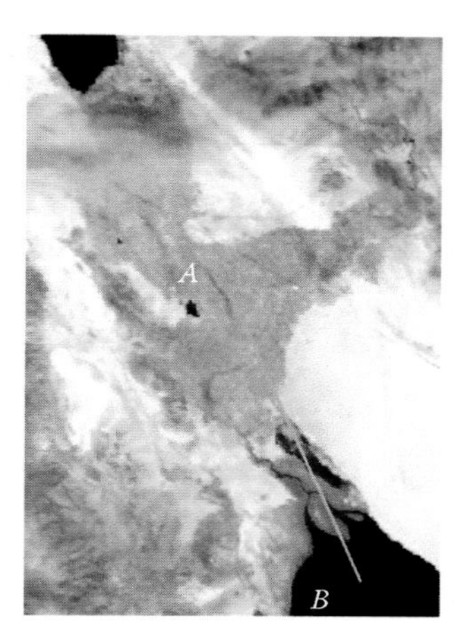


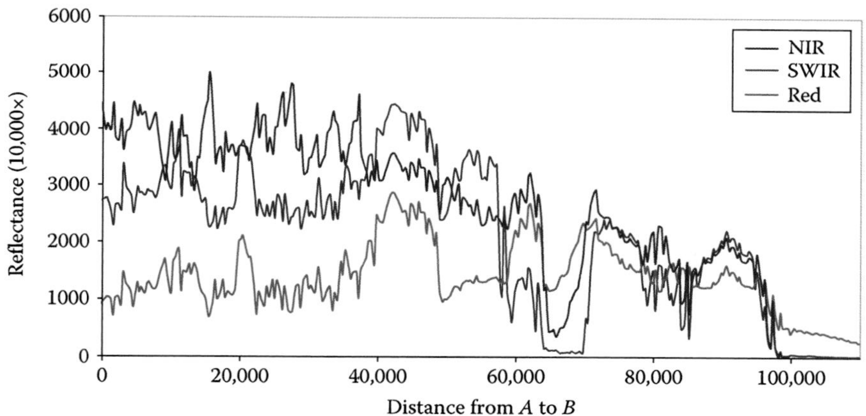


FIGURE 6.12 Radiometric profile over a window of the MODIS image.

## 6.5 VISUAL ENHANCEMENTS

In this section, we consider various techniques aimed at improving the visual quality of the image, including a discussion on contrast enhancement, color compositions, pseudocolor, and filters.

### 6.5.1 Contrast Enhancement

The process of contrast enhancement attempts to adjust the radiometric resolution of the image to the capabilities of the display system. It is required when the digital range of values does not match the number of brightness values available for display. Two opposite situations may occur: (1) the range of *DL*s in the image is less than the range of brightness display values and (2) the image has a greater number of *DL*s than can be visualized. In the first case, the interpreter should apply techniques for contrast stretching, while in the second one the data need to be compressed.

The idea of digital contrast can be illustrated with a photographic analogy. A photograph appears with low contrast when there is not much difference between light and dark tones. In the same way, we can define the digital contrast by the *DL* range or any other measure of image variability. The most common digital definitions of contrast are the ratio between the maximum and minimum value, the range of values, and the standard deviation (*S*) of the *DL*s in the image (Schowengerdt 2007):

$$C_1 = \frac{DL_{\max}}{DL_{\min}} \tag{6.5}$$

$$C_2 = DL_{\max} - DL_{\min} \tag{6.6}$$

$$C_3 = S_{DL} \tag{6.7}$$

Stretching operations should provide a higher contrast, increasing any of the three referred measures. The opposite operation to stretching is termed *color compression* and implies maintaining the visual quality of the color images with a significant reduction of the file sizes. This operation has been greatly advanced in the last few years with the increasing use of Internet-based image servers, which are favored by smaller file sizes.

#### 6.5.1.1 Color Lookup Table

A color lookup table (CLUT), or simply a lookup table (LUT), is a numeric matrix that shows the brightness values (*BV*s) on the screen that will be assigned to each of the *DL* values of the image. In other words, it defines the brightness intensity at which each *DL* is displayed.

In most IPS, the LUT is a numerical matrix of 3 columns × 256 rows (as BLs in computer systems usually have 8-bit resolution). The order of the row indicates the input *DL*, while the value in the LUT expresses the *BV* that will display that *DL*. The three columns correspond to the three primary colors (RGB: Figure 6.13). We should remember that IPS uses the additive process to create color images (see Chapter 5).

| ***BV* (blue)** | ***BV* (green)** | ***BV* (red)** |
|---|---|---|
| 1 | 1 | 1 |
| 2 | 2 | 2 |
| 3 | 3 | 3 |
| ... | ... | ... |
| 255 | 255 | 255 |

FIGURE 6.13 Tabular structure of a lookup table.

For this reason, the higher the *BV*s in each column, the more R, G, or B the pixel will be. In the case of displaying only one band, the image is commonly shown in gray levels, which implies that each *DL* has the same *BV* for the three primary colors. For instance, when the *BV* for a particular *DL* is 0, 0, 0, all pixels with that *DL* will be displayed as black, while a *BV* of 127, 127, 127 will show as medium gray and a *BV* of 255, 255, 255 as white. When the three values are different, then color tones appear on the screen, which are usually obtained through mixing three different spectral bands.

The LUT can also be represented as a line in a bivariate graph, where the *DL* of the image (input) is located along the *X*-axis and the *BV* (output) is represented along the *Y*-axis. If we displayed the original *DL*s, without any enhancement, the LUT would be described graphically as the diagonal of the square going from the origin (0, 0) to the maximum (255, 255) brightness (Figure 6.14a). In other words, the *BV*s would have the same values as the *DL*s. With the processes of contrast stretch and compression, this relation is modified in such a way that the full range of *BV*s is better represented. For instance, we can decide to assign all BL values to the range between the minimum and maximum *DL* (Figure 6.14b).

The utility of this numeric table can be easily understood with an example. Let us assume that we want to stretch the contrast of an image with *DL* distributed between 0 and 127. To adjust this range to the available number of *BV*s (0, 255), it will be sufficient to multiply by 2 each of the input *DL*s, which would expand the original range to between 0 and 254. Now, this option presents two problems: on the one hand, it transforms the original *DL*s to create an output image, which results in either losing the original information or having the need to increase the storage capacity, and on the other hand, the process involves a considerable volume of procedures—786,432 multiplications for a small 512 × 512 pixel, three-band image. If we consider that the enhancement is only an intermediate stage, and usually an ephemeral part of the process, it is reasonable to find an alternative to this process. Using an LUT, we can easily modify the relation between *DL* and *BV* using a simple

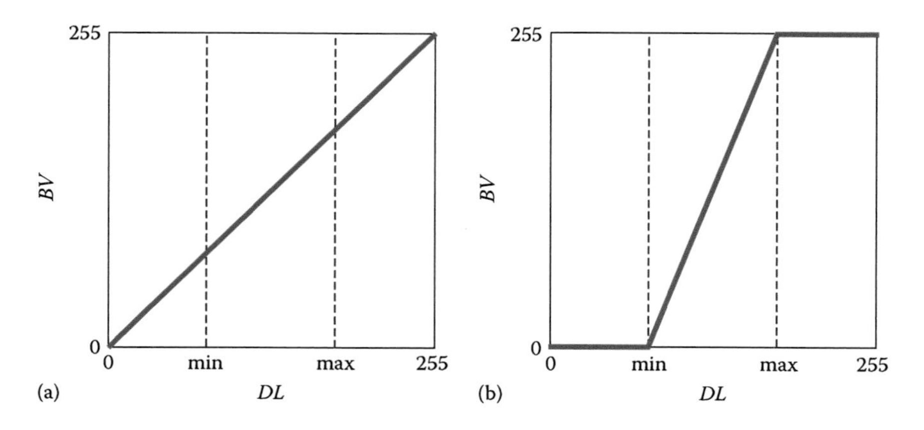


**FIGURE 6.14** Linear profile of the LUT: (a) original; (b) after linear stretching.

function ($BV = 2 \times DL$). This option would have the same visual effect as the multiplication of all original *DL*, but with two advantages: the original *DL* would be preserved and the transformation would be much quicker, since the modification through an LUT involves only 768 calculations (256 $BV \times 3$ color guns), independent of the size of the image.

The use of the LUT can be effective for both contrast enhancement and compression, although it has been more utilized in the former case. Pseudocolor and categorical color tables can also be created, as explained in the next sections.

#### 6.5.1.2 Contrast Compression

If the range of the sensor exceeds the number of available *BV*s, the interpreter should compress the original *DL* range. This adjustment is necessary when working with a sensor of high radiometric resolution, but it is also advisable when the image needs to be transferred at high speeds for visual interpretation, such as most image servers actually do.

The most recent sensors such as the L-8 OLI or Sentinel-1 radar systems provide high spectral sensitivity (16 bits), and therefore some sort of compression is required when trying to visualize them in 8-bit display systems. The most common approach is to reduce the original range of *DL* to a number of reduced intervals, selected from the histogram distribution of *DL*s. Intervals of equal width or equal frequency are two simple options, or the user can adjust the interval thresholds manually. In any case, a unique *BV* is assigned to each interval, thus eliminating its internal variation. When the contrast is reduced, the colors are less discernable, decreasing the quality of display. Nevertheless, if the selection of colors is adequate, a high-quality display can still be obtained.

There are more refined algorithms to compress data, which control the number of assigned intervals for each band used in the color composition through a process similar to an unsupervised classification that will be discussed later on (see Section 7.2). These algorithms first establish a series of partitions in a 3D space formed by the three bands used in the color generation. Based on the width of the histogram of each band, color partitions are assigned. To assign an interval of *DL* to each of these partitions, the most significant variation range is established on each of the bands, from their mean and standard deviation. Finally, the *DL*s of the image are classified into the closest partition in that space (Baudot 1990; Muller 1988b). The result of the process is a new image composed of a single band, which is associated with an LUT to represent, as close as possible, the chromatic variation of the multiband composition, using a fraction of the original stored file size (Figure 6.15).

Currently, the increasing use of Internet image servers has made color compression algorithms a very active field or research. Those algorithms generate high-quality versions of the original image at a fraction of its size (sometimes with compression ratios larger than 50:1). The most commonly used compression algorithms are the LZW, the JPEG, and the PNG. The Lempel–Ziv–Welch (LZW) was developed in 1984 and allows the creation of Graphic Interchange Format (GIF) images, widely used on web pages, although limited to 256 colors. The Joint Photographic Experts Group (JPEG) accepts 24 bits per pixel and guarantees a 20:1 compression ratio without noticeable quality loss, and it is more common in commercial digital cameras.

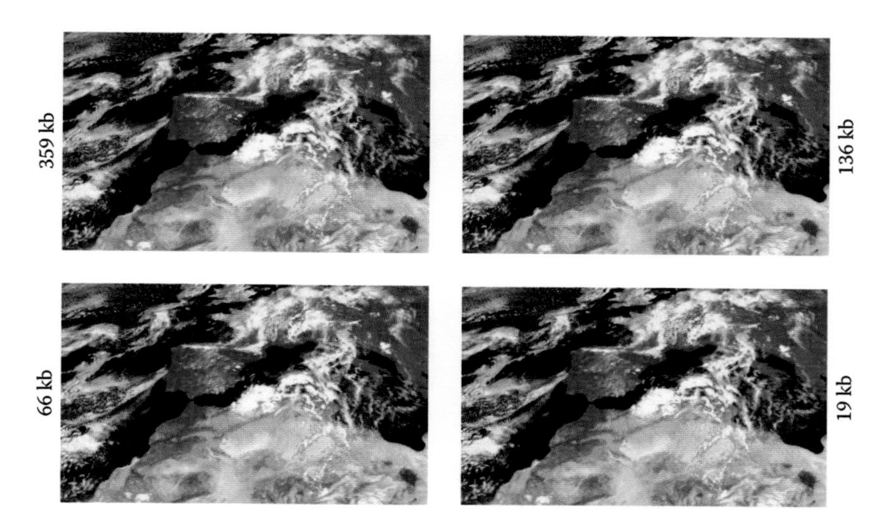


FIGURE 6.15 Three-band color Meteosat image of Southern Europe with different compression algorithms. Resulting file sizes are included at the margin of each image.

The Portable Network Graphics (PNG), developed by CompuServe and W3C, also accepts up to 24 bits per pixel as well as overlays, although the ratios of compression are not as high as in JPEG. Lately, new techniques have been developed for massive compression of images for distribution via the Internet. Among them, those based on wavelets are popular because they allow a high index of compression with low or null loss of information (Yocky 1996).

#### 6.5.1.3 Contrast Stretch

Contrast enhancing is widely used in both remote sensing and other image processing applications. It implies adjusting the original *DL* distribution to the capabilities of the display system, thereby improving image contrast and therefore the visual interpretation capability. When a sensor is designed, it is calibrated to acquire a wide range of radiances, from the very low of water to the very high of desert or snow areas. A single image will only very unusually include such a diversity of radiances, and therefore, part of the designed sensor sensitivity will not be used. In other words, even if the sensors have more than 8-bit radiometric resolution, in most images the original data distribution do not cover the lower and/or upper ranges of brightness values. Therefore, the original *DL* histogram never populates the 256 or higher possible *BV*s. Visually, this implies that there is a large range of gray tones that are not used, resulting in low-contrast images on the screen.

As an example, Figure 6.16a shows the original color composition (NIR, SWIR, R) of the OLI image, without any enhancement. The image has a low contrast, concealing many features of interest. The other sections of Figure 6.16 exemplify different contrast stretching methods for the same color composition. Figure 6.16b shows the impact of distributing the *BV*s linearly between the minimum and

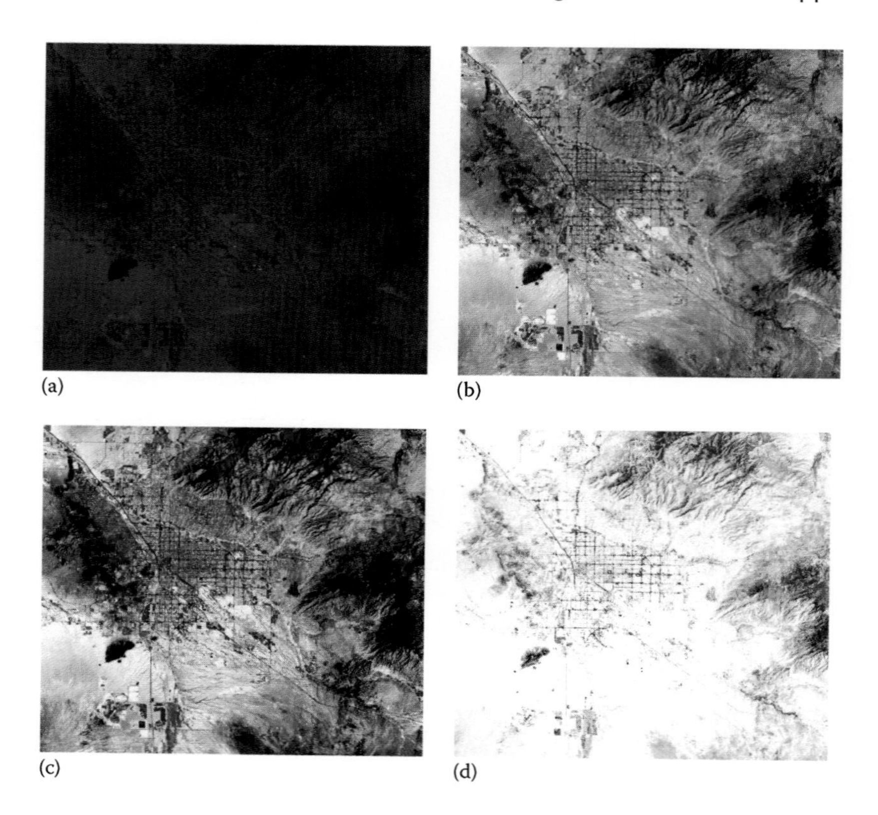


**FIGURE 6.16** Different types of contrast enhancement over the OLI image: (a) original data, (b) linear enhancement, (c) frequency enhancement, and (d) selected enhancement over areas of low radiance.

maximum *DL* of the image, Figure 6.16c shows a nonlinear stretching based on the frequency distribution of *DL*s, and Figure 6.16d shows an example of stretching based on a small range of *DL*s, in this case the darker tones. This option implies that the selected range of *DL* values is enhanced and therefore internal differences are more clearly seen, but at the cost of reducing contrast in other brightness ranges of the image, in this case the lighter ones.

In the following sections, these three contrast enhancement methods will be reviewed in more detail by considering the LUT of a single band. If a color enhancement is desired, the same procedures should be followed to stretch each of the bands that form the color composition.

#### *6.5.1.3.1 Linear Contrast Enhancement*

A linear stretch is the most basic way to adjust the contrast of an image and enhance its visual quality. In this case, we just need to design an LUT in which the minimum and maximum *DL* of the image correspond to *BV*s of 0 and 255, respectively, and linearly distribute the rest of the values between these two margins. Figure 6.16b shows

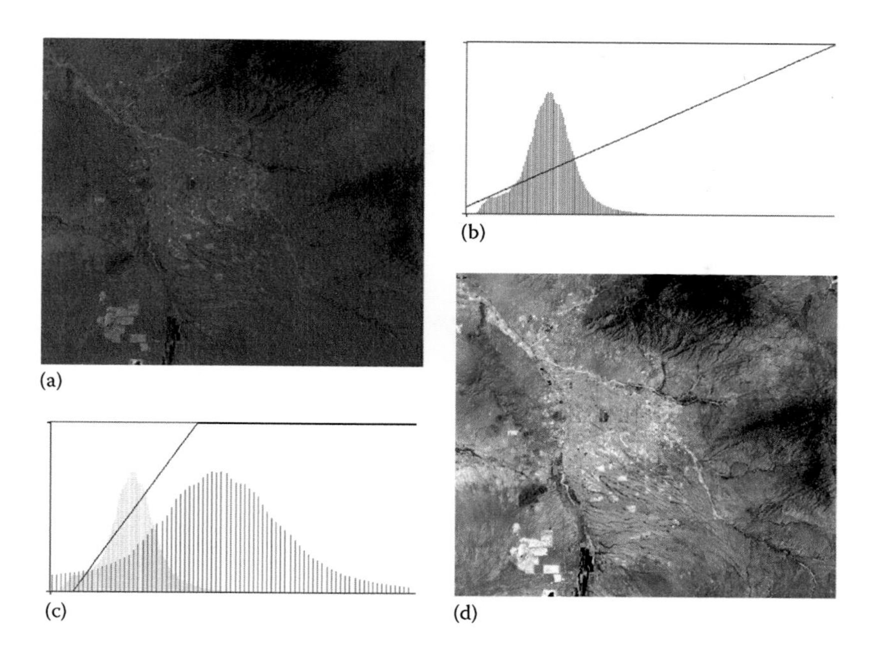

**FIGURE 6.17** Impact of linear stretch on image brightness (a, original; d, expanded) and *DL* histograms (b, original; c, expanded).

the visual quality of the new image after applying the linear contrast stretch. We will illustrate the linear stretching process using another Landsat image of the same area acquired by the Thematic Mapper (TM) sensor exactly 29 years earlier than the OLI image previously commented (on May 13, 1985). Since this image was acquired with 8-bit radiometric resolution, the display of LUT values is easier than for the 16-bit OLI images, but the process for both images is really the same.

Selecting the R band (band 3 for TM), the raw histogram of the target area shows poor contrast with medium to high brightness values missed (Figure 6.17a and b). The stretched image displays more information, as the brightness differences between covers are more evident (Figure 6.17d). The histogram of the stretched image is better distributed along the full range of brightness values. The LUT profile shows the graphic transformation applied to the original data, with greater slope than the original transformation, as now the expansion is focused between the minimum and maximum *DL* values of the image.

To accomplish this linear transformation, the interpreter should find a linear function that adjusts the *DL–BV* relation as follows:

$$BV = b + g\ DL \tag{6.8}$$

where each *BV* is a function of the original *DL*, after applying two coefficients, known as bias (*b*) and gain (*g*). The latter determines the slope of the LUT and the former the intersection of the LUT with the *BV* axis. To determine the value of these coefficients, we should solve a simple set of two equations that sets the minimum and maximum $DL$ values to 0 and 255, respectively, as follows:

$$0 = b + g\, DL_{min}$$

and

$$255 = b + g\, DL_{max}$$

Substituting in both equations, the coefficients are calculated as follows:

$$g = \frac{255}{DL_{max} - DL_{min}} \tag{6.9}$$

and

$$b = \frac{-255 * DL_{min}}{DL_{max} - DL_{min}} \tag{6.10}$$

Alternatively, $DL$–$BV$ relations can be expressed in a single term as

$$BV = \frac{DL - DL_{min}}{DL_{max} - DL_{min}} * 255 \tag{6.11}$$

but it is more appropriate to use Equation 6.8, which is the conventional equation for linear transformation of data. Instead of making $DL_{max}$ and $DL_{min}$ equal to the absolute maximum and minimum values, the interpreter may select different thresholds to fit the linear function. A common way is to select a certain percentile from the maximum and minimum values, such as 99% and 1%, respectively, or a given number of standard deviations above and below the mean. This choice will usually be more efficient than the absolute maximum and minimum, since the presence of anomalous high or low values would imply missing a certain range of $BV$s.

##### *6.5.1.3.2 Histogram Equalization*

A linear contrast stretch only considers the extremes of $DL$ distribution, and therefore all the $DL$ internal intervals are equally stretched. However, from a statistical point of view, it is more reasonable to stretch the values following the actual frequency distribution of the $DL$s. This requires the generation of an LUT in which each $BV$ has approximately the same frequency of $DL$s in the image. In other words, $DL$s with the higher number of pixels will occupy a proportionally larger range in the enhanced image than those with a lower frequency. This process is known as *histogram equalization,* and it offers better results than linear stretch, especially if the original image has a Gaussian distribution.

As can be seen in Figure 6.18a, histogram equalization results in a more balanced enhancement, showing less contrast between zones of high and low reflectance. As a result of this enhancement, the histogram of the image offers a better distribution of

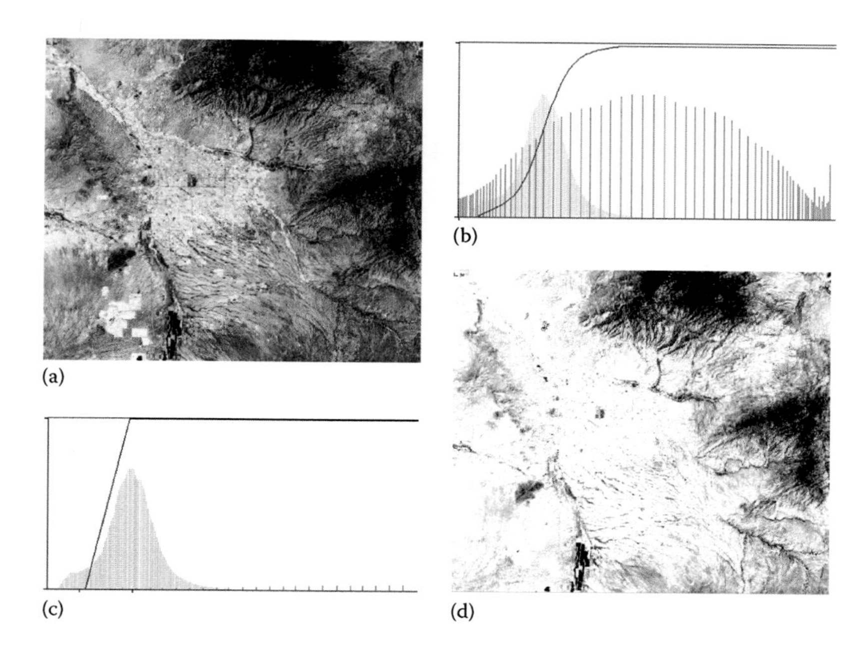


**FIGURE 6.18** Impact of histogram equalization on (a) image brightness and (b) resulting histogram. (d) Special contrast stretch and (c) histogram.

the data (Figure 6.18b). It can also be observed that the graphical profile of the LUT is not a straight but a curved line, similar to the cumulative distribution function of the original *DL*s. Other forms of nonlinear stretch can be applied, such as root or power functions, to better enhance the low or the high values, respectively. Table 6.2 shows the resultant LUT values using three different criteria: linear, histogram equalization, and root stretch.

#### *6.5.1.3.3 Special Contrast Stretch*

This is a specific case of the previous methods. The idea is to restrict the contrast to a specific range of the *DL* in order to enhance a specific land surface feature of interest. For example, in applications involving marine zones, it may be more convenient to enhance the lower end of the histogram to emphasize water differences, even though the contrast for the brighter ranges of the image is lost.

The special contrast stretch is based either on setting a maximum and minimum threshold selected by the user, considering the *DL*s of the target feature (linear method), or by restricting the histogram equalization to a certain range. In either case, the objective surface feature is enhanced while sacrificing the contrast of other areas in the image. Figure 6.18d shows an example of this type of enhancement, done over the lowest brightness values of the Tucson TM image, thus emphasizing internal differences in low reflectance (forest and water). These areas now show a better contrast than in the previous stretches, although the brightness range and visual content of the rest of the image is now poorly characterized, since the higher *DL* values are saturated (see the histogram, Figure 6.18c).

**TABLE 6.2**
**LUT Values after Different Stretches of the Red Band of Tucson 1985 Image**

| Original | Linear | Histogram | Root |
|---|---|---|---|
| 0 | 0 | 0 | 0 |
| 25 | 2 | 4 | 26 |
| 30 | 16 | 9 | 64 |
| 35 | 29 | 14 | 87 |
| 40 | 43 | 20 | 105 |
| 45 | 56 | 29 | 120 |
| 50 | 70 | 41 | 134 |
| 55 | 84 | 61 | 146 |
| 60 | 97 | 90 | 157 |
| 65 | 111 | 123 | 168 |
| 70 | 124 | 158 | 178 |
| 75 | 138 | 188 | 187 |
| 80 | 151 | 211 | 196 |
| 85 | 165 | 226 | 205 |
| 90 | 179 | 236 | 213 |
| 95 | 192 | 96 | 221 |
| 100 | 206 | 103 | 229 |
| 105 | 219 | 255 | 236 |
| 110 | 233 | 255 | 243 |
| 115 | 246 | 255 | 250 |
| 120 | 255 | 255 | 255 |

*Note:* Only values every five *DL* have been included for brevity.

A special case of this type of stretch is the creation of a binary image, in which the contrast of the image is reduced to just two *BV*s, white and black. In this case, the LUT has a stepping profile. The first step includes the range that will be black and the second step the one that will be assigned to white. This process is used to isolate a range of *DL*s for further analysis; for example, the *DL*s corresponding to a water body in an oceanographic study should first be isolated by masking out all land values in order to avoid the noise created by the coastal zone. Of course, in this case, there is no stretch or enhancement, but rather a drastic compression of the contrast as it is reduced to two levels.

### 6.5.2 Color Composites

As seen in Chapter 5, numerous color composite images can be made and displayed from remotely sensed multispectral datasets. The process is based on assigning each of the primary colors (RGB) to three spectral bands. The process of contrast stretch is the same as explained before, but in this case the three columns of the

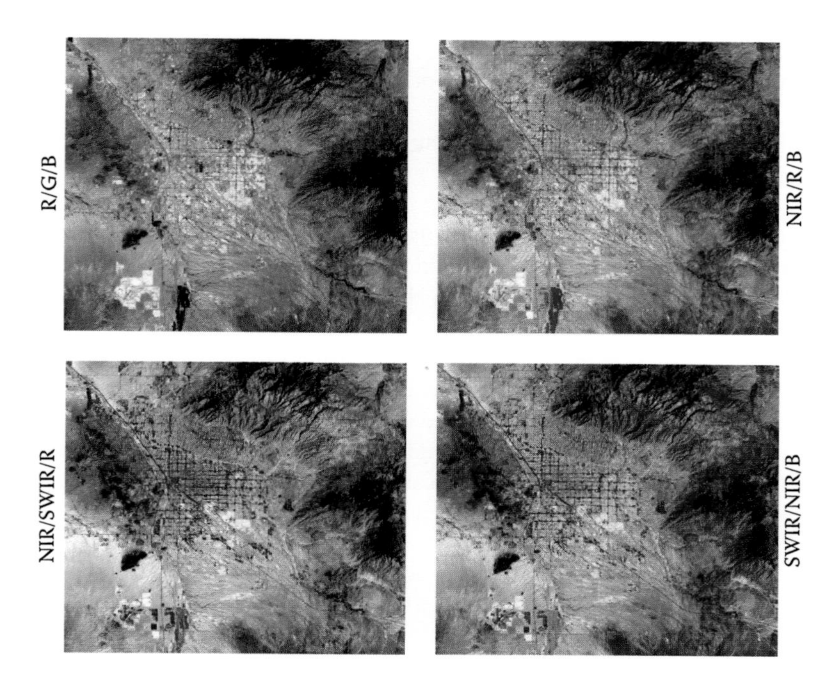


**FIGURE 6.19** Different color composites for the Tucson OLI image (order of band assignment is always RGB).

LUT (one for each primary color) will have different values and therefore the actual stretch parameters may be different in each band.

The selection of bands in a color composite image and the order of assigned colors for each band will vary with the sensor used and the type of application. As previously discussed, the most common color composite is called a *false color* or *infrared color composite*, in which the red, green, and blue primary colors are assigned to the near-infrared, red, and green bands, respectively. This type of color composite is widely used in a diverse range of applications and facilitates the mapping of vegetation, water bodies, urban areas, etc. (Figure 6.19).

When more spectral bands are available, and particularly in the case of hyperspectral sensors, the possibilities of creating color composite images are almost unlimited. Logically, only some will have interest for a particular application, and some of them will be almost identical. Besides the subjective visual perception, quantitative criteria to select the optimal band combination are available as well. Statistical and chromatic indexes, which retain the original variance of the data, have been developed. This is the case of the optimum index factor (*OIF*) that evaluates the variances of each band and their inter-band correlations with the aim of maximizing the information content of the three bands selected for compositing:

$$OIF = \frac{\sum_{k=1,3} s_k}{\sum_{j=1,3} \| r_j \|} \tag{6.12}$$

where

$s_k$ is the standard deviation of each of the three bands included in the color composite

$r_j$ is the correlation coefficient between each pair of the selected bands (Jensen 1996, pp. 97–100)

Other methods of selecting the input bands for the color composite image consider the maximum variation ellipsoid (Sheffield 1985) and the divergence value (Mather 1998). These techniques generally show a similar tendency; for example, in the case of Landsat TM images, nearly all of the various indexes will generally select one of the visible bands, another in the NIR band, and a third one in the SWIR, thus determining the main axes of spectral variation in a TM image (excluding the thermal).

Figure 6.19 includes a set of color band combinations for the Tucson OLI image, showing the impact on visual interpretation of using different sets of bands. The upper left image is a natural color composition, which shows little contrast among the different land covers, particularly green vegetation from other low-reflectance surfaces, such as dark soils. Contrast between irrigated crops and other vegetated areas is particularly clear in those color compositions using the NIR, while those using SWIR provide a better characterization of nonphotosynthetic vegetation and bare soils.

### 6.5.3 Pseudocolor

Several authors have shown that the human eye is more capable of discriminating color tones than brightness intensities. Therefore, the use of color can help in the interpretation of images, even when only a single band is available. In this case, we are not referring strictly to color, since three bands are required, but rather to pseudocolor. In a three-band color composite, we used different *BV* values for each of the three color guns (RGB). When a single image is displayed, the RGB values are the same, resulting in a gray-scale image. However, we can also display a single image in color if we create an LUT where a single *DL* is represented in different values of RGB. This is often used in two instances: (1) when a color scheme is needed for a classified image and (2) to enhance a single-band image, by substituting the gray levels with color tones. Pseudocolor is also used when a color image of 24 bits is compressed to 8, generating a color palette that captures the chromatic variation of the original image.

As an example, Table 6.3 shows an LUT designed for a classified image. The *DL*s represent various thematic classes (e.g., categories representing bare soil, vegetation, etc.) that will be displayed visually with different colors. $DL = 1$ (e.g., indicating urban zones) will appear in red, since it is represented by the highest intensity of red (scale from 0 to 255) and a minimum in green and blue. In the same way $DL = 2$, which could represent vegetation, will be seen as green, with a maximum value in this color and a minimum in the other two. Different colors will also be achieved with different intensity levels of red, green, and blue. For example, the color ochre is obtained from mixing maximum values of red and medium levels of green and blue. The available color palette will allow combinations of 256 levels of RGB, or

**TABLE 6.3**
**Example of a Lookup Table for a Classified Image**

| *DL* | *BV* Red | *BV* Green | *BV* Blue | Resulting Color |
|---|---|---|---|---|
| 0 | 0 | 0 | 0 | Black |
| 1 | 255 | 0 | 0 | Red |
| 2 | 0 | 255 | 0 | Green |
| 3 | 0 | 0 | 255 | Blue |
| 4 | 0 | 255 | 255 | Cyan |
| 5 | 255 | 255 | 0 | Yellow |
| 6 | 255 | 127 | 0 | Ochre |
| 7 | 127 | 127 | 127 | Gray |
| 8 | 255 | 0 | 127 | Pink |

16.8 million different tones ($256^3$). Obviously not all of the colors can be visually distinguished, and a small selection of colors (between 10 and 20) is normally enough to create an adequate color key for thematic classifications.

Pseudocolor can also be used to better visualize one of the original bands, enhancing its characteristics. Pseudocolor can be applied to a group of *DL*s in the scene, designing an LUT, for example, by combining numbers between 0 and 255 for each of the colors. Now, this will complicate our interpretation, because there is no scale of reference for the analyst. For this reason, it is more common applying a progressive color transformation, where the LUT is gradational: for example, blue tones for lower *DL* to red for the highest. Applications of this visualization technique would be the display of temperature variations or vegetation index images representing variations in vegetation amount or status. The thermal image can be represented with a blue-red gradient (Figure 6.20), while the vegetation image may be displayed by an ocher-green gradation scale. In these cases, instead of applying an LUT to the entire range of *DL*s (values), the range can be divided into a series of discrete intervals (*density slicing*), in which each of the resulting groups corresponds to a class of relatively similar radiance values. The thresholds that define each interval can be established arbitrarily or by statistical criteria. For instance, intervals of constant width or selecting quartiles are commonly selected.

### 6.5.4 Filters

#### 6.5.4.1 Digital Filters

Digital filters, similar to optical filters, are used to select certain features and eliminate or reduce others. When using a blue filter in a digital camera, for instance, the photographer tries to reject blue light from the incoming solar energy. In IPS, filters are used to select certain spatial features of the image, eliminating others. Therefore, using filtering techniques, the image spatial contrast can be smoothed or sharpened, so the original *DL*s become either more similar or more contrasting with their surrounding pixels.

FIGURE 6.20 Color coding of the brightness temperature for the Tucson OLI image (band 10). The reddish tones are the warmest temperatures.

As shown in Figure 6.21, when we draw a radiometric transect over an area of interest, numerous peaks and valleys are observed as a result of changes in radiometric characteristics of the different areas crossing the transect. Those spatial variations can be considered as the sum of two components: low-frequency information, which measures the global spatial variation, and high-frequency variations, which indicate local contrasts. The purpose of digital filters is to remove one or the other. They are referred to as *low-pass filters* when they select the global variations (smooth trends) and *high-pass filters* when they emphasize the local variations (sharp changes). Digital filters can be applied in the spatial domain, by considering values

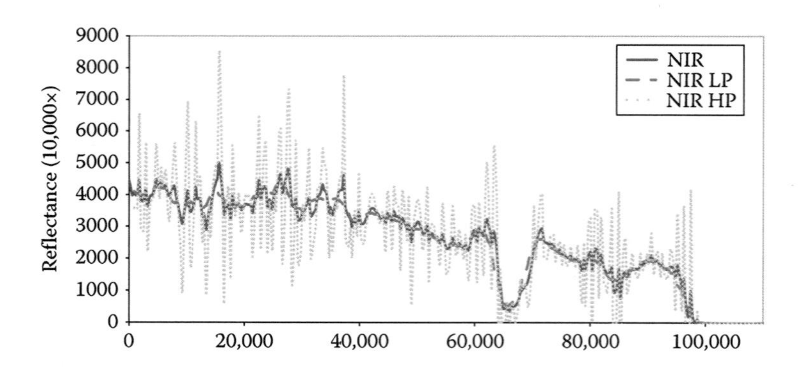


FIGURE 6.21 Radiometric profile over the NIR band (solid red), composed of low (green) and high frequencies (orange) of spatial variation.

of neighboring pixels, or in the frequency domain, by selecting specific components of the image frequency distribution.

Unlike contrast enhancements, spatial filtering modifies the original *DL*s, not only the *BV* stored in the LUT, although the process is commonly done only in the computer memory and the original *DL*s are preserved on the hard drive. In any case, the use of filtered images is not advisable when the final goal is to obtain physical variables or cover classifications, because the physical meaning of *DL*s will be lost after applying spatial filtering operations. However, when the objective of the user is obtaining a visually enhanced product, such as for generating image web servers, filters can be applied to improve the spatial features of the images.

Spatial domain filters are simpler, conceptually and operationally, while frequency domain filters involve more complex mathematical operations, such as Fourier analysis. Fourier transformation converts an image into a 2D function, with two frequency components (horizontal and vertical), assuming that the variation of the *DL* in the image can be approximated by a continuous complex function. Once the transformation is applied, the Fourier spectra are obtained, which define the image by their components of phase and amplitude. Filters and corrections can be applied on the Fourier spectra, especially when dealing with systematic noise (e.g., stripping effects), since they will appear in a fixed direction of the frequency (Richards 1993; Schowengerdt 2007). Once the operation is performed, it can be returned to the image domain by applying an inverse Fourier transformation.

Spatial filters can be grouped in two categories: low-pass filters, which enhance the homogeneity component of the image, thereby reducing spatial contrast, and high-pass filters, which select local contrast and therefore improve the visualization of highly variable areas. In both cases, the mathematical procedure to achieve the filtering effect is the same. A moving window with filter coefficients (*FC*s) is used to compute the filtered values for each pixel (Figure 6.22). This filter matrix (kernel) may have variable sizes, with 3 × 3 to 7 × 7 pixels being the most common. The larger

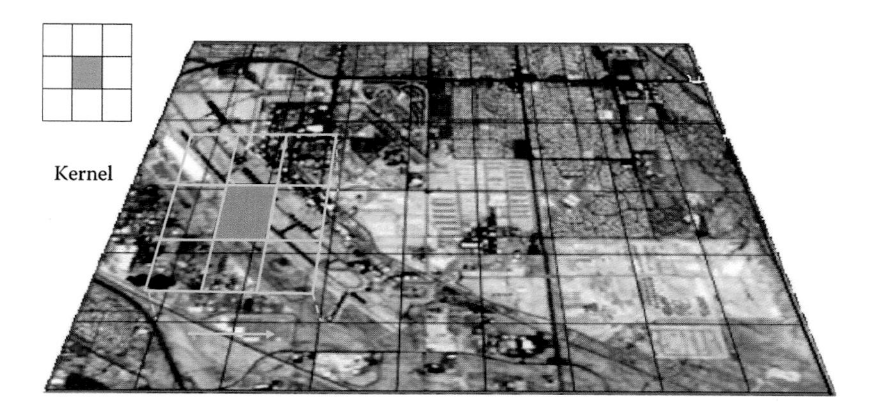


**FIGURE 6.22** Filtering through the use of a spatial moving window (kernel) that is used to define the neighboring pixels.

the kernel size, the more intense the smoothing or sharpening effect will be, since more distant neighboring pixels will be included in the computation.

In summary, the mathematical operations of a digital filter are based on relating each pixel with its neighbors through a user-specified *FC* matrix. In the case of a 3 × 3 matrix, the equation to obtain the filtered *DL* is (Mather 1998)

$$DL'_{i,j} = \frac{\sum_{p=-1,1} \sum_{q=-1,1} DL_{i+p,j+q} FC_{r+p,c+q}}{\sum_{p=-1,1} \sum_{q=-1,1} FC_{r+p,c+q}} \tag{6.13}$$

where

$DL_{i,j}$ is the original *DL* of pixel *i,j*
$DL'_{i,j}$ is the output *DL*
*FC* is the filter coefficient
*r* and *c* are the central row and column of the filtering matrix

The final output result is rounded to the nearest whole number. The results of a digital filter depend on the *FC* matrix designed by the user. If the *FC* tends to weigh the central pixel, the output *DL*s will be more diverse than the original image (high-pass filter). When the *FC* values the border pixels are higher than or equal to those of the central one, then the spatial variation of *DL*s will decrease (low-pass filter).

The process involved in filtering can be illustrated with the following example. Let us apply a 3 × 3 *FC* filter matrix to a 5 × 5 image using Equation 6.13. Observing the results (Figure 6.23), we first note that the first and last row and column are not filtered since they do not have the eight neighboring pixels that are needed to compute the operation. The operator can either repeat the original values or equal them to zero. This is common in both low- and high-pass filters. Obviously, in the case of a larger image (e.g., 3000 × 3000 pixels), the missing pixels will not be a relevant problem. The other values of the filtered image are computed from the original values and the *FC* of the filter matrix. In this example, the output *DL* of the center pixel is calculated by a weighted average of that pixel plus the eight neighboring pixels. The center pixel is weighted by a factor of two.

Original image

| | | | | | |
|---|---|---|---|---|---|
| 12 | 14 | 17 | 24 | 32 | 34 |
| 10 | 18 | 21 | 35 | 38 | 40 |
| 25 | 15 | 17 | 27 | 40 | 43 |
| 18 | 16 | 18 | 24 | 29 | 39 |
| 14 | 16 | 20 | 20 | 27 | 36 |

Filter matrix

| | | |
|---|---|---|
| 1 | 1 | 1 |
| 1 | 2 | 1 |
| 1 | 1 | 1 |

Filtered image

| | | | | | |
|---|---|---|---|---|---|
| 0 | 0 | 0 | 0 | 0 | 0 |
| 0 | 17 | 21 | 29 | 35 | 0 |
| 0 | 17 | 21 | 28 | 35 | 0 |
| 0 | 18 | 19 | 25 | 31 | 0 |
| 0 | 0 | 0 | 0 | 0 | 0 |

**FIGURE 6.23** Example of digital filtering.

For example, the pixel in row 3, column 3, with an initial value (*DL*) of 17 will have a *DL* in the filtered image of

$$DL'_{3,3} = \frac{18+21+35+15+(2\times17)+27+16+18+24}{10} = 20.8$$

which is rounded to the closest integer number, 21. The global contrast of the filtered image is appreciably reduced, from the original standard deviation (8.91) to that of the filtered image (6.77), resulting in a greater spatial homogeneity. A similar process is used in any other type of filtering.

#### 6.5.4.2 Low-Pass Filter

A low-pass filter reduces or smoothes the spatial contrast in an image by minimizing the spatial variability in *DL*s inside the filter window, thus adjusting the *DL* of the center pixel to be more similar to its neighboring pixels. Visually, the filtered image is more diffuse (blurred), since border transitions are smoothed by the filter operations. Figure 6.24 compares the results of applying two low-pass filters the Tucson OLI image. We observe that the filtered images are fainter, which can be seen in the airport or main highways. This spatial averaging effect becomes more evident at larger *FC* windows.

Low-pass filters can be used to clean up (restore) errors that can occur in the *DL*s of the image due to problems in image acquisition or data reception. Low-pass filters are also used to reduce the spatial variability over some land cover classes, as a prior step to classification. The goal is to attenuate "image noise" produced in highly heterogeneous areas in relation to the spatial resolution of the sensor. The most studied cases are from urban zones, where a multitude of different cover types are present in relatively small spaces (at local scale). To diminish this spatial variability, low-pass filters are used to smooth and generalize the urban class into a more homogeneous one. The effect of these filters has been very efficient for digital classification (Cushnie and Atkinson 1985; Haack et al. 1987). The low-pass filter can also be used to obtain the "background signal" when we want to compare a *DL* with its context in order to detect a specific feature of interest that is anomalous to the area, such as active fires (Flasse and Ceccato 1996) or clouds (Saunders and Kriebel 1988) or to

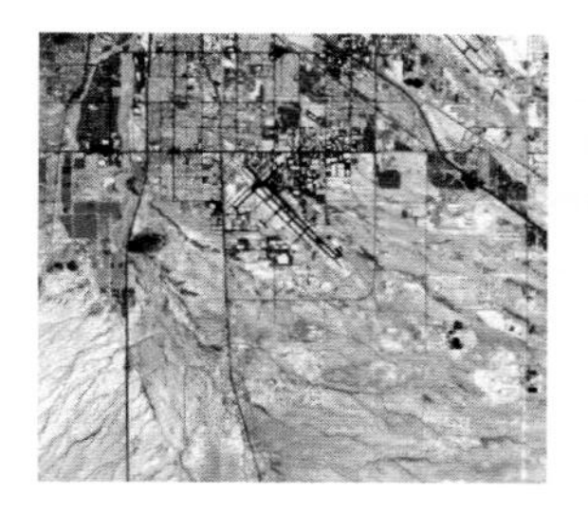
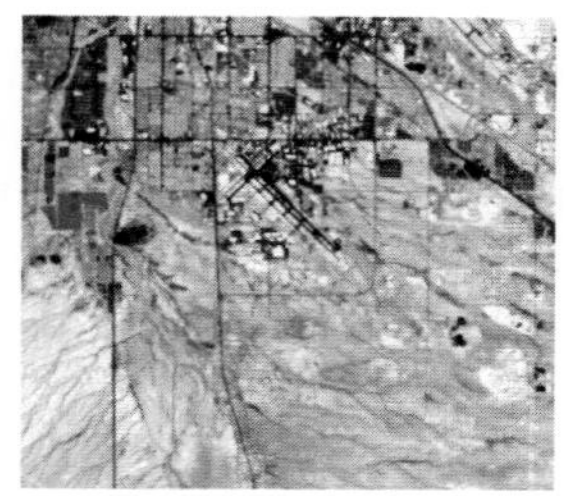
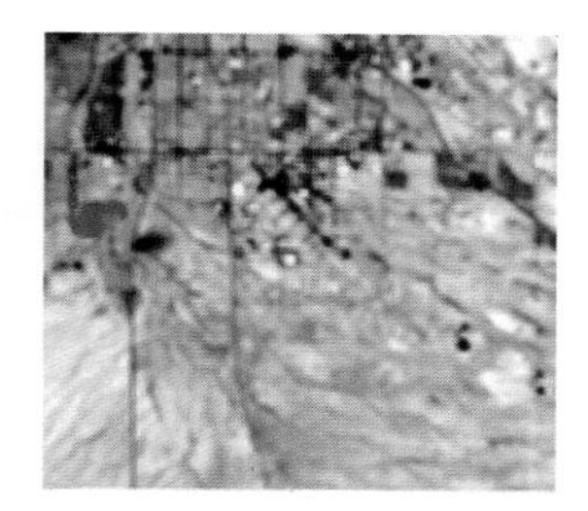

**FIGURE 6.24** Low-pass filters of a window of the Tucson OLI image. From left to right: original image, 3 × 3 filtered image, and 9 × 9 filtered image.

obtain the characteristic reflectance of an area in order to correct for the adjacency effect (Richter 1997).

There is a wide range of *FC* matrices for low-pass filtering. Some of the most common are as follows:

(1)

| | | |
|---|---|---|
| 1.00 | 1.00 | 1.00 |
| 1.00 | 1.00 | 1.00 |
| 1.00 | 1.00 | 1.00 |

; (2)

| | | |
|---|---|---|
| 1.00 | 1.00 | 1.00 |
| 1.00 | 2.00 | 1.00 |
| 1.00 | 1.00 | 1.00 |

; (3)

| | | |
|---|---|---|
| 0.25 | 0.50 | 0.25 |
| 0.50 | 1.00 | 0.50 |
| 0.25 | 0.50 | 0.25 |

The first one is a simple average of nine pixels within the kernel, while the other two filters include a weight for the central pixel to better preserve original features.

Low-pass filtering can also be achieved using the median value instead of the average. The median filter substitutes the center pixel value by the median value of the corresponding surrounding pixels. Since the median is less sensitive to extreme values (Shaw and Wheeler 1985), this filter results in an output image more similar to the original radiometry, and it has been recommended for the elimination of anomalous values caused by noise in data acquisition. Median filters are very common for removing the speckle in radar images (Blom and Daily 1982).

#### 6.5.4.3 High-Pass Filter

The goal of a high-pass filter is to preserve the local contrast of the image by keeping the high-frequency variation. These filters enhance the spatial contrast among neighboring pixels, and therefore they are very useful to emphasize linear changes, such as roads, agricultural fields, and geologic faults (Jong 1993).

The simplest method to obtain a high-pass filter is the subtraction between the original image and a low-pass-filtered image. In summary,

$$DL'_{i,j,k} = DL_{i,j} + (DL_{i,j} - DL_{L,i,j}) \tag{6.14}$$

where

$DL'$ corresponds to the filtered output $DL$ value
$DL$ is the original pixel value
$DL_L$ is the output value result of a low-pass filter

The application of this filter to the Tucson image can be observed in Figure 6.25 for two different filter matrix sizes. The process implies an enhancement of the linear features of the image, such as highways, railroads, the airport lines, and urban networks.

A high-pass filter can also be based on *FC*, similar to the low-pass filters. However, in this case, *FC* tries to reinforce the contrast between the center and neighboring pixels. Two commonly used high-pass *FC* matrices are as follows:

(4)

| | | |
|---|---|---|
| −1 | −1 | −1 |
| −1 | 9 | −1 |
| −1 | −1 | −1 |

; (5)

| | | |
|---|---|---|
| 0 | −1 | 0 |
| −1 | 4 | −1 |
| 0 | −1 | 0 |

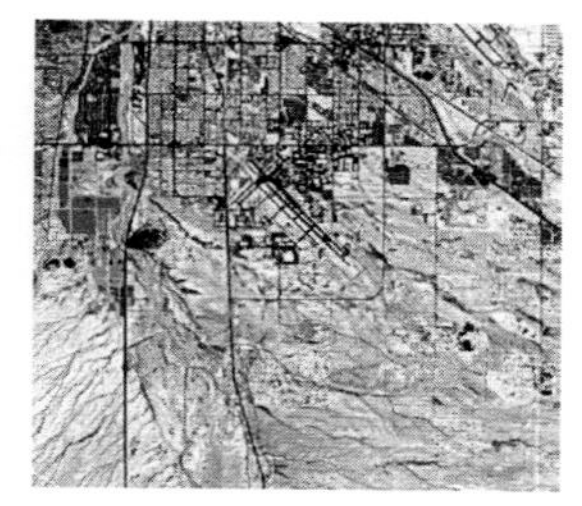

**FIGURE 6.25** High-pass filters on a window of the Tucson OLI image. From left to right: original image, 3 × 3 filtered image, and 9 × 9 filtered image.

Both are derived from gradient analysis. The second filter, known as Laplacian, is recommended to enhance linear features in urban areas or geological lineaments.

To further enhance the border pixels, several nonlinear filters can also be applied. One of the most common is known as Sobel (Gonzalez and Wintz 1977) and emphasizes the variations between columns and lines in the filtering matrix. In a 3 × 3 window, the *DL* value of the center pixel is calculated as follows:

$$DL'_{i,j} = \sqrt{C^2 + L^2} \tag{6.15}$$

where

$$C = (DL_{i-1,j+1} + 2DL_{i,j+1} + DL_{i+1,j+1}) - (DL_{i-1,j-1} + 2DL_{i,j-1} + DL_{i+1,j-1})$$

and

$$L = (DL_{i-1,j-1} + 2DL_{i-1,j} + DL_{i-1,j+1}) - (DL_{i+1,j-1} + 2DL_{i+1,j} + DL_{i+1,j+1})$$

In addition to the filters already discussed, a simple way to enhance the borders of an image is to make a copy of the image, shift it by one row and one column, and then subtract it from the original. The resulting pixels with values close to zero will indicate low variation, while those pixels exhibiting greater differences are indicative of greater contrast between neighboring pixels. In the same way, a directional contrast can be defined toward higher or lower values depending on whether the difference is negative or positive. Other methods have been developed for edge enhancement and specific applications, such as lineament detection, enhancement in water plumes, or delimiting of city streets (Gong and Howarth 1990).

The filter matrices discussed thus far allow for the enhancement of adjacent features, independent of its spatial orientation. The interpreter can also design directional filters that enhance only those linear features in a specific direction. The filter

**FIGURE 6.26** Directional filters applied to a window of the Tucson OLI image. From left to right: original, north, and east.

coefficients need only be appropriately distributed in the matrix. Some examples of these directional filters are as follows:

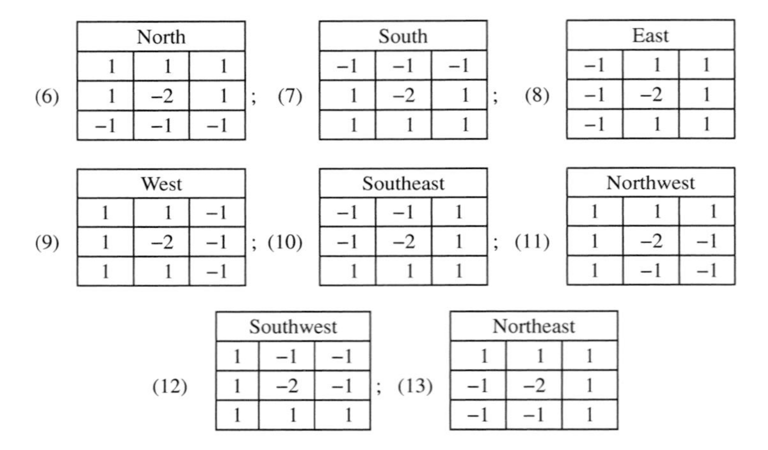

(6) North;

| North | | |
|---|---|---|
| 1 | 1 | 1 |
| 1 | −2 | 1 |
| −1 | −1 | −1 |

(7) South;

| South | | |
|---|---|---|
| −1 | −1 | −1 |
| 1 | −2 | 1 |
| 1 | 1 | 1 |

(8) East

| East | | |
|---|---|---|
| −1 | 1 | 1 |
| −1 | −2 | 1 |
| −1 | 1 | 1 |

(9) West;

| West | | |
|---|---|---|
| 1 | 1 | −1 |
| 1 | −2 | −1 |
| 1 | 1 | −1 |

(10) Southeast;

| Southeast | | |
|---|---|---|
| −1 | −1 | 1 |
| −1 | −2 | 1 |
| 1 | 1 | 1 |

(11) Northwest

| Northwest | | |
|---|---|---|
| 1 | 1 | 1 |
| 1 | −2 | −1 |
| 1 | −1 | −1 |

(12) Southwest;

| Southwest | | |
|---|---|---|
| 1 | −1 | −1 |
| 1 | −2 | −1 |
| 1 | 1 | 1 |

(13) Northeast

| Northeast | | |
|---|---|---|
| 1 | 1 | 1 |
| −1 | −2 | 1 |
| −1 | −1 | 1 |

The name assigned to these matrices follows the direction they indicate and not the direction they enhance, which is their perpendicular. For example, the north matrix emphasizes the northern front of lines oriented east–west, while the east enhances the eastern border of the N–S lines. Figure 6.26 shows the results obtained from these directional filters applied to the Tucson image. The differences from the original image and between the two directions are evident, particularly the contrast between the N and E filters, which enhances different streets and highways that follow the main direction of the edge enhancement.

## 6.6 GEOMETRIC CORRECTIONS

### 6.6.1 Sources of Errors in Satellite Acquisitions

The process of acquiring satellite images involves potential errors in both the radiance detected and the position of the observed pixels. We will cover in this section the geometric corrections and reserve the next one for radiometric transformations.

A satellite image cannot be directly overlaid on a cartographic projection, but rather includes several problems that prevent the direct georeferencing of the image pixels. The most frequent geometrical factors causing georeferencing problems can be grouped into three categories:

1. *Errors caused by the sensor platform.* Even though a satellite orbit is more stable than an airplane track, several geometric problems can occur as a result of small variations in orbit altitude, velocity, and orientation along the three axes (roll, pitch, and yaw). These variations will cause changes in the scale or the geometric position of observed areas (Figure 6.27). They are sporadic and are hard to predict and model (Welch and Usery 1984).
2. *Distortions produced by the Earth's rotation.* The effects of the Earth's movement are quite evident in satellite images when they have a large field of view (GOES or NOAA, for instance), but they also need to be considered even in more local observation sensors. For example, a Landsat image covering 185 × 185 km is acquired in around 30 s. During this period, the Earth moves by ~8 km from the first to the last line of the image at middle latitudes. This displacement, combined with the inclination of the orbit,

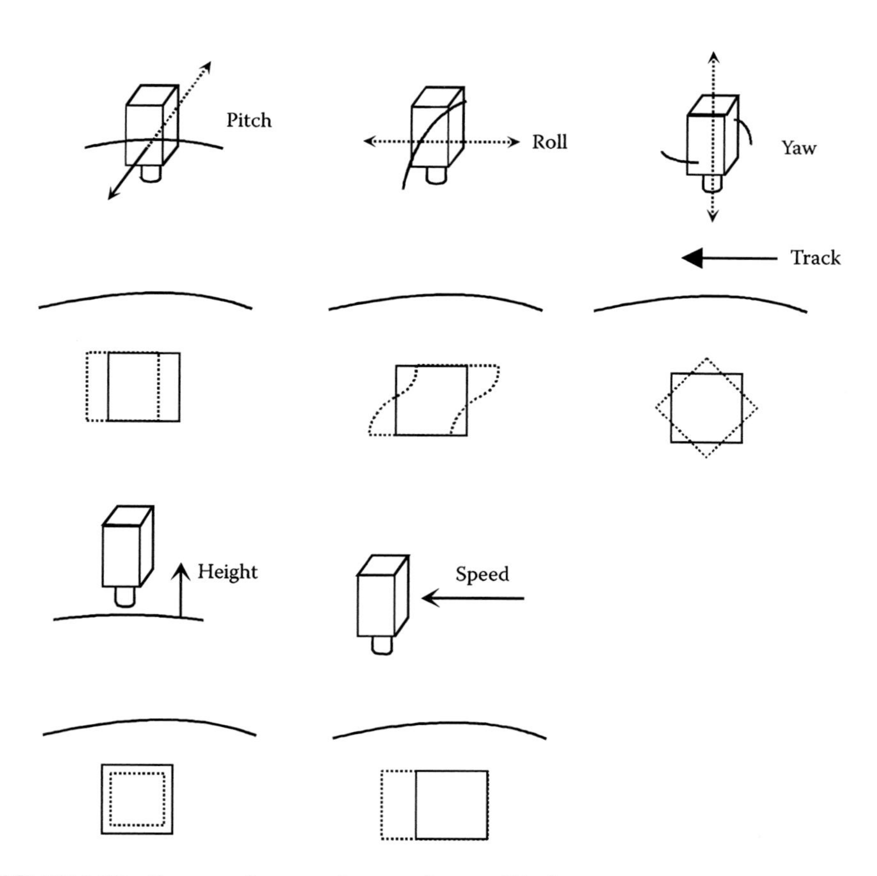


**FIGURE 6.27** Sources of geometric errors in a satellite image.

results in a NE–SW orientation of the image. At the same time, there are also variations of the pixel size due to the Earth's curvature.

3. *Problems caused by the sensor operation.* Several radiometric and geometric distortions can be attributed to sensor performance. For instance, across-track scanners require a very precise synchronization between the orbit and scan speed. Otherwise, lines may be duplicated or missed. As is shown in Chapter 3, an error in the compensating mechanism of the Landsat 7 ETM+ sensor is responsible for the presence of image gaps, areas that are not observed, which increase from the center to the image edges. A much more common problem is the geometric deformation caused by the sensor zenith angle. For those sensors with a wide FOV (e.g., AVHRR, Vegetation, or MODIS), the observation angle may change in daily image series, from nadir to very oblique observations, causing a significant change in the effective ground resolution. In AVHRR images, for instance, this panoramic distortion implies that the pixel size changes from 1.1 km at nadir to up $2.4 \times 6.5$ km at the edge of the scan (Goward et al. 1991). Variation of sensor zenith angle also implies different atmospheric effects across the image, since the optical thickness varies from nadir to the borders. In sensors with a smaller field of view, such as Landsat TM or SPOT-HRV, these effects are less critical.

Some of these problems are handled at the receiving stations, while others are corrected through image software programs. Other factors important to the retrieval of physical parameters, the generation of spectral indices, and addressing multitemporal studies need to be handled by the image analyst.

Geometric corrections include any change in the position of the image pixels. They imply generating a set of equations that relate image and reference coordinates, the latter being the cartographic projection to which we aim to locate image pixels. The general expression for these functions is

$$f(c') = f(x, y)$$
$$f(l') = f(x, y)$$

The column and line coordinates ($c'$ and $l'$) of the corrected image are a function of the coordinates, column and line, of the reference cartographic system to which the image ($x$, $y$) needs to be overlaid. Reference coordinates can also be another image from the same area, and therefore the same approach can be used to overlay two or more images for multitemporal change detection.

The georeferenced image can later be combined with ancillary information or integrated into a GIS (Chapter 9). In the second case, the geometric transformation of the image is aimed at multitemporal studies (Section 7.3) or to create mosaics to cover greater extensions of territory than a single scene.

The geometric correction of images can be approached in two ways. The first consists of mathematical modeling of known geometric errors and applying inverse transformations. To do so, it is crucial to have access to highly precise information on the platform's orbital characteristics and sensor specifications. These orbital corrections can solve systematic errors, such as those derived from the Earth's rotation or curvature and the inclination of the satellite orbit. The second approach is more empirical. It attempts to model the geometric error of the image from a series of points with known coordinates. In this case, the error is modeled inductively, since the transformation functions include, simultaneously, all the sources of error, assuming that these points are sufficiently representative of the geometric deformation of the image.

Both methods have their advantages and disadvantages. The first one is more automatic, since it does not require human input. It is very adequate when the sensor resolution is too coarse to find ground control points (GCPs), or in areas of difficult location (marine or desert areas, high cloud cover). This is the main procedure used to correct meteorological satellite data. Its main disadvantage is the poor precision when the telemetry of the sensor is not very exact or when the image includes random errors, such as unexpected platform movements or local relief, which can induce displacements of ±5 pixels (Ho and Asem 1986). The most recent satellites include refined navigations systems that offer greater accuracy from orbital corrections and are becoming the standard process. Most space agencies and image distributors offer a georeferenced version of the images (in the case of commercial satellites, at a higher cost), with usually high accuracy.

The empirical geometric correction method, based on GCPs, is rather tedious and requires significant human intervention to locate those points accurately, but it offers high precision, and therefore is still used for very detailed products or for correcting historical archives.

### 6.6.2 Georeferencing from Orbital Models

Orbital corrections are particularly adequate for coarse-resolution sensors. The growing interest in the use of global observation sensors has emphasized the need for these corrections and has become a mandatory step in image analysis. Usually while applying orbital corrections, the term *image navigation* is used (Emery et al. 1989) because we try to locate on a geographic grid each of the pixels.

Image navigation involves a complex mathematical process that exceeds the introductory character of this book. A complete reference can be found in more specific texts (Bachmann and Bendix 1992; Bordes et al. 1992; Emery et al. 1989; Ho and Asem 1986; Krasnopolsky and Breaker 1994; Rosborough et al. 1994). We will briefly review the most common orbital corrections.

#### 6.6.2.1 Image Inclination

Polar orbiting satellites frequently have inclined orbits with respect to geographic north. This results in an inclination of the axis of the image to the northeast. In addition, during the time of acquisition of the image, the Earth rotates to the east, with a variable distance depending on the latitude, the angular velocity of the satellite, and

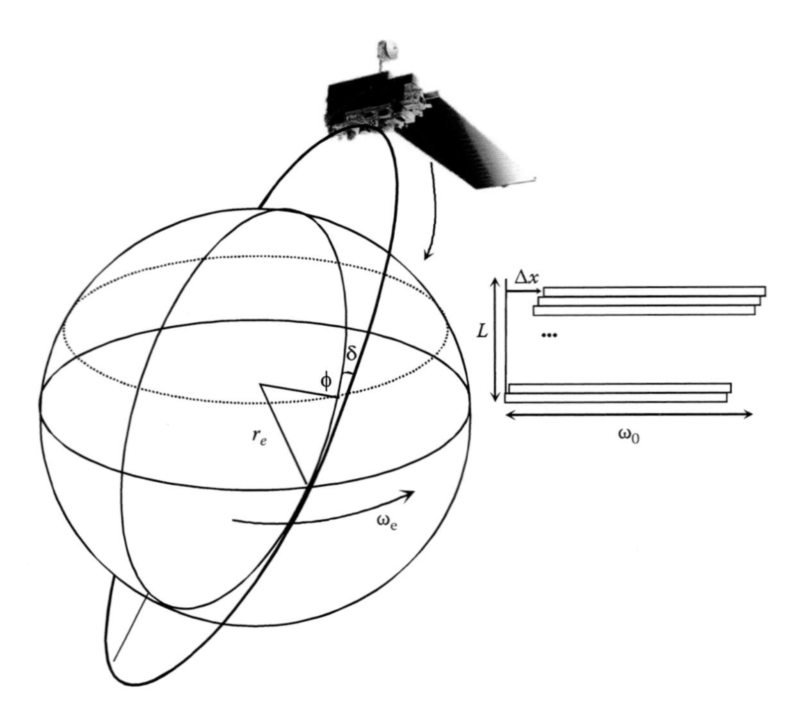


**FIGURE 6.28** Geometric errors due to Earth's rotation. (After Richards, J.A., *Remote Sensing Digital Image Analysis: An Introduction*, Springer-Verlag, Berlin, Germany, 1993.)

the area covered in the image. These problems an be solved with the following equations (Richards 1993, pp. 48–54; Figure 6.28):

$$\Delta x = \Delta x_e \cos \delta \tag{6.16}$$

$$\Delta x_e = \nu_e T_s \tag{6.17}$$

$$\nu_e = \omega_e r_e \cos \varphi \tag{6.18}$$

$$T_s = \frac{L}{r_e \omega_0} \tag{6.19}$$

From the first equation, the displacement of the image on the $x$-axis ($\Delta x$) can be derived to equal the displacement due to the Earth's rotation ($\Delta x_e$) multiplied by the cosine of the inclination angle of the orbit ($\delta$). From Equation 6.17, Earth's displacement ($\Delta x_e$) can be calculated from the angular velocity ($\nu_e$) and the time of acquisition of the image ($T_s$). The angular velocity can be calculated from the rotational speed of the Earth ($\omega_e = 72.72$ µrad s$^{-1}$), the Earth's radius ($r_e \approx 6378$ km), and the cosine of the latitude of the image ($\phi$). The time of acquisition is a function of the area covered ($L$), the Earth's radius, and the angular velocity of the satellite ($\omega_0$). Therefore, only the inclination of the orbit of the satellite, its angular velocity, and the area covered by the images are required to position the image in a northern direction.

#### 6.6.2.2 Panoramic Distortion

Any optical sensor obtains the images from an observation point, which makes the acquisition parameters (mainly the Earth–sensor distance) exact only at nadir. One of the most critical effects of this panoramic observation is the variation of the pixel size from nadir to the edges of the image, since the distance between the sensor and the observed area increases (Figure 6.29). In the case of images with a small field of view (Landsat TM/OLI, SPOT-HRG), this effect is low because images are acquired with a narrow field of view. This is not the case for global sensors, which have a greater scanning angle and, therefore, considerable errors toward the border of the image. A similar effect occurs for off-nadir pointing cameras, which provide oblique observations that are also affected by panoramic distortion.

A simple correction for the panoramic effect is given by the following equation:

$$p' = \frac{p}{\cos^2 \gamma} \tag{6.20}$$

on the $y$-direction of the pixel (latitude), and

$$p'' = \frac{p}{\cos \gamma} \tag{6.21}$$

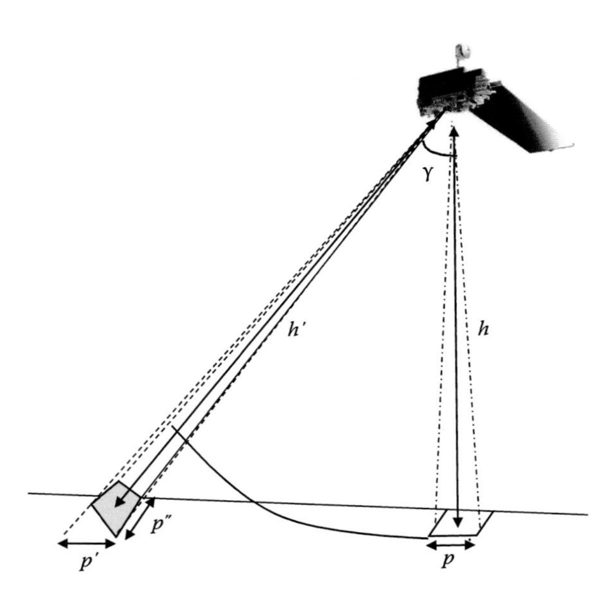


FIGURE 6.29 Effect of panoramic observation on the geometry of the image. (After Richards, J.A., *Remote Sensing Digital Image Analysis: An Introduction*, Springer-Verlag, Berlin, Germany, 1993.)

on the $x$-direction of the pixel (longitude), with $p$ as the size of the pixel at nadir and $\gamma$ the scanning angle (formed by the line of observation and the vertical to the sensor; Figure 6.29). In this scheme, it is assumed that the IFOV is constant. This correction allows the adjustment of the pixel size to the nadir distance, throughout each scanned line.

#### 6.6.2.3 Effect of Earth's Curvature

When the sensor has a wide field of view, the assumption of a flat Earth may create significant geometrical errors. Therefore, it is required to account for the Earth's terrestrial spherical character. By using nonlinear angular relationships, the correction becomes (Richards 1993; Figure 6.30)

$$p_c = \frac{\beta(h + r_e(1 - \cos\psi))}{\cos\gamma\cos(\gamma + \Psi)} \tag{6.22}$$

where

- $p_c$ is the pixel size corrected for the curvature effect
- $\beta$ is the angle of the instantaneous field of view
- $h$ is the height of the sensor
- $r_e$ is the Earth's radius
- $\gamma$ is the scanning angle
- $\psi$ is the angle formed by the center of the observed pixel and Earth's center

From these equations, knowing the geographic position of the satellite, we can navigate the image, generating a grid of coordinates $x$, $y$ with the $DL$ acquired by the

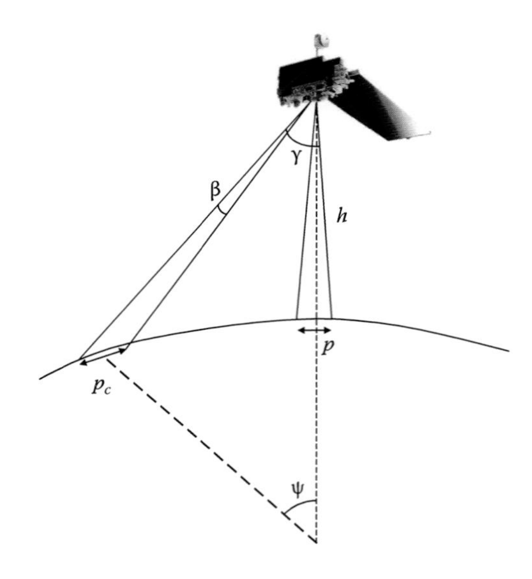


**FIGURE 6.30** Errors generated by terrestrial curvature. (After Richards, J.A., *Remote Sensing Digital Image Analysis: An Introduction*, Springer-Verlag, Berlin, Germany, 1993.)

sensors that are nearest to each position. This process of interpolation is relatively similar to the one later discussed for geometric correction with control points. The whole process requires a large number of calculations because it deals with complex equations. On occasions, orbital models are enhanced by locating a small number of GCPs that facilitate a better modeling of the satellite orientation (attitude).

### 6.6.3 Georeferencing from Control Points

Nowadays, orbital models are becoming the standard process for image georeferencing, as satellite position and attitude are now very precisely estimated. However, empirical methods based on GCPs are still much extended. In this approach, the interpreter assumes that the origin of the errors is unknown, but they can be modeled by adjusting some empirical equations to a set of points. The image coordinates to be corrected, as well as the map or image coordinates used as reference, are known. In both cases, the correction is performed in three phases (Figure 6.31): (1) location of common points to the image and the map (or other image of reference), (2) calculation of the transformation functions between the corrected and reference images, and (3) transfer of the original *DL*s to their new position defined by the previous transformation.

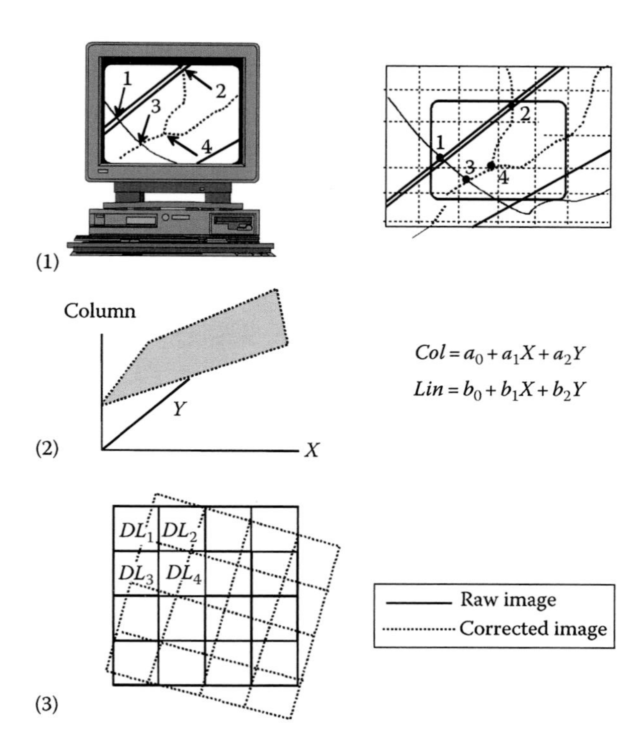


**FIGURE 6.31** Phases in the empirical correction based on ground control points: (1) selection, (2) generation of transformation functions, and (3) interpolation of the *DL* to the new position.

#### 6.6.3.1 Establishing Control Points

To build functions for transforming image and map coordinates, a series of common coordinates to both geometric systems need to be identified. Since the functions are fitted following those sampled coordinates, the quality of the georectification process will strongly depend on the precision and distribution of these common points (commonly named *ground control points* [GCP]). Imprecise location of these points, on the image or the map, or a biased spatial distribution (favoring some areas over others with different errors), will produce an erroneous geometric correction. For this reason, finding GCPs is the most crucial phase of the georeferencing process and the one that demands more human input. Three aspects are essential in the selection of GCP: number, location, and distribution.

The ideal number of control points depends on the size and geometric deformations of the image. In a flat area, acquired by a narrow FOV sensor, the image map transformation can be based on linear equations. In areas with rough terrain, second- or third-degree polynomials may be required. The number of GCP increases in parallel to the degree of the equation. Mathematically, only 3 points are needed to fit a first-degree equation, 6 for a second-order function, and 10 for a third-degree function. However, it is convenient to exceed these minimums to properly model the potential errors. For a small area (say, 1000 × 1000 pixels), 20 or 24 points and a first-order transformation are usually sufficient. If the elevation of the area is very diverse, a greater number of points and a more complex polynomial transformation should be used, although for rough terrain the use of digital elevation models (DEMs) are more adequate.

With respect to the location of the points, it is advisable that they are clearly identifiable in both the source and the reference products. For image-to-map transformations, human-made features that are not subjected to temporal changes, for example, crossroads or railroad tracks, are preferable. The interpreter should try to avoid points along the coastline, because the tide effect can modify the exact location of a geographic feature. The same reasoning can be applied to water bodies (rivers, dams) that can undergo important seasonal changes. An outstanding element of vegetation can be used as a control point as long as the appropriate thematic cartography is available and no major changes have occurred between image acquisition and map inventory. When two images are to be registered, the search for GCP is easier than for maps because there are common visual characteristics between images that are not included in basic cartography, as in the case of vegetation or agricultural patterns. Most of the IPS currently available facilitate the simultaneous visualization of both images to help the identification of common points.

With respect to the GCP distribution, it is convenient to select these points evenly over the image area. This will avoid errors due to the excessive weight of a certain area of the image at the expense of increasing errors in other areas. When correcting several images to a single one, the search for common points can be computer aided, based on a library of reference GCPs. The same points can be automatically searched in the other images of the same area using iterative searching algorithms (Gao et al. 2009).

We will present an example of georeferencing based on GCP for the same area we are using in this chapter. In this case, we have used an uncorrected image acquired by the Landsat TM sensor in 1986. Correction is based on the latest acquisition of the

**FIGURE 6.32** Ground control points selected to correct the Tucson image of 1986. Check points are displayed in blue.

same area by Landsat 8 OLI, which is used as reference image. This facilitates the location of control points and the analysis of the impacts of different resampling techniques. Figure 6.32 shows the location of selected control points for the 1986 image. Two sets of points have been selected: GCP, which are used for calibrating the regression model, and check points, which are only used for validation. This figure shows how GCPs are well distributed on the image and cover most sectors, although they are less represented in the Catalina Mountains (Eastern side). Table 6.4 includes the column and line coordinates of the image, and the $X$ and $Y$ from the map for those points.

### 6.6.3.2 Calculating the Correction Function

As mentioned previously, the digital correction of the geometry of the image is performed by establishing some functions that relate the coordinates of the image to those of the map. In this way, from the $X$, $Y$ coordinates of the map, the new column and row of the image can be estimated. This transformation can be expressed as (Mather 1998)

$$\hat{s} = \sum_{j=0,m} \sum_{k=0,m-j} a_{j,k} X^j Y^k \tag{6.23}$$

where $\hat{s}$ is the estimated coordinate of the corrected image obtained from the $X$ and $Y$ coordinates of the map. The subindex $m$ indicates the order of the polynomial. The simplest case would be a linear equation where

$$\begin{aligned} \hat{c}_i &= a_0 + a_1 X_i + a_2 Y_i \\ \hat{l}_i &= b_0 + b_1 X_i + b_2 Y_i \end{aligned} \tag{6.24}$$

where $\hat{c}_i$ and $\hat{l}_i$ are the estimated column and line, respectively, for a given $X_i$ and $Y_i$ map coordinates.

**TABLE 6.4**
**Coordinates of the Control Points Used to Correct the Tucson Image of 1986**

| GCP | Status | Col. | Lin. | X | Y |
|---|---|---|---|---|---|
| G0001 | Active | 1271 | 1330 | 510,335 | 3,559,693 |
| G0002 | Active | 586 | 717 | 494,073 | 3,580,203 |
| G0003 | Active | 689 | 672 | 497,221 | 3,580,986 |
| G0004 | Active | 909 | 499 | 504,243 | 3,584,768 |
| G0005 | Active | 995 | 409 | 507,061 | 3,586,820 |
| G0006 | Active | 1474 | 743 | 518,879 | 3,575,159 |
| G0008 | Active | 1625 | 1082 | 521,451 | 3,564,915 |
| G0009 | Active | 1644 | 1194 | 521,451 | 3,561,693 |
| G0023 | Active | 737 | 947 | 497,235 | 3,573,030 |
| G0011 | Active | 1703 | 1535 | 521,467 | 3,551,828 |
| G0012 | Active | 1726 | 1700 | 521,294 | 3547106 |
| G0013 | Active | 1853 | 1645 | 525,118 | 3,548,012 |
| G0014 | Active | 1459 | 1791 | 513,349 | 3,545,804 |
| G0015 | Active | 1290 | 1766 | 508,726 | 3,547,369 |
| G0017 | Active | 1002 | 1765 | 500,683 | 3,548,780 |
| G0018 | Active | 985 | 1537 | 501,300 | 3,555,250 |
| G0019 | Active | 847 | 1561 | 497,322 | 3,555,234 |
| G0020 | Active | 588 | 1434 | 490,701 | 3,560,067 |
| G0021 | Active | 860 | 1157 | 499,655 | 3,566,537 |
| G0010 | Check | 1663 | 1479 | 520,569 | 3,553,625 |
| G0016 | Check | 1200 | 1820 | 505,974 | 3,546,240 |
| G0007 | Check | 1749 | 731 | 526,618 | 3,574,203 |
| G0022 | Check | 861 | 984 | 500,517 | 3,571,385 |

As we can see, equations for geometric correction are simply multiple linear regressions, where $a_0$, $a_1$, $a_2$, $b_0$, $b_1$, and $b_2$ are the regression coefficients, $X_i$ and $Y_i$ the independent variables, and $c_i$, $l_i$ the dependent variables. The conversion can be performed in both directions: from the map coordinates to the image and vice versa.

As in conventional regression, these functions can be defined graphically by one plane that expresses the relation among the coordinates of the independent variables ($X$- and $Y$-axes) and the dependent variables ($c$- or $l$-axis). From a practical point of view, it is worth considering that one linear function is enough to approach a wide set of transformations of the image, such as to change the scale and origin, alter the image skew, modify the relation of the axes, and rotate them (Figure 6.33). These functions are adequate to correct small images as long as there is no contrasting topography. If a more rigorous modification is needed, second- or third-degree functions can be applied, where nonlinear geometric alterations are addressed (in other words, the function is defined not by a plane with linear axes but by a surface with curvilinear axes). For simplicity, we focus on first-order functions, although these comments can also be applied to more complex polynomials.

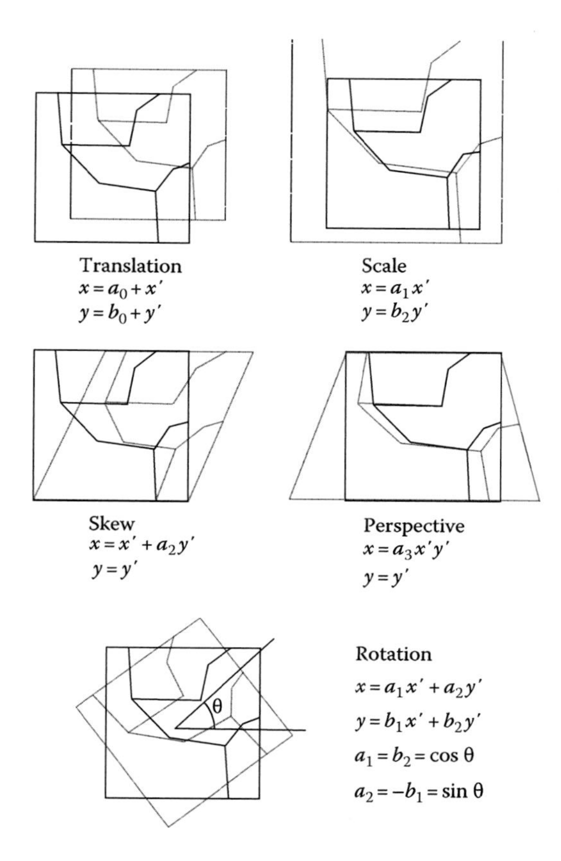


**FIGURE 6.33** Examples of geometric transformation functions with the effect they produce. (After Schowengerdt, R.A., *Techniques for Image Processing and Classification in Remote Sensing*, Academic Press, New York, 1983.)

The coefficients of the transformation functions ($a_0$, $a_1$, $a_2$, etc.) are calculated from the coordinates of the control points previously selected. The most common method to obtain the correction functions is by applying a minimum-square fitting, with identical calculation as those required for a conventional multiple regression (Snedecor and Cochran 1980). The quality of the fitting is measured by the importance of the residuals, which are the differences between the estimated and the observed values, for each of the sample points used in the process; the greater the value, the smaller the adjustment between the independent and dependent variables. In our case, the quality of the geometric correction can be evaluated by comparing, for each of the control points, the estimated coordinates by the regression with the real ones. The average of the residuals is commonly known as *root-mean-squared error* (*RMSE*). The *RMSE* is calculated with the square root of the differences between the observed values and those estimated by the regression for columns and lines. In our case

$$RMSE = \sqrt{\frac{\sum_{i=1,n} (\hat{c}_i - c_i)^2 + (\hat{l}_i - l_i)^2}{n-1}} \tag{6.25}$$

The *RMSE* for each point can also be calculated as the square root of the squared residuals at each point. That value is only the distance between its real coordinates and those estimated by the regression. As a result, we can use the term *longitudinal error* (LE) to refer to the *RMSE* for each particular point:

$$LE = \sqrt{(\hat{c}_i - c_i)^2 + (\hat{l}_i - l_i)^2} \tag{6.26}$$

Using this term, we can clean up a frequent confusion that considers the *RMSE* as applicable only to the set of points (because of the term mean error) and not to each individual point. In addition, it is more appropriate to express the variable we are estimating (position) and the types of errors we made (distance to the observed point).

The general quality of the adjustment is evaluated through the *RMSE* or the LE value for each point. If the mean exceeds a value previously set (usually equal to or less than 1 pixel), it will be necessary to apply one of the following solutions:

1. Verify the coordinates of the selected points, ensuring that the coordinates of the image and the map are accurate. At this point, the user should consider the degree of confidence of the point and how accurate its location is.
2. If the *RMSE* is still high after the coordinates have been modified, it is worth eliminating the point with the highest error, as long as it is one of the points considered as having doubtful reliability or is close to other points included in the equation. This will guarantee a good coverage of that sector in the final adjustment. Eliminating a point in an area poorly represented can create a fictitious reduction of the error that will result in that area of the image to be incorrectly rectified.
3. We can also decide to increment the order of the polynomial if the deformations of the image are severe, for example, in rough terrains. This solution must be approached with caution, especially if the number of control points is low, because an overadjustment can be obtained in these areas, introducing notable errors in less well-represented areas, even when the residual values are very low (Richards 1993, pp. 67–70).

For these and other reasons, to verify an accurate geometric correction, it is necessary to include check points that have not been used in the calculation of the equation and provide an independent estimation of the *RMSE* obtained in the process. This will avoid the bias that an incorrect distribution of points can generate, since the error is evaluated not on the points used to generate the model.

For the correction of the Tucson image, it was decided to select an adjustment equation of first order, with a *RMSE* no greater than 1 pixel (30 m). From the 23 control points previously selected, 19 were used to compute the polynomial model and 4 to check the error (Table 6.5). The equation that was derived provided a mean

**TABLE 6.5**
**Estimated Column and Line Coordinates and Residuals of the Correction of the Tucson Image of 1986**

| GCP | Status | $\hat{c}$ | Residual *c* | $\hat{l}$ | Residual l | RMS |
|---|---|---|---|---|---|---|
| G0001 | Active | 1270.96 | −0.46 | 1329.51 | −0.01 | 0.464 |
| G0002 | Active | 584.85 | 0.75 | 717.41 | −0.05 | 0.747 |
| G0003 | Active | 689.15 | 0.15 | 671.46 | 0.11 | 0.186 |
| G0004 | Active | 909.55 | −0.15 | 498.51 | 0.81 | 0.823 |
| G0005 | Active | 994.83 | −0.33 | 410.62 | −1.42 | 1.455 |
| G0006 | Active | 1473.98 | 0.35 | 743.24 | 0.12 | 0.371 |
| G0008 | Active | 1624.47 | 0.13 | 1082.23 | −0.03 | 0.128 |
| G0009 | Active | 1643.81 | 0.11 | 1193.71 | 0.39 | 0.400 |
| G0023 | Active | 737.34 | −0.03 | 946.60 | 0.51 | 0.511 |
| G0011 | Active | 1703.58 | −0.11 | 1534.89 | 0.50 | 0.510 |
| G0012 | Active | 1725.92 | −0.38 | 1699.30 | 0.24 | 0.452 |
| G0013 | Active | 1852.88 | −0.38 | 1645.02 | −0.52 | 0.646 |
| G0014 | Active | 1458.67 | 0.43 | 1791.96 | −0.86 | 0.964 |
| G0015 | Active | 1289.20 | 1.02 | 1765.50 | 0.55 | 1.163 |
| G0017 | Active | 1002.25 | 0.18 | 1764.89 | −0.05 | 0.185 |
| G0018 | Active | 984.79 | −0.19 | 1537.36 | 0.04 | 0.195 |
| G0019 | Active | 847.14 | −0.09 | 1561.75 | −0.36 | 0.370 |
| G0020 | Active | 588.94 | −0.64 | 1434.24 | −0.14 | 0.652 |
| G0021 | Active | 860.10 | −0.36 | 1156.75 | 0.16 | 0.392 |
| G0010 | Check | 1661.70 | 0.97 | 1478.12 | 0.44 | 1.061 |
| G0016 | Check | 1200.67 | −0.53 | 1821.06 | −0.90 | 1.043 |
| G0007 | Check | 1747.65 | 1.17 | 729.93 | 0.73 | 1.381 |
| G0022 | Check | 860.86 | −0.20 | 983.85 | 0.15 | 0.252 |

residual (*RMSE*) of 0.67 pixels, with 0.42 RMS in the *X*-direction and 0.52 RMS in the *Y*-direction. For the check points, the *RMSE*s were a bit higher: 0.94 and 0.72, respectively, but still acceptable. The equations computed were

$$\hat{c} = 4962.5165 + (0.034622 * X) - 0.00600076 * Y \quad (6.27)$$

$$\hat{l} = 127{,}539.647 - (0.005993 * X) - 0.0345961 * Y \quad (6.28)$$

### 6.6.3.3 Generation of the Georeferenced Image

The functions previously analyzed allow the translation of map to image coordinates. However, if we want to create an image that corresponds satisfactorily with those coordinates, it is necessary to somehow transfer the *DL* of the original image to the new position. The equations obtained from the regression model can calculate the correct position of each pixel, but do not move the pixels to that new position: an additional criterion to transfer the original *DL* values to the new position is required. In other words, correction functions create a new matrix, correctly positioned but empty. The process of "filling" this matrix is the last phase of the geometric correction.

This is a complicated process. Ideally, each pixel of the corrected image corresponds to only one pixel of the original one. This does not usually occur, but often the pixel of the new image is located between several pixels of the original image (Figure 6.34) because this process produces a considerable alteration of the original geometry of the scene. In addition, the size of the pixel can be changed in the corrected image, making it more difficult to find the *DL* that shows best the radiometric value detected by the sensor for that position.

The transfer of *DL* to the desired cartographic coordinates has been approached mainly with three methods: (1) nearest neighbor, (2) bilinear interpolation, and (3) cubic convolution. They can be divided in two groups based on whether they select a *DL* from the original image or they interpolate the output *DL* from the original *DL* values.

The nearest neighbor method is based on locating in each cell of the corrected image the *DL* of the nearest pixel in the original image. This is the fastest method and involves less transformation of the original *DL*s. Its main disadvantage is the distortion introduced in linear features of the image (geological faults, highways, and roads) that may appear in the corrected image as broken lines. Bilinear interpolation obtains the average of the four nearest *DL*s in the original image. This average is weighted depending on the distance of the original pixel to the corrected one: pixels that are closest in the original image have the greatest weight. This method reduces the effect in linear features but tends to blur the spatial contrast of the original image. Finally, cubic convolution considers the *DL*s of the nearest 16 pixels. The visual impact of this correction provides better quality but requires greater computation time and more alteration of original values.

The effect of these three transformation methods are shown in Figure 6.35, which includes a small area of the Tucson image, centered at the airport. The distortion introduced by the nearest neighbor method is evident in the airport tracks and some of the wider streets. As can be observed, the differences between the bilinear interpolation and the cubic convolution are nearly unnoticeable.

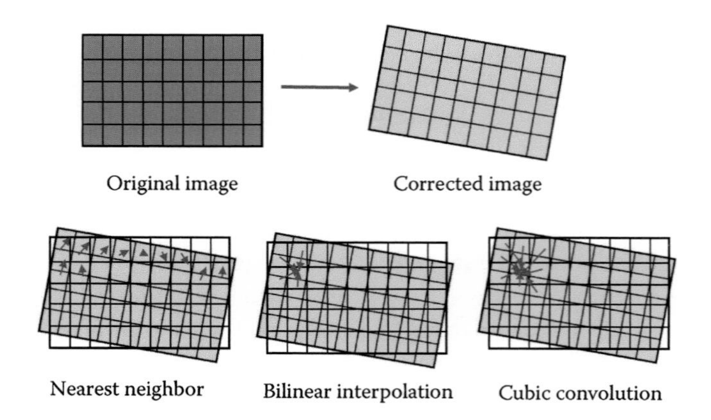


**FIGURE 6.34** Process to transfer the original *DL* to the corrected position.

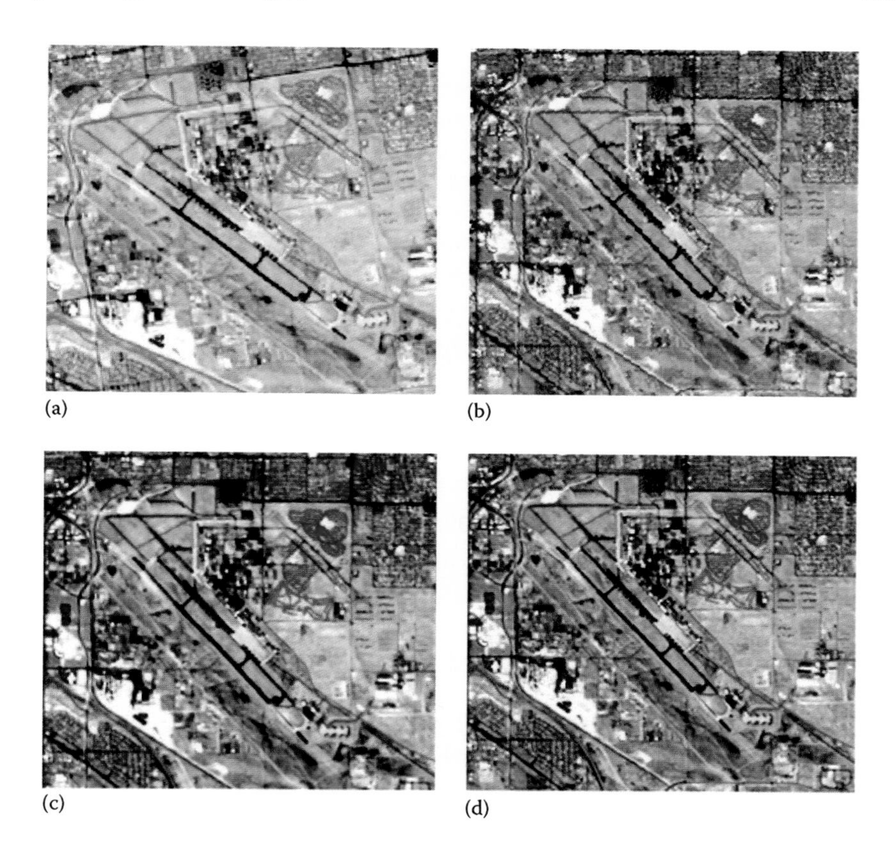


**FIGURE 6.35** Effects of the interpolation criteria in the Tucson image: (a) original, (b) nearest neighbor, (c) bilinear interpolation, and (d) cubic convolution.

The selection of the method depends on the purpose of the process and computer resources. If a classified image is to be corrected, the nearest neighbor method is the only option, because it is the only method that preserves the original values and avoids averages (which would not make sense in a thematic classification). If, however, we are trying to improve the visual analysis, more elaborate interpolation algorithms should be considered, especially cubic convolution if the resources are available.

To better illustrate the operations carried out by these methods of *DL* transformations, we will discuss the processes followed to correct an image via the nearest neighbor–bilinear interpolation methods. The example is based on the Tucson image, for which the adjustment equations have been discussed (Equations 6.27 and 6.28).

First, the interpreter has to define the area where the process is going to be applied. To do so, the $(X, Y)^*$ UTM coordinates of the corners and the pixel size in $X$, $Y$ need to be entered into the computer, or else the coordinates of the corners and

* To avoid confusion, when we refer to the coordinates of a given pixel, we refer to the coordinates at the center point of the pixel.

the size of the image in columns and rows. In the first case, the size of the image is calculated by dividing the difference between the coordinates by the pixel size, and in the second case the size of the pixel is obtained by dividing the difference between coordinates by the number of rows and columns. Usually the first option is used. In our example, the following coordinates were entered for the corners NW (UTM coordinates 481,395, 3,590,535) and SE (UTM coordinates 548,415, 3,531,255), using the WGS84 ellipsoid and datum definition. The output pixel size was selected as 30 × 30 m to match the Landsat 8 image.

With these data, the program operates as follows:

1. It estimates the coordinates, in columns and rows, of the corners of the corrected image, from the adjustment equations previously obtained (Equations 6.27 and 6.28). It usually starts at the NW corner, since it is the origin of the new image, using the functions previously fitted:

$$\hat{c} = 4{,}962.5165 + (0.034622 * 481{,}395) - (0.00600076 * 3{,}590{,}535) = 83.89$$

$$\hat{l} = 127{,}539.647 - (0.005993 * 481{,}395) - (0.0345961 * 3{,}590{,}535) = 435.93$$

2. The column and row coordinates estimated by the equations are used to extract the *DL* that will be situated in the desired position ($x$ = 481,395, $y$ = 3,590,535). The estimated coordinates are not integers, meaning that there are no original pixels perfectly matching these UTM coordinates. For this reason, we need to resample the original *DLs* to find out which is the most suitable to fill the desired coordinates. If we apply the nearest neighbor criterion, the output *DLs* will correspond to the pixel located in column 84 and line 436 of the original image (the closest integer numbers to the values estimated by the equation). If we use bilinear interpolation, the *DL* would be obtained by a weighted average of the four nearest pixels, using the following equation:

$$DL_{\text{out}} = \frac{\sum_{i=1,4} (DL_i / D_i^2)}{\sum_{i=1,4} (1/D_i^2)} \tag{6.29}$$

where $D_i$ is the distance between pixel $i$ and the one to be interpolated. In our case, the *DL* assigned to position (481,395, 3,590,535) would be a weighted average of the corresponding *DLs* 83–435, 83–436, 84–435, and 84–436. The distance $D$ would be calculated from

$$D = \sqrt{\Delta x^2 + \Delta y^2} \tag{6.30}$$

where $\Delta x$ and $\Delta y$ are the differences between the estimated value and the original column and row, respectively. For example, for the pixel located in the coordinates 83–435, $\Delta x = 0.89$ and $\Delta y = 0.93$, then $D = 1.29$. Consequently, that pixel would have less weight in the calculation than the one located at 84–436, which has a distance $D = 0.13$.

3. The rest of the *DL*s of the corrected image are computed in a similar way. Knowing the UTM coordinates of the NW corner and the pixel size, the UTM coordinates of the area to be "filled" with the *DL*s from the original image can be calculated. It is sufficient to add iteratively to the *X* coordinate of the NW corner the pixel size (30 m in this case) to obtain the *X* coordinate of the next pixel. When the program computes the last pixel of the first line, we should keep in mind that *Y* coordinates decrease to the South, and therefore, the *Y* coordinate of the second line is obtained by subtracting the pixel size from the *Y* coordinate of the first line, and so on. Once the UTM coordinates are calculated, they are included in the previous equations, obtaining the estimated column and row for such coordinates. From these estimated values, the *DL*s corresponding to those coordinates can be obtained. For example, the pixel located in column 220 and row 165 of the output image will have the following UTM coordinates:

$$X(220,165) = 481,395 + 220 \times 30 = 487,995$$
$$Y(220,165) = 3,590,535 - 165 \times 30 = 3,585,585$$

From these coordinates we can calculate the position in column and row of the original image, again using Equations 6.27 and 6.28, which gives an estimated value of column and row of 342.10 and 567.63, respectively. Therefore, the *DL* of the pixel 220–165 in the corrected image, with UTM coordinates 487,995 and 3,585,585, will be assigned the *DL* of the pixel located in column 342 and row 568 of the original image (using the nearest neighbor method). This process continues until the SW corner assigned by the user is reached, thus concluding the correction.

The final result of the process is shown in Figure 6.36. It includes the original and the corrected image of the Tucson city area. The geometric transformation

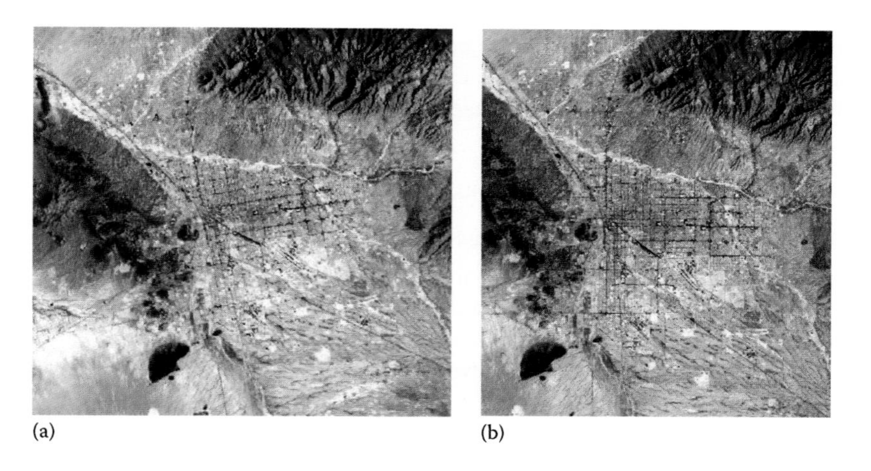


**FIGURE 6.36** Tucson image before (a) and after (b) geometric correction.

applied to the original image becomes evident in the direction of the main city streets and highways.

### 6.6.4 Georeferencing with Digital Elevation Models

Geometric corrections based on GCPs have been the most common procedure for satellite images of medium to high resolution, Landsat or SPOT, and is the one available in most of the commercial image processing programs. The results are in most of the cases sufficiently satisfactory. However, if working in areas of rough terrain or with aerial images where the geometric deformations are closely associated with the relief, it is necessary to incorporate topographic information. To perform a geometric correction with topographic effect, it is necessary to have a DEM of a similar resolution to the image to be corrected. A DEM is very similar in structure to an image, but instead of having radiance values, it includes elevation heights of each pixel. Using DEM for georeferencing implies usually a differential rectification, where each *DL* is transferred separately from the original image to the output image by correcting the position of the pixel based on its displacement due to the relief (Palà and Pons 1995). Each coordinate $X$, $Y$, $Z$ of the DEM is transformed onto the image through colinear equations. This correction can be simplified using a polynomial model where elevations are introduced as a new independent variable

$$\hat{c} = a_0 + a_1X + a_2Y + a_3Z + a_4ZX + a_5ZY \quad (6.31)$$

which offers better results than the basic linear adjustment, or even the second-order polynomial, particularly when the terrain is rough. The inclusion of elevation data are crucial when data are not acquired vertically (as can be the case of SPOT), where errors due to topography are very noticeable.

## 6.7 RADIOMETRIC CORRECTIONS

Within this section, we cover a series of operations intended to improve an image's radiometric properties, which serve to better prepare the data for extracting relevant information, either as biophysical values (reflectance, temperature) using empirical or simulation models or as thematic maps from classification techniques.

### 6.7.1 Solving Missed or Deteriorated Data

#### 6.7.1.1 Restoration of Missing Lines and Pixels

Random problems in sensor performance, communications, or reception systems may cause missing lines or pixels, resulting in images with anomalous lines or pixels highly contrasting with their surroundings. In both cases, the missing information was never acquired and thus is permanently lost, so the correction process is aimed at just enhancing the visual appearance of the image, but they do not provide a biophysical measurement for those missed pixels.

A simple way to estimate the values of missing or corrupt pixels is to rely on neighboring pixels. Variables in close geographic proximity tend to be spatially autocorrelated, and therefore closer areas tend to be more similar with one another than with those further away (Cliff and Ord 1973). The probability that two nearby weather stations have similar precipitation amounts is higher than for two weather stations farther apart. Similarly, radiance values tend to be correlated with distances (Woodcock et al. 1988a,b). Therefore, it seems a logical choice to estimate the *DL* of missing pixels from the neighboring pixels. There are many ways to use these vicinity criteria, of which the most common is the simple substitution of the *DL*s of each line/pixel by the previous values:

$$DL_{i,j} = DL_{i-1,j} \tag{6.32}$$

where

$DL_{i,j}$ corresponds to the digital value of the pixel in line $i$ and column $j$ (the corrupted one)

$DL_{i-1,j}$ indicates the digital value of the pixel in the preceding line. This can also be accomplished with the subsequent lines

A second method involves the substitution of the missing pixels by the average values of the previous and subsequent values:

$$DL_{i,j} = \text{INT}\left\{\frac{(DL_{i-1,j} + DL_{i+1,j})}{2}\right\} \tag{6.33}$$

where INT {} indicates that the resulting value is rounded to the closest whole number. According to Mather (1998), this method is more risky than the first approach, particularly when the assumption of spatial auto-correlation is less clear.

In case the problem occurs only in one band, a third method of recuperating lost information is to use another band that is highly correlated with the problematic one. The values of the defective pixels are estimated from the pixel values in the second band, as in the following formula (Bernstein et al. 1984):

$$DL_{i,j,k} = \left(\frac{s_k}{s_r}\right)\left\{\frac{DL_{i,j,r} - (DL_{i+1,j,r} + DL_{i-1,j,r})}{2}\right\} + \frac{DL_{i+1,j,k} + DL_{i-1,j,k}}{2} \tag{6.34}$$

where $s_k$ and $s_r$ are the standard deviations of the band to be restored ($k$) and the auxiliary band ($r$), respectively. As in the previous method, the result is rounded to the nearest integer.

In the three algorithms (Equations 6.32 through 6.34), it is assumed that the corrupted lines have already been identified. This is a trivial task in some cases, but can become very complex in others. Therefore, an automatic searching routine based on comparing the arithmetic mean of each line with the previous and subsequent lines can be used. If they differ significantly, that is, above a determined threshold, the *DL*s are analyzed in more detail to detect anomalous values (e.g., 0 or 255). If these extreme values are detected, it is considered to be a result of corrupt data, and the interpreter can apply any of the algorithms previously mentioned to those values.

The same can be said of pixels with anomalous values produced by an acquisition or reception error. In these cases, it will be harder to visually detect the defective pixels. Lillesand and Kiefer (2000) proposed an automatic search routine, based on detecting those pixels that have values significantly higher or lower than the surrounding values (calculated from a 3 × 3 moving window). If the *DL* of the central pixel in the window exceeds (above or below) a given threshold, the output *DL* is substituted by the mean of the neighboring pixels.

### 6.7.1.2 Correction of Striping Effects

Another common problem of sensor performance is calibration deterioration. In across-track scanning sensors, each scan collects several lines simultaneously to cope with the along-track speed of the satellite. Each line is detected by a different sensor, which should be well calibrated with the others to ensure uniform response along the image. The radiometric performance of the detectors may deteriorate, producing a striping effect in the image (Figure 6.37a), especially in areas of low radiance (shadows, marine surface). This effect is caused by intercalibration discrepancies in the sensor detectors, which provide a systematically lower or higher signal than the others. The effect is periodic, since each detector records 1 of every *p* lines (*p* being the number of detectors per band). In the case of Landsat MSS, 6 detectors were recorded simultaneously, while 16 were used for Landsat TM and ETM+.

In the case of along-track scanners, the potential calibration noises are even greater since these sensors have a large number of detectors per line (7000 for the multispectral bands of the Landsat 8 OLI, double for the panchromatic one). In this instance, a miscalibration between detectors would create a longitudinal striping on the image. An example of vertical striping is seen in Figure 6.37b with the ALI

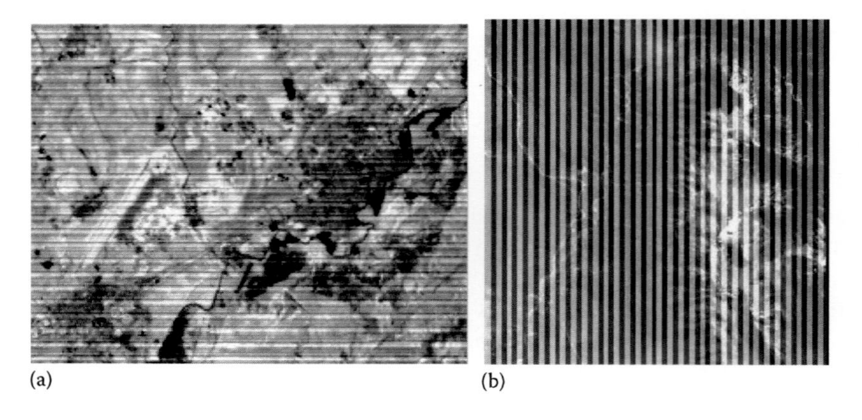


FIGURE 6.37 Effects of calibration problems among detectors: (a) thermal band of a Landsat ETM+ image acquired over Central Spain and (b) red band of an EO-1 ALI image acquired in the Mendoza province (Argentina). (Courtesy of Ruiliang Pu, University of California–Berkeley, Berkeley, CA.)

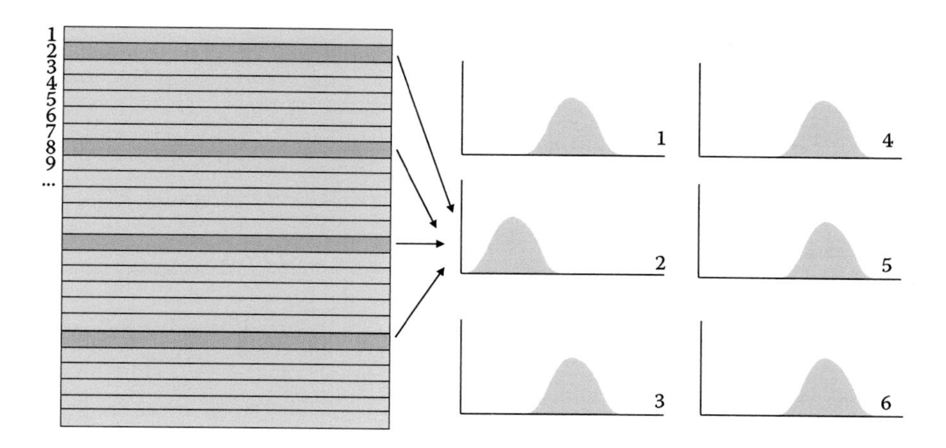


**FIGURE 6.38** Histograms of *DL* for values detected by each detector in a striped image. (Adapted from Campbell, J.B., *Introduction to Remote Sensing*, The Guilford Press, New York, 1996.)

sensor (onboard the EO-1 satellite) over the Mendoza region of Argentina. The calibration coefficients are specific for each column, or detector, in the image.

In order to adjust the signal for each of the detectors, it is assumed that all of the detectors should have a similar histogram for the entire image when all are properly calibrated. Therefore, one can restore the histogram of each detector by calculating the histograms of all pixels acquired by each detector (Figure 6.38). In the example of the Landsat TM/ETM+ image, with 16 detectors per band, the histogram of detector 1 would be computed using lines 1, 17, 33, 49, 65,...; the pixels acquired by the second detector would be lines 2, 18, 34, 50, 66,...; and so on until the histograms for all of the 16 detectors are calculated. The next step is to adjust all of the detector-specific histograms to a reference histogram. Most likely, the reference histogram corresponds to that of the entire image with all lines included. A simple way to perform this correction is to apply linear coefficients (bias and gain) to the histograms of each detector, in order to match their means and standard deviations to the reference histogram.

The process starts by computing the mean and standard deviation of the pixels sampled by each detector (partial histograms). Then, the histograms are adjusted to the reference histogram using

$$DL'_{i,j} = a_i + b_i DL_{i,j} \tag{6.35}$$

where $a_i$ and $b_i$ are the correction coefficients, computed as

$$b_i = \frac{s}{s_i} \tag{6.36}$$

and

$$a_i = \overline{DL} - b_i \overline{DL_i} \quad (6.37)$$

where

$s$ and $s_i$ indicate the standard deviation of the reference and partial histogram, respectively

$\overline{DL}$ and $\overline{DL_i}$ are the mean *DL* for both

To minimize the volume of calculations, this algorithm is usually applied in a similar way as a LUT previously addressed.

Another problem causing missing areas in across-track sensors occurred in 2003, when the scan line corrector (SLC) mechanism of Landsat 7 failed. This device compensated the mirror rotation and the forward motion of the satellite, thus aligning scan strips and avoiding either missing or duplicating observed areas. As seen in Chapter 3, the failure of this system causes the ETM+ images to have longitudinal strips of missing data that extend from the center to the image edges (Figure 3.30). Since ETM+ images were widely used in various land applications, different researchers proposed several methods to recuperate those missing areas, the most common being the use of geostatistical approaches (estimation of gap areas from nearby observed pixels) and the recourse to images acquired for the same area in other dates using regression models (Zeng et al. 2013).

### 6.7.2 Conversion from DL to Radiance

Most space agencies use a terminology that identifies satellite products according to their level of processing. The raw data are termed level 0, implying that no correction has been applied to the acquired data and therefore it may include different errors and artifacts. Level 1 implies that geometric and radiometric corrections have been applied. Depending on the agency processing the sensor, this level may imply different corrections. The most standard are those dealing with sensor equalization (thus solving stripping effects), geometric corrections (either with or without georeferencing, depending on the correction model), and sensor calibration. This last step implies converting the *DL* to radiance, which is the original physical measurement that a sensor registers.

Level 2 implies the calculation of a secondary physical variable associated to radiance, which is reflectance when working on the solar spectrum and temperature when working in the thermal wavelengths. This processing level implies maintaining the original sensor's spatial resolution.

Level 3 implies resampling the original resolution to a certain standard grid, with spatial and temporal resolution determined by product specifications. If this product proceeds from different orbits, it may also need further processing to ensure completeness and consistency. Finally, Level 4 implies a derived variable, such as evapotranspiration or land cover, probably obtained through a combination of different inputs.

This section covers the first radiometric correction, which aims to generate radiance products from the image *DL*s. This process is commonly named *calibration*, as the main goal is to convert back digital values to radiance units. This is the first step to work with physical variables instead of *DL*s, and therefore necessary for estimating both reflectance and temperature. Working with physical units has many advantages including the potential for comparing values over time and space from the same sensor, across different sensors, and between satellite measurements and other remote sensing platforms (from airplanes, UAV to ground and laboratory instruments), as all will be measuring in the same physical units. Implications of these corrections affect change detection, image fusion, scaling-up processes, and calibration and assessment of interpretation models.

Sensor calibration implies converting *DL*s to radiance. As explained in Sections 3.2 and 6.1, a satellite image is acquired by a sensor system that converts incoming radiance to digital values. The range of radiance values to which the sensor is sensitive and the number of different input radiances that a sensor can discriminate are part of the design specifications of the sensor. From these parameters, calibration coefficients can be obtained as (Figure 6.39)

$$a_{1,k} = \frac{L_{\max,k} - L_{\min,k}}{R_{DL}} \tag{6.38}$$

where

$a_{1,k}$ is the gain coefficient

$R_{DL}$ is the range of *DL* values of the image (256 or 2048 depending on sensor radiometric resolution)

$L_{\max}$ and $L_{\min}$ are the maximum and minimum radiances detected by the sensor (in W m$^{-2}$ sr$^{-1}$ μm$^{-1}$)

To fit better the equation for lower radiances, a bias coefficient ($a_{0,k}$) is commonly used and therefore the calibration formula becomes

$$L_k = a_{0,k} + a_{1,k} DL_k \tag{6.39}$$

Coefficients to obtain radiances from *DL* values should be provided by the sensor manufacturer, at least for pre-launch conditions. Commonly, sensors need to be

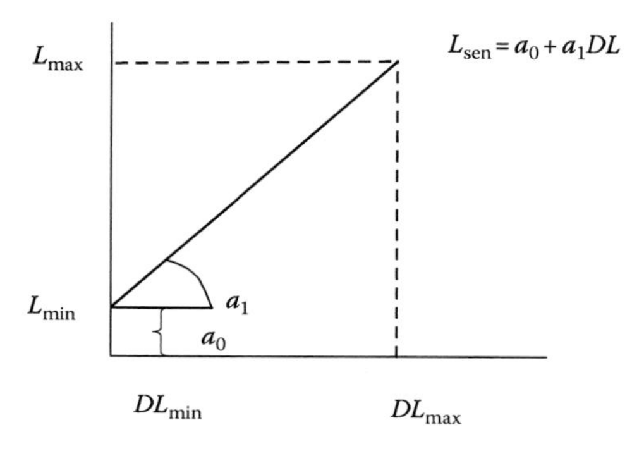


**FIGURE 6.39** Sensor calibration implies translating DL into radiances. $a_0$ and $a_1$ are calibration coefficients.

**TABLE 6.6**
**Parameters to Convert *DL* to Radiance for Landsat TM/ETM+/OLI Images**

| | Landsat 5 TM | | | Landsat 7 ETM | | Landsat 8 OLI |
|---|---|---|---|---|---|---|
| Band[c] | $a_{0,k}$ | $a_{1,k}$[a] | $a_{1,k}$[b] | $a_{0,k}$ | $a_{1,k}$ | $a_{1,k}$ |
| 1 (2) | −1.52 | 0.602 | 0.763 | −6.2 | 0.786 | 0.01256 |
| 2 (3) | −2.84 | 1.175 | 1.443 | −6.0 | 0.817 | 0.11576 |
| 3 (4) | −1.17 | 0.806 | 1.039 | −4.5 | 0.639 | 0.009762 |
| 4 (5) | −1.51 | 0.815 | 0.873 | −4.5 | 0.635 | 0.005978 |
| 5 (6) | −0.37 | 0.108 | 0.120 | −1.0 | 0.128 | 0.001485 |
| 7 (7) | −0.15 | 0.057 | 0.065 | −0.35 | 0.044 | 0.000501 |

*Sources:* Calibration coefficients are derived from Chander et al. (2004). For recent ETM+ images, the calibration coefficients were taken from the header files.

[a] Before May 5, 2003.

[b] After May 5, 2003.

[c] In parenthesis the OLI bands.

recalibrated after launching using a set of ground control areas that are well characterized radiometrically, such as deserts or deep lakes.

The calibration coefficients are usually included in the header of the image file or are provided by the sensor manufacturer. As an example, Table 6.6 shows the published calibration coefficients of the Landsat TM/ETM+/OLI sensors.

## 6.7.3 Calculation of Reflectance

### 6.7.3.1 Simplified Reflectance

In Chapter 2, we defined directional reflectance as the energy reflected from a surface in the direction of the sensor and normalized by the incident energy on that target. Thus, reflectance can vary between 0 (a perfect absorbing surface) and 1 (a perfect reflecting surface). The spectral reflectance of a target will depend on its physical, biological, and chemical properties, in addition to the conditions of the observation. Reflectance also varies with wavelength, resulting in *spectral reflectance signatures* that enable the discrimination of surface features and land cover types (see Chapter 2).

Assuming the target behaves as a Lambertian surface, the radiance reaching the sensor is a function of the solar irradiance, the surface reflectance, and the Sun–target–sensor observation geometry (Figure 6.28):

$$L_{\text{sen},k} = \frac{E_{0,k} \cos\theta_i \rho_k^*}{D\pi} \tag{6.40}$$

where

$E_{0,k}$ is the solar irradiance at the top of the atmosphere (ToA) in spectral band $k$ (see Table 6.7)

**TABLE 6.7**
**Top-of-the-atmosphere (ToA) Irradiance Values for the OLI and TM/ETM+ Spectral Bands**

| Landsat 8 OLI | | Landsat TM | |
|---|---|---|---|
| **Band** | **TOA*** | **Band** | **ToA** |
| 2 (B) | 2067 | 1 (B) | 1983 |
| 3 (G) | 1893 | 2 (G) | 1796 |
| 4 (R) | 1603 | 3 (R) | 1536 |
| 5 (NIR) | 972.6 | 4 (NIR) | 1031 |
| 6 (SWIR) | 245.0 | 5 (SWIR) | 220 |
| 7 (SWIR) | 79.72 | 7 (SWIR) | 83.44 |

$\rho^*_k$ is the apparent reflectance of the surface in band $k$
$\theta_i$ is the zenith angle of the incident flux, formed by the vertical and the solar rays (this is the complementary angle of the solar elevation angle included in the header of the image)
$D$ is a correction factor accounting for the variations of the Earth–Sun distance, obtained from

$$D = \left(1 + 0.01674\left(\sin\left(\frac{2\pi(J - 93.5)}{365}\right)\right)\right)^2 \tag{6.41}$$

where
$J$ is the day of the year
the sine term is in radians

This factor varies approximately between 0.983 (January 3) at the perihelion and 1.017 (July 4) at the aphelion.

In summary, the simplified reflectance can be computed from the simple equation

$$\rho^*_k = \frac{D\pi L_{sen,k}}{E_{0,k}\cos\theta_i} \tag{6.42}$$

Therefore, the conversion of *DL* to simplified reflectance requires sensor radiances, solar irradiance at the ToA, illumination angle, and the day of acquisition (alternatively, latitude, longitude, and day, from which $D$ and the solar zenith angle $\theta_i$ can be computed) (Figure 6.40).

An example of transforming *DL* values to apparent reflectances for three different surface features is shown in Figure 6.41. The reflectance values and spectral profiles differ from those using radiances as a result of band-specific differences in calibration factors and solar irradiances. However, the conversion of radiances to apparent reflectances does not consider atmospheric influences, nor adjacency effects, sensor

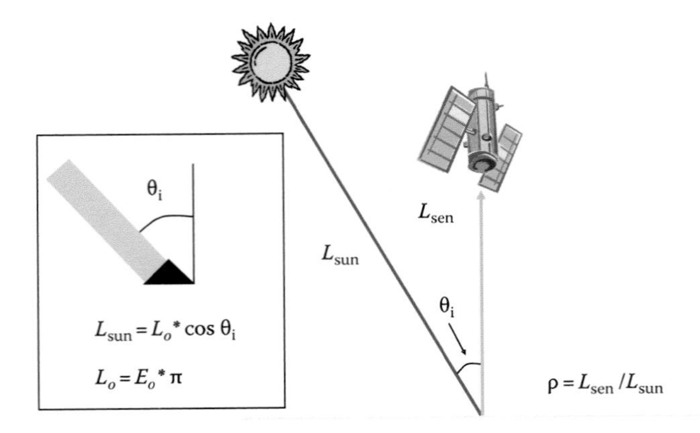


FIGURE 6.40 Calculation of simplified reflectances (ρ*).

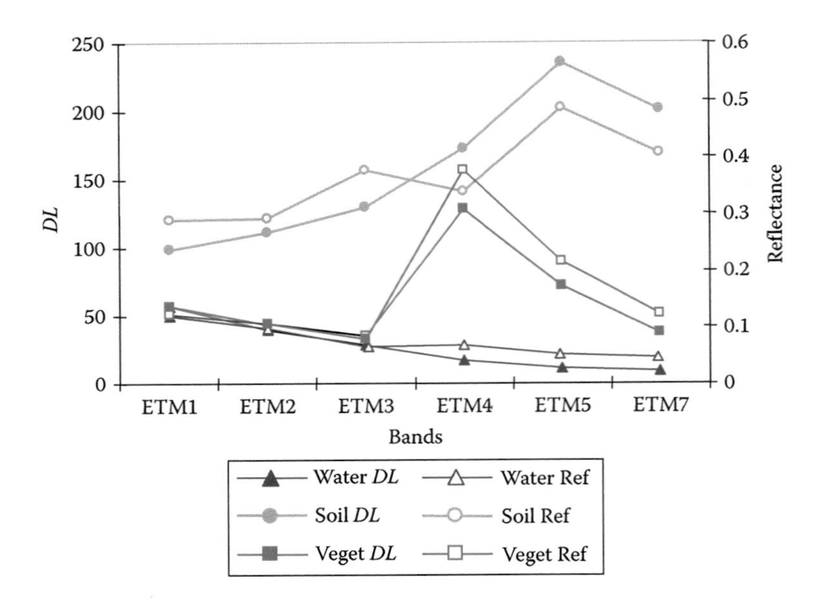


FIGURE 6.41 Comparison of the spectral curves of raw *DL* and simplified reflectances (ρ*) for water, bare soil, and vegetation.

observation angles, and illumination conditions. In summary, apparent reflectances are computed at ToA conditions and assume that the sensor observed Earth through a transparent atmosphere, at nadir view angle, and over flat and perfect Lambertian surfaces.

### 6.7.3.2 Atmospheric Correction

As mentioned in Chapter 2, the various constituents of the atmosphere modify the irradiances at the surface as well as the radiances leaving the surface, primarily through absorption and scattering processes (Figure 6.42).

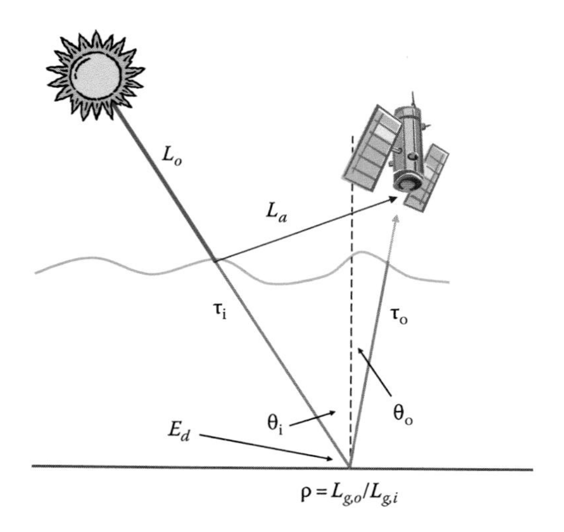


**FIGURE 6.42** Atmospheric effects on reflectance calculations.

Scattering effects are caused by aerosols and particulate matter from different sources and imply that part of the radiance received by the sensor does not come from the ground but actually from the atmosphere ($L_a$ in Figure 6.42). The most important scattering process (Rayleigh's) is particularly strong at shorter wavelengths, and therefore it needs to be taken into account even for computing transformations among bands of the same image.

Absorption implies that part of the incident solar energy is absorbed by the atmosphere, thus arriving at less radiance values than those measured at ToA. Similarly, the outgoing radiance is partially absorbed before reaching the sensor, and therefore the finally detected radiance is lower than at ground level. The opposite effect occurs from diffuse irradiance, which is energy illuminating a certain object from energy reflected by nearby objects, as they also reflect solar energy in different directions ($E_d$ in Figure 6.42). This effect causes, for instance, that objects in a shade are visible even if they do not receive direct solar energy. Also associated with the atmosphere is the "adjacency effect," in which a part of the signal from neighboring pixels influences the radiance of each pixel, thereby making the spectral contrast between neighboring pixels lower than expected and their discrimination difficult.

In summary, the radiance measured by a satellite sensor is not the same as that leaving the Earth's surface due to the complex role of the atmosphere. The radiance at the sensor is related to the following parameters (Gilabert et al. 1994):

$$L_{\text{sen},k} = L_{g,k}\tau_{k,0} + L_{a,k} \tag{6.43}$$

where

$L_{g,k}$ is the radiance coming from the ground in band $k$

$\tau_{k,0}$ is the transmittance of the atmosphere between the ground surface and the sensor

$L_{a,k}$ is the atmosphere path radiance contribution

Similarly, the incoming radiance arriving at the ground surface ($E_{g,k}$) is not the same as that measured at the ToA ($E_{0,k}$) due to atmospheric attenuation effects and the diffuse irradiance component:

$$E_{g,k} = E_{0,k} \cos \theta_i \tau_{k,i} + E_{d,k} \tag{6.44}$$

where

$E_{0,k}$ and cos $\theta_i$ have the same meaning as in Equation 6.42
$\tau_{k,i}$ is the atmospheric transmittance of the incoming flux
$E_{d,k}$ is the diffuse irradiance associated with downward atmosphere scattering (Mie and Rayleigh), dependent on atmospheric conditions

The transmittance in the upward direction depends on atmospheric scattering and absorption processes (attenuation), and includes the optical thickness of ozone ($\tau_{0z,k}$), aerosols ($\tau_{a,k}$), and the molecular or Rayleigh optical thickness ($\tau_{r,k}$) for a given band at the angle of observation ($\theta_0$):

$$\tau_{k,0} = \exp\left(\frac{-\tau_{0z,k} - \tau_{a,k} - \tau_{r,k}}{\cos \theta_0}\right) \tag{6.45}$$

The transmittance of incoming solar radiation through the atmosphere is similarly expressed as with the reflected radiance, except that the sensor angle of observation is replaced with the angle of incidence ($\theta_i$):

$$\tau_{k,i} = \exp\left(\frac{-\tau_{0z,k} - \tau_{a,k} - \tau_{r,k}}{\cos \theta_i}\right) \tag{6.46}$$

The angle of observation determines the path length of the atmosphere and becomes important in the case of sensors with off-nadir acquisition capacity (SPOT-HRG, GeoEye, WorldView), as well as sensors with wide swath angles, as in AVHRR, MERIS, and MODIS. For example, a satellite observation angle $\theta_0$ of 30° increases the path length of the atmosphere by a factor of 1.25 (1/cos 30°), thereby increasing the amount of attenuation between the ground and sensor.

Following the previous formulas, the true surface reflectance, which is defined as

$$\rho_k = \frac{D\pi L_{g,k}}{E_{g,k}} \tag{6.47}$$

requires solving the following equation, from Equations 6.43 and 6.44:

$$\rho_k = \frac{D\pi\left(L_{\text{sen},k} - L_{a,k}\right) / \tau_{k,0}}{E_{0,k} \cos \theta_i \tau_{k,i} + E_{d,k}} \tag{6.48}$$

In other words, to compute surface reflectance, we need to remove the atmospheric effects, which requires estimating atmospheric transmissivity (downward, $\tau_{k,i}$, and upward, $\tau_{k,0}$), diffuse irradiance ($E_{d,k}$), and atmospheric scattering ($L_{a,k}$).

The atmosphere correction process is quite complex and requires information on atmospheric conditions and atmospheric properties at the time of image acquisition. This information, however, is usually not available. Furthermore, atmospheric effects can be highly variable across an image, especially when caused by the spatial variability of aerosols or water vapor.

In general, atmospheric corrections have been approached in the following four ways (King et al. 1999; Robin 1998; Sobrino 2000):

1. Through *in situ* measurements, including radiosonde profiles of atmospheric composition and indirect estimates of atmosphere optical thickness through estimations of visibility and through measurements of the incident solar radiance or reflectance on the ground. These *in situ* methods are generally most accurate as they involve measurements of atmospheric conditions at the time of image acquisition (Figure 6.43), but such measurements are limited to local areas and specific time conditions. They are ideal for airborne or UAV campaigns.
2. Another way is by using satellite data and information from other sensors to provide estimates of the key atmospheric variables or properties at the time of acquisition of the images. The simultaneous use of two or more sensors, one measuring the ground surface and the other the atmosphere, makes it possible to better estimate atmospheric conditions. For example, the Terra platform includes sensors designed to specifically measure atmosphere properties (e.g., MISR) simultaneously with other sensors that measure surface reflectance and temperature (e.g., MODIS). Even though the spatial resolution of these sensors may not be the same, the integration of these data has provided quite useful in deriving calibrated products with adequate accuracies (King et al. 1999).

**FIGURE 6.43** *In situ* measurements for atmospheric correction of an airborne campaign.

3. Physically based radiative transfer models (RTMs), which usually include a series of standardized atmospheres, may be used in atmosphere corrections. The accuracy of these models depends upon the validity of the model assumptions for the specific conditions when the image was acquired, as well as the access to external information for model calibration, such as *in situ* measures of water vapor or aerosols.
4. Atmospheric correction may be performed from the image data and properties of the scene itself, such as areas of known surface optical properties. This would be the most operational method, since it does not require external information or *in situ* measurements.

Of the four methods of atmospheric correction, we will focus on the last two, as they are the most available to use. There are quite a few RTMs with standard atmospheres that are widely used in atmospheric correction with a high degree of accuracy, especially when there are good similarities between the atmospheric conditions of the image and the standard atmospheres included in the model. Two of the best known models are 6S (Second Simulation of the Satellite Signal in the Solar Spectrum: Kotchenova et al. 2008; Vermote et al. 1997) and MODTRAN (Berk et al. 2006). The 6S was initially developed by the Laboratory of Atmospheric Optics, Lille University, and is now part of the MODIS land surface reflectance products. The model simulates a flat land surface, considers elevation and a surface bidirectional reflectance distribution (BRDF), and models the influence of gases ($CH_4$, $N_2$, O, CO) in the calculation of gaseous transmission. It facilitates an estimation of the atmospheric effect for the whole optical spectrum, with bands up to 0.025 μm wide. MODTRAN was based on a model developed by the U.S. Air Force and a private company (Spectral Sciences, Inc.). It also considers surface BRDF effects and the adjacency effect. Modifications of this code are the basis for the FLAASH algorithm included in the ENVI IPS. It is also useful to mention the ATCOR model, developed by Richter (1996, 1997), and included in the modules of PCI Geomatics software. The ATCOR model centers the correction in a series of standard atmospheric profiles, included in the program, although they are easily adaptable by the user in case more detailed information exists. The definitions of atmospheric profiles include the water vapor contents and types of aerosols. The program requires an estimation of the atmospheric thickness, indicating a visibility range (in km), and an average height of the atmosphere (a DEM can also be incorporated) and the solar zenith angle. The software includes an optional estimation of the adjacency effect of neighboring surfaces, considering the differences between the reflectances of the pixel and the surrounding ones (in this case, an area of 1 × 1 km is considered). The authors of this model note that it works adequately only for sensors close to nadir (a tolerance of ±8° is accepted), in zones below 1500 m of elevation, and with surfaces with moderate slopes.

Naturally, the ideal case would be to derive the atmospheric correction from the information contained in the image itself. The alternatives can be summarized in four possibilities: estimate the atmospheric contribution from very dark surfaces with reflectances close to zero ("dark object" method), from the absorption and scattering differences in the different spectral bands of the same image, from a multitemporal comparison between images, and from images acquired at different angles. The first method, based on radiance observed for dark objects, was first proposed by Chavez in 1975, with successive improvements in 1988 and 1996 (Chavez 1975, 1988, 1996). This method assumes that any image has some areas with high-absorption materials, where the reflectance should be close to zero (water, shadows). However, the minimum *DL* of any image is commonly higher than zero, which is attributed to atmospheric scattering. Additionally, in most images this minimum value decreases as the wavelength increases, being close to zero in the SWIR. As we have seen, the most common atmospheric scattering (Rayleigh) affects mainly the shorter wavelengths, which confirms that this minimum value should be attributed to atmospheric effects. Therefore, a simple approximation to reduce the impact of atmospheric scattering consists of subtracting the minimum *DL* of each band from all *DL*s of that band, thus shifting the origin of the histogram at zero. In terms of radiance contribution, this expression leads to the following formula:

$$L_{a,k} = a_{0,k} + a_{1,k} DL_{\text{min},k} \tag{6.49}$$

which can be used as an input parameter in Equation 6.48. The Chavez method assumes there is always a high absorption area in the image, which can be reasonably accepted for shadowed areas or deep water bodies. However, it is assumed that atmospheric scattering is constant throughout the image, since it is computed from a single value. Some authors have proposed establishing a network of dark surfaces distributed in the image, which might be used to account for the spatial variation of the atmospheric optical depth (Ouaidrari and Vermote 1999).

The dark object method is very simple and is valid for calculating the atmospheric thickness of aerosols, but errors can be significant for high reflectance values (above 15%) when the transmittance of the incident and reflected flux are not considered. To avoid this problem, it has been suggested to complement this method with estimations of atmospheric transmittance on the ground or with ancillary climate parameters (mainly to estimate the ozone and water vapor contents, Ouaidrari and Vermote 1999). In fact, the last version proposed by Chavez suggests estimating the transmittance from the atmospheric thickness defined by the cosine of the zenith angle $\theta_i$ (Chavez 1996). In addition, similar to the previous version, the author suggests not to consider the diffuse irradiance. In other words, for vertical observations, the equation (Equation 6.48) should be

$$\rho_k = \frac{D\pi(L_{\text{sen},k} - L_{p,k})}{E_{0,k}\left(\cos\theta_i\right)^2} \tag{6.50}$$

where the downwelling transmittance is approached from the cosine of the Sun's zenith angle ($\theta_i$). The author assumes that the upwelling transmittance is 1, which is reasonable for vertical observations. This model has been shown effective only for TM bands 1, 2, 3, and 4 (B, G, R, and NIR, respectively).

According to Chavez's model, the value of transmittance would be independent of the wavelength, which is not appropriate. Consequently, TM1 is slightly corrected and TM4 is highly corrected with this method, although the author claims that this error is not significant. The transmittance is practically 1 for the SWIR bands: TM5 and 7 (Gilabert et al. 1994); thus, in these bands the correction of the cosine is not applicable because these bands would be overcorrected. Finally, the same author suggests the use of some standard values of transmittance instead of the cosine correction. The proposed equation in this case would be

$$\rho_k = \frac{D\pi(L_{sen,k} - L_{p,k})}{E_{0,k}\cos\theta_i \tau_{k,i}} \tag{6.51}$$

where $\tau_{k,i}$ = 0.70 (TM1), 0.78 (TM2), 0.85 (TM3), 0.91 (TM4), and 1 for the SWIR (TM5 and 7). These values of $\tau_{k,i}$ are very similar to the ones proposed by other authors. Gilabert et al. (1994) suggested, respectively, 0.73, 0.79, 0.85, 0.91, 0.95, and 0.97 for the same wavelengths. Pons and Solé-Sugrañes (1994) also recommended average values similar to the previous ones, although they are incorporated into a function of the solar zenith angle.

Figure 6.44 shows a graphical comparison between reflectances calculated with the Chavez method (with average transmittances) and ones obtained in the previous section (ToA reflectance). The effect of the correction is very obvious, mainly in the blue band, in which the values of corrected reflectances decrease notably. In addition, the standard signature of vegetation is now more clearly observed by the absorption in the blue and red bands, which were highly distorted by the atmospheric effect in the apparent reflectance. The water also displays the expected reflectance curve, with low values and descending towards longer wavelengths, although an overcorrection may have occurred in the blue band.

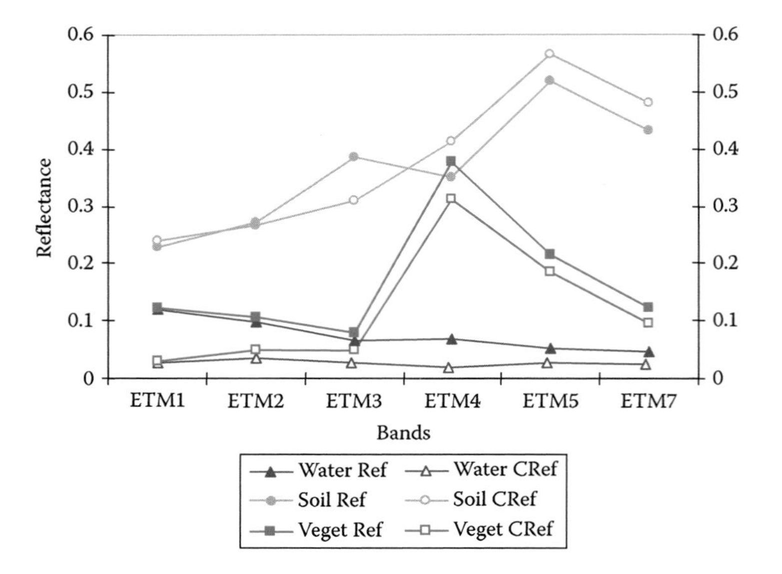


**FIGURE 6.44** Comparison of apparent reflectance (Ref) and atmospherically corrected reflectances for water, bare soil, and vegetation using Chavez method (CRef).

In Figure 6.44, the Chavez method is compared with the outputs of an RTM (included in the ATCOR model) for the same land covers. The correction effects of both methods are quite comparable, especially in water and vegetation. However, the Chavez method underestimates the transmittance in the red band, shown in the vegetation curve as a lack of an absorption band in relation to the green band, and the anomalous high value for soil. Both seem to overestimate the atmospheric effect in the blue band. Their comparison requires a more detailed analysis to be applied to images acquired on different dates. Both methods were applied to two Landsat TM images obtained during the summer in the central part of the Iberian Peninsula. The ATCOR method overestimates the correction in the shortwave bands (even generating negative numbers) and underestimates values in the NIR and SWIR bands (Riaño et al. 2003).

#### 6.7.3.3 Topographic Shadow Corrections

Another important factor to be considered in the calculation of reflectance is the effect of topography. Reflectances are dependent on both the incident and observation angles. Thus far, we have assumed that the incident angle depends only on the solar zenith angle, but this is true only for a flat terrain. In rougher terrains, slope gradient and aspect affect the incident energy, generating higher radiation in Sun-facing slopes and lower in the opposite ones. Figure 6.40 shows how incident energy is a function of the solar zenith angle ($\theta_i$) over a flat surface. The maximum energy incident on a flat surface is obtained when the solar zenith angle is 0° ($\cos \theta_i = 1$; i.e., the Sun is at nadir). Over rough terrain, however, the slope and aspect will modify the angle of the incident flux ($\gamma_i$), which will be no longer the same as the zenith angle as shown in Figure 6.45. The most severe impact of changing illumination conditions occurs in slopes facing away from the Sun, which do not receive direct solar energy at all (i.e., when the surface is in shadow).

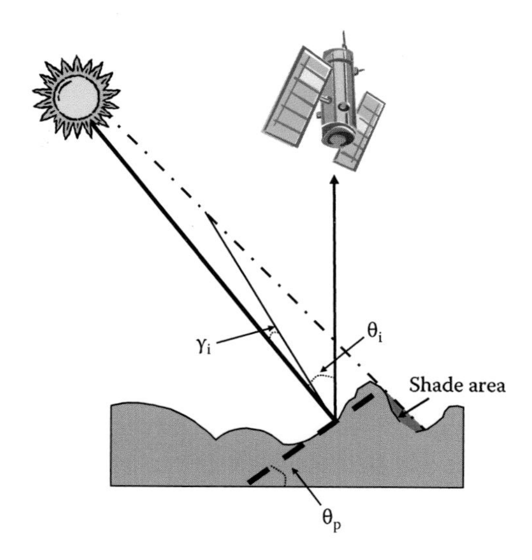


FIGURE 6.45 Effects of topography on reflectance calculations.

As the energy reflected by a rough terrain will vary proportionally with the incident energy, our computed reflectance values will not be the same for similar surface features unless topographic effects are considered and corrected. The reflectances derived for the same land cover characteristics, but over different slopes, will be very different, creating several problems in quantifying surface biophysical quantities and the discrimination and classification of land cover types.

Computing the ratio of two image bands is a simple way of reducing the topographic effects on reflectance. Since the solar incident angles ($\cos \theta_i$ in Equation 6.42) are constant for two simultaneously acquired bands, variations in reflected energy due to topographic influences are reduced, assuming that radiation variability is constant for different bands and that atmospheric effects have been previously removed (Colby 1991). These assumptions are generally not correct, given the strong spectral dependence of atmospheric influences that modify the direct irradiance at the surface, as well as atmospheric scattering contributions to diffuse radiation at the surface (shown in Equation 6.48), which are not removed by ratioing (Leprieur et al. 1988). In addition, the ratioing of bands reduces spectral information, potentially hindering land surface classification.

An alternative to image ratios is correcting the radiation–slope effects by modeling them using a DEM that needs to be previously registered to the image (or, more commonly, vice versa). In this way, variations introduced by topography can be estimated for each *DL* by modeling the illumination conditions at the time the image was acquired. For doing so, we first need to calculate the incident angle ($\gamma_i$) of a slope, which is a function of the solar zenith angle, the slope aspect, and gradient (Hantson and Chuvieco 2011; Riaño et al. 2003):

$$\cos \gamma_i = \cos \theta_i \cos \theta_p + \sin \theta_i \sin \theta_p \cos(\phi_a - \phi_0) \tag{6.52}$$

where

$\gamma_i$ corresponds to the incident angle
$\theta_i$ is the solar zenith angle
$\theta_p$ is the slope gradient
$\phi_a$ is the azimuth solar angle
$\phi_0$ is the aspect of the slope

These four angles can be obtained from a DEM and the date and time of the image acquisition. The $\cos \gamma_i$ term enables the estimation of the effect of slope on illumination. Additionally, this term is used to generate the shadowing effects that numerous GIS programs compute. Values of this parameter vary from –1 to +1, indicating for each pixel of the DEM the minimum and maximum condition of illumination, respectively. Figure 6.46 shows the Tucson image with clear examples of shadowing effects, along with an illumination image ($\cos \gamma_i$) for the same area and same conditions of image acquisition. The arrows indicate in which sectors the shadowing effect is more evident on the original image.

Once the incidence angles are derived, several methods can be used to account for the differences in incident radiation between slopes. A simple classification of

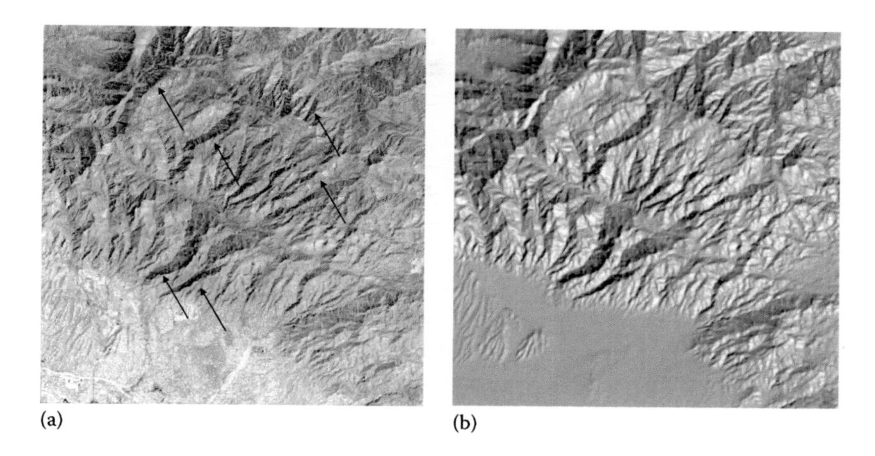


**FIGURE 6.46** Effect of relief on the Tucson image: (a) original band 4; (b) shaded relief derived from the DEM.

those methods group them on the basis of whether they assume Lambertian surfaces or they consider directional effects (Itten and Meyer 1993; Riaño et al. 2003). Among the Lambertian methods, the simplest is the one proposed by Teillet et al. (1982):

$$\rho_{h,i} = \rho_i \left( \frac{\cos \theta_i}{\cos \gamma_i} \right) \tag{6.53}$$

where

$\rho_{h,i}$ is the reflectance of pixel *i* in horizontal terrain
$\rho_i$ is the reflectance on a slope (corresponding to the image before correction)
$\theta_i$ is the solar zenith angle of the scene
$\gamma_i$ is the incident angle for that pixel

This method has been shown to overcorrect the image, especially in areas of low illumination (Duguay and LeDrew 1992; Holben and Justice 1981; Meyer et al. 1993), which is why it has been proposed to balance the correction as a function of the average conditions of illumination of the scene (Civco 1989):

$$\rho_{h,i} = \rho_i + \rho_i \frac{(\cos \gamma_m - \cos \gamma_i)}{(\cos \gamma_m)} \tag{6.54}$$

where $\cos \gamma_m$ indicates the average value of the illumination in the scene of interest. This model reduces the topographic effect, although not as much as it should. In this correction, as well as the previous one, it is assumed that the topographic effect is constant in all bands.

Within the non-Lambertian methods, one of the most cited was proposed by Minnaert (1941) to evaluate the rough surface of the moon:

$$\rho_{h,i} = \rho_i \left( \frac{\cos \theta_i}{\cos \gamma_i} \right)^l \tag{6.55}$$

where $l$ is introduced to model the non-Lambertian bodies. If $l = 1$, the surface behaves like an ideal Lambertian reflector. The problem with this method is getting a true value of $l$, since it varies from band to band and for each surface. A simple way to consider the importance of the Lambertian character on each band is to apply a semiempirical adjustment, as the one named on the C correction (Teillet et al. 1982):

$$\rho_{h,i} = \rho_i \left( \frac{\cos \theta_i + c_k}{\cos \gamma_i + c_k} \right) \tag{6.56}$$

where $c_k$ is an empirical constant for each band $k$, which is related to the average roughness of that band and is obtained from

$$c_k = \left( \frac{b_k}{m_k} \right) \tag{6.57}$$

where $b_k$ and $m_k$ are the constant and the slope, respectively, of the regression line between the reflectance image (band $k$) and the illumination image ($\cos \gamma_i$).

The problem of these simple corrections is the lack of consideration of the diffuse irradiance, which can be important on slopes in shadowed areas. Some alternatives have been proposed to solve this problem, but they are more complicated in their practical application. For example, Conese et al. (1993) suggests the use of a variation of the equation for the incident radiance that considers slope effects:

$$E^*_{su,k} = E_{0,k} \cos \gamma_i \tau_{k,i} + E^*_{d,k} \tag{6.58}$$

where

$E^*_{su,k}$ is the radiance that reaches the ground
$E^*_{d,k}$ is the diffuse radiance in a rough terrain

If the incident angle ($\gamma_i$) is greater than 90° (or if preferred, $\cos \gamma < 0$), the slope is in a shadowed area. Therefore, there is no direct illumination, and only the diffuse radiance ($E^*_{d,k}$) reaches the ground. Diffuse radiation is computed as a function of the incident angle, the slope of the terrain, and an anisotropic coefficient ($E_{g,k}/E_{0,k}$), dependent on the band and observation date. This topographic correction greatly enhances the computation of reflectances in rough terrains, reducing notably the spectral variability of cover types. Nevertheless, the authors (Conese et al. 1993) warn that the algorithm is

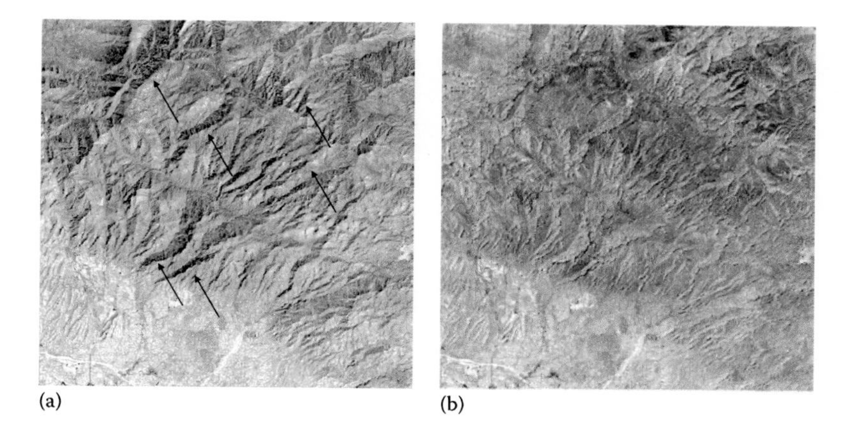


**FIGURE 6.47** Effect of the shadow correction on the Tucson area: (a) original band 4; (b) band 4 after correction.

not adequate for images with a large Sun zenith angle or with very rough topography. This same conclusion has been reached by other authors who have proposed similar methods for diffuse irradiance (Itten and Meyer 1993; Richter 1997).

The effect of topographic correction over the Tucson study zone is shown in Figure 6.47, which includes the same area as in Figure 6.46. In this case, band 4 is shown before and after the illumination correction, using Civco's method, that for this date confers good results, although registration problems between the DEM and the Landsat ETM+ image may cause an overcorrection of the upper mountainous areas. The arrows indicate the sections where the correction is more evident. One way to verify the effect of the correction is to calculate the correlation between the illumination image and the bands of the image, before and after correction. For our study area, the correlation decreases after correction from 0.57 to 0.28 in band 4, indicating that part of the topographic effect has been eliminated over the signal. Another verification method is to measure the difference in reflectances in a series of land covers along diverse slopes. If the correction is accurate, values should be homogeneous, eliminating their differences due to Sun exposure (Hantson and Chuvieco 2011).

#### 6.7.3.4 Correction of Bidirectional Effects

The last aspect that will be considered in the retrieval of surface reflectance is the effect of observation and illumination angles. Normally, the surface is assumed to behave in a Lambertian manner in the computation of reflectance. This implies that the radiance reflected by the surface is the same in all directions and largely independent of the sensor's viewing angle. This assumption is reasonable only as a first approximation in the computation of reflectance because most land surface conditions are anisotropic. Everyone has experienced directly how the appearance of an object varies according to the position of the observer in relation to the incident light (Figure 6.48); hence one must consider this factor for detailed computations of reflectance.

**FIGURE 6.48** Variations on the reflectance of a forest canopy by the effect of the observation angles. (a) The Sun is behind the camera; (b) the Sun is in front (Photo courtesy of Don Deering from https://www.umb.edu/spectralmass/terra_aqua_modis/modis, BRDF Explained, last access November 2015.)

Most medium- and high-spatial-resolution sensors have a narrow FOV and take images almost from the nadir. Therefore, it is generally assumed that the reflected radiances are derived from vertical observations. However, in the case of sensors with wider swaths, such as Terra-MODIS, SPOT-VEGETATION, or NOAA-AVHRR, there are large variations in viewing and Sun angles across the pixels of the image. These observations would not be perfectly comparable in space or time, thus creating potential noises in time series analysis. Compositing techniques may reduce, but not eliminate, bidirectional effects in the images, and corrections are necessary for the accurate retrievals of surface reflectance.

To address these corrections, we should consider the BRDF of each observed surface for trying to model a vertical observation for each image pixel. BRDF is a theoretical concept that describes the directional effects of reflected radiance for any given combination of viewing and solar illumination angles (Sandmeier and Itten 1999):

$$f(\theta_i,\phi_i;\theta_r,\phi_r,\lambda)=\frac{dL(\theta_r,\phi_r,\lambda)}{dE(\theta_i,\phi_i,\lambda)} \tag{6.59}$$

where

*dL* is the reflected radiance at a given view zenith ($\theta_r$) and azimuth angle ($\phi_r$)

*dE* is the incident irradiance at a specified solar zenith ($\theta_i$) and azimuth ($\phi_i$) direction (Figure 6.49)

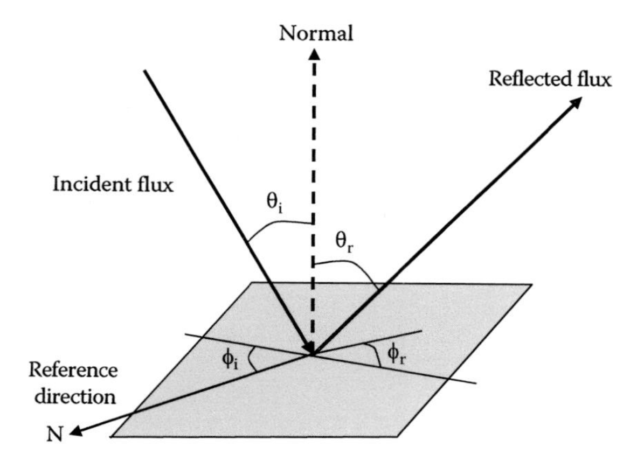


**FIGURE 6.49** Angles involved in the calculation of the BRDF.

The λ parameter has been included to indicate that BRDF is also wavelength dependent, or that the differences in the direction of observation will not have the same effect in all spectral wavelengths and is more pronounced at shorter wavelengths, such as in the visible in the case of vegetation (Sandmeier and Itten 1999). One can highlight BRDF effects more by dividing the directional reflectances by their corresponding nadir view observation:

$$ANIF\left(\theta_i,\phi_i;\theta_r,\phi_r,\lambda\right)=\frac{R\left(\theta_r,\phi_r,\lambda\right)}{R_o\left(\theta_i,\phi_i,\lambda\right)} \tag{6.60}$$

where *ANIF* is the anisotropic factor that measures reflectance variations relative to the nadir view direction. Some examples of anisotropic factors are included in Figure 6.50 in the case of herbaceous cover. One can see very different directional effects, depending on the wavelength. In the high absorption bands (B, G, and R), there is a strong directional component, with the higher intensities at angles closer to the vertical and also when observed in the backscatter direction (Sun backward). On the other hand, the NIR and SWIR bands have less marked directional effects.

To make bidirectional corrections to an image, the BRDF signatures of each land cover should be available. This information is not easily accessible, since few sensors provide multiangular observations. Most of the available studies are based on field goniometer data (Sandmeier and Itten 1999), but recent measurements by MISR or POLDER are very useful in this regard. In addition to performing the correction of nonvertical observations, the BRDF is very useful for retrieving relevant information of a particular cover, such as vertical structure or composition.

Since the BRDFs are not available, operational corrections of the angular effects need to rely on simplified models. In the case of AVHRR images, a simple correction based on a physical model was proposed by Dymond et al. (2001), which estimates a

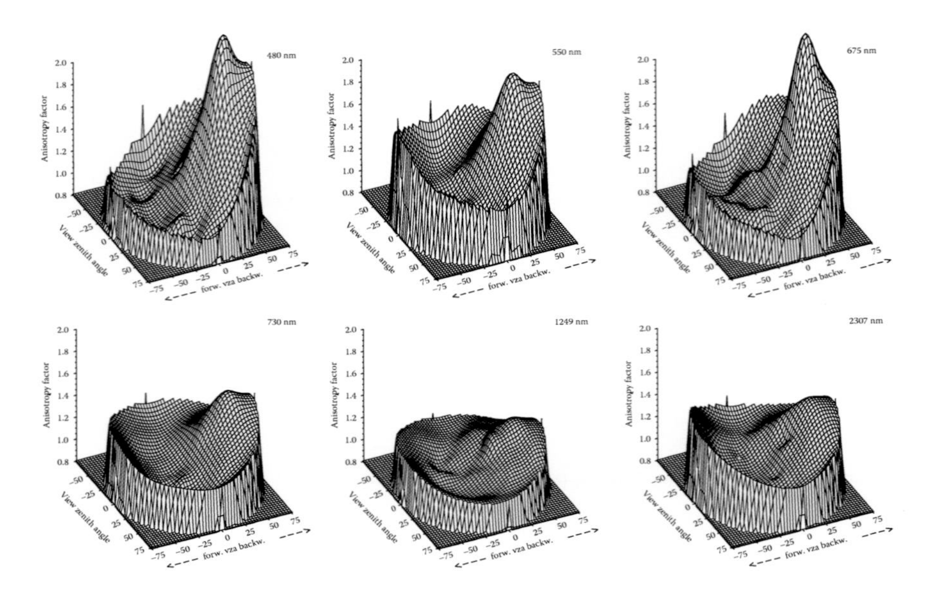


**FIGURE 6.50** Anisotropic graphs calculated from radiometric measurements obtained with a field goniometer over a grass cover. (Adapted from Sandmeier, S.R. and Itten, K.I., *IEEE Trans. Geosci. Remote Sensing*, 37, 978, 1999; Courtesy of Stefan Sandmeier.)

standardization factor of the reflectance, assuming reference observation conditions (vertical view and a 45° solar zenith angle). The conversion was based on

$$\hat{\rho} = \varphi\rho \tag{6.61}$$

where

$\hat{\rho}$ is the normalized reflectance
$\phi$ is the conversion factor
$\rho$ is the observed reflectance

The $\phi$ factor is computed differently for the R and NIR for AVHRR (bands 1 and 2, respectively). The authors of the method (Shepherd and Dymond 2000) suggest computing this factor from the following relationships:

$$\varphi_R = \frac{\cos\theta_i + \cos\theta_e}{\cos\theta_{io} + \cos\theta_{eo}}\left[\frac{a + (\pi - a)e^{-k\alpha_o^2}}{a + (\pi - a)e^{-k\alpha^2}}\right] \tag{6.62}$$

and

$$\varphi_{IRC} = \frac{\cos\theta_i + \cos\theta_e}{\cos\theta_{io} + \cos\theta_{eo}}\left[\frac{(8/3\pi)(a + (\pi - a)e^{-k\alpha_o^2}) + H(\theta_{eo}, w)H(\theta_{io}, w) - 1}{(8/3\pi)(a + (\pi - a)e^{-k\alpha^2}) + H(\theta_i w)H(\theta_e, w) - 1}\right] \tag{6.63}$$

where

$\theta_i$ and $\theta_e$ indicate the incident and reflection angle of the signal to be corrected

$\theta_{i0}$ and $\theta_{e0}$ are the angles corresponding to the reference direction of observation (usually 45° and 0°, respectively)

$\alpha$ and $\alpha_0$ are the phase angles (the ones formed by the solar rays with respect to the position of the sensor, that is, the azimuth difference between the Sun and the sensor) of the signal to be corrected and the reference direction of observation (in this case 45°)

The term $H$ considers the multiple reflections in the interior of the vegetation canopy and can be calculated as

$$H(\theta, w) = \frac{1 + 2\cos\theta}{1 + 2\cos\theta\sqrt{1 - w}} \tag{6.64}$$

for each of the angles considered (incident and reflection angle, for the detected signal as well as the reference). The rest of the parameters, namely, $w$, $a$, and $k$, are dependent on the observed cover and are calculated from their BRDFs. If they are not available, $w$ can be estimated from the leaf reflectance ($\rho$; $w = 2\rho$), and $a$ and $k$ from a regression analysis from data proceeding from two different orbits of the AVHRR over the same zone (i.e., taken with different angles), minimizing the residuals ($R_i$) in the following expression:

$$\frac{\rho_a}{\rho_b} - \frac{\cos\theta_{ia} + \cos\theta_{ea}}{\cos\theta_{ib} + \cos\theta_{eb}} \left[ \frac{a + (\pi - a)e^{-k\alpha_a^2}}{a + (\pi - a)e^{-k\alpha_b^2}} \right] = R_i \tag{6.65}$$

where $\rho_a$ and $\rho_b$ are the reflectances of a given cover extracted from two AVHRR images ($a$ and $b$) taken with different angles.

Figure 6.51 shows an example of the importance of these directional effects in NOAA-AVHRR images. It includes a series of reflectance values for bands 1 and 2, extracted from an area covered by evergreen forest, as well as the corresponding solar angles at the moment of acquisition. All of them are from images of the diurnal pass of the NOAA-16 acquired between 13:00 and 14:22 GMT. The dependence of the azimuth angle is quite clear, and a cyclic tendency of the reflectance closely related to the cyclic orbit of the satellite is observed.

For MODIS data, angular reflectance corrections are routinely derived using a BRDF model. The corrected reflectance is a standard product of the MODIS program (termed MOD43: Román et al. 2009; Schaaf et al. 2002).

### 6.7.4 Calculation of Temperature

Temperature is a critical variable to understand the Earth system, affecting both ocean and land processes. Location of ocean currents and blooms, evapotranspiration and crop's water demand, ice and glacier melting, and vegetation productivity are

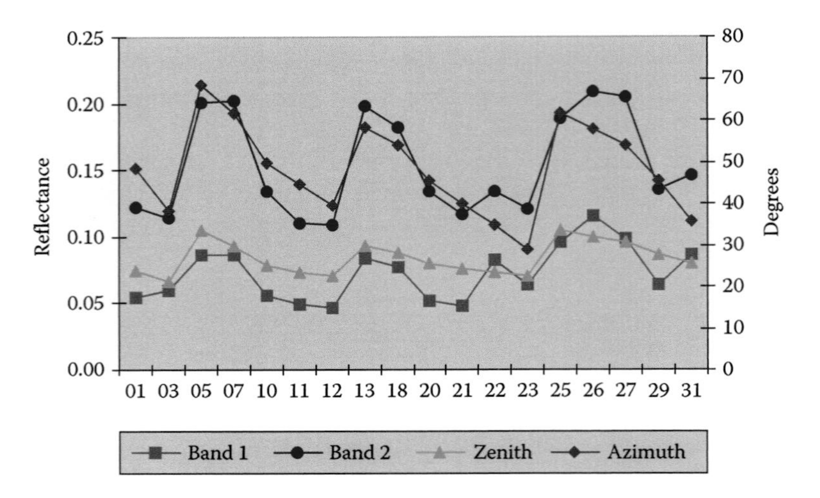


**FIGURE 6.51** Temporal daily series of reflectance values extracted from AVHRR bands 1 and 2. Azimuth and Sun zenith angles are included to observe their effect on the signal detected by the sensor.

associated with temperature changes. Climate studies commonly use air temperature to understand and monitor those processes. EO satellite retrievals of temperature are based on surface or radiant temperature, which are linked to thermal status of the upper part of the ground surface.

To calculate radiant temperature from EO sensors, first *DL*s are converted to radiances. Since radiance in the thermal infrared (10–12 μm) has little influence from the Sun's incoming energy, the energy detected by the sensor is considered as coming directly from the ground, and therefore Sun angles are not considered in this case. The radiance received at the sensor ($L_{sen}$) is the result of the flux emitted by the ground ($L_g$), minus the energy absorbed by the atmosphere, plus a minor emission of the atmospheric constituents. From the detected radiance, one can compute the temperature based on Planck's radiation law (Equation 2.8), which assumes that the radiant object is a blackbody. Since real covers are not blackbodies, we additionally need to estimate the emissivity to obtain an accurate estimation of the surface temperature (Figure 6.52).

Several models have been proposed to take into account both atmospheric and emissivity corrections. Methods have been derived for images from NOAA-AVHRR (Coll and Caselles 1997), Landsat TM (Sobrino et al. 2004), ATSR (Sobrino 2000), MODIS (Wan et al. 2002), geostationary satellites (Trishchenko 2006), and, more recently, the Landsat 8 thermal infrared sensor (TIRS: Jimenez-Munoz et al. 2014).

Since the AVHRR has a long historical series of data, the illustration of the process to compute surface temperature will be based on this sensor.

As in the case of reflectances, the first step in the calculation of temperatures is to convert the *DL*s into radiances. From radiances, brightness temperatures can be computed from the inverse of Planck's equation (Kidwell 1991):

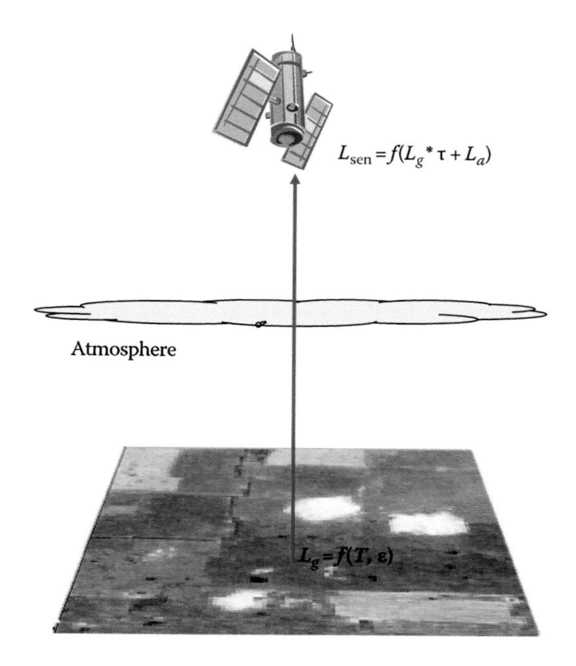


**FIGURE 6.52** Schematic retrieval of surface temperature.

$$T^*(L) = \frac{c_2 \nu}{\ln(1 + c_1 \nu^3 / L_{sen,\nu})} \tag{6.66}$$

where

$T^*$ is the brightness temperature in K (also termed $T$ at ToA) for a given radiance value $L$, and corresponds to the central wavelength of the thermal band under consideration ($\nu$ in $cm^{-1}$)

$c_1$ and $c_2$ are constants ($c_1 = 1.1910659 \times 10^{-5}$ mW $m^{-2}$ $sr^{-1}$ $cm^{-4}$ and $c_2 =$ 1.438833 cm K)

Brightness temperature refers to the thermal values at the ToA, assuming the emitting surface is a black body (emissivity = 1).

The main gas affecting atmospheric transmissivity in the thermal bands is water vapor, particularly in the 10–12 μm region, while $CO_2$, $O_3$, and $N_2O$ contribute also to absorption in the range 4–5, 9–10, and 14–16 μm. Atmospheric correction of thermal wavelengths is commonly based on the differences in brightness temperatures of two adjacent bands. These methods are known as *split window algorithms*, since at least two bands in the TIR window are needed. The generic equation for this type of correction is (Li et al. 2013)

$$T_s^* = a_0 + a_1 T_i^* + a_2 \left(T_i^* - T_j^*\right) \tag{6.67}$$

where

$T_s^*$ is the ground temperature

$T_i^*$ and $T_j^*$ are the brightness temperatures in two contiguous bands of the spectrum

Coefficients $a_0$, $a_1$, and $a_2$ are coefficients that depend on the spectral response function of the two thermal channels, the emissivities, and the atmospheric composition (particularly water vapor)

Other authors have proposed similar equations for biangular observations, that is, the acquisition of the same TIR band but with two different angles, as in the ATSR.

One of the proposed forms of the split window method, valid for middle latitudes, is (Coll et al. 1994)

$$T_s^* = T_4^* + \left(1 + 0.58\left(T_4^* - T_5^*\right)\right)\left(T_4^* - T_5^*\right) \tag{6.68}$$

where

$T_s^*$ is the brightness temperature on the surface

$T_4^*$ and $T_5^*$ are, respectively, the brightness temperatures of bands 4 and 5 for the AVHRR

The last step for obtaining the surface temperature would be to include the emissivity correction that assesses the radiative efficiency of a determined surface. The importance of this factor is incorporated in the following equation:

$$T_s = T_s^* + 0.51 + 40\ (1 - \bar{\varepsilon}) - 75\Delta\varepsilon \tag{6.69}$$

where

$\bar{\varepsilon}$ corresponds to the average emissivity of a pixel in bands 4 and 5 of AVHRR

$\Delta\varepsilon$ is the difference of emissivity for the same pixel in those two bands ($\Delta\varepsilon = \varepsilon_4 - \varepsilon_5$)

Emissivity of marine water varies with the observation angle and wind velocity, being 0.987–0.991 in the 10–12 μm channel at ±25° from nadir (Niclòs et al. 2005). For land applications, emissivity is more complex to obtain, as the surface may be covered by different materials with very diverse emissivity values. Since land cover is not available in many cases or it changes throughout the year (i.e., seasonal variation of vegetation covers), emissivity estimation may be estimated by assuming that a pixel is covered either by soil or vegetation, calculating an average emissivity as a function of the proportion of surface occupied by these two covers. To estimate these proportions, Valor and Caselles (1996) suggested a method based on the R and NIR reflectances of the same sensor, acquired simultaneously with the thermal channels. They estimate the pixel proportion occupied by vegetation from the normalized difference vegetation index (NDVI), which will be discussed in Chapter 7, and the minimum and maximum greenness detected in a sufficiently long series of images:

$$P_v = \frac{(1-(NDVI/NDVI_{min}))}{(1-(NDVI/NDVI_{min}))-k(1-(NDVI/NDVI_{max}))} \quad (6.70)$$

where

$NDVI_{max}$ and $NDVI_{min}$ correspond to the values of vegetation indices observed as a maximum and minimum vegetation cover, respectively

$k$ is a normalization factor, given by

$$k = \frac{\rho_{2v} - \rho_{1v}}{\rho_{2s} - \rho_{1s}} \quad (6.71)$$

$\rho_{2v}$ and $\rho_{1v}$ being the reflectances of vegetation in bands 2 and 1 of the AVHRR, respectively, while $\rho_{2s}$ and $\rho_{1s}$ are the reflectances of the soil in the same channels.

From these proportions of green cover, the emissivity of each pixel can be calculated, by multiplying that proportion by the average emissivity of vegetation ($\varepsilon_v = 0.952$ for dry grass, 0.983 for trees, and 0.985 for green grass and shrubs; an average global value of 0.985 can also be applied) and the rest by the average emissivity of the bare soil ($\varepsilon_s$ = 0.968 for sandy soil, 0.972 limy soil, and 0.974 for clays; here the average value would be 0.96). In this way, we would have (Caselles and Sobrino 1989)

$$\varepsilon = \varepsilon_v P_v + (1 - P_v)\varepsilon_s + d\varepsilon \quad (6.72)$$

and

$$\Delta\varepsilon = \Delta\varepsilon_v P_v + (1 - P_v)\Delta\varepsilon_s + d\Delta\varepsilon \quad (6.73)$$

The terms $d\varepsilon$ and $d\Delta\varepsilon$ evaluate, respectively, the effects of the cavity on $\varepsilon$ and $\Delta\varepsilon$, and can be determined from the geometric characteristics of the plants, although they have only a very small contribution (values vary from 0.004 to 0.0025 for the most common vegetation covers). Other methods to estimate the emissivity are based on field or lab measurements on empirical relationships, inversion techniques, day/night temperature–independent indices, and two temperature methods, among others (Li et al. 2013).

## 6.8 IMAGE FUSION METHODS

As we reviewed in Chapter 3, the interpreter of EO data has nowadays access to images with a wide range of spectral, spatial, and temporal resolutions. Many environmental phenomena are scale dependent, meaning that their characteristics vary depending at which level of spatial detail they are analyzed. For this reason, analyzing the same phenomenon at different spatial scales provide interesting insights to better understand its impacts. In addition, even though many environmental problems are global, they need to be studied at fine spatial detail as well. In this framework, scaling-up methods are required to integrate knowledge that is generated at different spatial resolutions. For instance, a better estimation of evapotranspiration flows requires understanding in detail mechanisms of water exchange between the soil–plant system and the surrounding atmosphere, as well as knowing the impact of landscape composition (species mixing, canopy cover), and the interactions of landscapes and biomes.

The integration of data and products at different spatial scales is greatly benefited by using fusion methods. The general definition of fusion is the generation of hybrid products by combining images acquired at different spatial resolutions, generally from the same spectral bands. The most common approach to image fusion has been the generation of improved resolution images by combining the multispectral and panchromatic bands of the same or different sensors (Pohl and Van Genderen 1998). As reviewed in Chapter 3, different EO missions include sensors that acquire simultaneously images from different spatial resolutions. For instance, the first SPOT satellites included a multispectral camera that acquired images at 20 m resolution and a panchromatic camera that acquired images at 10 m. Landsat 7 ETM+ had a similar system with 30 and 15 m, respectively, which is the same configuration as the new Landsat 8 OLI. The higher-spatial-resolution satellites (QuickBird, IKONOS, GeoEye, etc.) included similar sensor systems, which are able to acquire simultaneous images at 0.5–1 resolution in panchromatic mode and 2–4 m in the multispectral mode.

Fusing images acquired at two different pixel sizes implies the integration of the spatial detail provided by the panchromatic image and the spectral information included in the multispectral one (Bruzzone and Serpico 1998; Pellemans et al. 1993). Different methods have been proposed to carry out this merging. They can be grouped in two categories: those using algebraic operations and those based on image transformations. Before applying any fusion technique, the multispectral image needs to be resampled to the panchromatic resolution, so the number of pixels in both sets of images is the same. This can be easily done by reduplicating pixels in the multispectral image or interpolating them using nearest neighbor or cubic convolution algorithms (similar to those commented in Section 6.6.3.3). In addition, a radiometric equalization between the panchromatic and multispectral histograms may be carried out to improve color matching.

Algebraic methods rely on creating synthetic images by combining the panchromatic and multispectral images using simple mathematical operations. An example of this approach is the high-pass filter modulation method (Zhang and Moore 2015):

$$DL_{o,k} = \frac{DL_p}{DL_{LP,p}} DL_{ms,k} \tag{6.74}$$

where the output $DL$ in band $k$ is the result of multiplying the multispectral input data for that band ($DL_{ms,k}$) by the high-pass information of the panchromatic image, resulting from dividing the original data ($DL_p$) by the low-pass filtered ($DL_{LP,\,p}$).

The second group of fusion methods is based on image transformations, which create new images from the original spectral bands by maximizing the statistical or spectral contrast between input bands. We will comment on some of these transformations in Chapter 7, but here we should mention one of the most commonly used, which is the intensity, hue, and saturation (IHS) (Carper et al. 1990). This is an alternative way of representing colors. Instead of using the composition of the three primary colors (RGB) in the IHS space, each color is defined by its physical properties. Intensity refers to the brightness or average reflectance, hue to the dominant radiant wavelength, and saturation to the color purity, that is, the degree of combination with other tones. The IHS space represents the colors in a hexacone, where the length is the intensity, the angular direction the hue, and the proximity to the hexagon center the saturation (Figure 6.53).

Different algorithms have been proposed to convert the RGB to the IHS space (Pellemans et al. 1993). The user needs to select which spectral bands and with which color assignment will be used for the color composite. In the case of image fusion, the three bands to be merged should have a similar spectral range to the one covered by the panchromatic image. In the case of the Landsat or SPOT sensors, this implies selecting the NIR, R, G color composite, since the SWIR is not commonly covered by the panchromatic channel. The process starts by computing the three IHS components from the RGB space. Then the I channel is assumed to include the spatial variation of the image, while the H and S channels include mainly the spectral variation. Therefore, the I channel is replaced by the panchromatic image, and afterward the inverse transformation is computed (IHS to RGB) to obtain the new color composite. The new image should have a similar (ideally the same) color information of the multispectral bands and the spatial details of the panchromatic one.

Figure 6.54 includes an example of image fusion with the Landsat OLI image of Tucson. The new image shows much more spatial detail, particularly evident when objects are observed at larger scales, such as narrow roads, airport buildings, or vegetation features.

More elaborate methods to merge images of different resolutions are available in the research papers, the frequency transformations (Fourier and Wavelet analysis) being among those providing better results (Garguet-Duport 1997; Lillo Saavedra and Gonzalo 2006; Yocky 1996). Fusion techniques can also be used for other purposes, such as integrating information acquired by sensors working in different spectral regions or at very different spatial resolutions. For instance, radar and optical data can be merged for vegetation mapping (Simard et al. 2008), or active lidar and passive optical data for mapping forest parameters (Garcia et al. 2011b). In recent years, the growing availability of medium-resolution images (Landsat, SPOT, DMC,

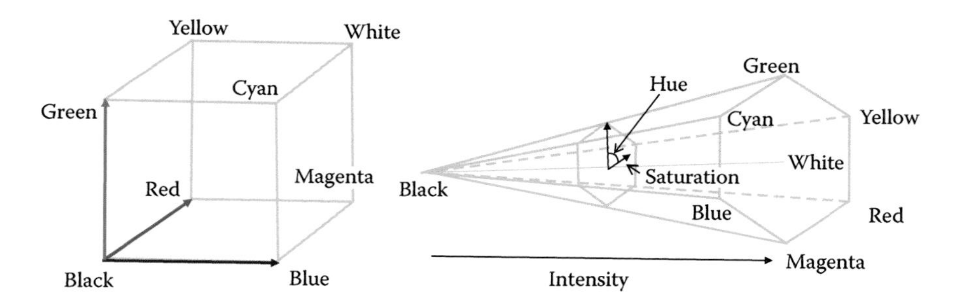


**FIGURE 6.53** Color representation using the RGB and the IHS color spaces.

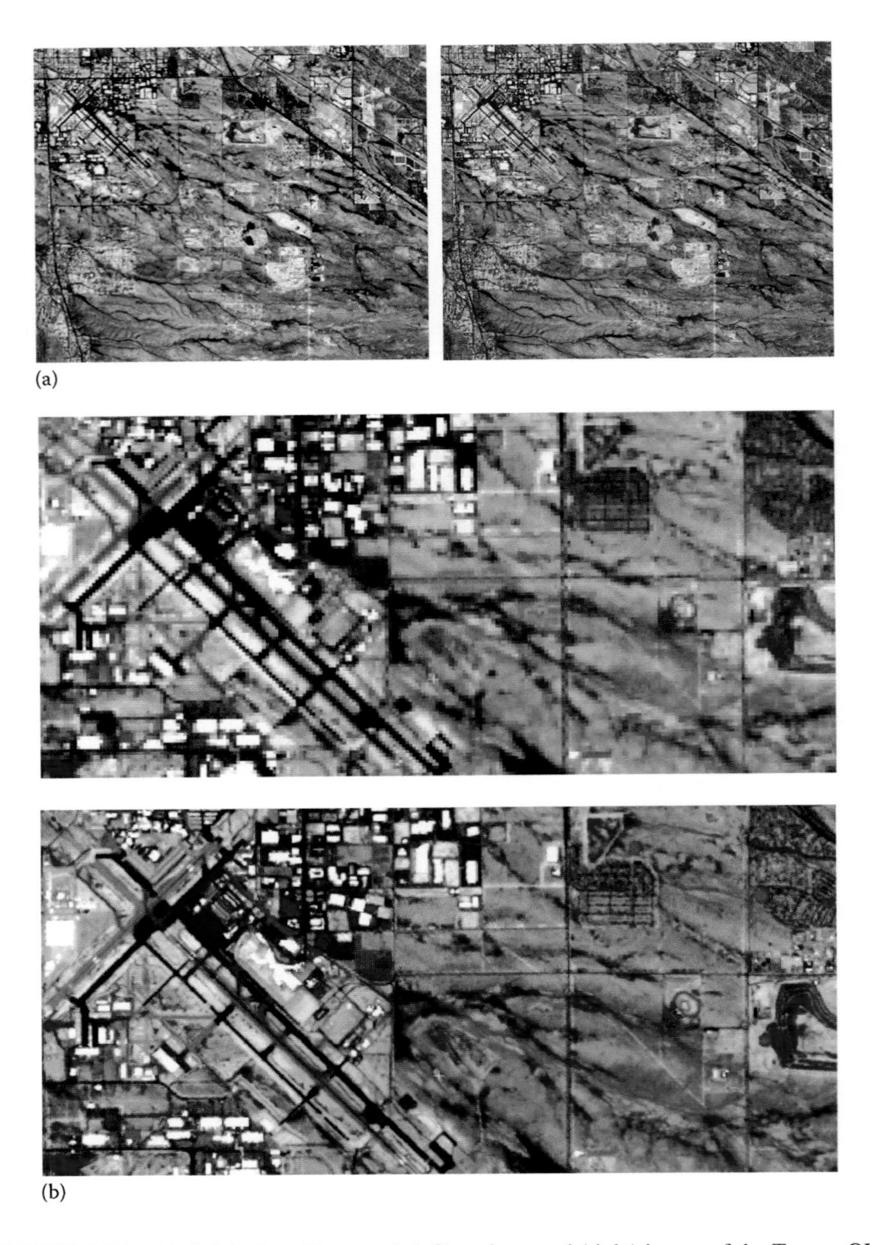


FIGURE 6.54 (a) Original multispectral (left) and merged (right) image of the Tucson OLI image, showing the impact of the fusion process. (b) Detail window that shows the original (up) and merged (down) images in the area surrounding the Tucson international airport.

Sentinel-2) has made their use more common in continental or even global studies, by fusing them with coarse-resolution sensors, such as MODIS-VIIRS, SPOT-VGT, or AVHRR data (Boschetti et al. 2015; Iverson et al. 1989).

## 6.9 REVIEW QUESTIONS

1. Raw digital levels in a remote sensing image are a numeric translation of
   a. Emittance
   b. Radiance
   c. Reflectance
   d. Radiant intensity
2. The image histogram provides relevant information on
   a. Spectral resolution
   b. Average conditions
   c. Variability
   d. Both average conditions and variability
3. When two bands have a high correlation, the interpreter should
   a. Accept the one with higher variance
   b. Reject both
   c. Accept both
   d. Accept the one with lower variance
4. Contrast stretch in a digital image involves
   a. Applying spatial filters
   b. Creating noncorrelated images
   c. Adjusting image histogram
   d. Transforming image geometry
5. Which of the following operations is most relevant for image enhancement?
   a. Emphasize spatial features
   b. Blur spatial features
   c. Apply atmospheric correction
   d. Apply geometric correction
6. What is the main purpose of digital filters?
   a. Enhancing spatial contrast
   b. Color stretching
   c. Histogram normalization
   d. Color composites
7. When the size of the filter window increases
   a. The filtering is more intense
   b. The filtering is less intense
   c. The color contrast is more evident
   d. The histogram is stretched
8. Which one of these problems is not accounted for in orbital corrections?
   a. Earth's curvature
   b. Satellite orbit
   c. Terrain characteristics
   d. Panoramic distortion

9. The geometric correction of a digital image is assessed using
   a. Training fields
   b. Ground control points
   c. Ground test points
   d. Image variance
10. Which one of the following geographical features would not be suitable as a ground control point?
    a. A road crossing
    b. A river and road crossing
    c. A lagoon
    d. A road and trail crossing
11. Which interpolation method can be used with thermal images?
    a. Nearest neighbor
    b. Bilinear interpolation
    c. Cubic convolution
    d. All of the above
12. The calculation of simplified reflectance does not consider
    a. Sensor calibration
    b. Atmospheric effects
    c. Solar angles
    d. Solar irradiance
13. The dark-object method for atmospheric correction identifies a low-reflectance image area with
    a. Atmospheric scattering
    b. Diffuse radiance
    c. Atmospheric transmissivity
    d. Ground reflectances
14. Which one of the following processes does not require atmospheric correction?
    a. Classification
    b. Relationship with spectroradiometer measurements
    c. Computation of spectral indices
    d. Multitemporal comparisons
15. A key parameter to obtain surface temperature from satellite data is the estimation of
    a. Solar angles
    b. Atmospheric scattering
    c. Emissivity
    d. Day of acquisition
16. When merging two images for creating hybrid resolution outputs, the user should first
    a. Ensure that both have atmospheric correction
    b. Ensure that both have the same frame and pixel size
    c. Perform image enhancement.
    d. Adjust the histograms of both images

# 7 Digital Image Processing (II)

## *Generation of Derived Variables*

A very important use of satellite remote sensing is the generation of spatially contiguous, thematic information and inventories of the Earth's surface, along with the detection of land cover changes. This chapter presents methods to retrieve different parameters from the primary variables detected by Earth observation (EO) sensors.

In the first decades of EO satellites, information obtained from satellite images was restricted to thematic maps (e.g., land covers) using different classification techniques. Even though this approach still attracts great interest, it is quite reductionist and does not take full advantage of the quantitative measurements acquired by EO sensors. We have reviewed in Section 4.2 other possible approaches to image interpretation: conversion to biophysical measurements to estimate a continuous-scale variable, such as fractional vegetation cover or leaf moisture content, and analysis of multitemporal trends and spatial patterns as monitored by satellite data. We will explain in this chapter different processing techniques that facilitate those approaches. First, the extraction of continuous biophysical variables is reviewed, then various classification schemes are discussed, also techniques for temporal analysis are reviewed, and finally different possibilities of extracting spatial information are presented.

## 7.1 GENERATION OF CONTINUOUS VARIABLES

Satellites can uniformly sample entire landscapes and regions with spatial coverages that are largely unobtainable with field measurements on the surface. By transforming such satellite measurements into certain biophysical or biochemical parameters, a comprehensive spatial distribution of a certain variable can be retrieved. In this regard, satellite observations complement existing ground sensor networks. Commonly, ground sensors provide a very sparse spatial observation, but highly frequent (seconds to minutes), while satellite data are spatially continuous with long temporal gaps (hours to days). This is the case of meteorological stations, which can measure temperature or rainfall at short intervals but with low density (one observation every cent or thousands of square kilometers), while satellite thermal or microwave information can estimate these variables every few hours to days at a relatively high spatial resolution (1–10 $km^2$). The two sets of measurements complement each other and provide a more accurate representation of how temperatures are spatially distributed and how they change with time, thus improving our understanding of surface–climate interactions in a region. The coupling of satellite measurements with ground-acquired measurements will be discussed in more detail in this chapter, as well as in Chapter 8 dedicated to validation methods.

### 7.1.1 Inductive and Deductive Models in Remote Sensing

Biophysical variables that can potentially be derived from remote sensing measurements are numerous. The basis for this estimation relies on the physical relationship that links the surface properties with the radiance detected by the sensors. A distinction between surface properties that are directly measurable by satellite sensors and those that are inferred or derived indirectly can be made in this regard. The former are primary variables, and the latter are secondary variables (see Section 4.1.1). Primary variables include the following:

- Spectral radiances in the whole spectrum
- Spectral reflectance in the solar reflective spectrum (bands in the visible [VIS], near-infrared [NIR], and shortwave infrared (SWIR) spectral regions)
- Land surface temperature (ST) in the thermal infrared (TIR) region
- Surface backscatter coefficients in the MW region
- Heights, from optical stereoscopy, radar interferometry, or lidar measurements

EO images can also be used to estimate biophysical variables derived from the primary acquired ones. Many of these variables are critical to our understanding of environmental change and landscape processes. Examples of variables that have been obtained from remote sensing satellite data include chlorophyll and water content in plant canopies, leaf area index (LAI), fraction of absorbed photosynthetically active radiation (fAPAR), soil moisture, sediments suspended in water, ozone or aerosol density in the atmosphere, surface albedo, evapotranspiration (ET), and marine ice thickness. Estimating these variables requires developing empirical, semiempirical, or physical models that relate the target variable with primary variables detected by the sensor. Auxiliary information, such as elevation or climatic products, may be used in combination with EO images to control some of the external parameters and obtain better estimations.

Any physical or empirical model should be based on a theoretical relationship between the parameter being estimated and the radiances measured by the sensor. For example, green vegetation cover can be estimated from red (R) and NIR reflectances, since vegetation pigments will absorb the R radiance, while green leaves are highly reflecting in the NIR. This behavior is pretty unique, and therefore a combination of R and NIR bands can be used to estimate the density of vegetation cover. The specific amount, however, requires models to convert reflectance to the proportion of cover. These models may be empirically derived, that is, generated for a particular site by taking ground measurements simultaneously with satellite acquisitions, or may be based on deductive models that are based on physical relations (Figure 7.1).

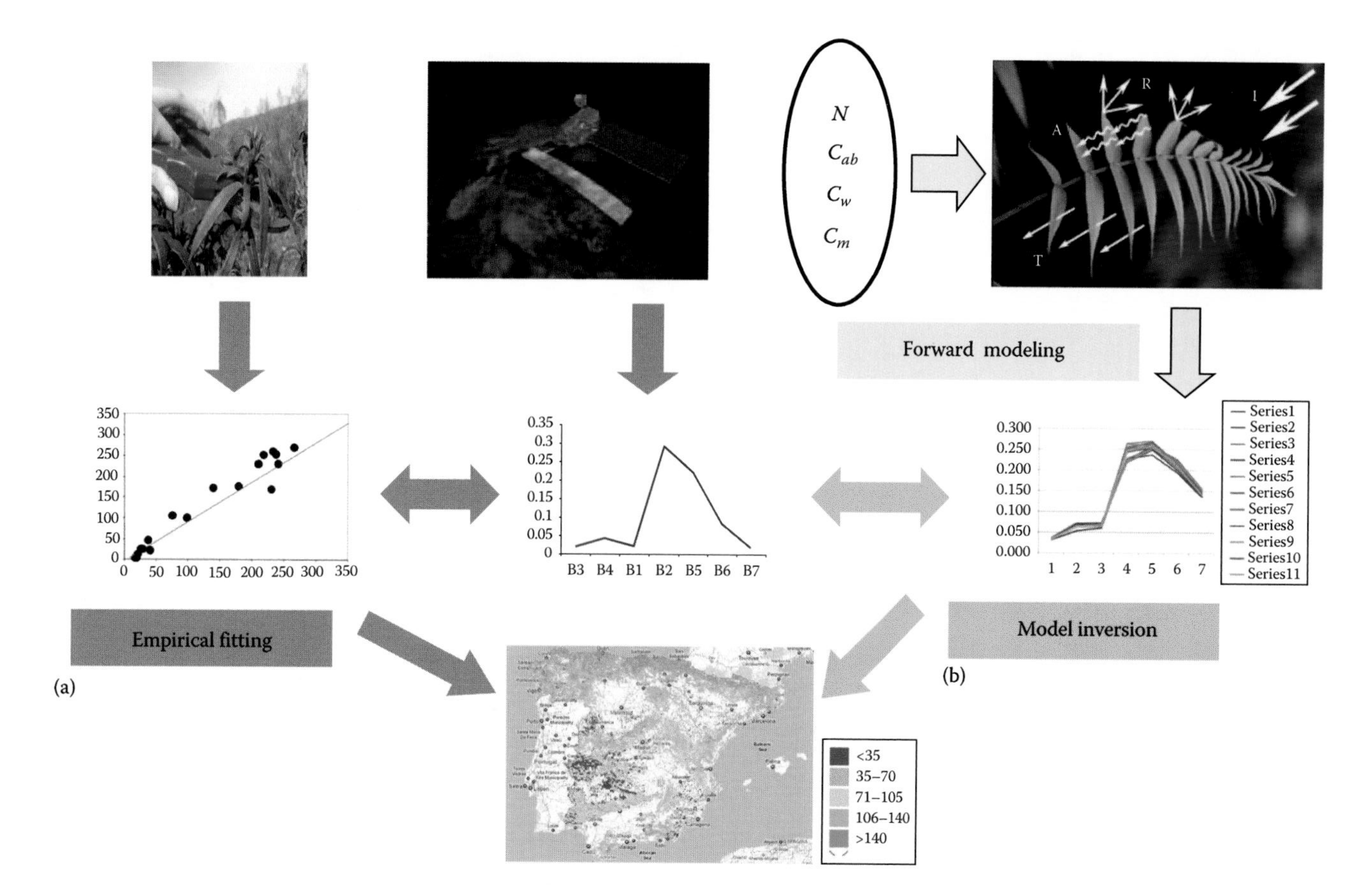


FIGURE 7.1 Framework to retrieve fuel moisture content from (a) inductive and (b) deductive models.

Inductive or empirical models fit a numerical relationship between the parameter to be estimated and the satellite measurements (raw data, radiance, or reflectance units), based on in situ observations taken at the time the image is acquired. In other words, the satellite image data is calibrated with the surface parameter of interest by means of local fits, usually based on regression techniques (Chuvieco et al. 2004; Koutsias and Karteris 1998; Morisette et al. 2005; Silva et al. 2004). There have also been adjustments using neural networks (Baret 1995; Fang and Liang 2003). In both cases, the equations are locally based, are simple to derive, and generate results of known accuracy. The main disadvantages of the inductive, or empirical, approach are that it requires field measurements simultaneously with the acquisition and the results only have local applicability, thus limiting their extension to other areas with different surface and observation conditions, as well as to different time periods and other sensors.

Deductive or theoretical models try to establish general relationships between the required parameter and satellite measurements. Therefore, the utility of the model extends beyond a single site, specific image, or specific time period (Strahler et al. 1986). There are many examples of land surface bidirectional reflectance models that describe how soils and plant canopies scatter radiation (Asner 2000; Gao et al. 2003; Verhoef and Bach 2003). Canopy reflectance models such as scattering by arbitrarily inclined leaves (SAIL) model physically the radiative transfer in the canopy as a function of vegetation structural properties, biophysical variables, and the optical properties of vegetation elements and soil (Goel 1988; Jacquemoud 1990; Verhoef 1984; Verhoef and Bach 2003). There are lidar-based models to retrieve canopy heights and estimate forest biomass (Nelson et al. 2000) and chlorophyll and canopy water content models (Zarco-Tejada et al. 2003, 2004). The surface energy balance algorithm for land (SEBAL) was developed to estimate ET from satellite data inputs of land ST and vegetation cover. Once the relationships are verified with some empirical test data, the physical models can be inverted to retrieve the biophysical parameters (e.g., LAI) from the surface reflectances contained in an image (De Santis and Chuvieco 2007). Remotely sensed data are increasingly being used in land surface atmosphere and climate models (Chuvieco 2008).

Physical models also have their limitations in their attempts to simplify highly complex land cover structures. As a result, many assumptions are made to simplify the real complexity, for instance, a uniform single- or two-layer vegetation canopy, an isotropic distribution of leaves, a flat surface, a Lambertian reflectance, and a certain soil background. For these reasons, physical and theoretical models are in constant need of being tested and constrained by actual data and observations in order to discover their weaknesses and inaccuracies and to improve their performance to be more comparable to the real world. However, even if the relationship between the estimated variable and the spectral data is consistent, it may not be easy to retrieve the variable because of extraneous influences that hinder or prevent a model inversion (Combal et al. 2002; Yebra et al. 2008). In spite of these difficulties, the deductive modeling approach is much stronger and is preferred to the inductive approach for the estimation of biophysical variables from remote sensing because such models are more extendable in space and time and they do not require simultaneous field data to be acquired with the satellite data.

There are also many intermediate approaches that involve both theoretical and empirical approaches, sometimes labeled as semiempirical models. These models assume a previous physical modeling of the estimated variable, but with empirical adjustments made by local regressions. An example of this is the model by van Leeuwen et al. (1994), who used local, site-based field measurements at the Hydrologic Atmospheric Pilot Experiment (HAPEX) field campaign to adjust and fine-tune the retrieval of biophysical parameters from the SAIL model. Other examples involve the use of vegetation indices (VIs) with an already demonstrated theoretical relationship with the variables being estimated (Danson et al. 1995). Neural nets and support vector machines (SVMs) may also be considered in this mixed class of retrieval methods, since they are essentially inductive machine learning techniques (Yang et al. 2007).

Given the introductory nature of this book, we will not review in detail the derivation and inversion of the large variety of physical models published in the literature. Instead, we will discuss and illustrate more about how these models are incorporated into satellite data processing workflows to generate biophysical products. In this section, we include the generation of image products from principal component analysis (PCA), spectral VIs, spectral mixture analysis (SMA), and advanced satellite products. For most of these image products, we assume that the original data or digital numbers of the image have been converted into physically calibrated radiances, reflectance units, or temperatures. These are essential prerequisite steps for model implementation in the retrieval of surface physical and biological variables useful in land surface characterization. This facilitates spatial and multitemporal comparability across different land surfaces and different time periods, and further enables integrated analysis among images from different sensors.

Spectral measurements made over terrestrial surfaces often can be treated as spatially variable mixtures involving two or more surface-reflecting features. The Earth's surface is rarely spectrally pure but contains complex vegetated canopies with leaves, stem, bark, litter, water, and soil background. Each of these surface elements, in turn, will have a certain portion illuminated directly by the Sun (sunlit) and a shaded portion that is primarily illuminated by diffuse or sky radiation in a manner dependent on canopy structure and the Sun–surface–sensor observation geometry. Over 75% of the Earth's terrestrial surface consists of open canopies (deserts, grasslands, savannas, open forests, etc.) with vegetation overlying significant amounts of exposed soil, litter, and even understory vegetation that will contribute to the overall spectral signal detected by a satellite sensor. These vegetated surfaces will further vary seasonally and with land cover disturbance. Resolving the presence and amount of the individual surface features, elements, and physical variables from the spectral information content present in satellite imagery is the primary objective of thematic-based digital image analysis. The extraction of key spectral features, such as rangeland forage, is key to effective natural resource management and for understanding hydrologic and biogeochemical processes within ecosystems.

### 7.1.2 Principal Component Analysis

PCA is a well-known mathematical technique for solving multivariate problems in the social and natural sciences. The primary objective is to characterize a large group of variables into a new smaller set, without losing a significant amount of original information. In the case of remote sensing, this usually involves the analysis of multispectral data structures to uncover complex interrelationships among spectral features, to discern independent sources of variability, and to compress the spectral data by removing redundant spectral information present in the multiband images (dimensionality reduction).

The origin of this technique may be traced to the field of psychiatry in the interpretation of human responses in intelligence tests. PCA was used to uncover an underlying series of factors that could explain the complex multivariate responses found within certain groups of variables. PCA is also widely used in the interpretation of complex multivariate data sets in the fields of chemistry and optics. A primary benefit of this approach is the ability to separate the multivariate data into independent sources of variability, which in linear combination can reconstitute the original data with minimal loss of information.

When the image has numerous spectral bands, they frequently contain redundant information either because those bands are in similar spectral regions or because some features have similar radiances across spectral regions. PCA allows us to remove the redundancies found in a spectral data set. This technique outputs a smaller data set from multiple bands, with minimal loss of original information. This is accomplished through a least-squares mathematical method called *eigenanalysis*, which decomposes the image spectral data into a new set of abstract eigenvectors with associated eigenvalues. This transformation is very useful, for instance, in selecting the three most suitable bands for a color composite (Green et al. 1988), or in change detection analysis of multitemporal data (Adams et al. 1995; Fung and LeDrew 1987).

On the other hand, and from a merely statistical point of view, PCA facilitates a first interpretation of image variability identifying the axes of variation, which allows one to identify those characteristics that are gathered in most of the bands and those that are specific to some group. Therefore, it provides a better distribution of the data, as opposed to those axes of variability, and can facilitate a more accurate identification of the different covers.

As is well known, the direction and strength of the linear correlation between two variables can be graphically represented in a bivariate axis plot. In our case, the cloud of points plotted indicates the location of the digital levels (*DL*s) or reflectances in the two bands considered. The ellipse that limits these points graphically expresses the strength of the correlation between the two bands. The more it resembles a straight line, or in other words, the greater contrast between the two axes of that ellipse, the greater the correlation. In Figure 7.2, there is a high positive correlation with two axes of variation: the first in the sense of the correlation (*A*) and the second in the sense of the residual (*B*). It is worth considering that this second component is perpendicular to the first, which indicates in statistical terms that the principal components are not correlated and that the information each one contains is unique.

Graphically, one can see that a rotation of *X*, *Y* axes toward the *A*, *B* directions will improve the original disposition of the data and probably also the separation

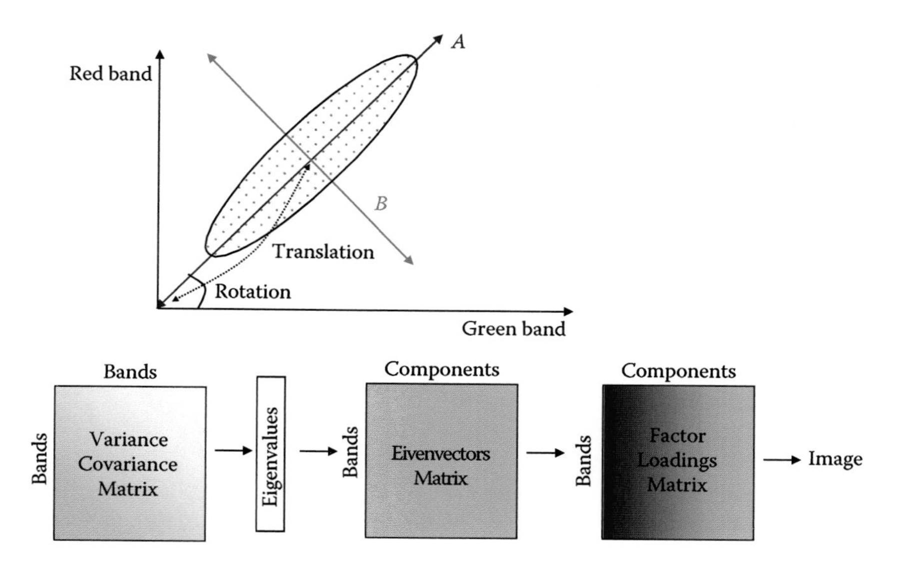


FIGURE 7.2 Process of obtaining principal components.

between the homogeneous groups of DLs in the image. That rotation is obtained simply by applying a function of the type

$$PC_1 = a_{1i} DL_i + a_{1k} DL_k \tag{7.1}$$

where $PC_1$ indicates the value corresponding to the first principal component, obtained from the DLs (or reflectances) of bands $i$ and $k$. From the geometric point of view, this new axis is only a rotation of the original ones (Figure 7.2). Sometimes, it can be useful to accompany the rotation with a displacement of the axes, locating the origin of coordinates at the resulting minimum values of the two bands. In doing so, a constant term should be added to the previous equation.

In more general terms, the principal components of an image are obtained according to

$$PC_j = \sum_{i=1,p} a_{i,j} DL_i + R_j \tag{7.2}$$

where

$PC_j$ indicates the DLs corresponding to principal component $j$

$a_{i,j}$ is the coefficient applied to the DLs of band $i$ to generate the component $j$

$R_j$ is a constant that is usually introduced in each component to avoid negative values

There are as many components ($p$) as bands, but they are extracted in such a way that only the former have significant information.

In simple terms, PCA can be summarized by the following steps. From the variance–covariance matrix of the original image, the eigenvalues for each of the components ($\lambda_j$) are extracted. The eigenvalues express the length of each of the new components and, consequently, the proportion of original information they retain. These data are useful in deciding which PCs are more interesting: they are usually those that retain the greatest amount of original information. The PCs are obtained in such a way that the eigenvalue decreases progressively from the first to the last ones, since the objective is to maximize successively the variance extracted in the analysis. The original variance explained by each component is calculated as the proportion of its eigenvalue, as opposed to the sum of all the eigenvalues:

$$V_j = \frac{\lambda_j}{\sum_{j=1,p} \lambda_j} \quad (7.3)$$

with $p$ being the total number of components.

To be able to interpret the meaning of the new variables (the principal components), we need to know their relationship with the original ones (the input bands). Similarly, to calculate the equations that will help obtain the new images, the coefficients of the transformation are needed. Both aspects can be deduced from the eigenvectors. The eigenvector indicates the weighting that must be applied to each of the original bands to obtain the new PC ($a_{i,j}$ in Equation 7.2). Briefly, it is equivalent to the regression coefficients in a standard linear transformation, the bands of the image being the independent variables and the PCs the dependent variables.

From the eigenvectors, the matrix of correlation between the PC and image bands is calculated, which allows us to know the spectral direction of these components. To do this, we must apply

$$r_{i,j} = \frac{a_{i,j}\sqrt{\lambda_j}}{s_i} \quad (7.4)$$

where

$r_{i,j}$ indicates the correlation coefficient between the component $j$ and band $i$
$a_{i,j}$ is the eigenvector of that component in the same band
$\lambda_j$ is the eigenvalue of the component $j$
$s_i$ is the standard deviation of band $i$

The last step is to obtain an image of the principal components using Equation 7.2. Once the image of the PCs is obtained, they can be represented in black and white or in color, enabling a new visualization of the study zone. To do this, we must scale the results to the most common range in visualization equipment (0–255). A simple method of carrying out this process would be to take a small sample of the resulting values in order to estimate the maximum and minimum values, and then apply some of the procedures to increase the contrast as previously mentioned (Section 6.5.1).

In order to illustrate the utility of the PCA in digital image interpretation, this technique was applied to the Tucson image that we used in Chapter 6.

The variance–covariance matrix for the seven bands of the image was computed (Table 7.1). The eigenvalues of the image were calculated on that matrix after performing a standardization, in order to reduce the effect of the different variabilities between bands. The eigenvalues ($\lambda_j$) extracted were 4477.44, 613.23, 112.82, 42.29, 14.77, and 4.38. Following Equation 7.3, the original variance associated with each of these components can be calculated as 85.04% for the first component, 11.65% for the second, 2.14% for the third, 0.8% for the fourth, 0.28% for the fifth, and 0.08% for sixth. According to these percentages, only the first three components would be selected, as they retained a total of 98.83% of the original variance of the image.

The spectral meaning of these components can be obtained from the eigenvector matrix (Table 7.2). In this case, the first component is a weighted average of all the bands and provides information on the general brightness of the image. The second component shows a contrast between NIR and SWIR bands and the visible ones, and seems more related to soils and building materials. Finally, the third component shows the contrast between the VIS and SWIR reflectance, which may be associated with dark objects and shadows.

The images of the resulting PCs seem to confirm this theoretical interpretation (Figure 7.3). The first is very similar in appearance to a panchromatic image, with low values in the water and vegetation, average values in the built-up areas, and high values over bare soils. The second component offers higher values for the urbanized area of Tucson, with a clear contrast with agricultural and vegetated areas. Finally, the third component displays in brighter tones the darker targets of the image, such as volcanic areas, asphalt, and water.

**TABLE 7.1**
**Variance–Covariance Matrix of the Tucson Image**

| | B1 | B2 | B3 | B4 | B5 | B7 |
|---|---|---|---|---|---|---|
| B1 | 1443.30 | | | | | |
| B2 | 443.07 | 243.85 | | | | |
| B3 | 1043.43 | 414.51 | 1166.48 | | | |
| B4 | 859.79 | 260.55 | 503.65 | 630.02 | | |
| B5 | 999.67 | 319.26 | 665.64 | 664.89 | 741.61 | |
| B7 | 1020.65 | 342.95 | 1065.25 | 492.80 | 642.28 | 1039.70 |

**TABLE 7.2**
**Eigenvector Matrix of the Tucson Image**

| | B1 | B2 | B3 | B4 | B5 | B7 | λ | Variance (%) |
|---|---|---|---|---|---|---|---|---|
| CP1 | 0.552 | 0.188 | 0.465 | 0.320 | 0.380 | 0.444 | 4477.44 | 85.04 |
| CP2 | 0.285 | −0.017 | −0.546 | 0.503 | 0.385 | −0.468 | 613.23 | 11.65 |
| CP3 | 0.250 | −0.847 | −0.252 | −0.033 | −0.063 | 0.389 | 112.82 | 2.14 |

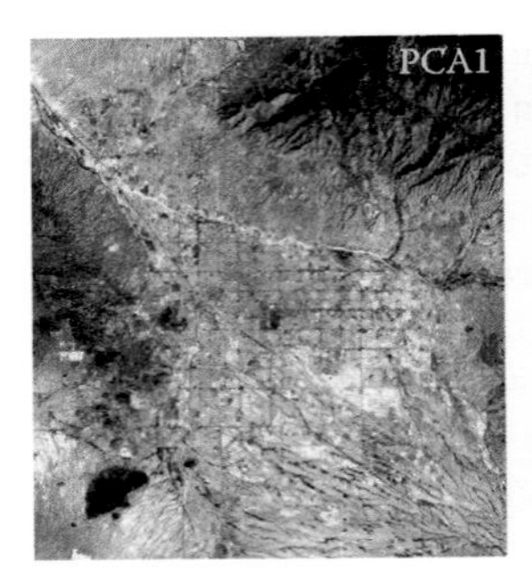

**FIGURE 7.3** Results of principal component analysis on the Tucson study area.

PCA is also a very appropriate technique for multiseasonal analysis when several images are used in a single classification. In that case, PCA serves to summarize the most significant information of every period. Later, the PCs of every date are combined to facilitate a better discrimination of land covers with distinct seasonal profiles. Several authors have taken advantage of this technique to perform multitemporal classification (Maselli et al. 1995; Siljeström and Moreno 1995), or to analyze a temporary series of images for global vegetation monitoring (Eastman and Fulk 1993). PCA provided in this case, in addition to the general patterns of the land covers, very useful residual information for detecting anomalies. Later, a more detailed use of the PCA as a technique for change detection between two images with different dates will be discussed (see Section 7.3.4.3).

In another context, the PCA has also been suggested as an alternative to merging data from sensors with different resolutions (Chavez et al. 1991). In this case, instead of replacing the SPOT panchromatic channel for the intensity component, it is replaced by the first PC, which is the one that gathers the brightness or general intensity of the image, returning later to the original variables with the inverse transformation. Also, other authors have used this technique to enhance the visual analysis of colored composites (Chavez and Kwarteng 1989).

An inherent problem of PCA is the difficulty of establishing an a priori interpretation of the components, since the transformation is scene dependent. In most cases, the first component is an indicator of the general brightness of a scene and the second one is related to the greenness in an image (Ingebritsen and Lyon 1985). However, in areas of low vegetation cover, the greenness component axis may not show up until the third or even fourth component. Actually, this is what happens in the Tucson image, which has a sparse vegetation cover, with significant influence of soils and lithological changes, and for that reason the greenness component was not found among the first three components.

Therefore, general physical rules for the interpretation of images derived from PCA cannot be applied. As will be discussed later in this chapter, PCA is a precursor step to unmixing techniques (SMA) and generally reveals the level of spectral information content, or dimensionality, of an image. In other words, PCA may reveal that there are *n* uniquely reflecting features in an image, and the task of SMA is to identify these unique landscape elements and assess their relative amounts in each pixel.

A particular application of PCA in the context of processing hyperspectral images is named *minimum noise fraction* (MNF) transformation (Pinzón et al. 1998). MNF is used to reduce the dimensionality of hyperspectral data by separating the noise in the data. The MNF transform is a linear transformation that is essentially two cascaded PCA transformations. The first transformation decorrelates and rescales the noise in the data. This results in transformed data in which the noise has unit variance and no band-to-band correlations. The second transformation is a standard PCA of the noise-whitened data.

### 7.1.3 Spectral Vegetation Indices

VIs are simple and robust techniques to extract quantitative information on the amount of vegetation, or greenness, for every pixel in an image. They typically involve spectral transformations of two or more bands, one in the chlorophyll-absorbing R spectral region (0.6–0.7 μm) and the other in the nonabsorbing NIR (0.7–1.1 μm), where there is significant leaf scattering. This feature is quite unique to photosynthetically active vegetation, since senescence vegetation tends to reduce NIR while increasing R reflectance (Figure 7.4). Soils regularly have little or no contrast between the R and NIR bands.

The R and NIR reflectance are combined in spectral VIs to enhance the vegetation signal while minimizing nonvegetation influences. In contrast to PCA and other techniques, VIs are designed to provide information on only one feature in an image, namely, vegetation greenness. As a result of their simplicity, VIs have proved to be among the most robust techniques in remote sensing, yielding consistent spatial and temporal comparisons of green vegetation at local to global scales.

VIs not only are used as a direct optical measure of greenness but are also useful as a proxy measure of biophysical variables, including LAI, fAPAR, green vegetation fraction (Fv), biomass, and photosynthesis (Kerr and Ostrovsky 2003;

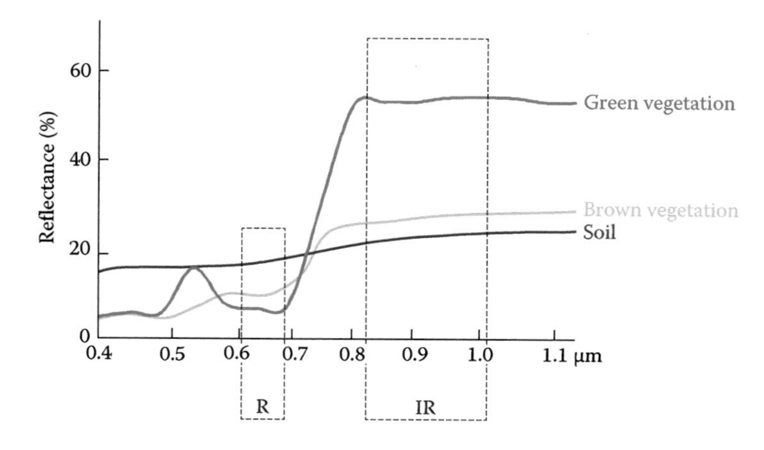


FIGURE 7.4 Vegetation indices, mostly based on the red and near-infrared reflectance contrast of green vegetation.

Pettorelli et al. 2005). The term *greenness* itself may be defined as a composite signal of leaf chlorophyll content, leaf area, canopy cover, and canopy structure. In effect, a VI is an "integrator" of many variables that together determine the photosynthetically active vegetation signal (Baret and Guyot 1991; Huete et al. 2002; Myneni et al. 1995). The term *greenness* recognizes both the strengths and limitations of VIs for directly retrieving specific biophysical variables. On the one hand, VIs exhibit high sensitivity to spatial and temporal variations in greenness. On the other hand, they generally cannot differentiate between the multiple causes responsible for a change in greenness as, for example, whether variations in the VI signal are due to a change in LAI or in leaf chlorophyll content. VIs are constrained in their ability to directly retrieve specific biophysical variables since they represent a composite of many vegetation and leaf canopy properties. Nevertheless, for local areas and specific land cover types, many have found VIs useful in constructing empirical and semiempirical models for the extraction of biophysical vegetation information.

When land surface pixels are plotted in red and NIR space, a triangular pattern of points is observed, which may be physically defined by (1) a lower baseline of substrate background spectra (with no green vegetation) that extends linearly outward from the origin, commonly known as the *soil line* (Figure 7.5), and (2) a green apex of maximum NIR reflectance and minimum red reflectance, indicative of very dense green vegetation. The soil line encompasses bright soil pixels (furthest away from the origin), dark soils (closer to the origin), and water (closest to the origin) and is defined as

$$\text{NIRsoil} = a * \text{Redsoil} + b \tag{7.5}$$

where $a$ and $b$ are the best fit soil line slope and intercept, respectively, for a range of soil backgrounds at different stages of wetting and drying, roughness, and

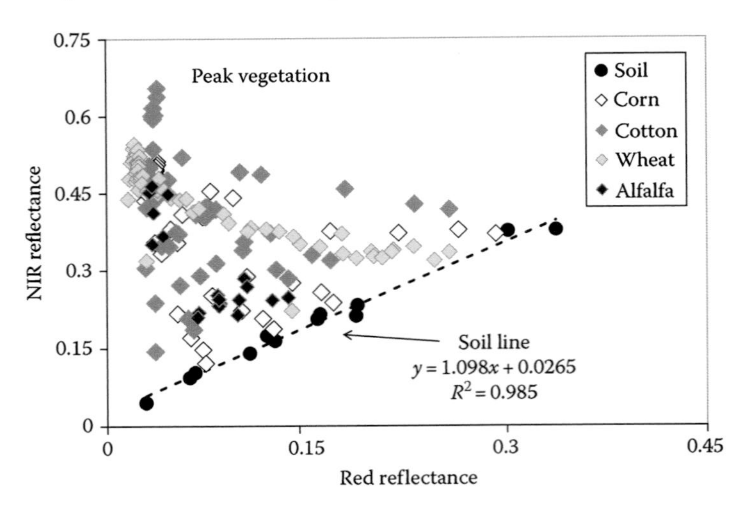


**FIGURE 7.5** Structure of vegetation spectra plotted in the red–near-infrared reflectance space. (Courtesy of Alfredo Huete.)

litter content. Different land cover types and biomes will follow the triangular pattern of pixels in the red–NIR space, but with slightly different soil lines or apices of maximum greenness. Note that the majority of pixels will plot inside the triangular cloud of spectra, indicating open canopies consisting of soil, water, and vegetation. The task of the VI is to quantify the amount of vegetation present, or greenness, in a pixel, based on its position in red–NIR space. For example, the closer a pixel is to the soil line, the lower will be the amount of vegetation present in that pixel.

There are a variety of ways one can model the cloud of pixels present in the red–NIR spectral space to estimate greenness or a specific biophysical variable. This has resulted in a multitude of VI equations, in which the red and NIR bands are combined differently, including band ratios (Gallo and Eidenshink 1988), normalized differences (Tucker 1979), linear band combinations (Crist and Cicone 1984b; Kauth and Thomas 1976), and optimized band combinations (Gobron et al. 2000; Huete 1988; Pinty and Verstraete 1992). Many of the VIs depict similar information and may be considered functionally equivalent to each other (Perry and Lautenschlager 1984), and so here we will discuss only a few examples of the different types of VIs available for satellite vegetation studies, namely, the (1) ratio-based indices, as in the simple ratio (SR), normalized difference vegetation index (NDVI), and vegetation water indices; (2) optimized VIs, such as the soil-adjusted vegetation index (SAVI), the global environment monitoring index (GEMI), and the enhanced vegetation index (EVI); and (3) linear orthogonal indices, which include the perpendicular vegetation index (PVI) and the tasseled cap green vegetation index (GVI). Although all VIs have been related to some extent with various vegetation canopy properties, there are significant differences in how they depict greenness and are, therefore, potentially useful and complementary.

#### 7.1.3.1 Ratio-Based VIs

The first VIs were formed by spectral ratioing of the NIR with red band using field radiometers (Jordan 1969; Tucker 1979). The greater the amount or vigor of green vegetation present in a given pixel, the larger the contrast between NIR and red values and the higher the resulting NIR-to-R ratios (Figure 7.6). This index is often called the *simple ratio*:

$$SR = \frac{\rho_{NIR}}{\rho_R} \tag{7.6}$$

where $\rho_{NIR}$ and $\rho_R$ indicate the reflectances of a pixel in the NIR and R bands, respectively. The SR values may be calculated from bands 2 and 1 in the Advanced Very High Resolution Radiometer (AVHRR) and MODIS instruments, bands 4 and 3 with thematic mapper (TM) and ETM+, bands 5 and 4 with OLI, and bands 3 and 2 in the SPOT-HRV, SPOT-VGT, and ASTER sensors.

The SR values in a digital image allow one to separate healthy vegetation from soil, snow, mulch, water, and other nonvegetation cover types. The SR values also estimate, pixel by pixel, the amount of vegetation present, from sparse to dense foliage cover (Figure 7.6). As vegetation undergoes stress (drought, disease, nutrients), its NIR reflectance decreases while the red reflectance increases because of lower

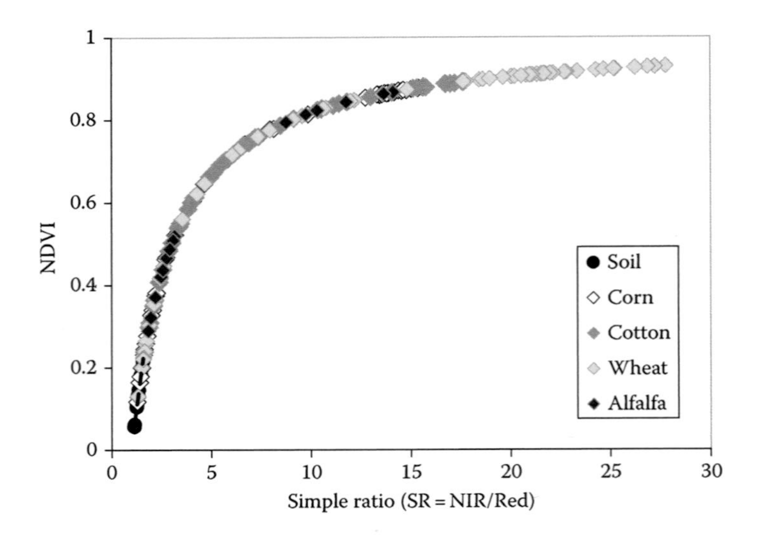


**FIGURE 7.6** Functional and nonlinear relationship between the normalized difference vegetation index and the simple ratio. (Courtesy of Alfredo Huete.)

chlorophyll absorption, thus reducing the overall ratio or spectral contrast of the pixels containing the stressed vegetation. A lower NIR/R spectral contrast, however, may indicate either a lower quantity of vegetation present in a pixel or a higher amount of vegetation but in stressed condition. Quite often, the cause of a change in the NIR/R ratio may be deduced from additional knowledge such as the temporal pattern of SR values, in which, for example, the reduced spectral contrast may be a result of less vegetation, senescing vegetation, or vegetation under stress.

Ratios are useful in their ability to reduce many forms of multiplicative noise, such as illumination differences, atmospheric attenuation, cloud shadows, and topographic variations (slope, aspect), all of which influence multiple bands similarly and, hence, may be minimized through band ratioing. Ratios, however, are sensitive to the soil background underneath vegetated canopies with values that are positively biased when soils are dark or wet (Major et al. 1990).

NDVI is a functional variant of the SR, with a dynamic range from −1 to +1:

$$NDVI = \frac{\rho_{NIR} - \rho_R}{\rho_{NIR} + \rho_R} \tag{7.7}$$

or

$$NDVI = \frac{SR - 1}{SR + 1} \tag{7.8}$$

SR has no upper bound, since as the denominator approaches zero, the SR value would increase without bound, approaching infinity. NDVI is a normalized but functionally equivalent version of the SR. By functional equivalence we mean that for any SR value, there can be only one predefined value of NDVI in accordance with Equation 7.8 (Figure 7.6). However, note that this nonlinear transformation results in NDVI being highly sensitive to low vegetation amounts and sparse vegetation conditions over which the SR is largely insensitive, while the opposite is true for the SR in which most of the dynamic range encompasses the high biomass vegetation range, with very little variation reserved for the lower biomass regions (grassland, semiarid, and arid biomes). Thus, NDVI values from 0 to 0.5 (half the positive dynamic range) only translate into SR values from 1 to 3, while the NDVI values from 0.5 to 0.9 translate into SR values from 3 to 19.

NDVI is successful as a vegetation measure in that it is sufficiently stable to permit meaningful comparisons of seasonal and interannual changes in vegetation growth and activity. As a ratio-based index, NDVI maintains the benefits of band ratios and normalizes multiplicative sources of extraneous noise to produce more stable values. Figure 7.7 shows the NDVI of the Tucson image. In spite of the low vegetated cover of this region, this VI clearly emphasizes the most vigorous green sectors of the city, such as the Reid Park, the irrigated areas near the Santa Cruz River, and the numerous golf courses, with the Randolf Golf course in the city center being depicted with special clarity. The forested areas of the Santa Catalina Mountains to the north of the city and the irrigated fields to the south (Sahuarita sector) are also clearly displayed. By contrast, the mining sectors to the south of the image and the industrial area near the airport show the darkest tones (lowest NDVI values).

FIGURE 7.7 Normalized difference vegetation index of the Tucson OLI image.

NDVI is used extensively to estimate several biophysical and ecologic variables of particular value in landscapes and for global analysis. Among the parameters that have been related to NDVI are the following:

1. LAI, which has a positive relationship with the NDVI, especially at low LAI values between 0 and 2. The sensitivity of NDVI to LAI, however, becomes increasingly weak with increasing LAI beyond values typically between 2 and 3, and becomes saturated (Baret and Guyot 1991; Tian et al. 2002).
2. Fractional vegetation cover, Fv (Carlson and Ripley 1997; Hansen et al. 2002; Jiang et al. 2006). In contrast to LAI, there is a much higher correlation between NDVI and percentage of green cover with minimal saturation issues.
3. Fraction of photosynthetically active radiation absorbed by the plant ($f_{APAR}$), given by

$$f_{APAR} = \frac{APAR}{PAR} \tag{7.9}$$

   where
   $APAR$ is the absorbed photosynthetically active radiation
   $PAR$ is the incident photosynthetically active radiation

   This radiation involves only the VIS portion of the spectrum (400–700 nm), and the amount absorbed is strongly related to vegetation productivity. NDVI has been shown to be strongly linearly correlated with $f_{APAR}$ (Asrar et al. 1984; Goward and Huemmrich 1992), especially when the leaves are planar and the soil substrates are dark (Sellers 1989). In a multibiome study across North America, Sims et al. (2006) estimated a global NDVI and $f_{APAR}$ relationship as

$$f_{APAR} = 124 * NDVI - 0.168 \tag{7.10}$$

4. Chlorophyll content in the leaf, through direct and linear relationships with NDVI (Blackburn 2002; Broge and Leblanc 2001).
5. Water content in the leaf, through indirect relationships between water deficit and amount of foliage (Yebra et al. 2013b). In addition, VIs, including NDVI, have also been found useful in estimating landscape biophysical processes, such as photosynthesis, ET, and phenology events.
6. Net $CO_2$ flux, at local and global levels (Hall et al. 1991a; Tucker et al. 1986).
7. Gross and net primary productivity of the vegetation, as related to APAR through a light use efficiency (LUE) term, specific to vegetation type (Sellers 1989; Sims et al. 2006), following

$$LUE = \frac{GPP}{APAR} \tag{7.11}$$

   where $GPP$ is gross primary productivity (photosynthesis) and $GPP$ and $APAR$ are expressed in the same molar units.
8. ET, related to NDVI directly or through leaf vigor, and leaf water stress (Yebra et al. 2013b).

Global coarse-resolution (8 km resolution) satellite NDVI datasets derived from the NOAA-AVHRR have been operationally implemented since 1981 and are routinely used for drought monitoring, for assessments of ecosystems and agricultural areas, and as key variables in global change studies (Myneni et al. 2002). The Global Inventory Modeling and Mapping Studies (GIMMS) and Pathfinder AVHRR Land (PAL) data sets are derived from imagery obtained from the AVHRR instrument onboard the NOAA satellite series 7, 9, 11, 14, 16, and 17 and have been available since 1981. These AVHRR-NDVI datasets have been corrected for sensor calibration, view geometries, volcanic aerosols, and other effects not related to vegetation change (Tucker et al. 2005). Later, the U.S. Geological Survey made available AVHRR-NDVI datasets at 1 km resolution over the United States (Eidenshink 2006; Millington et al. 1994).

Annually averaged NDVI time series images, acquired over several years, have also enabled improved discrimination of land cover types for use in global land cover classifications. These classifications are based on NDVI measurements of seasonal phenologic profiles and are supported with ancillary information, such as topography, soil maps, and biogeographical data (Hansen et al. 2000; Townshend et al. 1991). Phenomena such as desertification and the advance of deserts (Eva and Lambin 1998; Tucker et al. 1991), tropical deforestation (Eva and Lambin 1998; Shimabukuro et al. 1994), impacts of forest fires (Chuvieco et al. 2008a), and biomass characterization at continental scales (Goward and Dye 1987) have been studied effectively with the AVHRR VI images.

The long-term time series NDVI data record has also been recently extended to include newer sensor systems such as MODIS and SPOT-VEGETATION (Tucker et al. 2005). AVHRR and MODIS VIs are used in near-real-time and interactive monitoring tools for Famine Early Warning Systems (USAID, http://www.fews.net, last access November 2015). Zha et al. (2005) used MODIS NDVI to monitor locust infestation and outbreaks in East China. MODIS VI data are used for operational early warning deforestation detection in the Amazon (Ferreira et al. 2007) and invasive species forecasting systems. An increasing number of natural resource managers are using web-based geospatial decision support tools that utilize both historical and current NDVI data for planning and management decisions.

#### 7.1.3.2 Optimized VIs

Despite the usefulness of NDVI for vegetation studies, as with most remote sensing data, it is subject to errors and uncertainties. There are also many disadvantages to the use of ratio-based indices that limit their extension, scaling, and use in quantitative studies. These include the nonlinear behavior of ratios, their sensitivity to soil background and additive atmosphere influences, and saturation of the NDVI signal at moderate to high vegetation densities (Huete et al. 2002; Jiang et al. 2006). Some of these influences can be best handled with corrections to the spectral reflectances prior to computation of the NDVI, and some of the corrections can be designed into the equation itself, based on physical theory. Optimized VIs are an example of the latter case, in which the VI equation itself is designed to measure the NIR and red spectral contrast of a pixel while removing unwanted nonvegetation or extraneous influences.

The NDVI, although a vegetation measure, also responds to spatial and temporal variations in the soil underneath open vegetation canopies, rendering it difficult to reliably distinguish vegetation changes from those caused by the soil. Soils are very heterogeneous spatially but vary to a smaller extent temporally (usually through drying and wetting events). In field studies, NDVI–biophysical relationships have been shown to vary significantly with the brightness of the underlying soil background, with darker soils and wet soils resulting in much higher NDVI values (up to twofold) than brighter or drier soils for equivalent vegetation conditions (Huete et al. 1985; Figure 7.8).

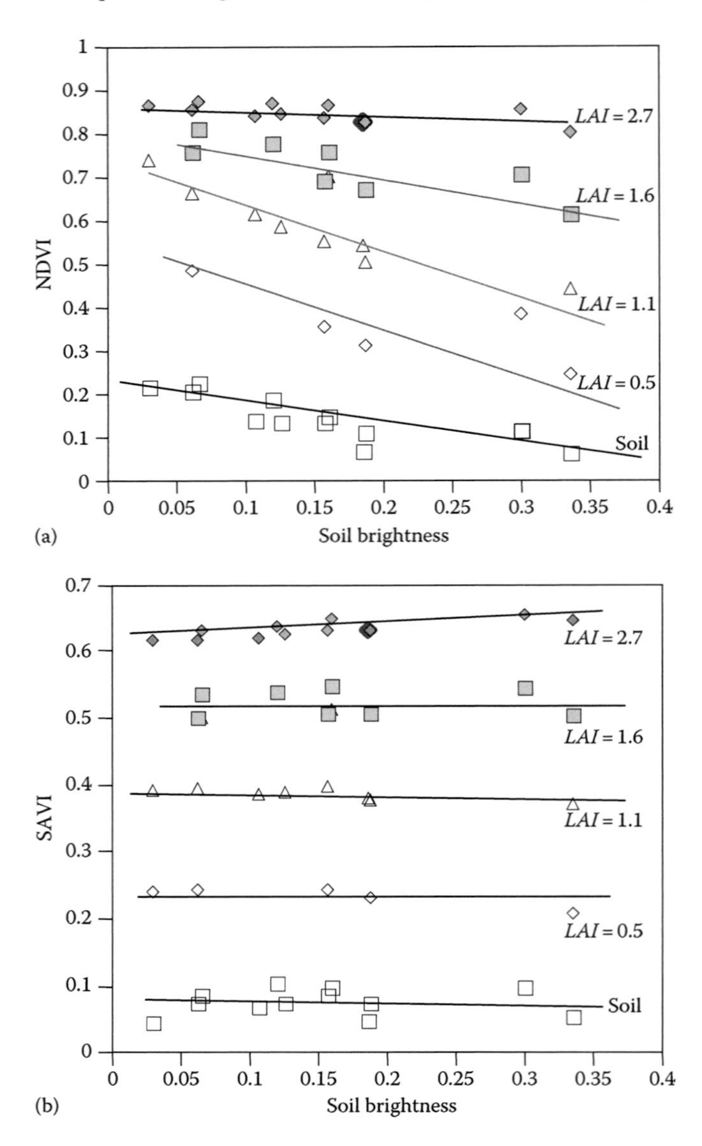


**FIGURE 7.8** Influence of soil substrate brightness on (a) normalized difference vegetation index and (b) soil-adjusted vegetation index. (Adapted from Huete, A.R., *Remote Sens. Environ.*, 25, 295, 1988.)

This effect is a result of first-order soil–vegetation optical interactions, which are most pronounced over intermediate vegetation amounts. Huete (1988) found that such interactions can be approximated with a first-order Beer's law application, which allows the influence of the backscattered soil signal to be removed. Consequently, the NDVI equation could be corrected by incorporating a parameter ($L$) related to the differential optical penetration of R and NIR light through a vegetation canopy, resulting in the SAVI:

$$SAVI = \frac{\rho_{NIR} - \rho_R}{\rho_{NIR} + \rho_R + L}(1 + L) \tag{7.12}$$

where $L$ is the soil adjustment factor. In field experimental studies over agriculture crops and grasslands, an $L$ value of 0.5 was found to best model soil influences across all vegetation types and amounts (Figure 7.8b). In contrast to NDVI, SAVI provides the opportunity to more accurately measure greenness through the use of local-, regional-, or global-based soil lines as zero vegetation reference baselines. In the case of NDVI, the soil line is simply taken to be NIR = R (slope = 1, intercept = 0, and $NDVI = 0$). Baret et al. (1989) developed the transformed SAVI (TSAVI) based on SAIL model simulations (Verhoef 1984), which explicitly use the soil line:

$$TSAVI = \frac{a(\rho_{NIR} - a\rho_R) - b}{a\rho_{NIR} + \rho_R + ab} \tag{7.13}$$

where $a$ and $b$ are the coefficients of the soil line as in Equation 7.5. Other variants of SAVI include the generalized SAVI (GESAVI; Gilabert et al. 2002) and the modified SAVI (MSAVI; Qi et al. 1994).

Another source of noise that can greatly influence NDVI values is atmospheric contamination, primarily associated with aerosols, water vapor, and molecular scattering. As a ratio, the NDVI is particularly sensitive to additive noise influences, such as atmospheric path radiances. Overall, atmospheric effects lower NDVI values by reducing the contrast between the red and NIR reflected signals. Atmospheric water vapor attenuates the reflected NIR signal from vegetation, particularly with the AVHRR-NIR bandwidth that overlaps with water absorption regions. Rayleigh and aerosol scattering tend to add and increase the very low red reflected signals from vegetated surfaces. Thus, NIR attenuation and red reflectance increases lower the NIR to red contrast and will lower VI values.

The most accurate and rigorous way to compute VIs is by using atmosphere-corrected surface reflectance values. An alternative method of minimizing atmospheric influences on VIs is to design the VI equation to self-correct for atmospheric effects. The atmospherically resistant vegetation index (ARVI) was developed to remove the influence of aerosols through the use of the blue band (which is the most sensitive to atmosphere) to correct for aerosol effects in the red band (Kaufman and Tanré 1992):

$$ARVI = \frac{\rho^*_{NIR} - \rho^*_{RA}}{\rho^*_{NIR} + \rho^*_{RA}} \tag{7.14}$$

where

$\rho^*_{NIR}$ indicates the top-of-the-atmosphere (ToA) reflectance in the NIR

$\rho^*_{RA}$ is a factor that considers the difference of reflectance between blue and the red bands, as defined by

$$\rho^*_{RA} = \rho_R - \gamma(\rho^*_B - \rho^*_R) \tag{7.15}$$

where

$\rho^*_B$ and $\rho^*_R$ indicate the ToA reflectances in the B and R bands, respectively

$\gamma$ is a calibration parameter that depends on the type of atmosphere, although in most cases it is equal to 1

In the case of ARVI, the ToA reflectances are corrected for Rayleigh scattering and ozone absorption, which are relatively straightforward. Another self-corrected VI is the GEMI, which was designed to be resistant to atmosphere effects in NOAA-AVHRR data sets (Pinty and Verstraete 1992):

$$GEMI = (1 - 0.25) - \frac{(\rho_R - 0.125)}{(1 - \rho_R)} \tag{7.16}$$

where

$$\eta = \frac{2(\rho^2_{NIR} - \rho^2_R) + 1.5\rho_{NIR} + 0.5\rho_R}{(\rho_R + \rho_{NIR} + 0.5)}$$

GEMI has been found particularly useful for burned land mapping with coarse-resolution data, being more robust than NDVI to AVHRR radiometric noises (Chuvieco et al. 2008a; Pereira 1999).

The EVI was formulated as a new satellite VI product for the MODIS sensor. The EVI was designed to minimize soil and aerosol influences and render the VI signal more sensitive in high-biomass areas. The EVI gains its heritage from the SAVI and the ARVI and is an optimized combination of blue, red, and NIR bands, designed to extract canopy greenness, independent of the underlying soil background and atmospheric aerosol variations. The blue band in the EVI is primarily used to stabilize aerosol influences in the red band resulting from residual and miscorrection of aerosols (Xiao et al. 2003):

$$EVI = 2.5\frac{\rho_{NIR} - \rho_R}{\rho_{NIR} + C_1\rho_R - C_2\rho_B + L} \tag{7.17}$$

where

$\rho_{NIR}$, $\rho_R$, $\rho_B$ are reflectances in the NIR, R and B bands, respectively

$L$ is the canopy background adjustment factor

$C_1$ and $C_2$ are the aerosol resistance weights (Liu and Huete 1995)

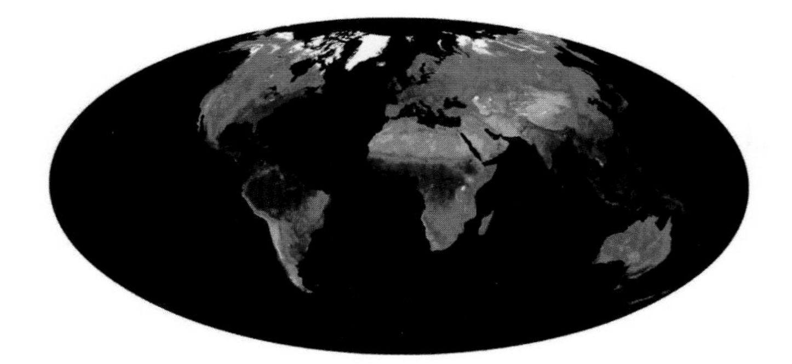

**FIGURE 7.9** MODIS enhanced vegetation index global image computed from the Terra satellite. (From http://visibleearth.nasa.gov, last accessed November 2015.)

The coefficients of the MODIS EVI equation are $L = 1$, $C_1 = 6$, and $C_2 = 7.5$ (Huete et al. 2002). The EVI becomes increasingly sensitive to the NIR band at moderate to high vegetation amounts with a greater optical depth penetration into canopies. Thus, the EVI better depicts biophysical canopy structural variations and is less prone to saturation in high biomass areas. Figure 7.9 shows an example of the global MODIS EVI product, showing the spatial differences in vegetation amount and productivity.

The EVI, along with NDVI, is a standard MODIS vegetation product on both Terra and Aqua platforms, and has been selected as a standard product for the VIIRS sensor on board the NPOESS mission (Murphy et al. 2006). A good example of EVI sensitivity in high biomass conditions was encountered in Brazilian Amazon and Southeast Asian tropical forests, in which pronounced seasonal patterns in satellite EVI data were observable and were highly seasonally correlated with seasonal flux tower measurements of photosynthesis or gross primary productivity (GPP; Huete et al. 2006). Other studies involving the coupling of flux tower measurements of GPP and ET with MODIS satellite EVI have also revealed strongly linear relationships (Sims et al. 2006).

Reduction of saturation effects and improvement of linearity add to the observed accuracy in estimating biophysical parameters from VI values and provide a mechanism for multisensor scaling of VI values. There have been several other studies aimed at mitigating the saturation problem of NDVI. One approach is to use alternative narrow bands, away from maximum chlorophyll absorption wavelengths at 680 nm, for the R band in the NDVI (Gitelson et al. 2002). However, this approach requires sensors designed with additional narrow bands or hyperspectral sensors, which are not yet available for operational applications. In fact, the saturation problem in MODIS NDVI is more serious than that encountered with the AVHRR instrument, partly due to the much narrower MODIS red band, which is more sensitive to chlorophyll absorption and hence saturates more quickly. Gitelson (2004) noted the lack of sensitivity of NDVI to the NIR with increasing levels of vegetation and suggested the use of weighting coefficients to enhance NDVI sensitivity. He proposed the wide dynamic range vegetation index (WDRVI)

$$WDRVI = \frac{aN - R}{aN + R} \tag{7.18}$$

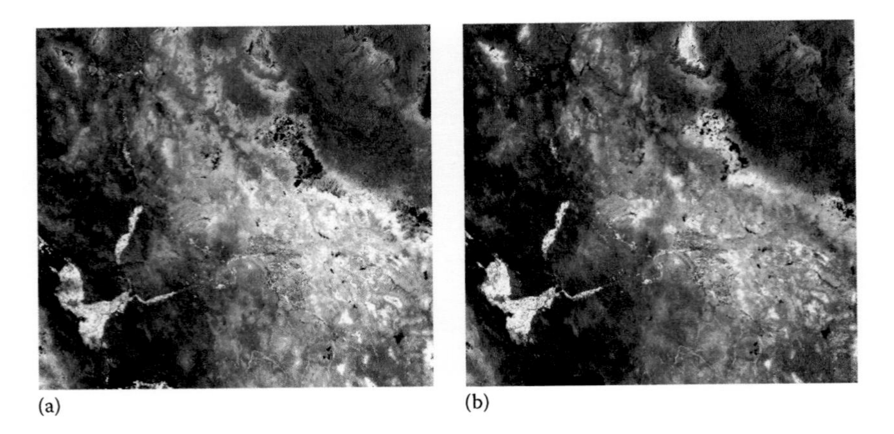


**FIGURE 7.10** Image of normalized difference vegetation index (a) and enhanced vegetation index (b) for the MODIS image of the Southwestern United States.

and suggested $a = 0.2$ as the value of the weighing coefficient. WDRVI was found to have a more linear relationship with vegetation fraction, the $f_{APAR}$, and much higher sensitivity to change in LAI over crops in comparison with NDVI.

An example of the two standard MODIS VI products for the Southwestern United States and California is shown in Figure 7.10. The agricultural regions along the United States–Mexico border display large cross-border differences in VI values. The Colorado River and Gila River riparian corridors show up as bright areas (lighter-color tones) with much more vigorous vegetation that has access to year-round water. The forested mountain areas show up as brighter areas (higher VI values) surrounded by desert soils that are fairly dark (low vegetation conditions). Comparing the NDVI and the EVI, the former appears brighter overall because of high NDVI sensitivity in sparsely vegetated regions.

### 7.1.3.3 Orthogonal-Based VIs

Besides ratios, one can also design a VI that measures the contrast between the R and NIR spectral regions by subtracting or linearly combining the R and NIR bands ($C1 \times R + C2 \times NIR$). These VIs include the PVI (Richardson and Wiegand 1977) and the "tasseled cap" transform (TCT) GVI (Crist and Kauth 1986; Kauth and Thomas 1976) and are sometimes referred to as *orthogonal indices* because they measure the perpendicular distance of a pixel in the R–NIR space to the soil line (Figure 7.11). The simplest orthogonal VI is the difference vegetation index, $DVI = NIR - R$, which is a measure of the perpendicular distance of a pixel in R–NIR space to the 1:1 line. In contrast, the PVI calculates the orthogonal distance of a pixel to the "soil line":

$$PVI = \frac{(\rho_{NIR} - a * \rho_{R}) - b)}{(1 + a^2)^{1/2}} \tag{7.19}$$

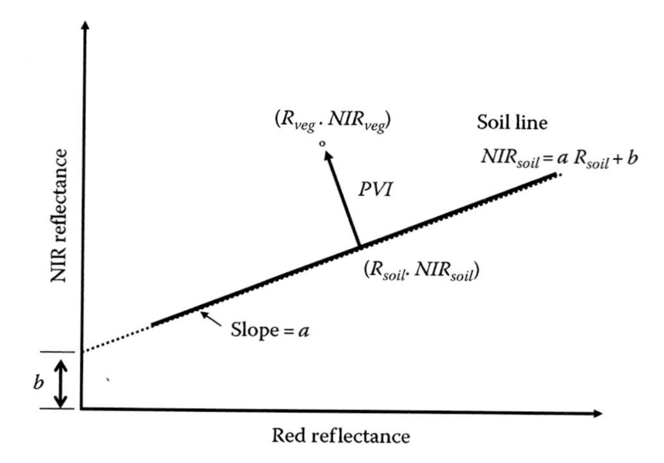


**FIGURE 7.11** Concept of orthogonal indices in near-infrared–red reflectance space. In this example, the perpendicular vegetation index is illustrated as the orthogonal distance of a vegetated pixel, *o*, from the soil line. (Courtesy of Alfredo Huete.)

where $a$ and $b$ are the slope and intercept of the soil line, respectively, as defined in Equation 7.5. The soil line generally has a slope that is slightly greater than one (~1.1 to 1.2). As with SAVI, PVI provides the opportunity to more accurately measure greenness through the use of local-, regional-, or global-based soil lines, whereas in the case of DVI, the soil line is simply taken to be $NIR = red$ (slope = 1, intercept = 0, and $DVI = 0$). Orthogonal indices such as DVI, PVI, and GVI are generally more linear and scalable and have less saturation problems than NDVI; however, they do not ratio out extraneous sources of noise and thus require accurate surface reflectance inputs.

TCT is a multidimensional interpretation of the spectral structure of pixels in an image incorporating additional spectral bands besides the R and NIR. This transform was developed by Kauth and Thomas (1976) in the framework of the joint NASA–USDA (U.S. Department of Agriculture) Large Area Crop Inventory Experiment (LACIE) project in the 1970s, with the aim of improving the prediction of grain harvests. Using a temporal series of Landsat MSS images over agricultural areas, the fundamental spectral components of crop development were modeled to characterize green-up, senescence, and crop fallow periods (Figure 7.12). As such, the spectral–temporal structure of crop development resembled a 3D tasseled cap shape represented by a soil brightness axis, a greenness axis, and a yellowness axis. The base of the tasseled cap consists of a reference "soil plane" in which bare soils with variable colors and brightness are found (Figure 7.13). Crop green-up and development proceeds from a specific location on the soil plane, based on the soil type, along the greenness axis away from the soil plane, and toward the point of maximum greenness. Finally, with crop senescence and yellowing, the crop pixels migrate back toward the soil plane, but along a separate yellowness axis. TCT extends the 2D PVI framework into four dimensions in the case of Landsat MSS data, and six dimensions

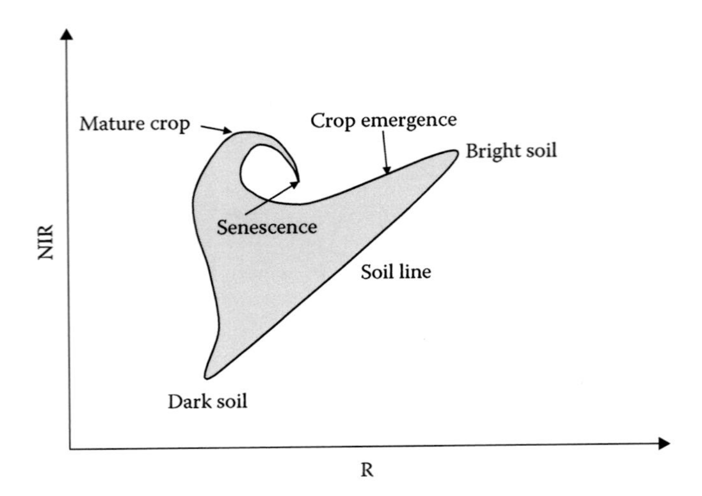


FIGURE 7.12 Temporal trends of crop reflectance plotted in near-infrared and red band space.

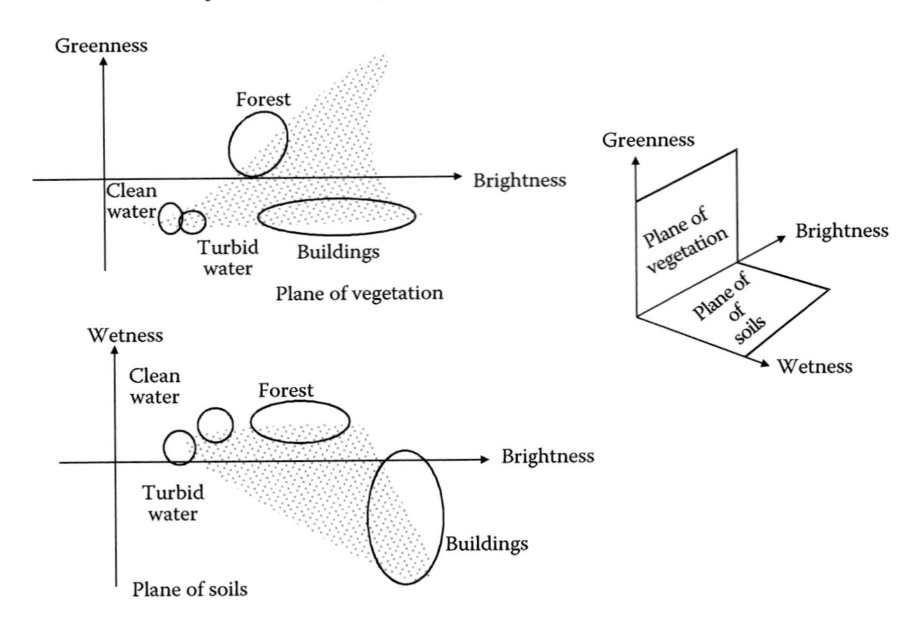


FIGURE 7.13 Spectral planes defined by the tasseled cap transformation.

for Landsat TM data, and enables simultaneous extraction of multiple biophysical features. The yellowness feature has been studied with interest for atmosphere haze analysis as well as for estimating vegetation senescence.

As in PCA, this transformation is used to obtain a multiple set of component features derived from linear combinations of the original bands, to quantify the unique features in a landscape. However, in contrast to PCA, the TCT generates a set of unique spectral feature components with more precise physical meaning, which are largely independent of the type of image that is being analyzed. These features typically include brightness, greenness, wetness, and yellowness. The TCT tends to show the spectral behavior of the vegetation and the soil well, by creating new axes that are better adjusted to that physical space. In that new system of coordinates, the separation between both covers is meant to be clearer. In the case of MSS images, this transformation was developed, with new axes obtained from

$$u_j = R_i' x_i + c \tag{7.20}$$

where

$u_j$ indicates the vector corresponding to the transformed image
$x_i$ is the input vector
$R_i'$ is the vector of transformation coefficients
$c$ is a constant to avoid negative values (the authors suggested a value of $c = 32$)

At first, this transformation was not widely accepted by the user community, in contrast to NDVI and other ratios. In the 1980s, however, it began to be used more, as Crist and Cicone adapted this approach to the Landsat TM bands (Crist and Cicone 1984a,b; Crist and Kauth 1986) as well as to AVHRR data. Table 7.3 includes the coefficients for Landsat TM images. The three main axes of variation are named brightness, greenness, and wetness:

1. *Brightness* reflects the changes in the total reflectance of the scene. As seen in Table 7.3, it is a weighted sum of all the channels, excluding the thermal one.
2. *Greenness* indicates the contrast between the VIS and the NIR bands. Band 4 stands out and there are negative weights in the VIS bands and a mutual cancellation of the VIS and NIR bands in the SWIR region.
3. *Wetness* is related to the water content in the vegetation and the soil and is marked by the contrast between the SWIR, where the absorption of water is strong, and the VIS and NIR bands. This is a new component that was

**TABLE 7.3**
**Coefficients to Obtain the Tasseled Cap Transform for Landsat Thematic Mapper Images**

| | B1 | B2 | B3 | B4 | B5 | B7 |
|---|---|---|---|---|---|---|
| Brightness | 0.3037 | 0.2793 | 0.4743 | 0.5585 | 0.5082 | 0.1863 |
| Greenness | −0.2848 | −0.2435 | −0.5436 | 0.7243 | 0.0840 | −0.1800 |
| Wetness | 0.1509 | 0.1973 | 0.3279 | 0.3406 | −0.7112 | −0.4572 |

*Source:* Crist, E.P. and Cicone, R.C., *IEEE Trans. Geosci. Remote Sens.*, GE22, 256, 1984.

not available in the original Landsat MSS data due to the lack of an SWIR band. Other authors have proposed calling the third TCT component *maturity* instead of *wetness*, since it has been related to the age and density of the vegetation cover (Cohen and Spies 1992). In addition, it was observed that this component was the least affected by the topography.

The combination of the TCT components provides several planes of variation (Figure 7.13), which are interesting for a detailed study of soils and vegetation. In this sense, three planes are usually distinguished: the vegetation plane, formed by the greenness and brightness axis; the plane of soils, formed by the brightness and the moisture; and the plane of transition, formed by the moisture and the greenness. Lobser and Cohen (2007) formulated the TCT for use with the seven primary land bands in MODIS data.

One of the advantages of this transformation compared to the PCA is that it can be physically interpreted. The components deduced from TCT have a precise meaning, regardless of the conditions of the scene, since they are supported by the characteristics of the spectral bands considered. Therefore, the proposed coefficients can be applied to different images from the same sensor. However, we should take into account that they were extracted from the agricultural characteristics of the American Midwest and may not be suitable for our particular target region. Another factor to be considered would be the scaling of the resulting image, which remarkably varies depending on the time of year and the conditions of illumination. For this, some of the procedures applied previously for ratios and VIs can be used.

Using the coefficients proposed by Crist and Cicone, the TCT components were calculated for the Tucson image (Figure 7.14). The first of the output components, brightness, displays spatial variations of albedo, and it is similar to a panchromatic image. The second one, greenness, is similar to the NDVI previously commented on. It also shows the vegetated areas of the city and surrounding mountains, with reduced contrast between built-up areas and bare soils. Finally, the wetness component shows mainly shade effects, since there are very few open water areas. The higher values for this component are observed in the volcanic regions of the western and southern sectors of the city, as well as paved areas, while the lower values correspond to bright soils.

**FIGURE 7.14** Tasseled cap transform components of the Tucson image.

#### 7.1.3.4 Fluorescence Indices

Most vegetation indices are usually interpreted as estimating chlorophyll activity. However, recent studies have shown that NDVI, EVI, and the most common vegetation indices are in fact associated with canopy structure but have little sensitivity to short-term physiological changes, particularly to those associated to heat and water stress (Zarco-Tejada et al. 2013). For this reason, new indices are based on chlorophyll fluorescence, which have been shown to be more directly associated to diurnal changes in photosynthetic activity.

Fluorescence is a biophysical process that implies that part of the absorbed photosynthetically active radiation (APAR) occurring mostly in the 0.4–0.7 μm spectral region is reradiated at longer wavelengths (0.66–0.8 μm). Several studies have shown the relation between fluorescence and gross primary production and $CO_2$ assimilation by plants, and therefore this information may be very useful to monitor global vegetation productivity (Frankenberg et al. 2011).

The detection of plant fluorescence requires sensors with very narrow spectral bands. Several methods to retrieve plant fluorescence from hyperspectral sensors have been published (Meroni et al. 2009), the most common being those based on the reduction of irradiance occurring at different narrow spectral bands in the region between 0.66 and 0.76 μm (commonly referred to as *Fraunhofer lines*), caused by hydrogen and oxygen absorption.

A specific satellite mission funded by the European Space Agency, called FLEX, intends to measure fluorescence at the global scale using high-resolution spectra. Previous estimations based on the narrow red bands recorded by the TANSO Fourier transform spectrometer (FTS), on board the Japanese satellite GOSAT, have shown promising results in analyzing global patterns of vegetation productivity (Frankenberg et al. 2011).

### 7.1.4 Other Spectral Indices

In addition to VIs, other combinations of input bands can be used to emphasize different image features. Spectral vegetation indices are always based on the radiance/reflectance contrast in two or more spectral bands between a certain target feature and others. One of the most widely used spectral ratio is the normalized difference snow index (NDSI), arithmetically similar to NDVI but using two different spectral bands (Hall et al. 1995):

$$NDSI = \frac{\rho_G - \rho_{SWIR}}{\rho_G + \rho_{SWIR}} \tag{7.21}$$

The G and SWIR bands are used in this case, as the snow has a high reflectance in the former and a low value in the latter, which makes it possible to discriminate snow from clouds (visually very similar), as clouds have high reflectance in both bands. NDSI has been extensively used in snow monitoring (Kelly and Hall 2008), and it is still the basis for the snow mapping algorithm of the MODIS sensor (Hall and Riggs 2007).

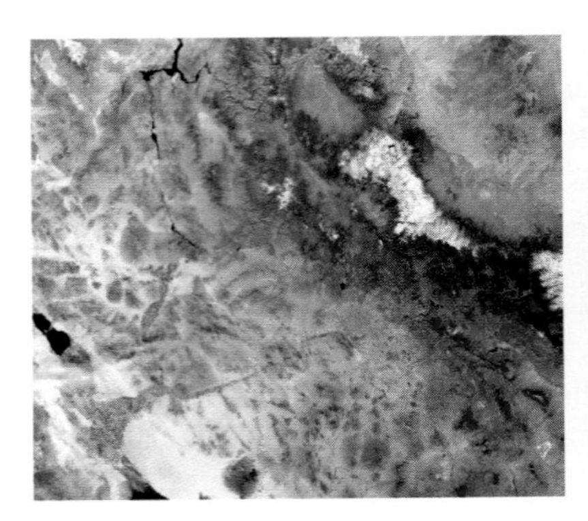

**FIGURE 7.15** Color infrared image and normalized difference snow index of the MODIS image of Southwestern United States.

Figure 7.15 shows the calculation of the NDSI over the MODIS image of SW USA, along with the infrared color image of the same date. Using the NIR, R, and G bands, it is difficult to discriminate between snow, clouds, and bright soils, while NDSI provides a clear separation. Visually, however, this index provides certain confusion with the water bodies (particularly the coastal areas), probably due to the impact of the atmospheric correction on low reflectances.

Other spectral indices have been applied to detect burned areas (Chuvieco et al. 2002; Martín and Chuvieco 2001) or improve the discrimination of water content on plants (Ceccato et al. 2002). The latter may even combine the solar spectrum with thermal bands, to estimate the water content by monitoring variations in air and surface temperatures (Sandholt et al. 2002; Yebra et al. 2013a).

### 7.1.5 Extraction of Subpixel Information

Spectral measurements made over terrestrial surfaces can be modeled as spectral mixtures involving two or more surface-reflecting features. The ground surface is rarely spectrally pure; most commonly, it contains complex vegetated canopies with leaves, stem, bark, litter, soil background, and shadows. Approximately, 70% of the Earth's terrestrial surface consists of open canopies (deserts, grasslands, savannas, open forests, etc.) with vegetation overlying significant amounts of exposed soil, litter, and even understory vegetation that will contribute to the spectral signal observed by a satellite sensor. Even in areas containing only soil or only dense vegetation, the satellite-observed reflected signal may be considered as mixtures, with the soil consisting of mineral and organic components and the dense vegetation containing sunlit and shaded leaves and young and old leaves. Thus, remotely sensed spectral responses of the Earth's surface are related to the number and type of reflecting elements, their individual spectral signatures, and their relative proportions within a given pixel. These mixtures will further vary seasonally and with land cover disturbance. Unmixing the spectral response and resolving the presence and amount of the individual components, such as vegetation and soil, are important in understanding hydrologic and biogeochemical processes within ecosystems.

Traditionally, the classification of satellite images has the objective of assigning each pixel to one, and only one, category as defined in the legend; for example, a pixel may be labeled *soil, water, grass, or forest.* Most commonly, the end users want to have such assignment, as the standard land cover maps include a single category. However, for certain applications, this "hard classification" may be quite limited. For instance, when trying to survey a certain vegetation species, the user may be interested to find out how much dense that species is in the study area, instead of mapping just only those areas where it is the most extended. In this case, the output of the classification analysis would be a map with the density of occurrence of that species (from 0% to 100%).

In addition to this advantage, a "soft classification" may be the most convenient alternative for heterogeneous areas, whereas most pixels are a mixed of two or more land covers. A *hard classification* assumes categorization of each pixel of the image with an exclusive code and assumes that such pixel is homogeneous, and therefore, the category covers the whole pixel. When in fact a mixture exists and it is beyond the minimum map unit at a certain scale, the mixture is simplified either by eliminating the least significant category or by generating a new mixed category, acknowledging the difficulty in separating its components. The denominations *crop mosaic* or *mixed forest* are examples of this second approach.

An alternative to both approaches is the use of techniques that make it possible to estimate the degree of mixing present in each pixel. These techniques output the proportion of each basic category in each pixel, assuming that the pixel reflectance is a combination of reflectances from those pure categories. The retrieval of that subpixel information has been extensively developed in the last decades. The pioneering studies on unmixing models were proposed in the late 1970s (Marsh et al. 1980), although they were extended during the 1990s with the massive use of coarse-resolution images such as those required by AVHRR, VEGETATION, or MODIS, where mixed pixels are very common.

One of the most extended techniques for subpixel classification is the SMA, which tries to obtain images with the proportion of each pixel occupied by a certain category. In other words, as a result of this technique, we obtain as many output images as there are categories under study. In each one of them, the DL of a certain pixel indicates the percentage of surface being occupied by that category (Figure 7.16).

SMA assumes certain conditions that are worth considering in assessing the full potential of this technique, since it will also help us to evaluate its limitations. The most relevant assumption states that the reflectance of a mixed pixel is a direct combination of the reflectances of the basic covers of that pixel, in proportion to the area they occupy. This also implies that radiation from neighbor covers does not affect the reflectance of the target pixel (Settle and Drake 1993). The linearity of the mixture is assumable only in the bands of the optical spectrum, not in the thermal ones. It is also assumed that there are pure spectral signatures valid for all the analyzed zones (Quarmby et al. 1992).

If these assumptions are met, we can state that the reflectance of a mixed pixel is a linear combination of the reflectance of individual pure covers (usually named

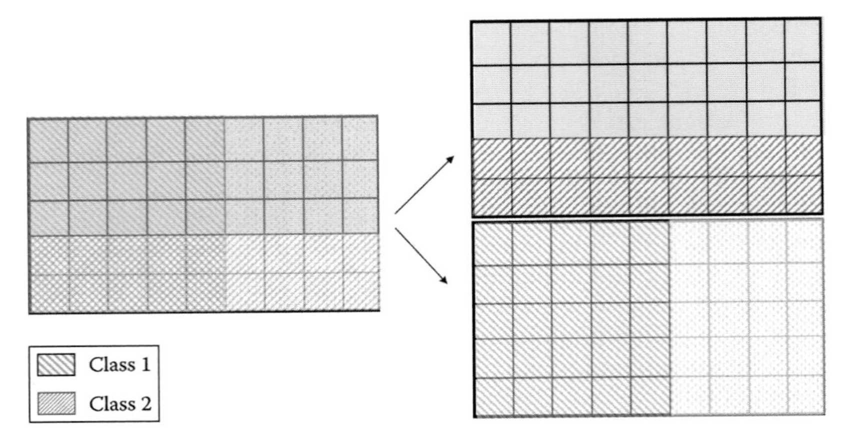


**FIGURE 7.16** Fundamentals of spectral mixture analysis. Two pure classes (reds and greens) are mixed in the images in different proportions. The challenge is extracting those proportions to obtain unmixed images.

*endmembers*), in a proportion equivalent to the surface areas they occupy. Therefore, the following formula is fulfilled (Shimabukuro and Smith 1991):

$$\rho_{i,j,k} = \sum_{m=1,p} F_{i,j,m}\rho_{m,k} + e_{i,j,k} \tag{7.22}$$

where

- $\rho_{i,j,k}$ indicates the reflectance observed in the pixel *i,j* in the *k* band
- $F_{i,j,m}$ is the fraction or proportion of the *m* component in that pixel for each of the *p* pure categories
- $\rho_{m,k}$ corresponds to the reflectance of the *m* component in the *k* band
- $e_{i,j,k}$ is a term of the error associated with the estimation of the proportions in each pixel *i,j*

The proportion of each pure component ($F_m$) must be as follows:

$$0 \leq F_{i,j,m} \leq 1 \tag{7.23}$$

that is, there cannot be negative proportions, and

$$\sum_{m=1,p} F_{i,j,m} = 1 \tag{7.24}$$

that is, the sum of the proportions for all the components is equal to the total surface of the pixel.

Equation 7.22 can also be expressed in terms of raw DLs, instead of reflectances, as long as the linearity of the information initially detected by the sensor is not modified. The interest in using physical values is related to the first and most important part of SMA: the accurate definition of the endmembers that will be the pure components to unmix. These endmembers are the basic covers present in the image area. It is assumed that all pixels of the image have different proportions of those endmembers. The pure components of an image could be healthy vegetation, soil, and shade for deforestation studies (Adams et al. 1995; Holben and Shimabukuro 1993); analysis of forest plantations, recently cut and with active regeneration (Hlavka and Spanner 1995); analysis of healthy vegetation, involving burned areas and shade classes for burned cover mapping (Caetano et al. 1996); or crop identification for crop inventories (Quarmby et al. 1992). An important limitation of SMA is that the number of endmembers of an image cannot surpass the number of bands plus the one used for the calculation of the fractions. This is explained by the mathematical requirements of the method to solve the problem, as we will soon see.

The basic input of SMA consists of calibrated image reflectances and endmember reflectances ($\rho_{m,k}$). These can be obtained from laboratory spectral measurements, from spectral libraries, from the image, or from simulation models. Ideally, they should come from spectral measurements, to avoid noise factors such as topography or substrate conditions. Nevertheless, this option has some inconveniences in assuming that atmospheric and topographic effects have been properly corrected in an image, which is not a trivial task (see Section 6.7.3.3). For this reason, in published studies, the extraction of endmembers is done from the image itself. The interpreter chooses areas that may be considered fully homogeneous representations of the different covers to be unmixed. When working with coarse-resolution images, it may be difficult to extract homogeneous pixels. In this case, some authors recommend extracting the endmembers from higher-resolution images, once they are registered and calibrated with the coarser-resolution images. The endmember reflectances can be generated by using regression techniques or degrading artificially the high-resolution images until the pixel size equals that of the coarse-resolution image (Foody and Cox 1994; Kerdiles and Grondona 1995).

Once the spectral signatures (in reflectance or DLs) of each endmember are obtained, the proportion images can be generated ($F_{i,j,m}$) by solving Equation 7.23 for each pixel. Normally, the proportions are obtained by minimizing the residual error of each pixel ($e_{i,j,k}$). Hence, Equation 7.23 can be rewritten based on the error

$$e_{i,j,k} = \rho_{i,j,k} - \sum_{m=1,p} F_{i,j,m}\rho_{m,k} \tag{7.25}$$

This criterion is known as *minimum square fitting* and has two modalities of solution: the adjustment with restriction and the weighted adjustment (Shimabukuro and Smith 1991).

Let us illustrate the process by using a very simple example (Figure 7.17). Consider that we have two bands (R and NIR) and try to deduce for each pixel the proportions of two categories ($c_1$ and $c_2$). The reflectances for those categories are 0.35 and 0.4 (R and NIR, respectively) for class 1 and 0.75 and 0.7, respectively, for class 2.

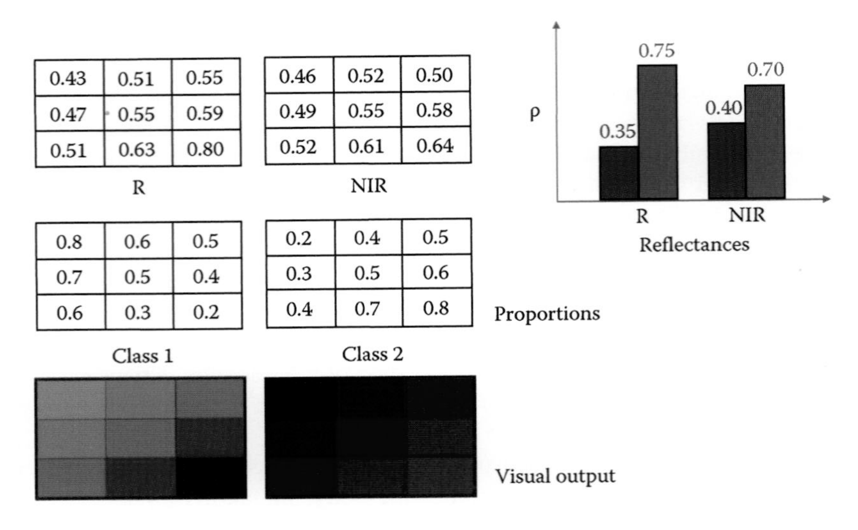


**FIGURE 7.17** Example of obtaining unmixed images of two endmembers.

With these data, two images with the proportion of $c_1$ and $c_2$ for each pixel have to be generated. Assuming the estimations have no error term, it is necessary to solve for each pixel a simple system with two equations (i.e., two bands) and two unknowns (i.e., two classes). For example, for the first pixel (top, left), we would have

$$0.43 = 0.35F_1 + 0.75F_2$$
$$0.46 = 0.40F_1 + 0.70F_2$$

that is, the reflectance values observed for that pixel in the two bands (0.43, 0.46) are produced by a linear combination of the reflectances of the $c_1$ and $c_2$ in proportion ($F$) to the area they occupy in those two bands. Since the proportions ($F$) are the same in all bands (as they account for the area occupied by each category), we can easily obtain $F_1$:

$$\frac{0.43 - 0.35F_1}{0.75} = \frac{0.46 - 0.4F_1}{0.7}$$

from which we deduce that $F_1$ is 0.8. Since there are only two categories, $F_2$ is 0.2. We then proceed in a similar way with the rest of the pixels of this image, obtaining the resulting matrix of proportions (Figure 7.17). In this case, assuming that there is no residual error in the estimation of the proportions, a single equation would be sufficient since if the restrictions of nonnegative values and exhaustivity that the SMA require are fulfilled, $F_2$ can be cleared from $F_1$ ($F_2 = 1 - F_1$). In sum, we could obtain the value of $F_1$ from a single band:

$$0.43 = 0.35F_1 + 0.75(1 - F_1)$$

Since it is necessary to have as many equations as there are unknown quantities, we select as many endmembers as available bands. Nevertheless, if we include the exhaustivity requirement (the sum of the proportions is equal to 1), an additional equation is available; this is why we can include up to a maximum of $k + 1$ endmembers, $k$ being the number of bands. All this is applicable as long as we assume the error is null. Should this not be the case, another unknown quantity would be added, and it would be necessary to limit the number of endmembers to the number of bands.

The method of solving those systems of equations is logically more complicated than in our example. Consequently, it is necessary to apply more refined criteria to solve the system of equations previously mentioned. Among the several procedures that have been proposed, the simplest estimator is the minimum squares fitting without restrictions. In matrix annotation, this estimator is expressed as follows (Gong et al. 1994; Settle and Drake 1993):

$$\hat{f} = (M^T M)^{-1} M^T X \tag{7.26}$$

where

$\hat{f}$ indicates the vector with the proportions calculated for each pixel
$M$ indicates the vector with the average values of the endmembers in the bands of analysis
$X$ is the vector with the values of the pixel for those same bands
superscript $T$ indicates the transpose of the matrix
$-1$ is the inverse one

This algorithm is relatively fast, but it does not satisfy the restrictions mentioned previously. In order to solve this, we must make the negative values of $\hat{f}$ equal to zero and scale the rest so they add up to 1. For other solution criteria that include these restrictions directly, consult Settle and Drake (1993) and Shimabukuro and Smith (1991).

The techniques of SMA were initially proposed in the 1970s but did not receive much attention until the early 1990s, initially with coarse-resolution sensors (mainly AVHRR) and then with the first hyperspectral sensors (both airborne and satellite-borne, such as Proba CHRIS or EO-1 Hyperion). The perspectives of the new sensors with high spectral resolution can lead to a remarkable growth in the use of this technique. The SMA has already shown excellent results in agriculture (Kerdiles and Grondona 1995; Quarmby et al. 1992), deforestation studies (Adams et al. 1995; Shimabukuro et al. 1994; Souza et al. 2005), estimation of the fraction of vegetation cover density (Gong et al. 1994; Hlavka and Spanner 1995), vegetation mapping (Roberts et al. 1998), and monitoring of burned areas (Caetano et al. 1996; Martín and Chuvieco 1998; Riaño et al. 2002b). In addition, it has been widely used in soil and mineral mapping (Palacios-Orueta and Ustin 1998; Smith et al. 1990).

The spectral unmixing of three basic endmembers for the Tucson image is shown in Figure 7.18. The endmembers delimited are soil, vegetation, and shade using six reflective bands of the Landsat OLI sensor. The comments of these subpixel images are quite similar to those provided by PCA and TCT, since they reflect the three

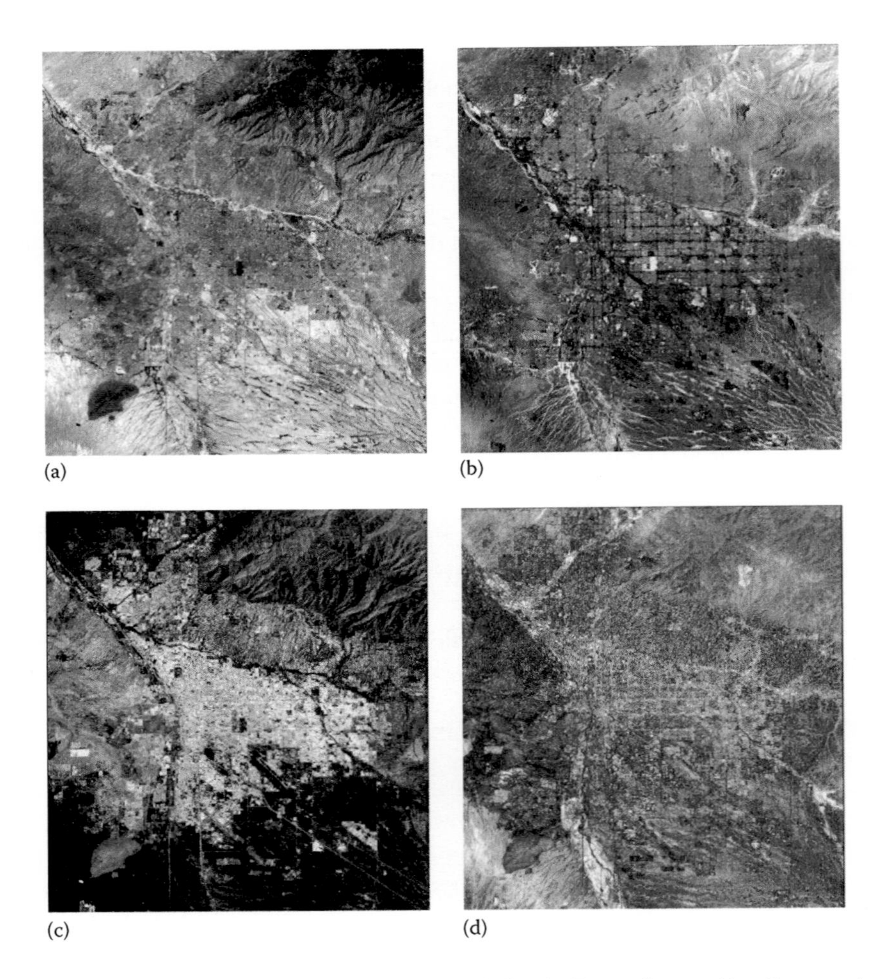


**FIGURE 7.18** Results of spectral mixture analysis for the Tucson image: (a) soil proportion, (b) vegetation proportion, (c) shade proportion, and (d) error term.

dimensions of information included in the image, although since the vegetation cover has a low global proportion in this area, this component was not extracted in the PCA. Comparing the SMA results and those of the PCA, in this case each component has a physical sense—the proportion of soil, vegetation, and shade—in a continuous way from the maximum amount (100%) that appears in areas where the pure member was defined to those in which they show a practically null proportion. The shade component shows the highest values in built-up areas and volcanic regions. The vegetation component is evident in irrigated land crops, golf courses, and riparian vegetation, as well as in gardens. The soil offers the highest values in arid areas of the center south and north of the image, with very sparse vegetation. Finally, the error component is higher in areas that share different covers, being very low in irrigated crops and water bodies where the endmembers were collected.

### 7.1.6 Lidar Data Processing

As previously mentioned (see Section 3.4.2), lidar systems emit laser pulses to the ground and record both the intensity and the time delay of the reflected signal. The former element provides similar information to other remote sensors working on the solar spectrum, while the time recordings provide ranging measurements to extract vertical information that is quite unique from other sensor systems. These measurements on the vertical structure of observing targets make lidar systems particularly useful for a wide range of environmental applications, such as estimating vegetation biomass and geometrical properties, terrain slope and aspect analysis, coastal transformations, or ice accumulation, as well as monitoring infrastructures and urban areas.

Even though the intensity of the lidar signal can also be useful for detailed ground characterization, we will cover in this section only those methods dealing with the coordinates of the laser returns. While lidar intensity is not exactly the same as reflectance, digital methods to process lidar intensity have many similarities to other sections of this chapter, while those oriented toward extracting information from lidar ranges are more distinctive, as they deal with irregularly spaced *X,Y,Z* point coordinates. This section is based on the analysis of current airborne lidar systems but includes some references to the ICESat-GLAS sensor, which is so far the only lidar ranging space-borne system. The main differences are the use of discrete returns (airborne systems) versus full waveform recordings (ICESat, as well as some airborne systems) and obviously the footprint size (10–30 cm airborne versus 70 m for ICESat).

Once the lidar campaign has been carried out, the survey company should provide different files with the raw and corrected acquisitions, making sure that the target area is properly covered (no gaps), that flight lines are well registered to a common reference, and that repeated pulses caused by flight-line overlapping are removed. A common format for distributing lidar data is the LASer file exchange format (LAS), proposed by the American Society for Photogrammetry and Remote Sensing (ASPRS) Lidar Subcommittee. This format is acceptable by most commercial and public domain software and includes the data coordinates (*X,Y,Z*) along with the intensity, return number (first, last, or intermediates), scan angle, a classification of the cover type, and GPS time, as well as a public header to record metadata.

Depending on the flight conditions and lidar specifications, the target area will have been covered at a certain density of observations (number of pulses per square meter, for instance). However, that density will not be uniform throughout the area, since some sections may have more observations in certain sections or less in rough terrain, for instance. A simple way to better understand spatial differences is to compute the return density within a certain grid size. For instance, Figure 7.19 shows the area observed by a lidar survey carried out in western Washington near Olympia.* One can notice the impact of the flight lines in the density distribution (long green

* This dataset is provided by the developer of a freely accessible lidar analysis software (FUSION/LDV: http://forsys.cfr.washington.edu/, last accessed: February 2015). This section includes examples extracted from this dataset.

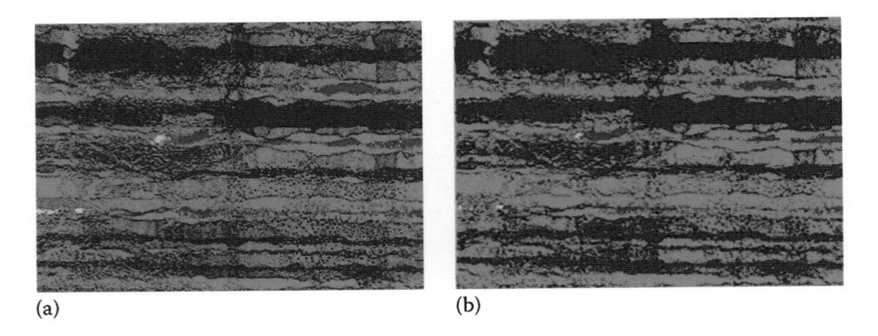


FIGURE 7.19 Density of observations for the lidar survey of Olympia (Washington): (a) grid size 4 $m^2$ and (b) grid size 16 $m^2$. Yellow, cells with no points; red, cells with less than 1 point; green, cells with 1–3 points; and blue, cells with more than 3 points.

or blue strips, indicating different point density), along with small sectors without observations (in yellow).

Displaying 3D representations of *X,Y,Z* lidar measurements greatly helps in understanding the vertical structure of the target area. In forested areas, this vertical structure may be quite complex, as different tree stands may have different heights, with different understory stratification. To simplify the analysis, most authors recommend discriminating three vertical layers (Figure 7.20):

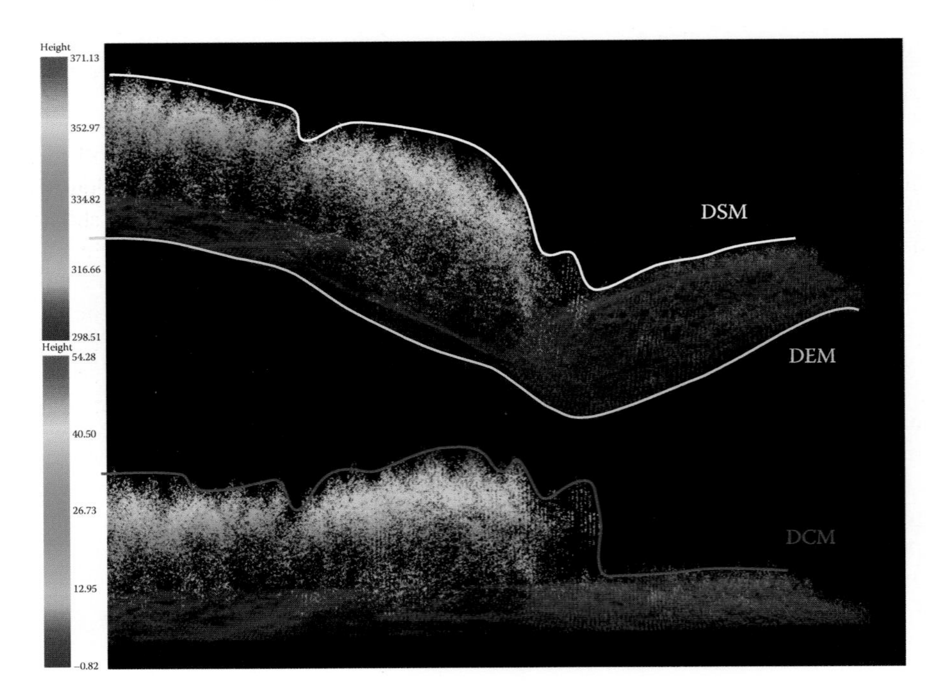


FIGURE 7.20 Digital elevation model, digital surface model, and digital canopy model over a section of the Olympia dataset.

- Digital elevation model (DEM), which is the underlying surface that represents the elevation of the ground at any particular area. This would be the elevation of the surface after removing all objects (vegetation, buildings, etc.).
- Digital surface model (DSM), which includes elevations of the upper part of the canopy, that is, the maximum elevation at any particular area.
- Digital canopy model (DCM), which gives the height of any particular area from the ground and is computed as the difference between the DSM and the DEM.

Figure 7.20 includes a section of the Olympia dataset that shows the distinction between these three surfaces. DEM corresponds to the standard definition of elevation, and it is the common dataset provided when the ground elevation is required, as most topographic maps do not take into account vegetation heights. It can be considered as the base line for measuring heights of any object in the target scene, since all estimations will be based on the *Z* differences between any return and the ground *Z* value for the same *X,Y* coordinate. For this reason, generating accurate estimations of the DEM is critical to obtain reliable estimations of object heights.

There are different approaches to estimate the ground baseline from lidar measurements. The simplest way would be to keep the last return from a system collecting several discrete pulses for the same position. Assuming the target cover is not opaque, the pulses will hit the top of the tree canopy, will be then transmitted to the branches, and finally will hit the ground. However, when the vegetation is dense or the target is opaque (such as a building roof), the first and last pulses will be in the same *Z* position (the canopy), and the ground elevation will be confused with the canopy height and therefore overestimated.

A more elaborate method consists in finding the lowest elevations for a coarse grid size, which is later used to estimate ground elevations at a denser grid, using different interpolation algorithms (Bolstad 2008, chapter 12). First, the frequency distribution of *Z* values in the coarse grid is computed from the last returns. This grid size should have sufficient returns so as to be very likely that some of them hit the ground. For instance, for a lidar flight acquiring an average of 4 pulses $m^{-2}$, this auxiliary grid may be 25 $m^2$, so an average of 100 pulses will be available at each location. The elevation found for 1%–2% of the lowest elevations is considered to hit the ground and it is retained as the baseline for that particular cell. Later, the grid is densified by spatial interpolation to a better spatial resolution (1–4 $m^2$ in this case) to improve estimations of ground baseline variability. This simple method provides good results if the area is diverse and the lidar pulses are dense enough. Information from lidar intensity or from a simultaneously acquired ortophoto may also be used to better discriminate ground from vegetation hits (Riaño et al. 2007). More sophisticated methods are available in the literature (Ackermann 1999; Kraus and Pfeifer 1998; Magnussen and Boudewyn 1998; Pfeifer et al. 1999; Vosselman 2000; Wehr and Lohr 1999).

The DSM is simple to derive, as the first return pulse should hit the surface canopy. In this case, the critical parameters are the pulse density and the scanning pattern. Depending of both, some areas may not have any pulses, and therefore the DSM for these areas will be just an estimation from nearby elevations. Similar to the DEM, the DSM surface is created by spatial interpolation. Special care should be taken in urban areas, as standard interpolation algorithms typically assume smooth gradients, while buildings have sharp elevation changes. The use of break lines and modeling surfaces (cubes, cylinders, cones) may be very useful to properly approximate building shapes (Vosselman and Maas 2010).

Once the DEM and DSM are obtained, the DCM can be easily computed as the elevation difference between the two. The output will be the heights of each cell from the ground. If the DCM is computed from the second, third, or fourth return, it will provide information on the vertical characterization of that particular cell (height distribution), providing the object has enough transmittance. Alternatively, the analysis of height distribution can be based on plots of a certain radius (e.g., 5 m for a 4 pulses $m^{-2}$ campaign) or those that have been measured on the ground. From these plots, a full waveform may be simulated by computing the frequency distribution of the returns. From them, several parameters may be computed: average, standard deviation, percentiles, bias, density, etc. These parameters are very useful in estimating different vegetation parameters. Some of those parameters are estimated directly, for example, maximum height, while others are modeled through statistical analyses, such as biomass. The specialized literature on using lidar data for forest inventory, carbon estimation, or fuel properties is very vast (Drake et al. 2002; Garcia et al. 2010; Morsdorf et al. 2004; Riaño et al. 2003, 2004b; Sun et al. 2008). Reliable estimations of tree crown height, crown base height, fraction of tree cover, LAI, crown bulk density (Figure 7.21), biomass or biomass fractions, and analysis of single tree have been obtained. In addition, lidar data have been combined with those from optical sensors to improve classifications of urban and forested areas, by including elevation or derived parameters as additional variables in image classification (Garcia et al. 2011; Wulder et al. 2009).

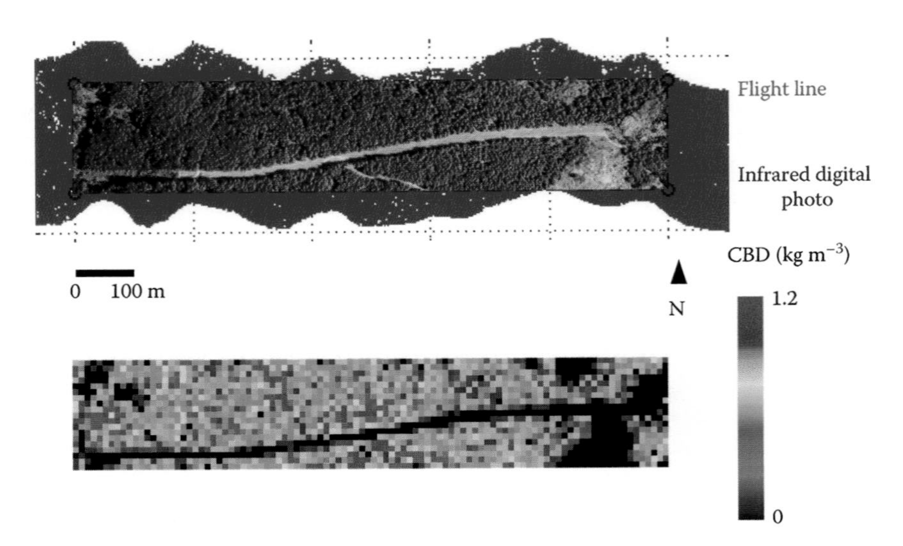


**FIGURE 7.21** Crown bulk density generated from a lidar flight in Central Spain. (After Riaño, D. et al., *Remote Sens. Environ.*, 92, 345, 2004.)

## 7.2 DIGITAL IMAGE CLASSIFICATION

### 7.2.1 Introduction

In spite of the growing interest in retrieving biophysical variables from EO images, the generation of thematic maps is still a very common objective of digital image processing. The goal is assigning each pixel to one of the target categories (crop or vegetation types, urban covers, etc.). Although recent developments in fuzzy classification techniques make it possible to assign more than one category to each pixel, most commonly the classification results provide a single assignment for each pixel, since most thematic maps still consider the space as homogeneous entities.

Classifying land covers has been traditionally carried out using visual image interpretation, either from aerial photography or satellite images. As previously mentioned (Section 5.3), visual analysis is a powerful tool to incorporate a wide diversity of interpretation criteria. However, digital classification techniques provide more robust and spatially consistent tools to generate thematic maps, particularly when working over large areas. Classification techniques can be applied to raw data, but most authors recommend first carrying out geometric and radiometric corrections, to avoid the impact of external factors of noise (atmospheric influences, topographic shades, etc.). Moreover, when working with physical values (e.g., reflectance or temperature), classification rules can be extended beyond the specific sensor or period of that particular image, and provide methods that can be generalized to a wider range of situations.

Once the classification method is completed, the output image will have the same spatial structure as the input one, although the DLs will no longer represent a quantitative measurement but a numeric label that identifies a category (evergreen forest, grass, corn, wheat, etc). Therefore, even though the pixel values will still be numeric, they should be considered a qualitative rather than a quantitative measurement scale, and so the classified images cannot be used for analytic operations such as filtering or PCA, for instance. However, statistical metrics designed for categorical variables can be properly applied to classified images, such as mode or majority filters. In the case of classifying a single variable in grades (e.g., low, medium, and high density of forest), some ordinal-scale techniques can be used in the output images.

The classified images can be easily transformed to thematic maps by assigning a proper patch or color key to each DL, while the histogram of those DLs will provide an inventory of the number of pixels (and the area if it is geometrically corrected) assigned to each class (Figure 7.22).

The first experiences of digital classification of satellite images were based exclusively on the spectral values of the input images. Spectral information is not always able to separate end-user categories, since different target classes may exhibit similar spectral behavior. For this reason, classification methods are also based on the spatial and the temporal dimensions of the images. The former refers to the spatial properties that may identify a target category, such as its shape or size (large or small patches), its spatial pattern (arrangement in rows, such as olive trees), or which categories are nearby. The temporal variability involves using the seasonal trends of each class, which is especially relevant for discriminating between vegetation covers. For instance, evergreen and deciduous trees can be easily discriminated in winter or early spring.

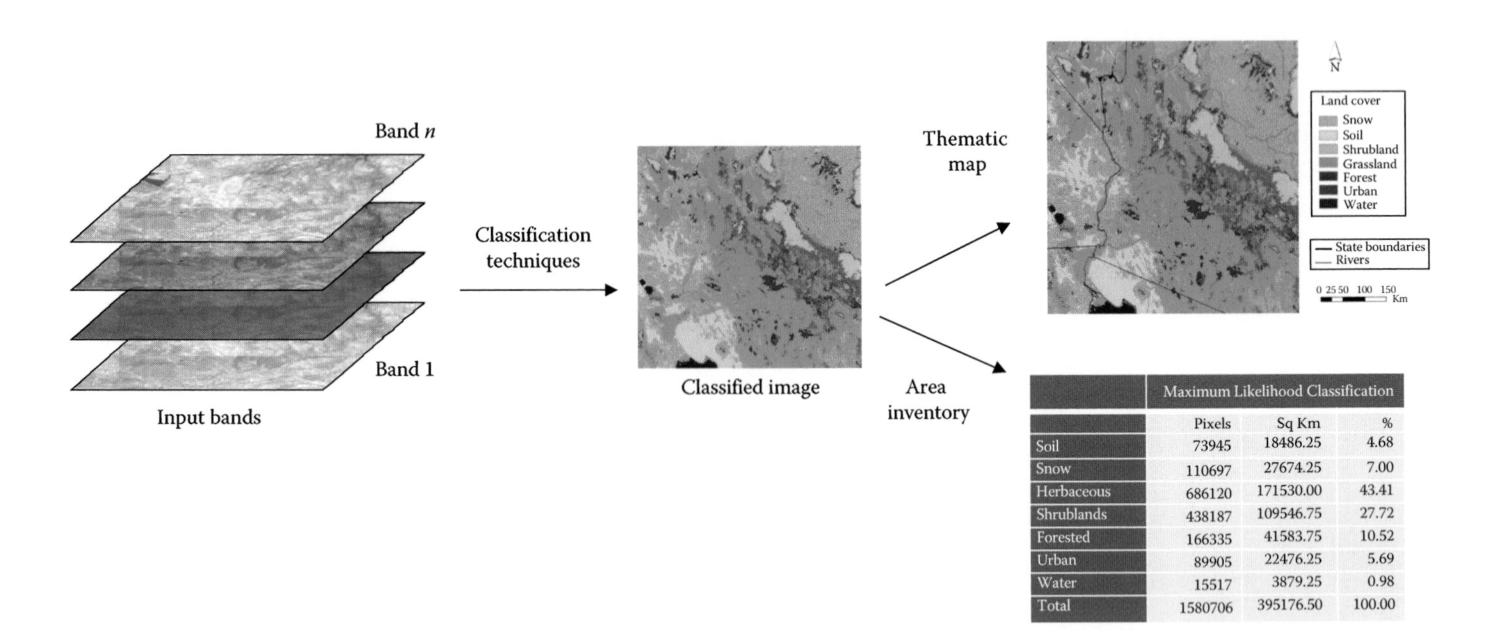


| | Maximum Likelihood Classification | | |
|---|---|---|---|
| | Pixels | Sq Km | % |
| Soil | 73945 | 18486.25 | 4.68 |
| Snow | 110697 | 27674.25 | 7.00 |
| Herbaceous | 686120 | 171530.00 | 43.41 |
| Shrublands | 438187 | 109546.75 | 27.72 |
| Forested | 166335 | 41583.75 | 10.52 |
| Urban | 89905 | 22476.25 | 5.69 |
| Water | 15517 | 3879.25 | 0.98 |
| Total | 1580706 | 395176.50 | 100.00 |

FIGURE 7.22 Fundamentals of digital classification.

If none of these three dimensions (spectral, spatial, and temporal) provides a clear discrimination of a target category, auxiliary information may be used, such as elevation, aspect, or soil maps. In this approach, the synergy between remote sensing interpretation and geographical information systems (see Chapter 9) is greatly beneficial.

The process of classifying digital images is analogous to that used in visual analysis. For photo interpretation, the analyst identifies the visual pattern associated with each cover from a set of control areas. Once the interpreter is trained to recognize the different covers, the learnt experience is extended to the whole study area, based on their similarity with the previously identified visual patterns. Finally, this visual interpretation needs to be assessed from groundwork or other external sources. In digital classification, the phases are similar, but the analog criteria are translated into digital criteria. Therefore, three phases can be distinguished: (1) digital definition of the categories based on sample pixels (training phase), (2) allocation of the whole image to one of the previously defined categories (assignment phase), and (3) assessment of results. We will deal here with the first two, while the third one will be covered in another chapter (Chapter 8), since verification techniques are applicable to both digital processing and visual analysis.

## 7.2.2 Training Phase

### 7.2.2.1 Basic Concepts

The first and most important phase of digital classification is the proper characterization of the different categories to be discriminated. Since the classification is based on pixels' numerical values, that definition must also be numerical. Therefore, the purpose of the training phase is obtaining the DLs, or better still the range of DLs that identifies each category for all the bands that will be used in the classification.

Since the radiometric response of each cover is not unique because of the impact of external factors (illumination conditions, density, phenological state, etc.), the digital characterization of each class cannot be achieved with average DL values. Rather, the dispersion trends around the mean should be considered as well, which will be more or less significant depending on the class and landscape complexity.

The goal of the training phase is to properly characterize that variability, based on sampling representative pixels within the study or similar areas. As in any other sampling, the objective of this phase is to obtain the most accurate results with the minimum sample size. An incorrect selection of the sample will lead inexorably to poor results in the final classification (Gong and Howarth 1990a; Hixson et al. 1980; Story and Campbell 1986). Therefore, the training phase constitutes the backbone of the numerical classification.

Traditionally, the classification methods have been divided into two groups: supervised and unsupervised, based on the methods used to collect training statistics. The supervised method implies a previous knowledge of the area that helps in selecting the samples for each category. The unsupervised method makes automatic searches to find the clusters of homogeneous values within the image. Later on, the user needs to find correspondences between those spectral groups and the target categories.

In this regard, we could consider two types of categories: those that are required by the end user and those that correspond to unique spectral characteristics.

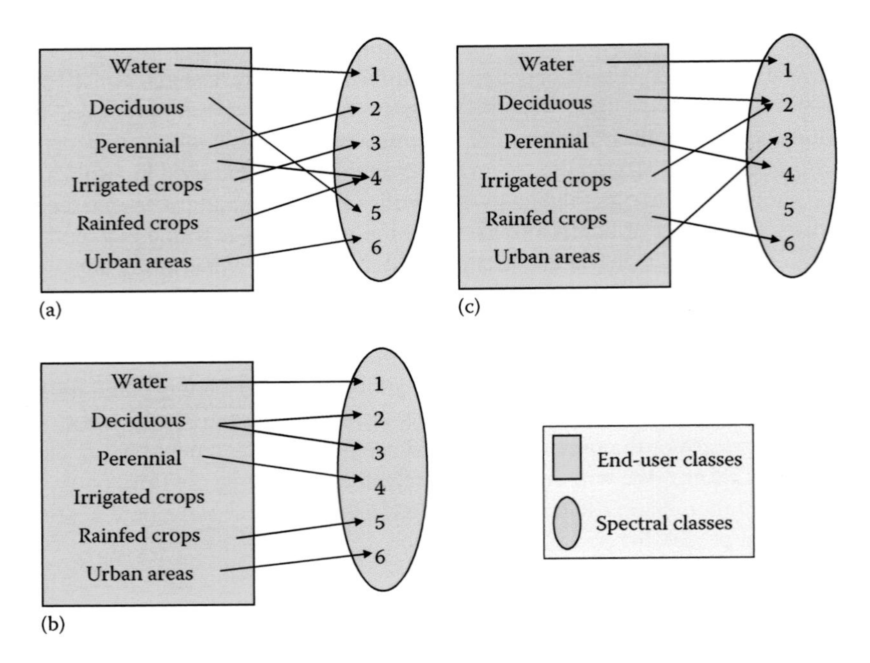


**FIGURE 7.23** Relationships between spectral and end-user classes. (a) One spectral class for each end-user class, (b) two spectral classes in one end-user class, and (c) two end-user classes in one spectral class.

The former can be named *end-user categories*, while the latter are called *spectral categories*. Ideally, these two groups should have a two-way correspondence: that is, each end-user category should correspond to just one spectral class, and vice versa (Figure 7.23a). Unfortunately, this two-way relation is infrequent, and normally either of the following situations occurs:

1. A target category could be defined by several spectral classes (Figure 7.23b). For example, pine trees in the Sun-facing slopes have a different spectral category than those in shaded slopes. Water with different depth or turbidity may also imply two or more spectral groups. When this situation occurs, the solution depends on the interpreter's ability to identify those contrasting spectral properties and create different spectral groups that will be fused after the classification ends. Otherwise, some end-user categories would have very heterogeneous spectral signals and be easily subject to confusion with other classes.
2. The opposite situation occurs when two or more informational categories share a single spectral class (Figure 7.23c). This implies that some target categories cannot be discriminated spectrally from the input data set. In this case, the interpreter may decide to unify the problematic categories into a single, more general one, if that is agreeable to the end user. For example, a coniferous cover category may replace two named "pine trees"

and "fir trees." When the end user cannot accept that combination or the confusion is created between two categories that have completely different thematic meanings (urban gardens and forest, for instance), then other options should be considered, such as adding spatial criteria or auxiliary variables (relief, soils, etc.) or imagery from other dates.

Following the previous groups of training methods that were commented on, the supervised method tries to define end-user classes while the unsupervised method tends to identify the spectral classes of the image. From previous paragraphs, we can conclude that neither of the two methods provides an immediate solution to all the problems that digital classification involves. On the one hand, the supervised method can be criticized for being biased and artificial, since the computer can be *forced* to discriminate between categories that do not have a clear spectral meaning. On the other hand, the unsupervised method may provide results that are difficult to interpret or have little connection with the end user's needs.

#### 7.2.2.2 Supervised Classification

A supervised classification assumes previous knowledge of the study area acquired either by external sources or fieldwork. This familiarity with the study zone allows the interpreter to identify on the image sufficiently representative areas of each one of the target categories. These areas are known as "training fields." The term indicates that such areas are used to *train* the computer in recognizing the different categories. In other words, from pixels in the training fields the computer calculates the DLs that define each of the classes, and later assigns the rest of the pixels of the image to one of those categories based on their DLs. The selection of training fields may be based on known optical properties of each category, field observations at known locations, and aerial and ground photographs, among other sources of information. It is strongly recommended to have the reference information as close as possible to the time of image acquisition, to avoid misallocation of pixels based on outdated reference data. Whenever possible, it is also recommended to make field radiometric measurements (Figure 7.24), which give the interpreter relevant information regarding potential factors of noise or to help properly select the most suitable bands for the classification (Barret and Curtis 1999; Milton et al. 1995).

Most image processing software (IPS) facilitate the interactive location of training fields. Using a digitizer or an electronic mouse, the interpreter introduces on the screen the vertices that define each of those areas, assigning them to one of the categories defined previously in the key. The auxiliary fieldwork and other documents—such as aerial photography or conventional cartography—can greatly help a more precise location of these areas, bearing in mind that they must be sufficiently representative and homogeneous examples of the class that we are trying to define. Once the boundary of a training field is completed, other fields for the same category can be selected. It is advisable to select several areas per category in order to suitably reflect its variability in the study zone.

Another alternative to select training fields is called "seeding." The interpreter inputs single pixels, and a region-growing algorithm creates the field by adding up the neighboring pixels that have similar spectral values with the one initially selected.

**FIGURE 7.24** Portable radiometers are very useful in the training phase.

The direction and distance to which new pixels will be added, as well as the degree of heterogeneity of those new pixels from the *seed pixel* can be controlled by the user. This method guarantees that the training fields are sufficiently homogeneous and avoids including border pixels between two categories, which would imply an artificial increase of the variance (Figure 7.25).

Once the selection of training fields is completed, the basic statistics of each category, mean, rank, standard deviation, variance–covariance matrix, etc. are computed from the DLs of all the pixels included in the training fields assigned to that class. These statistics will be used later on for categorizing the rest of the image pixels. Therefore, the measurements extracted from the training fields are assumed to be good representatives of each class. Otherwise, the interpreter would be forcing the computer to classify heterogeneous areas, and the classification would be inaccurate.

Since the training phase is the most relevant in image classification, the selection of training pixels should be done with great detail and accuracy. When the categories are homogeneous, a single training field per category may be enough to avoid confusion with other covers (Figure 7.26a). Special attention should be paid to avoid selecting fields in transitional areas, as the resulting class statistics may not be good

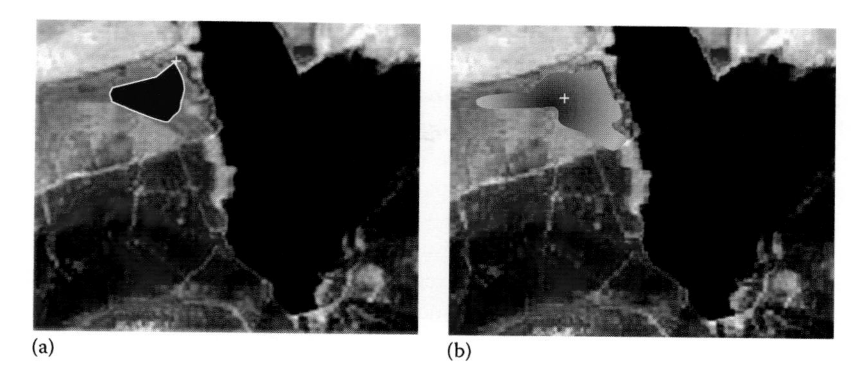
(a) (b)

**FIGURE 7.25** Training pixels, selected by either (a) digitizing polygons or (b) automatic seeding algorithms.

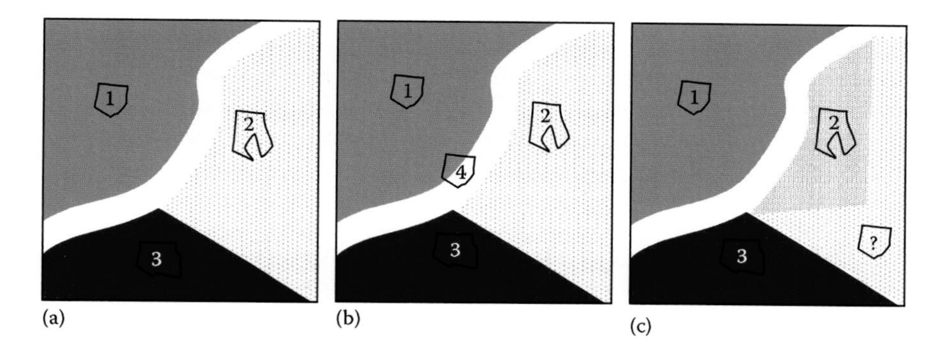


**FIGURE 7.26** Problems of selecting training fields: (a) homogeneous areas, (b) heterogeneous areas in field 4, and (c) heterogeneous categories not properly sampled.

representatives of the target cover (Figure 7.26b). The opposite error would be to select training sites only in those places where the category is very homogeneous, missing the real spectral variation of that cover (Figure 7.26c). For example, when trying to classify oak trees, areas with different canopy cover should be selected, either as a single class or as two or more spectral *oak* classes, to join them in the final map. From this we can infer that the supervised method requires a previous knowledge of the features of the study area, as well as the categories we want to discriminate between.

The last aspect to consider in the selection of the training sites is the statistical requirements that ensure they are good representatives of the different categories. As mentioned, this process implies a peculiar form of spatial sampling. For that reason, it is better to use some of the usual criteria in this type of technique to choose the size and the most suitable distribution of the sample, as well as to make estimations from it.

Regarding the size, it is required to select a minimum of $m + 1$ pixels per category, $m$ being the number of bands that integrate the analysis. However, it is very convenient to exceed this minimum limit. Several authors recommend selecting a number of pixels between 10*m* and 100*m* per category (Mather 1998). This minimum number is related closely to the degree of spatial association between the DLs of the image. Conventional sampling considers the selected samples to be random and independent. This assumption is not valid in this case because training fields have neighboring pixels affected by spatial autocorrelation (Congalton 1988b). Therefore, it is more advisable to choose several small-sized fields than one larger one, since the latter will tend to underestimate the variability of that category. Some authors recommend selecting random pixels, within fields of a greater size, instead of including them all (Campbell and Wyme 2011).

Regarding the distribution, it is best to consider the characteristics of the image itself and to try to include the spatial variations of each category: slope, aspect and gradient cover, density, moisture content, type of soil, etc. Thus, several authors have emphasized that the variability of a cover is directly proportional to the resolution of the sensor (Cushnie 1987). In other words, the greater the spatial detail registered in the image, the greater its sensitivity in detecting the internal variations of a category, which would be hidden under a larger pixel. Since greater heterogeneity means greater mixing with similar classes and greater risk of confusions in the later assignments, an increase in spatial resolution can complicate digital classification. This fact, which seems to contradict common sense (greater spatial resolution, greater accuracy), has been proven by several authors and is known as "scene noise" (Markham and Townshend 1981). The effect has been verified in the more heterogeneous cover types, as in the case of urban areas, where a larger pixel makes it easier to acquire an average signal of different components, whereas the more the pixel is reduced, the more intense the border effects become (Toll 1984). However, in agricultural and forest categories, that effect is less clear, and thus, higher-resolution images are more advisable for digital classification (Irons and Kennard 1986). As a result of greater spatial complexity of high-resolution images, new object-oriented classification systems have been developed that take better advantage of spatial features (Ferro and Warner 2002; Tso and Mather 2001). We will later review those methods (Section 7.2.3.8).

To illustrate the classification process, we will use in this section the MODIS image of Southwestern United States, acquired on February 19, 2001. The first seven bands of this sensor will be used in this exercise. All have 500 m of spatial resolution and cover from the blue to the SWIR bands. Figure 7.27 includes the training sites that were used for supervised classification. To simplify the procedure, only seven categories have been defined: soil, snow, water, herbaceous green vegetation, shrubs and brown vegetation, forested areas, and built-up areas. The location of the training sites was aided by higher-resolution images.

#### 7.2.2.3 Unsupervised Classification

This method tries to define the spectral classes present in the image by finding clusters of pixels with similar DLs. No prior knowledge of the study area is required, and the work of the interpreter is more focused on the labeling of the resulting groups than on providing input information to the clustering algorithm. The basic assumption behind unsupervised methods is the existence of unique clusters of pixels with

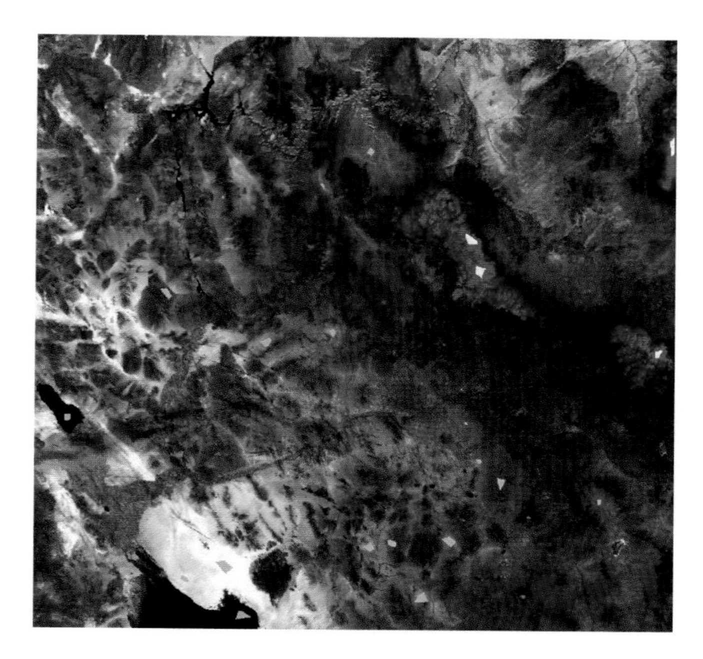

FIGURE 7.27 Training sites selected for classification of the MODIS study area.

similar spectral characteristics and, therefore, similar cover. As stated earlier, however, the spectral similarity does not necessarily imply homogeneous cover because the same cover may have different spectral responses, while the same spectral signal may refer to more than one cover. Consequently, the interpreter needs to make sense of the resulting classes, mainly by labeling them, but also by combining them or splitting them according to the end-user needs.

The unsupervised training is based on selecting the discrimination space (set of bands that will be used for the grouping) and the clustering algorithm. The variables used for the clustering may be the original bands or the result of some transformation, such as PCA or VIs. Eventually, auxiliary variables such as DEMs or rainfall maps could also be used in clustering, provided these variables are measured in a quantitative scale.

The clustering method of finding those homogeneous clusters can be illustrated with the simplest case, that is, when working with just two bands (NIR and R, e.g., Figure 7.28). In this bivariate graph, clusters of pixels with similar DLs in those two bands are clearly observable. Those groups may be related to thematic classes of interest. Finding the clusters of the image requires deciding on two factors: how to measure the similarity between pixels and how to group them. The former has been commonly based on the Euclidean distance:

$$d_{a,b} = \sqrt{\sum_{i=1,m} (DL_{a,i} - DL_{b,i})^2} \tag{7.27}$$

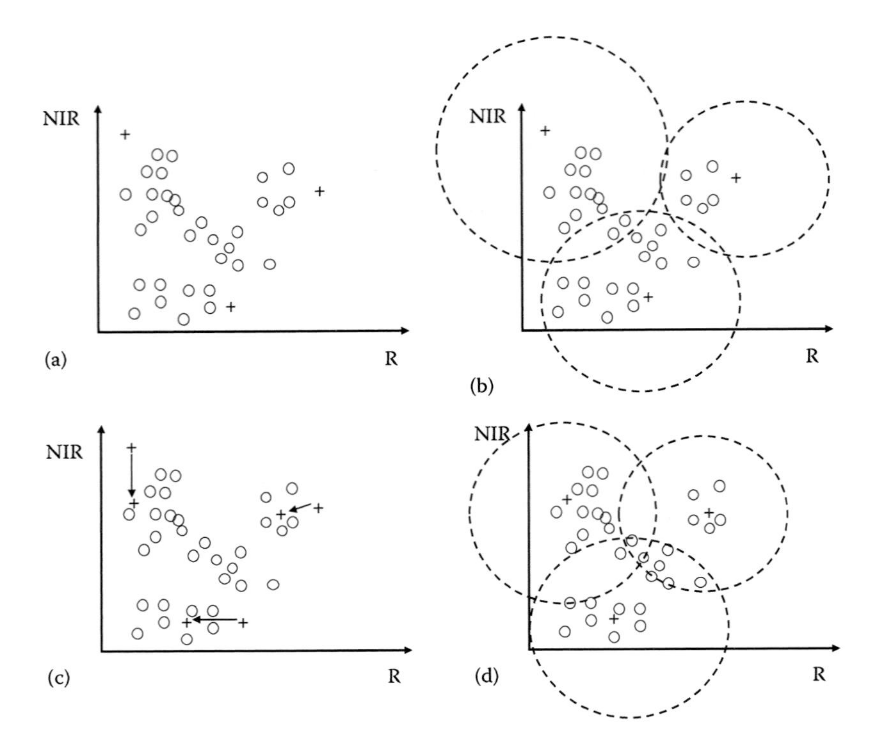


FIGURE 7.28 Phases of the ISODATA algorithm: (a) random selection of cluster centers; (b) pixels are assigned to the closest cluster center; (c) cluster centers are recalculated from the pixels assigned in the first iteration; (d) A second iteration changes pixel assignments as a result of center displacements in (c).

where

$d_{a,b}$ indicates the distance between two given pixels $a$ and $b$

$DL_{a,i}$ and $DL_{b,i}$ are the digital values of those pixels in band $i$

$m$ is the number of bands

Other criteria, such as the block distance or the Mahalanobis distance (Mather 1998), can also be used to measure the distance between pixels, although the Euclidean distance is the most common.

Regarding the clustering algorithm, the options also are very numerous. One of the most extended is known as ISODATA (Duda and Hart 1973), which provides an iterative scheme that is very flexible and robust. This algorithm can be described by the following phases (Figure 7.28):

1. A series of cluster centers are initiated randomly, with a number commonly specified by the interpreter.
2. The pixels of the image are assigned to the cluster with the closest center, following the distance criterion previously selected. This clustering may be performed with all image pixels or, most commonly, with a systematic sample (1 out of every 10, for instance) to reduce processing time.

3. Once the first iteration is completed, the class centers are recalculated, considering the DLs of all the pixels that have been incorporated in the previous phase.
4. A second iteration starts. Again, all pixels are classified under the nearest center class. Because of the displacement registered in step 3, the classification of many pixels will change from the previous cycle.
5. Again, the centers are recalculated and a new iteration starts. The process continues until the number of pixels that changed categories from one iteration to the previous one is below a certain threshold indicated by the interpreter.

Before each iteration starts, the centers of classes may be arranged following a set of parameters indicated by the interpreter (Figure 7.29). The most common are the minimum distance between clusters to be distinguished (below that threshold, the clusters are combined), the minimum size of a cluster (below that threshold, the cluster is eliminated), the maximum variance of a cluster (above that limit, the group is split), and the maximum number of clusters (above that number, the closest groups are joined).

The most critical point of this algorithm is the decision on the control parameters that are appropriate for each classification problem, since there are no universal rules to follow. Commonly, the interpreter does not know the number of spectral clusters present in the image, or the convenient thresholds of maximum variance, or the

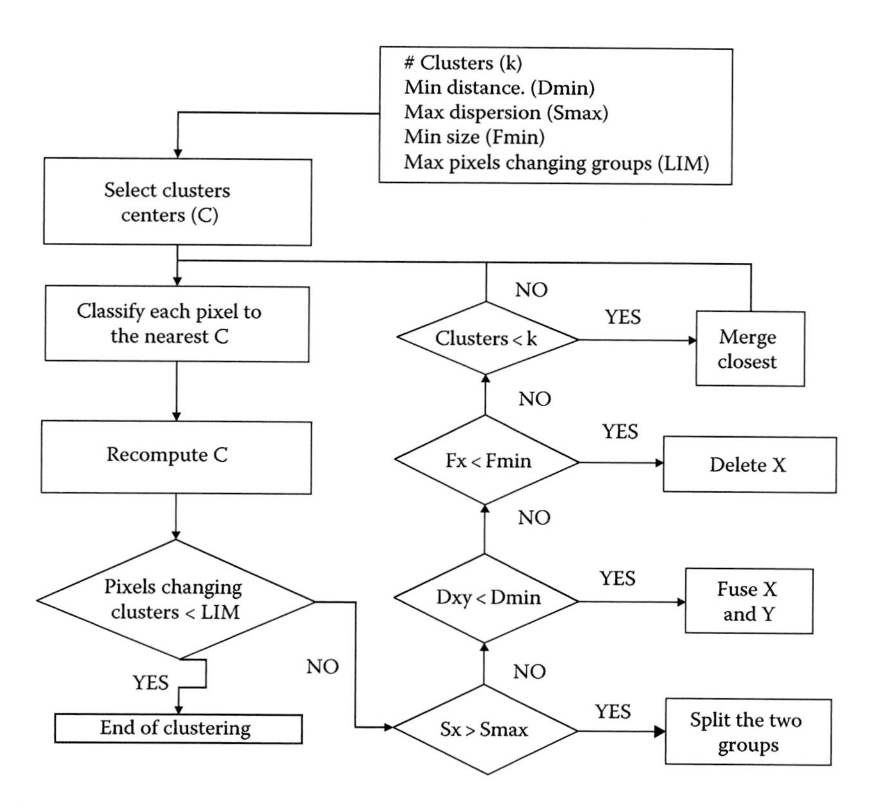


FIGURE 7.29 Control parameters for the ISODATA algorithm.

appropriate distance between clusters. In this way, the unsupervised analysis may be considered an exploratory process in which the user selects arbitrary values that are modified after seeing the results (Tso and Mather 2001). The unsupervised analysis may indicate, for instance, whether the end-user classes are spectrally separable or not, or their spectral variability, or whether they will have a bimodal or multimodal distribution.

In the case of the MODIS image, the unsupervised analysis was made on the seven reflectance bands. The following control parameters were used: 20 iterations, between 12 and 20 clusters, 5 as the minimum size threshold of each group, 2 standard deviations as the maximum dispersion in a group to be divided, and 1 as the minimum distance between clusters to be united. With this scheme, 13 thematic classes were generated and compared with the statistics generated by the supervised process to proceed with the assignment phase.

#### 7.2.2.4 Mixed Methods

Both the supervised and unsupervised methods have advantages and disadvantages (Thomas et al. 1987a; Tso and Mather 2001). The former method is often biased because the user decides a priori categories without considering the spectral characteristics of the image. The latter method does not guarantee that the output classes have a meaning for the end user. In order to offset the disadvantages of both methods, several alternatives have been proposed that combine both methods. These mixed methods can be grouped into two blocks:

1. Methods that use supervised analysis to "guide" the unsupervised analyses. In this case, the knowledge of the interpreter is used to lead the search for those homogeneous spectral clusters. For example, the unsupervised method can be applied to areas with a well-known cover, which would facilitate the spectral definition of this cover (Justice and Townshend 1981). Another way to guide the unsupervised analysis would be to introduce the average DLs derived from supervised training as the initial centers of clusters. Thus, the iterative process of the search is remarkably reduced and at the same time influences the formation of clusters with a thematic meaning.
2. Methods that combine statistics generated from supervised and unsupervised training. Categories deduced by the two methods can be complementary, since they will include both thematic and spectral meaning. Some authors, for example, use the unsupervised analysis to stratify the image, leaving the supervised one for the areas that were not classified previously. Categories can also be combined using automatic classification routines such as hierarchical clustering (Chuvieco and Congalton 1988). This joint classification facilitates three types of clusters: those consisting of supervised and unsupervised categories; those consisting of only supervised categories; and those consisting of only unsupervised categories. The first group identifies those classes selected by the interpreter that have a clear spectral meaning. The second group corresponds to thematic classes that are with little spectral meaning and are therefore difficult to discriminate.

Finally, the third group indicates spectral classes that cannot be assimilated by the classification key proposed by the user. From the analysis of these last two clusters, the interpreter can refine the scheme of classification or the selection of training statistics.

#### 7.2.2.5 Analysis of the Training Statistics

Regardless of the method used in defining the training phase, before carrying out the classification of the whole image, it is recommended to analyze the training statistics to determine whether the selected categories are really discriminable. In the negative case, the interpreter should find out whether the training process needs revision or the input variables or categories are relevant to the final goal. In the case of the MODIS image that we are using as a classification example, Table 7.4 displays the mean and standard deviation of the selected categories, based on the pixels included in the training classes. The mean provides information on the average spectral behavior of each category, while the standard deviation contains its internal homogeneity. In this case, the highest values of standard deviation are for the snow cover, while the lower values are for water. In spite of the great spectral complexity of urban areas, they do not show high standard deviations, probably due to the homogenization effect produced by having a coarse-resolution pixel size.

**TABLE 7.4**
**Mean and Standard Deviation of Classes Selected in the MODIS Image**

| | Mean | | | | | | |
|---|---|---|---|---|---|---|---|
| Center Wavelength (nm) | Soil | Snow | Herbaceous | Shrublands | Forested | Urban | Water |
| 645 | 3830.22 | 5311.49 | 1101.05 | 1972.30 | 614.49 | 1717.44 | 49.83 |
| 858.5 | 4574.02 | 5560.14 | 4504.79 | 2731.78 | 2511.85 | 2851.25 | 20.10 |
| 469 | 1780.79 | 5122.43 | 624.37 | 969.00 | 343.08 | 788.01 | 208.76 |
| 545 | 2799.26 | 5400.31 | 1150.96 | 1488.57 | 627.10 | 1441.01 | 240.66 |
| 1240 | 5097.45 | 1896.74 | 4043.67 | 3276.94 | 2478.55 | 2882.10 | 14.46 |
| 1625 | 5595.74 | 485.26 | 2797.17 | 2923.14 | 1595.88 | 2340.04 | 43.91 |
| 2130 | 5397.28 | 229.46 | 1629.87 | 3430.46 | 842.72 | 1826.89 | 25.79 |
| | Standard Deviation | | | | | | |
| 645 | 174.94 | 1939.56 | 241.35 | 394.40 | 278.28 | 210.61 | 53.83 |
| 858.5 | 210.80 | 1271.24 | 472.64 | 398.34 | 248.68 | 216.17 | 43.03 |
| 469 | 164.66 | 1920.62 | 130.45 | 270.48 | 196.52 | 103.07 | 143.73 |
| 545 | 180.76 | 1936.10 | 146.35 | 329.58 | 253.82 | 166.99 | 120.86 |
| 1240 | 206.99 | 398.49 | 319.46 | 587.30 | 220.77 | 242.97 | 36.52 |
| 1625 | 213.10 | 106.68 | 276.43 | 702.23 | 234.75 | 175.16 | 37.69 |
| 2130 | 209.87 | 67.54 | 325.44 | 806.42 | 210.30 | 191.95 | 25.59 |

*Note:* Values are expressed in reflectance × 10,000.

Both graphical and numerical methods have been developed to assess training statistics. Among the former, the most basic is a spectral signature plot, which shows the bands in the $x$-axis and the average DLs of each category in the $y$-axis (Figure 7.30). When classifying with corrected reflectances, this plot will be similar to the spectral curves shown in Chapter 2. This is the case of our example based on the first seven reflective bands of the MODIS sensor. The spectral signature plot provides a first assessment on the spectral tendencies of each category. Parallel and close lines indicate probable overlaps between categories, while the intersections between lines show those bands that are more likely to separate between categories.

The spectral signature graph shows only the average tendency of each category. To find out whether they can be discriminated from others, it is necessary to consider the dispersion values as well, since they will yield information about potential overlaps with similar categories. A graphical tool to evaluate those potential overlaps is the spectral dispersion graph. This is a bar chart that includes the digital range covered by each category, commonly expressed as the average ±1 or 2 standard deviations from the mean. In Figure 7.31, four examples of these graphs are included, showing the category overlaps between four contrasting bands (1, 2, 5, and 7) for the MODIS image. We observe how certain covers show clear overlaps in some bands (for instance, soil and snow in band 1), but are clearly separate in others (those two covers in band 7, especially), while others show a greater similarity in all of them (urban and herbaceous), and water is clearly discriminated from the rest in all four bands.

In addition to the graphical procedures, there are some quantitative criteria that provide more information to discriminate between the input categories. The simplest

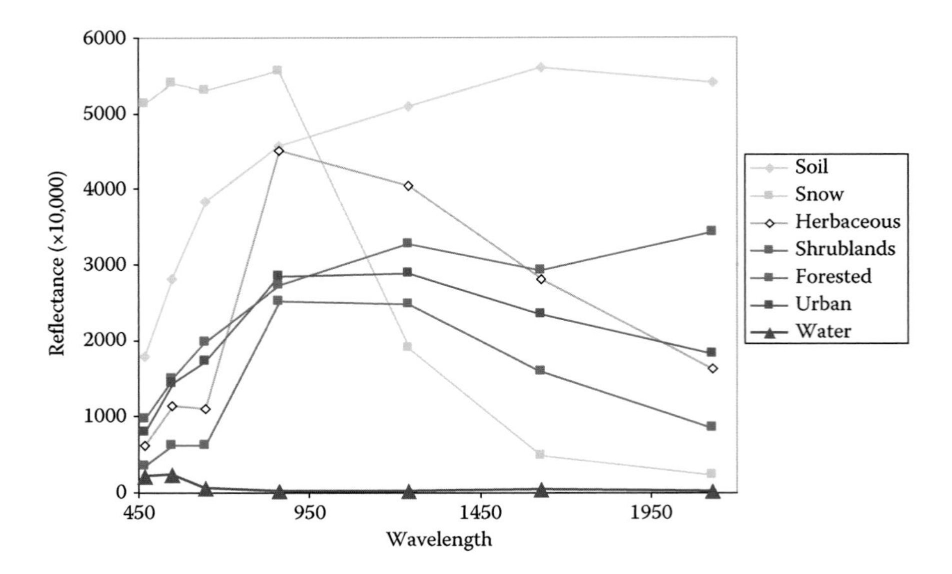


**FIGURE 7.30** Spectral signatures plots of the training classes. MODIS bands have been arranged by wavelength.

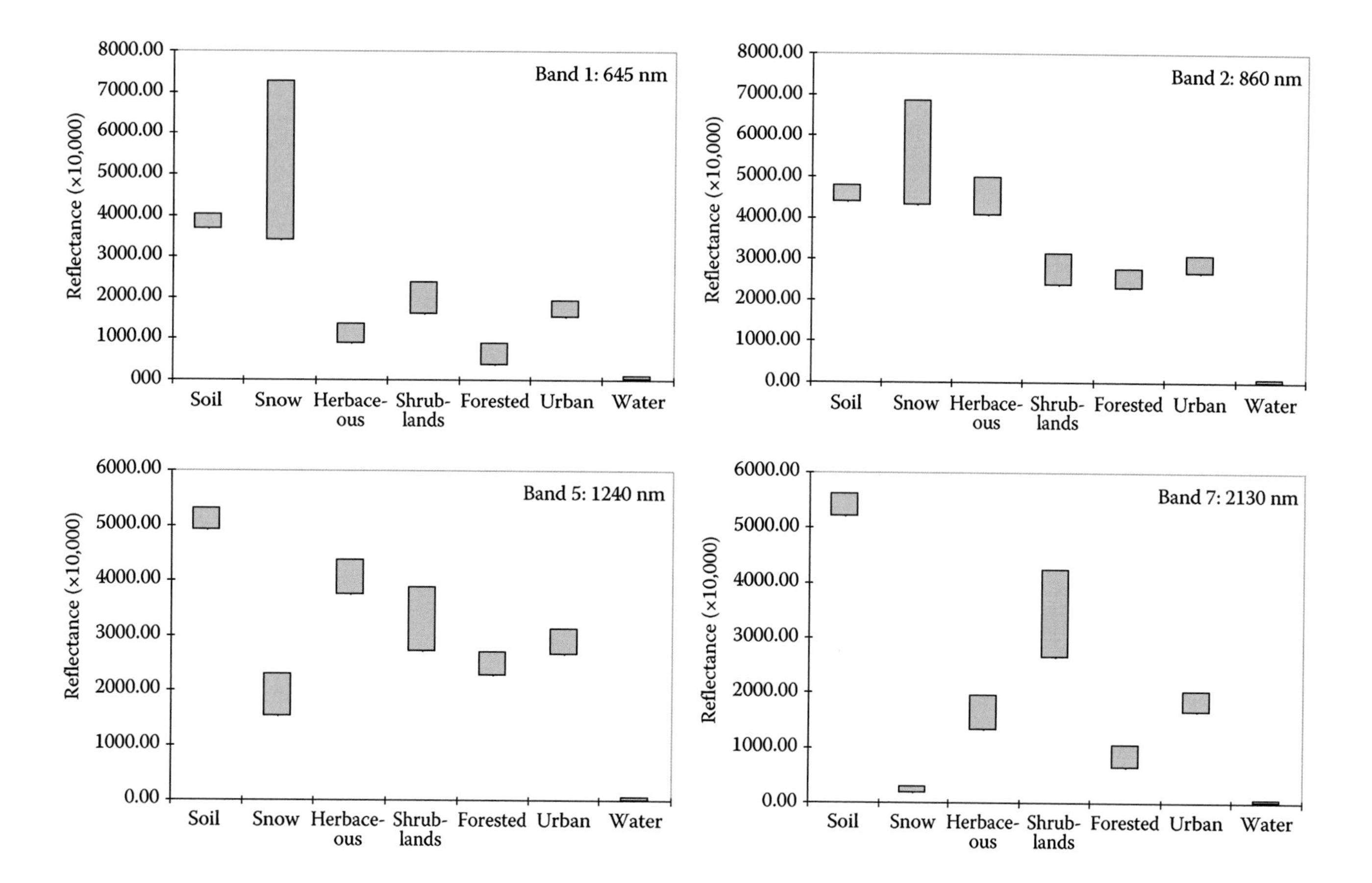


FIGURE 7.31 Spectral dispersion graphs.

index is the normalized distance (ND), which is calculated as the absolute difference between the averages of two categories, as a fraction of its standard deviation (Swain and Davis 1978):

$$ND_{A,B} = \frac{\left|\overline{DL_A} - \overline{DL_B}\right|}{s_A + s_B} \quad (7.28)$$

This calculation is applied to each pair of bands that take part in the classification, averaging its value to obtain a separability matrix.

A little more complicated, but also more versatile, is the statistical divergence (Swain and Davis 1978; Thomas et al. 1987a). It assumes that the DLs are normally distributed in each category and considers the separability as a measure of overlap between neighboring categories. Since it is a question of calculating that value for a multivariate space (as many variables as bands taking part in the process), the vector of means and the matrix of variance–covariance between pairs of categories need to be considered. The calculation is done as follows:

$$\begin{aligned} Div_{AB} &= 0.5\,Tr\left\{(\mathbf{V}_A - \mathbf{V}_B)\left(\mathbf{V}_B^{-1} - \mathbf{V}_A^{-1}\right)\right\} \\ &\quad + 0.5\,Tr\left\{\left(\mathbf{V}_A^{-1} + \mathbf{V}_B^{-1}\right)(\mathbf{M}_A - \mathbf{M}_B)(\mathbf{M}_A - \mathbf{M}_B)^T\right\} \end{aligned} \quad (7.29)$$

where

$Div_{AB}$ indicates the divergence between the categories $A$ and $B$
$Tr$ refers to the trace or sum of the elements of the diagonal of the indicated matrix {}
$\mathbf{V}_A$ and $\mathbf{V}_B$ are the variance–covariance matrices of $A$ and $B$
$\mathbf{M}_A$ and $\mathbf{M}_B$ are the vectors of means for the same categories
$T$ indicates the transposed matrix of the one indicated in parenthesis

The greater the value of the divergence, the greater the separability between classes, based on the input bands. This metric increases with the number of input bands, which implies that the discrimination ability of the classification would constantly increase when increasing the number of bands. Since this is not very realistic, the divergence is scaled within known margins, using a simple transformation (Kumar and Silva 1977):

$$TD_{AB} = c\left(1 - \exp\left(-\frac{Div_{AB}}{8}\right)\right) \quad (7.30)$$

where $TD$ is the transformed divergence, $c$ indicates the maximum value of the divergence, which usually takes the value of 2, although some authors recommend using a value of 100, so that the results can be interpreted in percentages.

The divergence has also been used as a measure to select the set of bands that provides the best separation between classes. This way, an average value of divergence is calculated, according to Richards and Xia (1999):

$$\overline{div} = \sum_{i=1,m} \sum_{j=1,m} p(w_i)p(w_j)Div_{i,j} \tag{7.31}$$

where

$p(w_i)$ and $p(w_j)$ indicate the a priori probabilities of classes $i$ and $j$, respectively
$m$ is the number of classes
$Div_{i,j}$ is the value of the divergence between those classes

The average divergence is calculated for any combination of $m$ bands, the ideal combination being the one that maximizes the global value of separation between classes. In addition to divergence, other measurements of separability have been proposed, such as the Bhattacharyya or Mahalanobis distances (Mather and Coch 2011; Thomas et al. 1987b):

$$MD_{AB}^2 = \left(\mathbf{M}_A - \mathbf{M}_B\right)^T \left(\mathbf{V}_B^{-1} - \mathbf{V}_A^{-1}\right)\left(\mathbf{M}_A - \mathbf{M}_B\right) \tag{7.32}$$

with the same meaning as in Equation 7.29.

Table 7.5 includes the calculation of the statistical separability for the classes selected on the MODIS image. This table offers high values for all categories, meaning that a clear separation was observed in the training statistics. The maximum limit was obtained in most cases, with a minor reduction in the urban-forested and herbaceous-forested categories.

As mentioned earlier, the divergence can also be used to indicate the suitable combination of bands for the classification. In our case, it was calculated for all the combinations of three bands, and what was obtained was that bands 1, 2, and 5 offered the greatest average separability between categories.

**TABLE 7.5**
**Transformed Divergence for Classes of the MODIS Image**

| | Soil | Snow | Herbaceous | Shrublands | Forested | Urban |
|---|---|---|---|---|---|---|
| **Snow** | 2.0000 | | | | | |
| **Herbaceous** | 2.0000 | 2.0000 | | | | |
| **Shrublands** | 2.0000 | 2.0000 | 2.0000 | | | |
| **Forested** | 2.0000 | 2.0000 | 1.9999 | 2.0000 | | |
| **Urban** | 2.0000 | 2.0000 | 2.0000 | 2.0000 | 1.9989 | |
| **Water** | 2.0000 | 2.0000 | 2.0000 | 2.0000 | 2.0000 | 2.0000 |

The analysis of the training statistics should provide some ideas regarding the convenience of the classification key used, the training methods adopted, or the input bands to discriminate between the target categories. Once these aspects are solved, different classification algorithms can be used to group the complete image, but results should not be very different from those deduced in the preceding analysis.

### 7.2.3 Assignment Phase

This phase aims to classify the whole image by assigning each pixel to one of the previously selected categories. This categorization is carried out on the basis of the DL values of each pixel, for all input bands selected by the interpreter. As a result of this process, a new image will be created, where the DL will identify the thematic category assigned to each pixel.

Several algorithms have been proposed to carry out the categorization process. From a statistical point of view, they all define discriminant functions that provide a rule to assign each pixel to any of the previously trained categories. The most common approaches to establishing these functions have been the minimum distance, parallelepiped, and maximum likelihood (ML), but many other methods have been suggested in the last couple of decades. These methods may be grouped in two categories, parametric and nonparametric, based on whether or not they assume normally distributed classes. A more detailed analysis of most common algorithms follows.

#### 7.2.3.1 Minimum-Distance Classifier

The simplest approach to classify a pixel is to allocate it to the closest category, that is, the one that minimizes the distances between that pixel and the class center (Figure 7.32). The most common definition of similarity is the Euclidian distance:

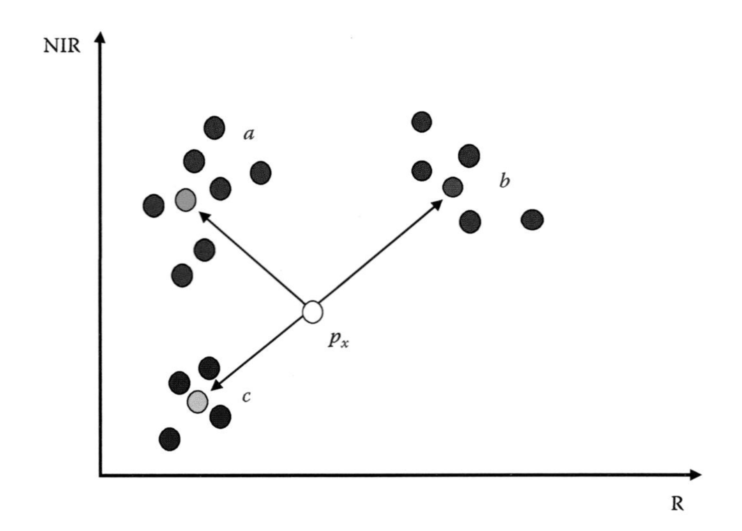


FIGURE 7.32 Minimum-distance classifier.

$$d_{x,A} = \sqrt{\sum_{i=1,m} \left( DL_{x,i} - \overline{DL}_{A,i} \right)^2} \tag{7.33}$$

where

$d_{x,A}$ is the distance between the pixel $x$ and the category $A$
$DL_{x,i}$ is the value of the target pixel in the band $i$
$\overline{DL}_{A,i}$ is the mean DL of category $A$ in the same band
$m$ is the number of input bands

The pixel is assigned to the category that minimizes $d_{x,A}$. Since the goal is to compare between categories, the previous formula can be simplified by eliminating the square root. This algorithm does not consider the dispersion of each category or the correlation between bands. Therefore, the discriminant functions have circular shapes around the class centers.

This algorithm is quite simple and fast, and offers good results when the categories have little overlap. The method ensures that all pixels are classified (always there is a nearest class, even if it is far away). This may cause important commission errors, particularly when the training phase has not identified a relevant category in the study area.

Some authors have proposed modifications in the calculation of the distance in a way that is considered the variance of the classes (Lo et al. 1986; Schowengerdt 2007).

#### 7.2.3.2 Parallelepiped Classifier

This algorithm is based on defining a dispersion area around the center of each category, which accounts for its spectral variability. Therefore, this assignment approach establishes multidimensional polygons of parallel sides (that is the reason for its name) around the center of each category. The classification criterion is based on including a pixel in a certain category when its DLs are within the limits of that class, in all the input bands (Figure 7.33). Schematically

$$\overline{DL}_{A,i} - R_{A,i} \leq DL_{x,i} \leq \overline{DL}_{A,i} + R_{A,i} \tag{7.34}$$

where $R_{A,i}$ indicates the dispersion range for the class $A$. The criterion is run for all input bands $i$. In summary, a pixel is assigned to a certain class if its DLs are within a certain distance of the center of that class. That distance is established by the interpreter and is commonly a multiple of the standard deviation. Other approaches such as the interquartilic deviation or class range can be used as well. In any case, it is worth considering that the broader the dispersion value, the more likely it is that pixels of neighboring categories (commission error) will be included, and the narrower the dispersion value, the more likely it is that pixels of the actual category will be missed (omission errors).

The parallelepiped algorithm is easy and quick to implement, since it requires only simple logical operations (IF … THEN …). It has frequently been used as the

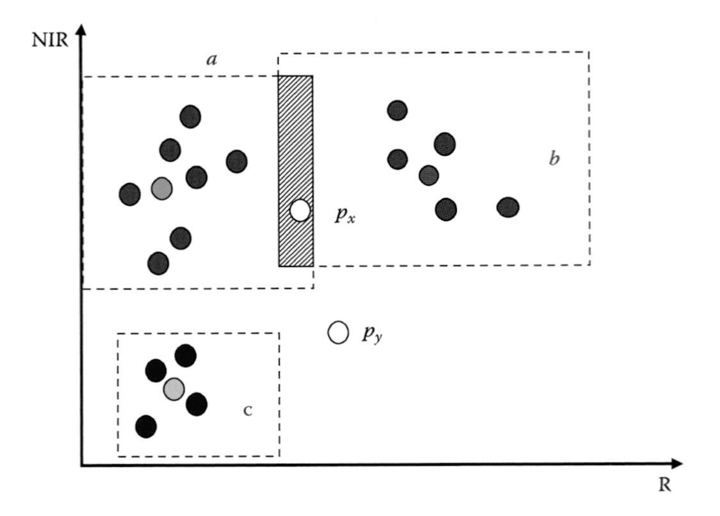


**FIGURE 7.33** Parallelepiped classifier: $p_x$ can be assigned to two categories, while $p_y$ will remain unclassified.

first approach to class variability or to obtain a first stratification of the image, isolating the most unique categories and then applying another algorithm on the most problematic pixels.

The main problems of this algorithm are related to the way the domain areas are established. First, there might be overlapping areas and, therefore, pixels can be classified in two or more categories. If the program does not have other criteria, it will include the pixel in the first category that fulfills the requirements given in Equation 7.34. As a consequence, the order of categories may have an important impact on the results. Some authors have suggested modifying the criterion for these conflict areas by introducing a minimum-distance rule, by adjusting the borders interactively (Mather 1998) or by creating mixed classes.

The second problem is unclassified areas, which will occur for those pixels whose DLs are outside any of the defined boxes. In this case, the interpreter may decide to extend the dispersion range, establishing a balance among the unclassified pixels and creating overlapping regions. Another option is to improve the definition of input classes, including new ones adapted to nonassigned areas.

#### 7.2.3.3 Maximum Likelihood Classifier

This algorithm assumes that DLs are normally distributed within each class, and so the actual frequency can be estimated using the Gaussian probability function, based on the mean vector and variance–covariance matrix. This method is more complex than the two previously reviewed and requires more computing time. However, it is the most widely used in remote sensing applications, since it provides a more consistent approach to data variability.

The basis of this algorithm can be illustrated using an example proposed by Swain and Davis (1978). Let us consider that the interpreter needs to discriminate between

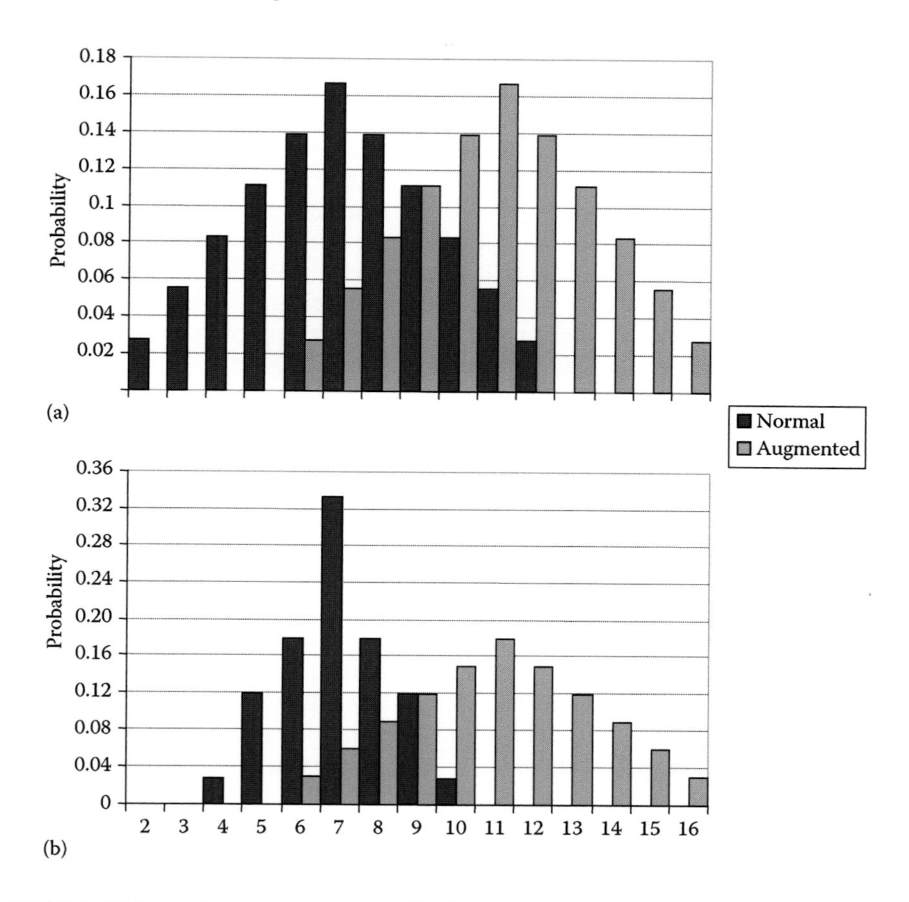


FIGURE 7.34 Basis of the maximum likelihood classifier. (a) Distribution functions for a normal and augmented pair of dice and (b) distribution functions for a modified and augmented pair of dice.

a specific outcome obtained by throwing a normal pair of dice, numbered from 1 to 6, from another result obtained from dice with a couple of additional points added to each face (numbered from 3 to 8). A reasonable way of solving the problem would be to calculate the probability that any specific outcome was obtained with the normal or the modified dice, assigning the unknown results to the most probable pair of dice. For example, with a throw of 7, the probability of having used the normal dice is 6/36 (6 + 1, 1 + 6, 3 + 4, 4 + 3, 5 + 2, and 2 + 5). For the augmented dice, the probability is smaller, just 2/36 (3 + 4 and 4 + 3), and therefore the "result" is assigned to the normal pair of dice. The same philosophy would apply to other outcomes, and the criterion to assign to one or the other pair would be the value of the expected probability. For the normal pair of dice, this function would be centered about the value 7 (maximum probability), with a minimum of 2 and a maximum of 12 (Figure 7.34a). For the modified pair, the most probable value would be 11, being the probability distribution from 6 to 16. From this graph, we can conclude that the probability of having used the normal dice is larger than the augmented one when outcomes are equal to or below 8. It is smaller when they are equal to or higher than 10, and it is identical when they are equal to 9. In other words, the threshold to assign any specific outcome to each pair of dice lies at the intersection between their respective probability functions.

This criterion does not imply that outcomes larger than 9 cannot be obtained from the normal dice, or lower than 9 with the augmented one. The probability is lower, but it is not zero. In other words, the ML rule is not fully accurate when there is overlap in the probability functions. It is completely reliable only when values are lower than 6 or higher than 12, since these values can be obtained only with one pair and never with the other. In summary, if natural overlap exists among classes, it is not possible to separate them with total reliability.

Extending this example, let us suppose that the normal pair of dice are also modified, but in this case to reduce the range of outcomes from 2 to 5, eliminating 1 and 6 and repeating the punctuations 3 and 4. This would not modify the mean value of the probability distribution, which would still be 7. However, this action would change the spread of the distribution, since the punctuations would tend to be closer to the average value. Consequently, the assignments to this pair of dice would be safer. The assignment threshold would continue to be 9, but the probability of having lower or higher outcomes to that value decreases with respect to the previous distribution (Figure 7.34b). Therefore, the more homogeneous a distribution (i.e., the lower the variance), the more accurate the discrimination of categories will be.

Applying this example to the context of digital classification, the interpreter needs to estimate the probability functions of the different thematic categories to be discriminated. The probability of obtaining a certain value when playing with two dices can be estimated before the actual results are obtained, since we know the number of possible combinations for each outcome and the total number of combinations. In other words, we can compute the a priori probability of obtaining each outcome. In the case of digital classification of thematic categories, we should find out a procedure to estimate their probability distributions. Assuming that the categories are normally distributed, the Gaussian function can be used to estimate the probability that a pixel with a certain DL is assigned to each category, providing we know the mean and standard deviation of that category following

$$p\left(\frac{x}{A}\right) = \frac{1}{\sqrt{2\pi\sigma_A^2}} \exp\left\{-\frac{\left(DL_x - \overline{DL_A}\right)^2}{2\sigma_A^2}\right\} \tag{7.35}$$

where

$\overline{DL_A}$ is the mean $DL$ for class $A$

$\sigma_A^2$ is the variance of that class

Both the mean and the variance are taken from the training statistics. Once this function is computed for all categories for a single pixel, that pixel is assigned to the category that maximizes the function

$$p\left(\frac{x}{A}\right) \geq p\left(\frac{x}{B}\right) \tag{7.36}$$

for any $B$ category. To speed up the calculation of the exponential function, some authors suggest using a logarithm transformation to obtain lineal functions (Schowengerdt 1983):

$$\ln p\left(\frac{x}{A}\right) = -0.5\ln(2\pi) - 0.5\ln\sigma_A^2 - \left\{\frac{(DLx - \overline{DL_A})^2}{2\sigma_A^2}\right\} \tag{7.37}$$

The expression can be simplified even more if we eliminate the first term ($-0.5\ln(2\pi)$), which is constant for every category. For the same reason, the second term ($0.5\ln\sigma_A^2$) can be calculated initially for each category, instead of computing it for each pixel.

Returning to the previous example, it could also occur that the outcomes were obtained with two pairs of normal dice and one pair of augmented dice. In this case, we should also consider in our classification criterion that it is more likely to get certain outcomes from the normal pair of dice than the augmented one, simply because they are more numerous, independent of the actual results. This difference in weight of two categories is also applicable to digital image classification because some categories may be more abundant in a certain area and therefore are more likely to occur in the image than others. This criterion of "a priori" probability can also be considered in the ML rule, by modifying the decision rule of Equation 7.36, which becomes (Strahler 1980)

$$p\left(\frac{x}{A}\right)p(A) \geq p\left(\frac{x}{B}\right)p(B) \tag{7.38}$$

where $p(A)$ and $p(B)$ are weight factors (or a priori probabilities) that take into account the abundance of each category in the study area. The threshold to discriminate between categories moves toward one side or another of the original value, as a function or a priori probability of each category. Remember that the intersection between curves marks the boundary in the area assigned to each category; this factor can modify the results significantly.

Although of great interest, the use of this option involves practical difficulties, since commonly there is no previous information on the land cover of the area to classify. When this is available, for instance, from a previous inventory, those weights may help solve confusing problems between spectrally similar categories. Otherwise, the interpreter may just introduce the same weights to all.

So far, we have considered the ML criterion with just a single band. Most commonly, this algorithm is applied to several bands. Therefore, we should extend the formulas previously presented to a multivariate space:

$$p\left(\frac{x}{A}\right) = (2\pi)^{-m/2}\left|V_A\right|^{-0.5}\exp\left\{-0.5\left(M_x - M_A\right)^T V_A^{-1}\left(M_x - M_A\right)\right\} \tag{7.39}$$

where

$m$ is the number of bands
$|V_A|$ is the determinant of the variance–covariance matrix for class $A$
$M_x$ is the vector of DLs for pixel $x$
$M_A$ is the vector of mean values for class $A$
$T$ and $-1$ indicate the transpose and inverse of the matrix, respectively

This formula can also be expressed in logarithmic form:

$$\ln p\left(\frac{x}{A}\right) = \left(-0.5\ln V_A\right) - \left(0.5\left(M_x - M_A\right)^T \ V_A - 1\left(M_x - M_A\right)\right) \quad (7.40)$$

once the constant terms are eliminated from Equation 7.39. For the case of two bands, the limits of the categories have an ellipsoidal shape (Figure 7.35). As in other classification criteria, the more the overlap between probability functions, the more the potential confusion between categories.

In principle, the ML algorithm classifies every pixel into one category, even when the probability is low. Since classification with low likelihoods commonly is associated with high commission errors, a postclassification filter can be applied to reject the assignment of those pixels with low probability of membership (Figure 7.36). In case of many pixels with low probability, the interpreter should consider whether the selected classes are really representative of the study area, or whether the training statistics were properly generated.

The ML method assumes that the categories follow a normal distribution, which is frequently the case in remotely sensed data. However, Swain and Davis (1978) showed that the results of this algorithm are consistent even if this condition is violated.

For the MODIS image used in this section, the ML criterion provides good results as was expected from the separability analysis. However, problems were found in distinguishing between irrigated crops and grasslands, which have also photosynthetic activity at this time of the year, and therefore a common herbaceous category was selected. Minor problems were also observed in distinguishing water in the Colorado River delta area from soil because of water turbidity. Urban areas were

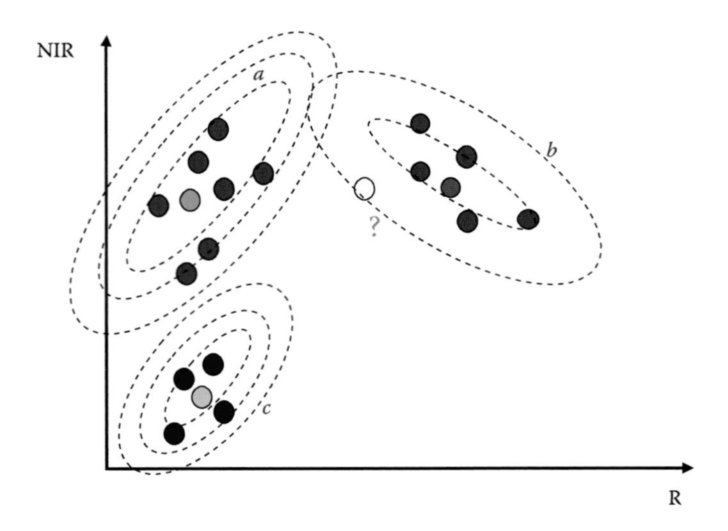


FIGURE 7.35 Isolines of the Gaussian probability function. The unknown pixel is assigned based on probability distribution functions.

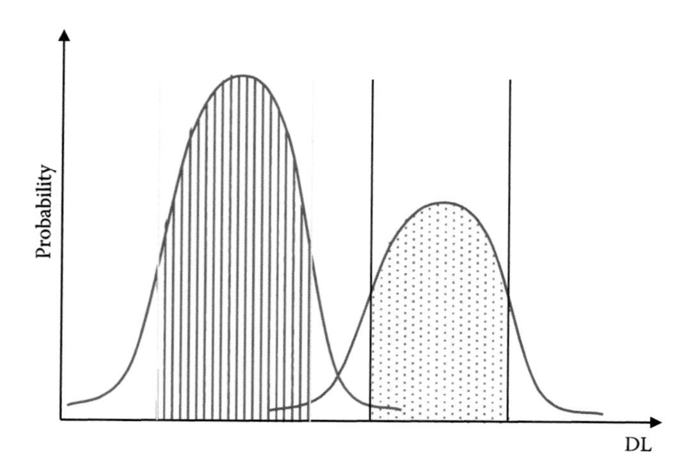


**FIGURE 7.36** Impact of applying postclassification probability filtering.

mixed with semiarid vegetation, even though the large metropolitan areas (Phoenix, Tucson, Las Vegas, Mexicali, etc.) were well classified. In addition, frequent confusion between grassland and shrubland was observed (Figure 7.37a).

Comparing these results with the unsupervised ISODATA training, an improvement in the definition of transitional areas was observed, especially in semiarid vegetation. Soil background effects were also noticeable in the definition of spectral classes. Urban areas were not identified in this case, as they were confused with irrigated crops or grasslands (Figure 7.37b).

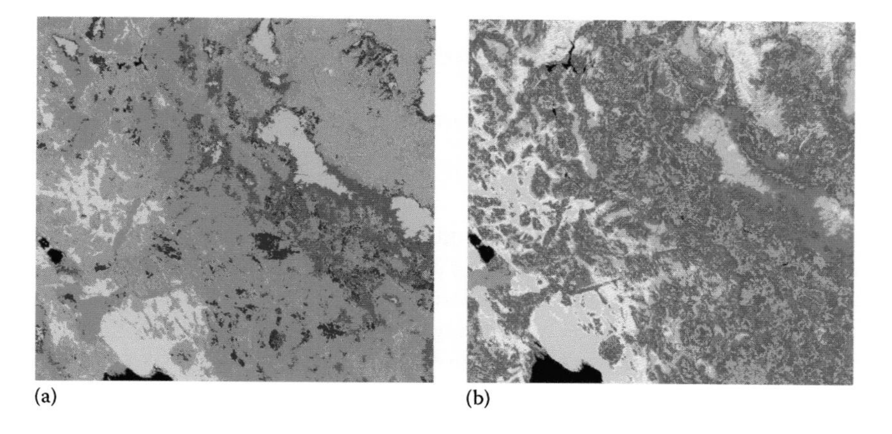


**FIGURE 7.37** Results from the (a) maximum likelihood classifier and (b) unsupervised ISODATA methods: blue, water; yellow-orange, soil, shrub, and sparse vegetation; light green, grassland and irrigated areas; dark green, forested; red, urban areas.

#### 7.2.3.4 Decision Tree Classifier

One of the simplest alternatives to traditional classification systems is decision tree classification. This approach is based on establishing a set of binary rules that are applied sequentially to discriminate between different target categories. Those rules include thresholds on spectral bands, but also on auxiliary information, such as soil maps, slope, or elevation, and therefore are very adaptable to different types of input data. The algorithm is built after identifying the most sensitive bands for category discrimination. Each category may be classified using different input variables, instead of using all simultaneously. For example, water can be separated in the NIR or SWIR reflectance, green vegetation using the normalized ratio of R and NIR bands plus elevation and slope, and residential areas using a combination of texture and NIR reflectance. This classification approach is very robust, easy to interpret, and flexible enough to handle a broad range of response types (De'ath and Fabricius 2000). For these reasons, the decision tree classifiers have been extensively used in the last few years, especially in global land cover classification studies (Friedl and Brodley 1997; Friedl et al. 1999; Hansen et al. 1996; Zhan et al. 2002). Algorithms developed to discriminate between single categories based on a set of sequential rules can also be considered tree classifiers. This is the case with the different cloud detection (Saunders and Kriebel 1988), active fires (Giglio et al. 2003), or burned area discrimination algorithms (Bastarrika et al. 2011; Chuvieco et al. 2008a).

The set of rules may be created from the interpreter's experience or based on statistical approaches, which try to find the linear functions that better discriminate between specific input categories. The former are difficult to generate, since the interpreter needs to identify the best discrimination space for each category. However, once the system is properly trained, it provides a quick and robust way to classify a broad range of categories, provided there is at least an input band that provides a unique response for each category. In this regard, the tree classifier can be considered a simple example of an expert system (Skidmore 1989). The decision rules are defined in terms of logical operators (IF condition1 AND condition2 OR condition3 … THEN action). These conditionals can involve all kinds of input variables, including spectral, textural, and auxiliary information:

$$\text{IF } DL_{NIR}/DL_R > 2 \text{ AND Texture} = \text{Rough THEN Forest}$$
$$\text{IF } DL_{NIR}/DL_R > 2 \text{ AND Texture} = \text{Smooth THEN Grasslands}$$

To create the decision rules, the user should consider the spectral, spatial, and temporal characteristics of the different target categories. In the case of vegetation covers, in addition to the spectral values, the seasonal trends may be critical in discriminating among different species, as well as elevation ranges or soil types. Following this logic, Skidmore (1989) carried out an inventory of eucalyptus forests in Australia, based on satellite images and auxiliary information. He observed a significant improvement in the classification after considering the ecological gradients of eucalyptus in the region, which were introduced as a set of discriminant rules. Other authors have used contextual rules, considering the vicinity or each cover as a decision criterion (Middelkoop and Janssen 1991). Decision trees can also be integrated

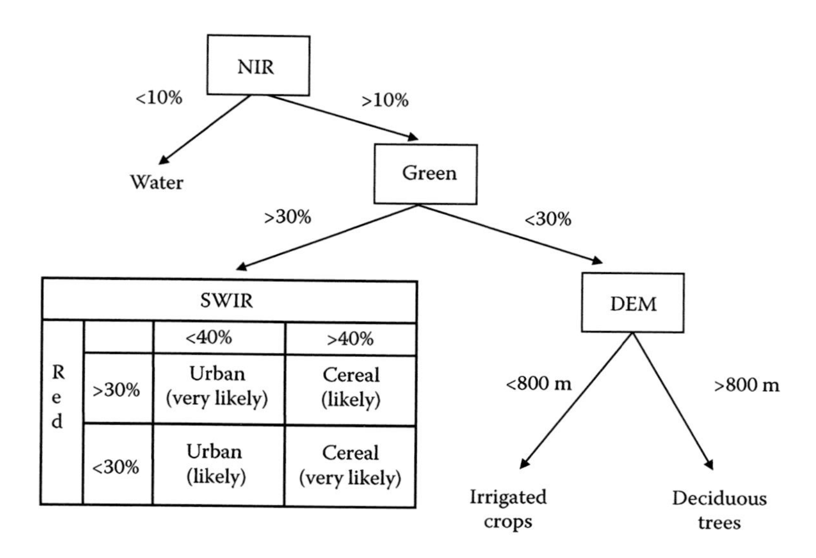


**FIGURE 7.38** Example of a simple decision tree classifier.

with other classifiers to obtain a more automatic discrimination of thematic classes (Garcia et al. 2011).

To illustrate the creation of a decision tree, a simple classification structure has been included in Figure 7.38. In this case, the exploratory analysis was carried out with four spectral bands: G, R, NIR, and SWIR, as well as elevation data (DEM). The first branch of the tree addresses the separation of water, which is easily discriminated in the NIR band. Those pixels with low NIR reflectance ($\rho_{NIR} \leq 10$) were labeled as water and eliminated from further analysis. The second step discriminates healthy vegetation (irrigated areas and deciduous trees) based on the green reflectance ($\rho_{Green} \leq 30\%$). Urban land and cereal crops were more difficult to differentiate because they present a significant overlap. For this reason, two bands were used for this threshold, based on the SWIR and red reflectance. Finally, the classes deciduous and irrigated crops were discriminated using elevation data, assuming that crops will occur only at low elevations.

Automatic algorithms for building decision trees have been proposed in the last few years. One of the most versatile is named classification and regression tree (CART) analysis (Lawrence and Wright 2001). This algorithm has been implemented in several commercial statistical packages and, therefore, is easily available for remote sensing applications. CART operates by repetitively splitting the data based on the input variables until a certain degree of homogeneity is found in the terminal nodes. Each branch of the tree is created after specific functions of the input variables to ensure maximum intragroup homogeneity and intergroup separability. To improve the generalization power of the final result, the splitting criteria are cross-validated with samples that are not used in the building of the tree. This assessment is commonly known as "pruning" and involves comparison of different sets of decision rules created from the training pixels against validation pixels, selecting the most accurate and simple set as possible. The CART algorithm has been used in several remote sensing applications, such as land cover (Lawrence and Wright 2001), forest fire risk assessment (Amatulli et al. 2006), and burned area classification (Stroppiana et al. 2003).

Another variation of decision tree classifiers is named *random forest* (RF), which uses an iteratively random set of the input variables to build a large set of trees (a forest) (Breiman 2001). Each tree is built so as to maximize separation between categories. Instead of using the best variables for each node of the trees, RF uses a random subset, thus minimizing the generalization error. In addition, the algorithm uses different samples of the input dataset, and therefore trees are generated from different subsets. Averaging tree outputs provides the most suitable set of rules to discriminate the input samples, part of which are not used to create the trees and, therefore, they can be used as unbiased predictors of classification error. This is an important feature of RF, as the traditional decision trees may be affected by overfitting. In the last years, RF has been extensively used in classification of EO images (Gislason et al. 2006; Lawrence et al. 2006; Rodriguez-Galiano et al. 2012).

#### 7.2.3.5 Neural Networks

Artificial neural networks (ANNs) have been extensively used for pattern recognition and classification analysis in a wide range of applications. Within the remote sensing community, ANNs have been mainly used for digital image classification, but there are also examples of using these systems for estimating physical parameters such as the LAI, water, or chlorophyll content (Fang and Liang 2003; Fourty 1997; Riaño et al. 2005), as well as for fitting prediction models (Vega-García et al. 1996) and multitemporal change detection (Liu and Lathrop 2002).

In essence, ANNs are used to predict a complex behavior (output) based on a set of input variables that can be multiply interrelated. They try to simulate the behavior of the human brain (hence, their name) by means of multiple interconnections among activation units (called "neurons"). An ANN is a set of connections between simple elements (neurons, processing units) distributed in layers (Figure 7.39). The units of a layer are connected with those of the neighboring layer through a certain

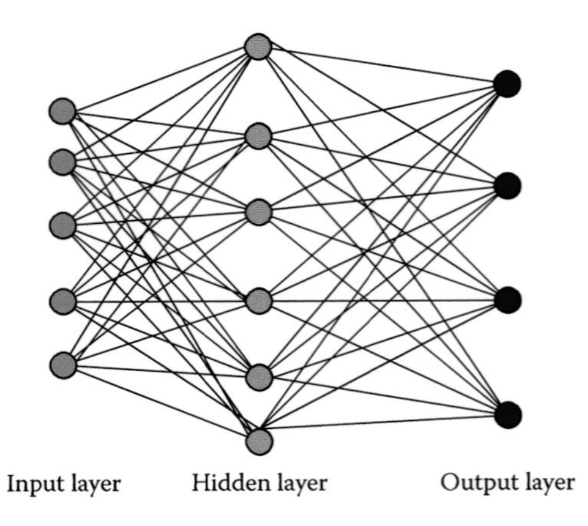


**FIGURE 7.39** Structure of an artificial neural network.

function. These equations (named "activation functions") can be linear or nonlinear (most commonly, sigmoid). They are used to estimate a value in the output layer from the values of the input layers.

Since the neurons are connected to other neurons through the intermediate layers (named "hidden layers"), the value of the output units depends on the values of all input variables. However, those input values may have more or less importance in the estimation of the output, according to the weights ($w_{i,j}$) calculated for each activation function. The calculation of those weights is carried out through an iterative process called "training" (or "learning") the network. In this process, the weights are changed iteratively from a set of random values, comparing in each step the estimated output with the actual output value with the final goal of reducing the estimation error. In summary, the outputs values are calculated as (Richards 1993)

$$o = f(wx + \theta) \tag{7.41}$$

where
- $o$ indicates the output vector
- $\theta$ is a threshold (in many occasions it is zero)
- $w$ is the vector of weights
- $x$ is the input vector

The output vector can in turn be the input vector of other layers, depending on the network structure. This structure can be very diverse, and the level of complexity is adapted to the classification needs. In image classification, the ANNs are usually structured in three layers (input, hidden, and output). The hidden layer serves to provide more flexible discrimination functions. The number of the input units depends on the input bands for classification and the radiometric resolution of those input bands (ranging code) (Benediktsson et al. 1990). The number of output units commonly is the number of categories to be classified, while the number of units in the hidden layer is usually at least similar to that of the input layer (Richards 1993).

The most critical point in the ANN is the *learning* or training phase, which includes the definition of the network structure and finding out the weights of the activation functions. Among the diverse learning algorithms, one of the most common is named "backpropagation." The basis of this algorithm is comparing the output of the ANN with the actual output values provided by the interpreter, and finding out which set of weights of the different activation functions provides the least errors. In summary, the algorithm proceeds as follows (Rumelhart et al. 1986):

1. First, it assigns random weights to the initial network. The computed outputs (in our case, the predicted membership, $t_{k,m}$) using those weights are compared to the actual category of the training sample (introduced by the user). As a result, an initial error is computed, following

$$E = \sqrt{\sum_{k=1,n} \sum_{m=1,c} (t_{k,m} - o_{k,m})^2} \tag{7.42}$$

for all expected outputs $m$ (in our case, for the $c$ categories) in all the input units $k$ (in our case, the number of input bands, $n$).
2. The weights are modified iteratively in the different layers to reduce the final error. This process starts from the output layer to the input layer, and therefore, the error is spread to the previous stages (backpropagated). Once the weights are readjusted, the estimated output values are compared again to the actual output values and a new error is computed.
3. The process is repeated until the error is very small or null (in that case, all the samples would be well classified), or until the number of iterations indicated by the interpreter has been completed. That number of iterations can vary widely depending on the complexity of the problem (in land cover classification problems, they may be as high as 250,000 iterations; Civco 1993).

To illustrate how the learning phase evolves, a simple classification example proposed by Richards (1993, pp. 219–223) can be used. In this example, a simple ANN will be derived for six elements classified in two categories ($c_1$ and $c_2$) that we want to discriminate using linear functions (Figure 7.40). The values of these six elements are

$$c_1 = (-4,3)(-4,-3)(-1,0)$$
$$c_2 = (-2,-3)(-2,3)(1,0)$$

We can conclude from Figure 7.40b that these two classes cannot be discriminated using linear equations without errors. To solve that problem, a simple ANN with

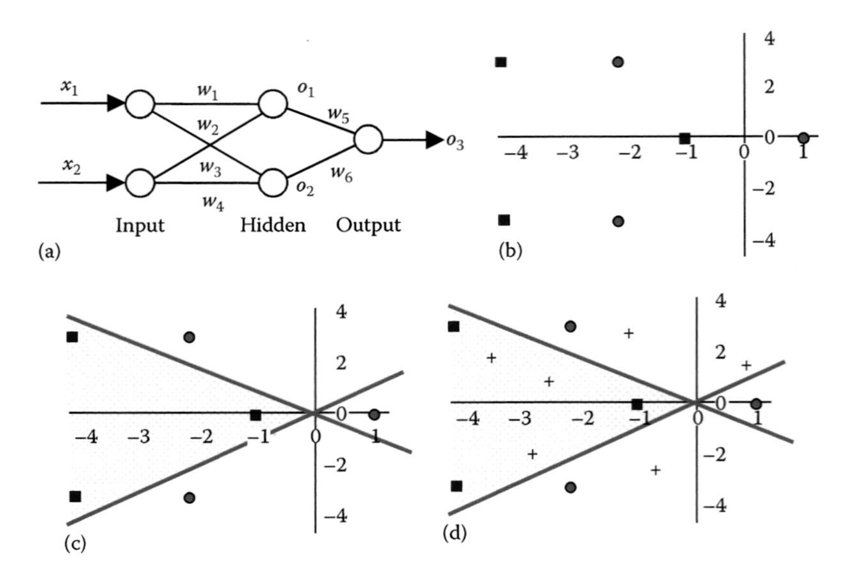


**FIGURE 7.40** Example of using an artificial neural network for classification of two categories. (a) Basic ANN structure, (b) units to discriminate, (c) discriminant function, and (d) new cases to classify. (Adapted from Richards, J.A., *Remote Sensing Digital Image Analysis: An Introduction*, Springer-Verlag, Berlin, Germany, 1993.)

**TABLE 7.6**
**Variation of Weights of the Neural Network Alters Different Iterations**

| Iteration | $w_1$ | $w_2$ | $w_3$ | $w_4$ | $w_5$ | $w_6$ | Error |
|---|---|---|---|---|---|---|---|
| 0 | 0.050 | 0.100 | 0.300 | 0.150 | 1.000 | 0.500 | 0.461 |
| 1 | 0.375 | 0.051 | 0.418 | 0.121 | 0.951 | 0.520 | 0.424 |
| 5 | 0.606 | 0.007 | 0.570 | 0.117 | 1.240 | 0.909 | 0.391 |
| 10 | 0.642 | −0.072 | 0.641 | 0.196 | 1.464 | 1.034 | 0.378 |
| 50 | 2.224 | −2.215 | 2.213 | 2.216 | 3.302 | 3.259 | 0.040 |
| 150 | 2.810 | −2.834 | 2.810 | 2.835 | 4.529 | 4.527 | 0.007 |
| 250 | 2.901 | −2.976 | 2.902 | 2.977 | 4.785 | 4.784 | 0.005 |

*Source:* After Richards, J.A., *Remote Sensing Digital Image Analysis: An Introduction*, Springer-Verlag, Berlin, Germany, 1993.

three layers (input, hidden, and output) is created. Since the problem only includes two classes, two units will be used in the input and hidden layer and just a single one in the output layer. Therefore, the problem requires computing six weights: four to connect the two input units to the two units of the hidden layer, and two more to connect the hidden layer to the single unit of the output layer. Following the backpropagation criterion, the process starts by assigning random weights and computing the final error (0.461: Table 7.6). As the iterative process progresses, the weights are recalculated and the final error is reduced. In the last step, the weights are used to calculate the final linear equations that will be used in this problem. Three equations are computed, two from the input layer to the hidden layer

$$o_1 = 2.901\, x_1 - 2.976\, x_2$$
$$o_2 = 2.902\, x_1 + 2.977\, x_2$$

and one from the hidden layer to the output layer

$$o_3 = 4.785\, o_1 + 4.784\, o_2$$

The final classification is based on the value of the output unit ($o_3$). When this value is lower than 0.5, it is assigned to class 1, and when it is higher, it is assigned to category 2. The two equations of the hidden layer serve to separate the discrimination space, which is finally applied in the output layer. Those intermediate equations can be conceived as a means of transforming the space of the original measure to achieve a suitable separation among the members of each class (Figure 7.40c). Linear equations are not commonly used to obtain the output values of exit of a neuron, but more complex functions such as sigmoids (the output value ranges from 0 to 1), or hyperbolic-tangent (output value from −1 to +1; Civco 1993). Among the many methods available for training an ANN, some do not require a previous definition of output classes, and therefore, ANN can also be used as an alternative to unsupervised training.

Once the learning is completed, new unknown cases can be easily classified, using the previously calibrated equations. In this example, some cases are added in the vicinities of those initially introduced, with good classification results (Figure 7.40d).

ANN classification has been used in different applications of remotely sensed images (Benediktsson et al. 1990; Foody 1995; Trombetti et al. 2008). They have been found to be robust when the normality assumption cannot be met or when the input data are very heterogeneous. Several authors have proved the robustness of this technique, which is not very sensitive to extreme values, and has the ability to include textural elements (Bischof et al. 1992; Civco 1993; Hepner et al. 1990).

Although ANNs have advantages, this classification approach involves some practical difficulties. First, there are no clear criteria for selecting the most appropriate network structure (number of layers, units in each layer) or training algorithm for a particular problem. Frequently, these choices are based on trial and error because universal rules are not generally available. The second problem is more connected with the training procedure. Since the activation functions can be very sophisticated, ANNs may be overtrained; that is, they fit the training dataset well but perform poorly in other situations. To avoid this, the training is commonly performed with two samples: one used to modify the function weights and the other to validate the generalization power.

#### 7.2.3.6 Fuzzy Classification

In our previous discussion of spectral unmixing methods, we discussed the interest in considering a pixel as a mix of several land covers, which is especially true in medium- and coarse-resolution sensors. Traditional classifiers assign one single category to each pixel, by assuming that either the pixel is homogeneous or at least has a highly dominant class. This approach can lead to remarkable errors when mixing is observed in similar proportions. For example, if 90% of the pixel is occupied by forest and 10% by water, the assignment of that pixel to forest might be considered accurate enough. However, if the proportions are 55% and 45%, the error would be significant. For these situations, a multiple-assignment classification approach would be more appropriate. These approaches are generally termed as *soft* classifiers; *hard* classifiers allow only one category for each pixel (Mather 1999; Schowengerdt 2007).

Fuzzy classification is one of the most common approaches to soft classifiers. This approach makes it possible to assign a degree of membership to each category instead of a single value for each pixel (Figure 7.41). Conventionally, the function of membership is binary (0/1), but it may also be a real number between 0 (absence of a certain category) and 1 (full pixel covered by that category). Therefore, the output image may include a simultaneous assignment to several categories, with different grades of membership (Townsend 2000; Wang 1990b):

$$U_{i,j} = f(DL_x) \tag{7.43}$$

where $U_{i,j}$ would indicate the grade of membership of the pixel $x$ to the class $i$ and can be a real number between 0 and 1.

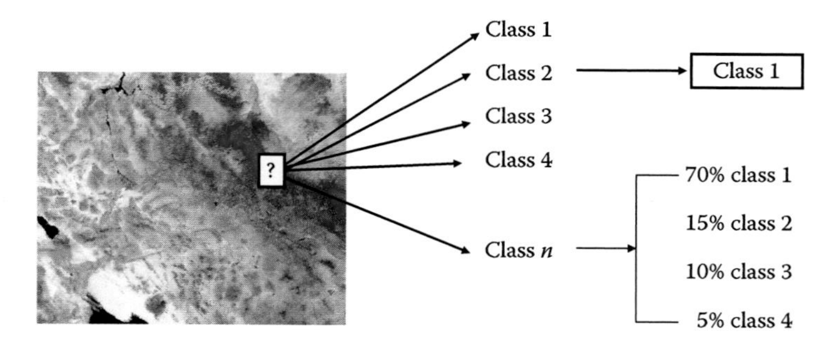


**FIGURE 7.41** Basis of fuzzy classification system.

In remote sensing applications, the most common methods of obtaining those values of ownership are the fuzzy variant of ML (Maselli et al. 1995; Wang 1990b) and a variety of the unsupervised analysis, called fuzzy c-means (Fisher and Pathirana 1990; Foody 1992). To simplify the text, we will focus on the first approach, since it is based on the ML algorithm previously analyzed. For a complete formulation of the c-means algorithm, the reader may refer to the article by Fisher and Pathirana (1990).

The ML algorithm can be used to compute the degree of membership in a fuzzy classification algorithm by changing the definition of the vector of means and the variance–covariance matrix. The alternative formulas are (Wang 1990a)

$$\mathbf{M}_A^* = \frac{\sum_{i=1,n} \mathbf{U}_A(\mathbf{M}_i)\mathbf{M}_i}{\sum_{i=1,n} \mathbf{U}_A(\mathbf{M}_i)} \tag{7.44}$$

and

$$\mathbf{V}_A^* = \frac{\sum_{i=1,n} \mathbf{U}_A(\mathbf{M}_i)(\mathbf{M}_i - \mathbf{M}_A^*)(\mathbf{M}_i - \mathbf{M}_A^*)^T}{\sum_{i=1,n} \mathbf{U}_A(\mathbf{M}_i)} \tag{7.45}$$

where

- $\mathbf{U}_A$ is the degree of membership to class $A$
- $\mathbf{M}$ is the vector of DL values for the $n$ sampled pixels
- $\mathbf{V}_A$ the variance–covariance matrix for the same class
- the asterisk (*) indicates that the $\mathbf{M}_A$ vector and the $\mathbf{V}_A$ matrix are calculated for a *fuzzy* space

From those two metrics, one can estimate the membership probability, using formulas similar to the ML algorithm:

$$p^*\left(\frac{x}{A}\right) = (2\pi)^{-m/2}\left|\mathbf{V}_A^*\right|^{-0.5} \exp\left\{-0.5\left(\mathbf{M}_x - \mathbf{M}_A^*\right)^T \mathbf{V}_A^{*-1}\left(\mathbf{M}_x - \mathbf{M}_A^*\right)\right\} \tag{7.46}$$

This equation provides for each pixel the probability of being a member of category $A$, and it is output to the resulting image. This image will have as many bands as the number of categories considered. For each of those classified bands, the pixel value would indicate the degree of expected membership for a particular category.

The training areas for fuzzy classification can be homogeneous, as in hard classifiers, or heterogeneous, with mixed components, provided their mix proportions are known. Fuzzy logic is especially suitable when dealing with applications involving frequently mixed covers, such as forest or vegetation inventories (Maselli et al. 1995) and maps (Townsend 2000), fuel classification (Nadeau and Englefield 2006), or analysis of urban areas (Fisher and Pathirana 1990).

The results of fuzzy classification can be compared with those of SMA. Even though both techniques have a different basis, they share the idea of mixing covers, by considering either a degree of membership (fuzzy classifiers) or proportions of cover (SMA). A few studies have compared the advantages and inconveniences of both techniques, considering their reliability in solving mixtures, computer time, and data limitations. When they have been compared, the results were found to be similar (Foody and Cox 1994). Fuzzy techniques can also be used for multitemporal change detection, making it possible to detect different intensities of change (Foody 2001).

On the other hand, fuzzy classifications require special methods for accuracy assessment, as multiple memberships are assigned to each pixel (Stehman et al. 2007).

Figure 7.42 shows an example of fuzzy classification of the MODIS image. Four categories have been included in this figure to illustrate the process. The output values provide an estimation of the membership function of each pixel to the user categories. Soil and vegetation are very well defined, but water and snow show some coincidences, in spite of their strong spectral contrast, especially in the VIS bands.

#### 7.2.3.7 Hyperspectral Classification

Similar classification methods can be applied to both multispectral and hyperspectral data, although some algorithms may require previous selection of most relevant variables, since they may have limitations in the number of input variables.

However, the use of standard classification techniques with hyperspectral data may reduce the potential of these images to identify special features of interest. For this reason, alternative methods of classification have been proposed, making better use of the hyperspectral dimensions. While in multispectral systems reference statistics are computed from training fields or unsupervised training derived from the same image to be classified, the hyperspectral reference data are commonly collected from spectral libraries or spectroradiometer measurements. If image values are to be compared with these spectral measurements to train the classifier, DL to reflectance conversion needs to be performed before the classification. Otherwise, the comparison between pixel values and reference spectra will be meaningless.

A simple way of assigning an unknown pixel to a reference category is by comparing how both the pixel and the reference reflectances change throughout the spectrum. This is the basis of the classification named "binary encoding," which converts the actual reflectance of a continuous spectrum to a set of 0/1 values, depending on whether the reflectance of a single band is above or below the average reflectance for the whole spectrum. The band is coded with 1 when reflectance is higher than the

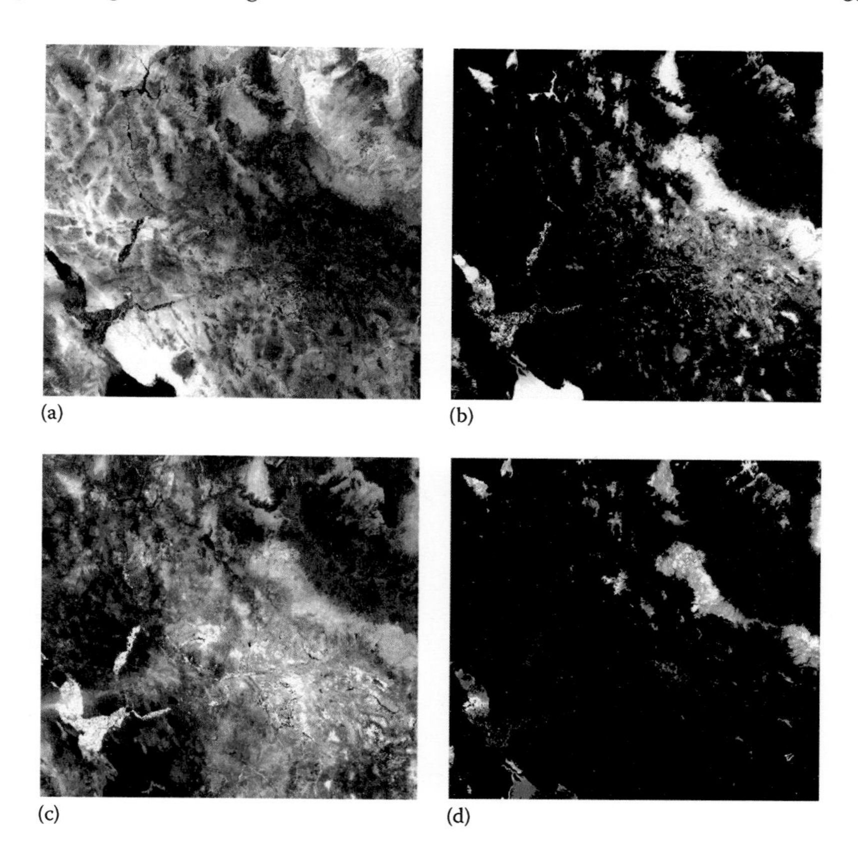

FIGURE 7.42 Examples of fuzzy classification of the MODIS image: (a) soil, (b) water, (c) green vegetation, and (d) snow.

average, or with 0 otherwise. When two spectra have similar covers, we can assume that the sequence of 0s and 1s along the spectrum will be similar. To obtain a more quantitative evaluation of this similarity, the index of spectral agreement (ISA) is computed as follows (Mazer et al. 1988):

$$ISA = \frac{\sum_{k=1,m}\left(BC_{i,k} - BC_{j,k}\right)^2}{m} \quad (7.47)$$

where

$BC_{i,k}$ is the binary encoding of spectrum $i$ (the one to be classified) in band $k$
$BC_{j,k}$ is the binary encoding of the reference spectrum for the same band
$m$ is the number of bands considered

Whenever $ISA$ is close to 0, the spectra are more similar, and when it is close to 1, they are more different. In Figure 7.43, we can observe three spectra extracted from a

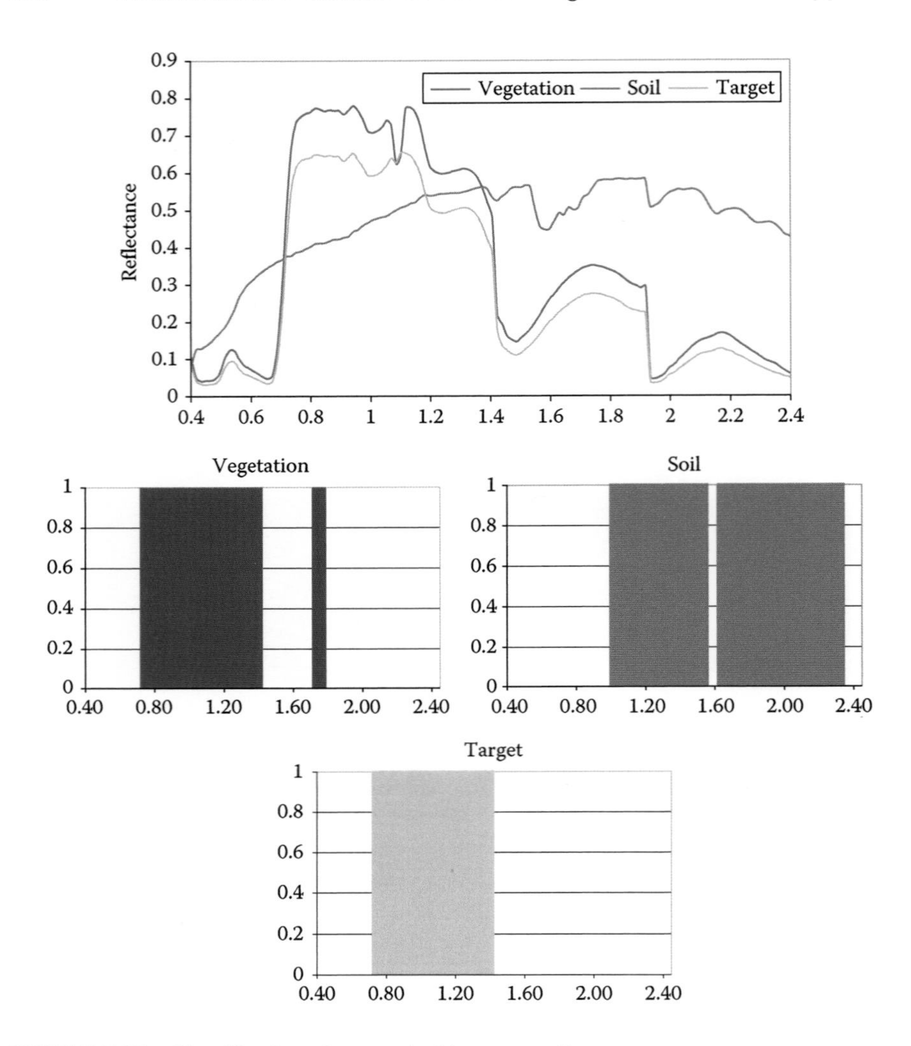


FIGURE 7.43 Classification of spectra by binary encoding.

hyperspectral image acquired by the Hymap airborne sensor. The target spectrum is compared with two known spectra: vegetation and soil. Even though visually it is clear that the unknown spectrum is more similar to the vegetation spectrum, we need to quantify this similarity. Using the ISA metrics, the value between the target spectrum and vegetation is 0.039 while that between the target and the soil spectrum is 0.5625. Therefore, the greater similarity to the vegetation spectrum is numerically confirmed.

A second technique of classifying hyperspectral images is based on comparing the differences in two spectra between the reflectance of each band and the average trend of that spectrum, defined by the local maxima. The analysis of these differences has been named "continuum removal" (Kruse et al. 1993a) and can be considered as the study of the absorption features that are present in a single spectrum.

These features characterize the specific absorption properties in a material such as chlorophyll, water, iron, or yeast. To analyze these absorption features, we need first to find the maximum locals (the continuum) and then compute the difference of each band reflectance against that line (Figure 7.44). Each absorption feature can be defined by the central wavelength where it occurs, its width, depth, and shape (Kruse et al. 1993b). Absorption features of an unknown pixel can be compared with those of a reference category, and therefore the target pixel can be classified (Tsai and Philpot 2002) or its biophysical properties extracted (Tian et al. 2001).

Finally, a widely used method of classifying hyperspectral images is the spectral angle mapper (SAM). SAM bases the classification on the angular distance between a target and a reference spectrum, assuming that each spectrum is a vector in the $m$-dimensional spectral space ($m$ being the number of available bands; Figure 7.45). The formula is

$$\theta_{i,j} = \arcos \frac{\sum_{k=1,m} DL_{i,k} DL_{j,k}}{\sqrt{\sum_{k=1,m} DL_{i,k}^2} \sqrt{\sum_{k=1,m} DL_{j,k}^2}} \tag{7.48}$$

$DL_{i,k}$ being the values of the target spectrum and $DL_{j,k}$ the values of the reference spectrum. In this case, $DL$ should be expressed in reflectance, since most commonly

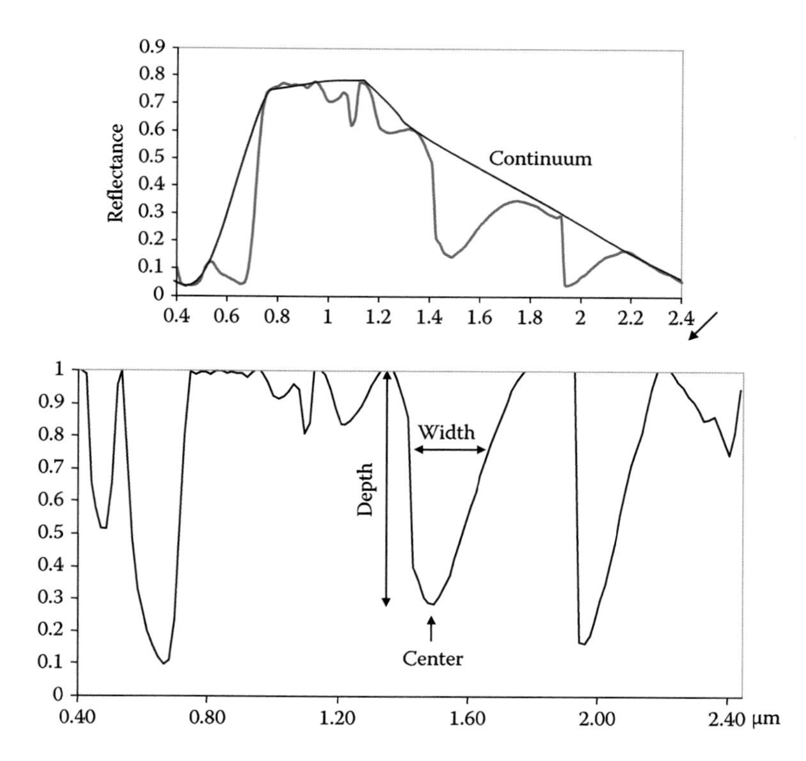


**FIGURE 7.44** Analysis of absorption features.

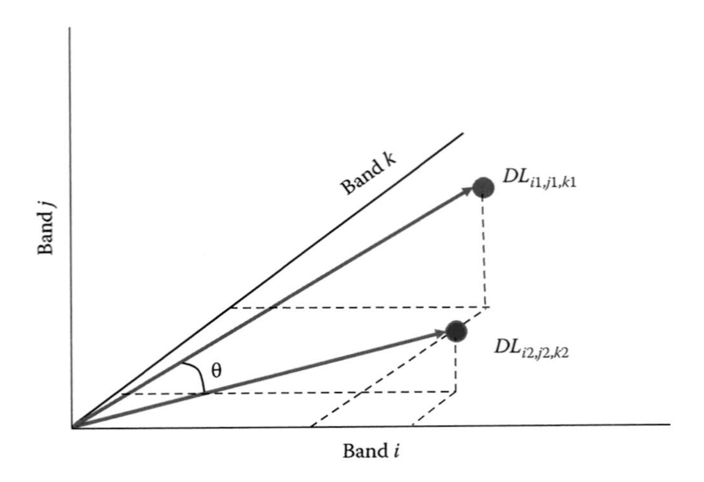


**FIGURE 7.45** Calculation of the spectral angle mapper.

the reference spectrum will be extracted from an external source (spectral measurements). The target spectrum will be assigned to the category that minimizes the angle θ.

This technique measures the similarity between spectra by measuring the spectral contrast (direction) rather than their absolute differences (intensity). Therefore, the algorithm is fairly insensitive to different illumination conditions. This property is very useful not only to classify hyperspectral images but also as an inversion strategy to assign biophysical values to pixels by comparing observed reflectances with simulated data. This approach has been successfully tested in the estimation of burn severity and water content from medium-resolution sensors (Chuvieco et al. 2007; De Santis and Chuvieco 2007; Jurdao et al. 2013).

#### 7.2.3.8 Object-Oriented Classifiers

In several parts of this book, we have stated that land covers do not present unique spectral signatures, but rather, there are external and internal factors that introduce a certain variability within the spectral signature of a particular cover and create potential confusion with other categories. Consequently, it will be difficult to discriminate them based just on the spectral values. Additional variables may need to be taken into account to complement spectral information.

Among those additional variables, the textural and contextual factors are very relevant. As in visual interpretation, the spatial properties of covers are very helpful in differentiating them from similar categories. The use of spatial variables is less frequent in digital than in visual interpretation, but there is growing interest in using these features, especially in the context of classifying high-resolution images.

The distinction between the concepts of texture and context refers to the area that is used to measure the spatial relationships (Gurney and Townshend 1983). Texture should be associated with the spatial variations inside contiguous groups of pixels, while context refers to the spatial relations between a certain group of pixels and the rest of the image. To simplify the discussion, we will consider both aspects here, since both aim to include spatial variables in digital image interpretation.

Three groups of spatial classifiers can be identified: those algorithms that include textural bands as inputs into traditional classification systems, those procedures that use the spatial context in the definition of the classification algorithm, and those methods that consider the spatial context after the classification to improve the results.

The first group will be covered in a later section (Section 7.4.2), where different methods to compute texture will be reviewed. At this point, it should be enough to say that those methods generate new bands by measuring the spatial contrast of DL within certain distances and directions (Ferro and Warner 2002; Maillard 2003). They have been extensively used in classifying urban areas (Barr and Barnsley 2000; Gong and Howarth 1990b; Moller-Jensen 1990) and in forest inventories (Drake and Weishampel 2000; Souza et al. 2005).

The use of texture in the classification phase was explored during the late 1970s, but has received greater attention more recently. One of the first attempts of contextual classification was the extraction and classification of homogeneous objects (ECHO) algorithm (Ketting and Landgrebe 1975), which proposed classifying on the basis of groups of pixels rather than using single pixels. Those groups should be areas of contiguous and homogeneous pixels that would serve as single entities for classification purposes. The algorithm first created those groups based on the spatial and spectral metrics of the image. After the groups were created, all pixels within each group were classified into a single category following one of the standard classification algorithms (Hoffer and Swain 1980).

These pioneering proposals have been extensively developed within the framework of the so-called object-oriented classification systems (Gitas et al. 2004; Zhang et al. 2005). An object in this context is a group of homogeneous and contiguous pixels. These systems are especially suitable for high-resolution images, since real-world objects are commonly formed by a spatial cluster of pixels. In addition, high-resolution images may have a significant amount of spatial noise, which makes identification of heterogeneous land covers difficult when considering per-pixel classification methods (Schiewe et al. 2001).

As in the case of the ECHO classifier, the object-oriented methods operate in two phases: in the first one, the image pixels are spatially grouped to form the objects (spatial segmentation: Figure 7.46), while in the second one, those objects are assigned to any of the target categories. A wide range of algorithms for spatial segmentation are available (Lobo 1997; Schiewe et al. 2001). Some of the control parameters for segmentation include the average size of objects (smaller, larger), their uniformity (homogeneous, heterogeneous), and their shape (circular, elongated). Once the objects are defined, all pixels within each object will be assigned to the same cover, and therefore the final classification will have a more coherent spatial structure. In addition to this advantage, unlike pixels, which all have the same size and shape, objects can vary in shape, size, and position, and therefore the classification may use not only spectral but also spatial properties as input variables to discriminate covers.

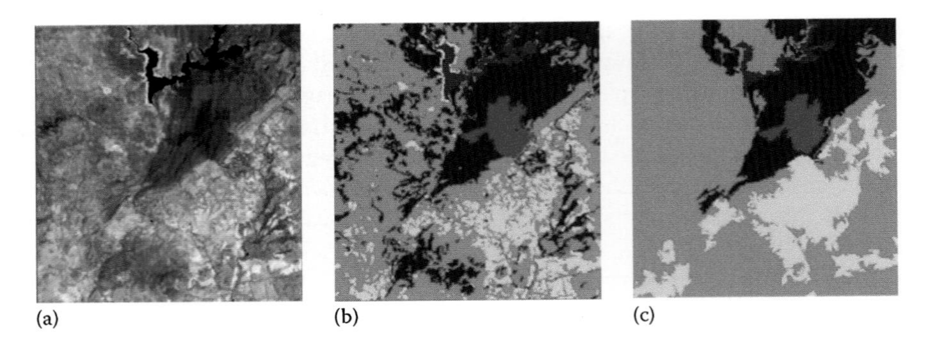


**FIGURE 7.46** (a) Original image split into (b) smaller or (c) larger objects depending on the segmentation criteria.

#### 7.2.3.9 Contextual Classifiers

Contextual classifiers comprise those methods that incorporate the context in the classification algorithm itself but still using a pixel-to-pixel approach. Most commonly, those methods assume that the assignment of a single pixel is a function not only of its DL but also of the DLs of the neighboring pixels. A simple way to include this criterion is by modifying the ML criterion to include the probability of the nearby pixels (Richards 1993):

$$p'\left(\frac{x}{c_i}\right) = \frac{p(x/c_i)q(x/c_i)}{\sum_{i=1,m} p(x/c_i)q(x/c_i)} \qquad (7.49)$$

where

- $p(x/c_i)$ indicates the probability that the pixel $x$ belongs to the category $c$ in band $i$, for all the $m$ categories
- $q(x/c_i)$ is a vicinity function, defined by the user, which is commonly related to the distance between pixels

This approach was successfully used in different thematic classifications (Danjoy and Sadowski 1978; González and López 1991).

The context may also be considered to discriminate objects that show a clear spatial difference with the background. For instance, the thermal difference between an active fire and the surrounding forested area is a key variable to discriminate a hot spot (Giglio et al. 2003), or the greenness change between recently cut trees and surrounding forest is relevant to monitoring deforestation processes (Souza et al. 2005). Context criteria are very useful for cloud detection (Khlopenkov and Trishchenko 2007; Latifovic et al. 2005) or discriminating burned scars (Alonso-Canas and Chuvieco 2015; Chuvieco et al. 2008b).

#### 7.2.3.10 Postclassification Generalization

The spatial context can also be considered after image classification. In this case, the processing aims to refine classification results based on spatial properties.

In traditional per-pixel classification approaches, results commonly show an excessive fragmentation. Pixels labeled with a certain category may be surrounded by other nonrelated categories, missing the geographical consistency that is found in thematic maps. This phenomenon is colloquially named "salt and pepper" and creates excessive discontinuity in the output classified image. Even though the assignment of those isolated pixels is accurate, it does not have a real cartographic meaning because it does not make much sense to map spatial units below a certain size.

Postclassification filtering techniques can be used to reduce the spatial fragmentation of classified images. Since the output images are measured in a categorical scale (the DLs are only labels of categories), analytical filters cannot be used in classified images. Instead, filters based on categorical statistics are used, most commonly the mode and majority (Thomas 1980). As in the case of arithmetic filters, they are applied to mobile windows of a certain size (3 × 3 to 7 × 7 pixels are the most common). The assignment of the central pixel of that window is compared with the assignments of the surrounding pixels (Figure 7.47). In the case of the mode filter, the central pixel changes its class to the most frequent class within that window. The majority filter changes its assignment only when there is one category with at least half plus one of all the pixels within the window. Otherwise, the central pixel maintains its original assignment.

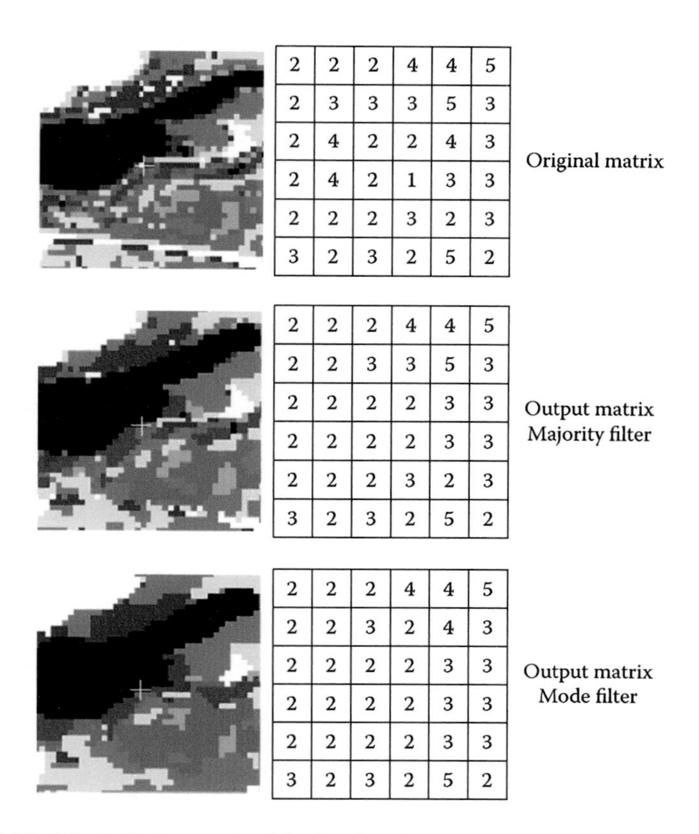


**FIGURE 7.47** Methods for postclassification filtering.

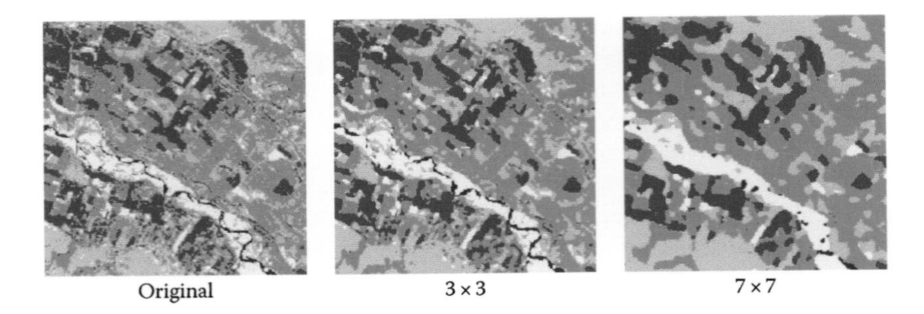


**FIGURE 7.48** Classes with small and elongated patches, such as rivers, may be removed as a result of postclassification filtering. The effects are more severe as the filter window is larger.

The selection of the postclassification filter depends on the desired generalization level and the complexity of the target area. For homogeneous landscapes, the spatial noise will be low and only minor transformations should be required. The mode filter implies a more intense transformation of original assignments and should be used with very complex landscapes or at regional or global scales. Several authors have proved that postclassification filtering increases the accuracy of results, by eliminating errors caused by border pixels (Booth and Oldfield 1989). However, this method can also create errors for categories that have a very limited spatial representation. For instance, rivers or small lakes can disappear after mode filtering, especially when window sizes are large (Figure 7.48). To cope with this problem, some categories may be excluded from filtering, or some polygons filtered according to a certain spatial threshold (Barr and Barnsley 2000).

### 7.2.4 Classification Outputs

The results of digital classification are typically stored in a new image. Though similar in structure and size to the original one, the main difference with the original data is the meaning of the DL, which now will code the class assigned to each pixel. This file can be converted to a thematic map or exported to a GIS to be integrated with other spatial variables. The former process requires some transformations of the output file (Figure 7.49):

1. *Geometric correction.* In case the input image was not previously corrected, the output file needs to be georectified so that the classification results have cartographic coordinates. Since the classified image will lack some of the spatial features of the original bands, it is more advisable to select the control points for geometric correction from the original bands and then apply the transformation to the classified image. It is important to note that the resampling phase should be done using the nearest neighbor algorithm; otherwise, the assignments will be changed without user control.
2. Once the image is geometrically rectified, the classification should be generalized to eliminate very small polygons. This generalization can be based on modal or majority filters, as previously explained, or on eliminating groups of pixels smaller than a certain size.

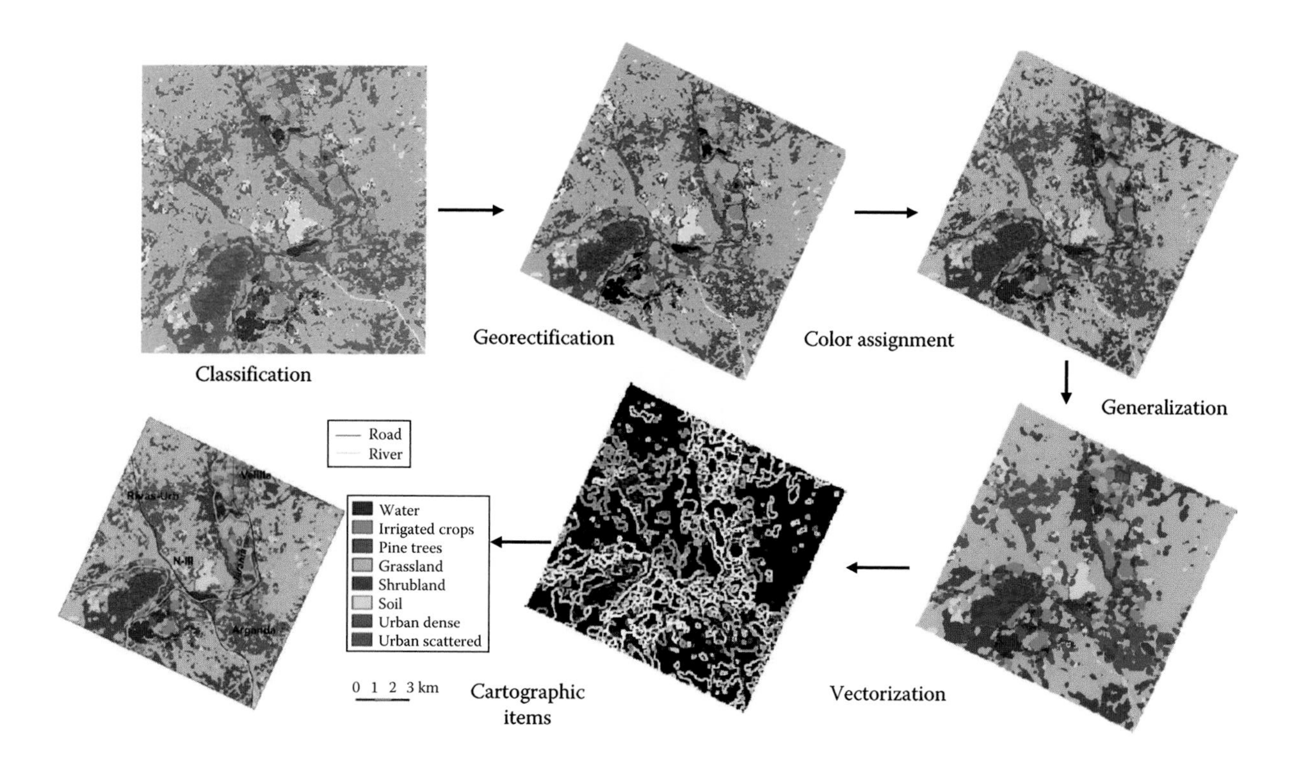


**FIGURE 7.49** Process to obtain thematic maps from classification files.

3. The third phase focuses on generating a color code for each category, by creating a lookup table (see Section 6.5.3). Colors have a cartographic meaning, following semantic rules. Therefore, the assignment of colors to the categories should consider those conventions: for example, red for urban areas, green for forested ones, and yellow for crops (Anderson et al. 1976; Coleman 1968), instead of assigning arbitrary colors. In thematic classifications, 15 or 20 colors, clearly distinguishable from one another, will be enough to visualize the output file. A larger number of categories would complicate the legibility of the final document.
4. When working with a vector-based GIS, the classification may be converted from raster to vector format. Different automatic algorithms are available for this task (see, e.g., Bolstad 2008).
5. For cartographic reference, the map should have a scale bar, a symbol to indicate northern orientation, and a class color key. Names of rivers, roads, or cities can also be added.

The area covered by each category is an important output of digital classification, especially in some applications such as forest or crop inventories. In this regard, digital classification is more powerful than visual analysis, since area inventory is as simple as computing the number of pixels included in each category. Given that the pixel size is known (after geometric correction), the area covered by each category is easily calculated. Area statistics may be obtained for the whole image or for sets of polygons extracted from external databases. In a crop inventory, for example, it will be useful to obtain area statistics for certain administrative boundaries (counties, states) to compare with other sources of agricultural data. Table 7.7 compares the area covered by each category in the classification of our study area, using ML and neural networks. The differences are significant, especially in the shrubland category, which is overestimated in the neural network classifier.

**TABLE 7.7**
**Area Covered by the Different Categories Following Two Different Classification Criteria**

| | Maximum Likelihood Classification | | | Neural Networks | | |
|---|---|---|---|---|---|---|
| | Pixels | km² | % | Pixels | km² | % |
| Soil | 73,945 | 18,486.25 | 4.68 | 160,282 | 40,070.50 | 10.14 |
| Snow | 110,697 | 27,674.25 | 7.00 | 30,550 | 7,637.50 | 1.93 |
| Herbaceous | 686,120 | 171,530.00 | 43.41 | 141,361 | 35,340.25 | 8.94 |
| Shrublands | 438,187 | 109,546.75 | 27.72 | 969,185 | 242,296.25 | 61.31 |
| Forested | 166,335 | 41,583.75 | 10.52 | 15,593 | 3,898.25 | 0.99 |
| Urban | 89,905 | 22,476.25 | 5.69 | 68,377 | 17,094.25 | 4.33 |
| Water | 15,517 | 3,879.25 | 0.98 | 195,358 | 48,839.50 | 12.36 |
| Total | 1,580,706 | 395,176.50 | 100.00 | 1,580,706 | 395,176.50 | 100.00 |

## 7.3 TECHNIQUES OF MULTITEMPORAL ANALYSIS

### 7.3.1 Temporal Domain in Remote Sensing Studies

Monitoring dynamic processes is one of the most important contributions of EO satellites to environmental studies. Since images are acquired by a sensor mounted on a spacecraft with a stable and repetitive orbit, they constitute a precious source of information about how Earth landscapes are changing, related to either seasonal trends, climatic cycles, human intervention, or catastrophic events.

To better understand the components of temporal changes, we can consider two temporal dimensions in environmental dynamics (Robin 1998): on the one hand, the speed with which the phenomenon takes place and, on the other hand, the persistence of the resulting change (Figure 7.50). The first aspect refers to the duration of any particular phenomenon, while the second affects the length at which impacts will still be observed (persistence). From the duration perspective, the required frequency to monitor environmental phenomena ranges from a few hours (tropical hurricanes, earthquakes, volcanic eruptions, etc.) to several years (desertification, urban growth, etc.). The persistence is related to the actual intensity of the transformation caused by that event. Persistence can also vary from a few days (low-intensity hurricane, for instance) to several years or even centuries (soil degradation, landslides, etc.). The temporal footprint of a natural disaster is closely related to its negative impacts: a flood can be brief when it is superficial or have permanent effects when it causes

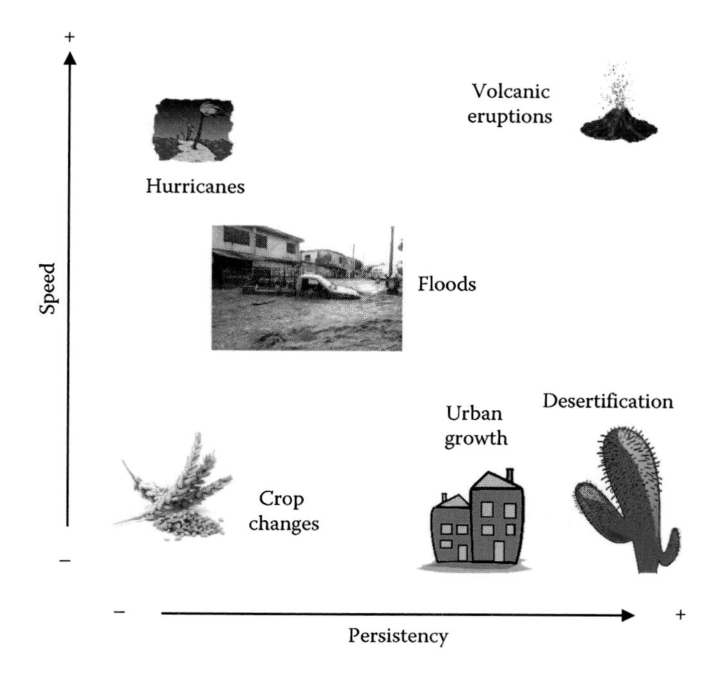


FIGURE 7.50 Temporal dimensions in environmental monitoring. (Adapted from Robin, M., *La Télédétection*, Nathan, Paris, France, 1998.)

changes in geomorphologic features (alluvial fans, for example). Both aspects, that is, the speed of the phenomenon and the persistency of its effects, are not necessarily related, since a quick event may have a long impact (volcanic eruption), and vice versa (Figure 7.50).

These thoughts may be useful when deciding on the most appropriate frequency to observe a phenomenon of interest. The most suitable temporal resolution will depend on both speed and persistency. Some natural events require continuous observation (weather monitoring), while for others an annual observation (urban growth) or even longer cycles (5 years for land cover) are sufficient.

As reviewed in Chapter 3, the observation frequency of a satellite observation system depends mainly on the orbital height and speed and the sensor's field of view. In practical terms, the nominal temporal resolution of a sensor is not the same as the actual observation cycle, since cloud coverage reduces the time frequency of useful observations when working with optical sensors. Other factors may also hinder multitemporal observation, such as sensor acquisition problems or lack of ground downloading coverage. Current sensor systems provide different observation frequencies, from a few minutes for geostationary satellites to several weeks for those oriented toward observing natural resources. During the first years of EO, spatial resolution was more appreciated, but nowadays temporal resolution is also considered critical to monitor many environmental variables, particularly when updated monitoring is decisive. For example, observing hurricanes with very fine spatial resolution imagery every week would be virtually worthless. Since clouds are very dynamic, their monitoring requires very high observation frequency even at the cost of losing spatial detail. The same will apply to other natural hazards: fire, floods, icebergs, etc.

To ensure multitemporal coherence in satellite observation of dynamic phenomena, the sensor should provide periodic images of the same spectral bands and under similar observation conditions (height, hour, angle of acquisition, etc.). These requirements are met by most natural resource satellites (Landsat, SPOT, IRS, etc.). Manned mission of the Space Shuttle or the International Space Station does not offer systematic observation, but their photographs can be used for visual–qualitative comparisons. Airborne sensors do not provide systematic observations either, but they can be used to detect changes between fixed dates, and they are of particular interest in analyzing long-term changes, since aerial photographs are a much older technology than satellite observation (Adeniyi 1980).

Two dimensions of temporal change can be analyzed from EO images: seasonal trends and land cover variations. In the first case, the interpreter tries to follow the phenology changes associated with vegetation dynamics throughout the year (Willis 2015). This seasonal contrast may be very useful in improving the discrimination of certain target categories, by including images from different periods of the year in the classification process (Latifovic et al. 2004; Lo et al. 1986). A seasonal separability analysis will help in identifying which season is more appropriate to discriminate particular land covers (Schriever and Congalton 1995). Seasonal analysis can also be used to monitor vegetation regeneration after a major disturbance (fire, wind storms, drought, etc.), or as a result of climatic variations (Alcaraz-Segura et al. 2010; Fraser et al. 2003; Goetz et al. 2005).

The second temporal approach aims to detect land cover changes, caused either by short-duration catastrophic events (floods, earthquakes, volcano eruptions, hurricanes, fire, etc.) or by slow-duration phenomena (urban sprawl, crop change, desertification, etc.). Images acquired in two reference periods (before and after the sudden event or separated by a fixed time span) are compared visually or digitally (Gopal and Woodcock 1996; Lucas et al. 1993; Sader and Winne 1992; Zhan et al. 2002). Most commonly, images for change detection analysis are acquired under similar observation conditions (close in the same year, or in the same season of different years) to avoid identifying as cover change those transformations caused by different phenological conditions.

### 7.3.2 Prerequisites for Multitemporal Analysis

Most multitemporal analyses involve pixel-to-pixel comparisons of DLs from different dates. One of the critical issues in temporal analysis is the discrimination of actual ground changes from those caused by other factors of potential variations between input images, such as calibration or georectification issues. For this reason, radiometric and geometric corrections are particularly critical for temporal analysis, as otherwise comparisons would be affected by external factors of change. In other words, the interpreter needs to be sure that the same variable and the same area are being compared for all the dates involved, or else the comparison would be meaningless.

#### 7.3.2.1 Multitemporal Matching

Multitemporal comparisons require that input images be precisely coregistered to ensure that the exact area is being compared. Otherwise the interpreter would identify changes between images that are in fact displacements. This effect can be quite serious for the case of heterogeneous landscapes with great spatial variability, such as in urban areas (Gordon 1980), especially with high-spatial-resolution sensors (Nelson and Grebowsky 1982; Roy 2000). Even in the case of coarse-resolution sensor imagery, it has been shown that up to 50% of the change observed between two images can be due to displacements when the mean registration residual among images exceeds 1 pixel (Townshend et al. 1992). Several studies have shown that to ensure errors less than 10% in multitemporal image comparisons, geometrical matching accuracy among them should be better than 0.2 pixels (Dai and Khorram 1998; Townshend et al. 1992).

To illustrate the importance of geometrical matching, we carried out a simple exercise to simulate the change in the classification of land cover types following an intentional displacement of two images. A Landsat TM image was classified into four categories. Then, the output classification was shifted a pixel toward the west and the south. A cross-tabulation was performed between the two classifications. Pixels along the diagonal of this matrix are unchanged areas, and outside the diagonal are pixels marked as change (Table 7.8). The percentage of cover change is quite high, reaching 34% of the total area. Obviously, in this case all changes are exclusively due to the systematic displacement that was introduced, since the same classification is being compared. In other words, more than 30% of change is spurious, not caused by actual cover change but simply by misregistration.

**TABLE 7.8**
**Impact of Misregistration on Multitemporal Comparisons of the Original Classifications**

| | Displaced Image | | | | |
|---|---|---|---|---|---|
| **Original Image** | **1** | **2** | **3** | **4** | **Total** |
| 1 | 94 | 247 | 42 | 6 | 389 |
| 2 | 251 | 20,653 | 9,315 | 897 | 31,116 |
| 3 | 43 | 9,512 | 51,449 | 15,337 | 76,341 |
| 4 | 3 | 755 | 15,560 | 27,897 | 44,215 |
| Total | 391 | 31,167 | 76,366 | 44,137 | 152,061 |

**TABLE 7.9**
**Impact of Misregistration on Multitemporal Comparisons after Both Classifications Were Smoothed Using a 5 × 5 Modal Filter**

| | Displaced + Filtered Image | | | | |
|---|---|---|---|---|---|
| **Original Image** | **1** | **2** | **3** | **4** | **Total** |
| 1 | 98 | 100 | 30 | 41 | 269 |
| 2 | 132 | 23,556 | 4,954 | 387 | 29,029 |
| 3 | 39 | 5,196 | 66,982 | 8,681 | 80,898 |
| 4 | 0 | 259 | 8,962 | 32,361 | 41,582 |
| Total | 269 | 29,111 | 80,928 | 41,470 | 151,778 |

The amount of change caused by geometric rectification is reduced when the comparison is made after previous generalization of classification results, since this process eliminates small isolated groups that are more likely to change by displacements. In fact, when comparing the original and the displaced classifications after generalizing the original with a 5 × 5 pixel modal filter, the detected change decreases by almost half from the previous detection, reaching 19% of the image (Table 7.9). Although real displacements in images are not as systematic as those introduced here, this exercise may provide a better awareness of the importance of ensuring a precise multitemporal matching between images to avoid noise in multitemporal comparisons. The exercise also shows the interest in generalizing the classifications before comparing them. Some authors have even suggested comparing classifications after segmenting the images in homogeneous patches instead of using individual pixels (Bruzzone and Fernández-Prieto 2000).

Multitemporal matching of images is based on the same principles as ordinary geometric correction. In this case, the reference coordinates are those of the image with which we try to compare. That image may have been previously rectified or not, since multitemporal comparisons do not need to be in cartographic coordinates. It is enough that both employ the same reference system. The selection of common control points between the target and reference images is simpler than using a standard base map as a reference, since there are many common features between two images that will not be clearly observable in topographic maps (vegetation shapes, parcel borders, rivers or streams, etc.). Some authors have suggested semiautomatic methods of selecting control points in multitemporal images, such as dynamic correlation analysis based on the detection of common features (Eugenio and Marqués 2003; Gao et al. 2009).

#### 7.3.2.2 Radiometric Calibration

The second factor to be controlled in multitemporal comparisons is radiometric consistency. If DLs are compared, we need to be sure that they are equivalent in the different input images. Obviously, reflectance values cannot be compared with raw digital values, for instance, or digital values from different sensors or even from the same sensor through time, because calibration changes between sensors and dates. As a consequence, the conversion of DLs to physical units (reflectance or temperature) is a critical requisite of multitemporal analysis, to ensure radiometric consistency between images. When using a long time series of satellite data, such as those built from NOAA-AVHRR or Landsat TM/ETM+/OLI images, calibration coefficients need to be adapted to the different platforms. In some cases, even the spectral characteristics of a sensor may change throughout time (e.g., the AVHRR sensor was modified from the NOAA-14 to the NOAA-15 satellites). Therefore, correction algorithms need to be adapted to the different spectral bands (Khlopenkov and Trishchenko 2007; Latifovic et al. 2005; Trishchenko et al. 2002).

An alternative to using reflectance or temperature values involves a relative correction between input images. In this case, the goal is not to perform absolute correction but just to make the images comparable. In doing so, a radiometric matching between dates may be performed, based on a series of control pixels that are assumed invariant through time. These points should preferably be chosen from high-contrast reflectance areas (low- and high-reflectivity targets), such as shades, clear and deep waters, bare soils, or parking lots. A regression line can be computed from those points, yielding the following transformation:

$$DL_{o,k} = b_k DL_{i,k} + g_k \tag{7.50}$$

where

$DL_{o,k}$ is the output pixel value in band $k$
$DL_{i,k}$ is the input value in the same band
$b_k$ and $g_k$ are the bias and gain coefficients, respectively, of the fitting function

Bias and gain coefficients may be obtained from dark and clear pixels on the target and reference images, following

$$b_k = \frac{DL_{cl,r,k} - DL_{d,r,k}}{DL_{cl,c,k} - DL_{d,c,k}} \tag{7.51}$$

and

$$g_k = \frac{DL_{d,r,k}DL_{cl,c,k} - DL_{d,c,k}DL_{cl,r,k}}{DL_{cl,c,k} - DL_{d,r,k}} \tag{7.52}$$

where

$DL_{d,r,k}$ and $DL_{cl,r,k}$ indicate the mean $DL$ of the darkest and brightest pixels, respectively, of the reference image

$DL_{d,c,k}$ and $DL_{cl,c,k}$ indicate the same for the target image

The method has been proven to be efficient at homogenizing the atmospheric effect between images. Additionally, it reduces the atmospheric contribution of the target images (Hall et al. 1991b). When invariant areas are not clearly observable, some authors suggest using pixels with lower and higher NDVI (Hill and Sturm 1991). More sophisticated is the method proposed by Heo and Fitzhugh (2000), which includes algorithms for selecting invariant areas with similar topographic conditions, preferably smooth and sparsely vegetated.

### 7.3.3 Methods for Seasonal Analysis

When multitemporal analysis aims to improve the discrimination of some land covers, the interpretation is based on images acquired at different climatic seasons, preferably from the same year. The goal is to emphasize the seasonal contrast between input images to complement spectral with phenological information. For instance, to discriminate between crops that may be similar spectrally in one particular date, images from other periods of the year when those crops have a greater contrast in terms of density or degree of maturity may be used.

Multiseasonal interpretation may be carried out by comparing classification outputs from the two seasonal images, or by including two or more seasonal images in the same classification run. In the former approach, the process is the same as in traditional classification methods. After generating the results, the interpreter needs to build a seasonal table to account for transitions of the different covers of interest. For instance, when comparing spring and summer images (Figure 7.51), grasslands would have green vegetation in the former and dry vegetation in the latter, while bare soils would have soils in both dates and evergreen trees would have green signal in both. Consequently, spectral and temporal curves may be drawn for the same pixel.

When the analysis is done with two or more images simultaneously, the number of input bands for classification increases substantially. Therefore, before the classification, it is common to apply a data reduction technique such as PCA. In this way, only the most significant information of each date is included in the classification, saving time and avoiding data redundancy problems (Maselli et al. 1996).

The high temporal sampling provided by meteorological or global observation sensors, such as NOAA-AVHRR, SPOT-VEGETATION, or Terra-MODIS, offers many opportunities for seasonal analysis. These sensors acquire daily images over the whole planet. Therefore, the interpreter may follow subtle changes in surface

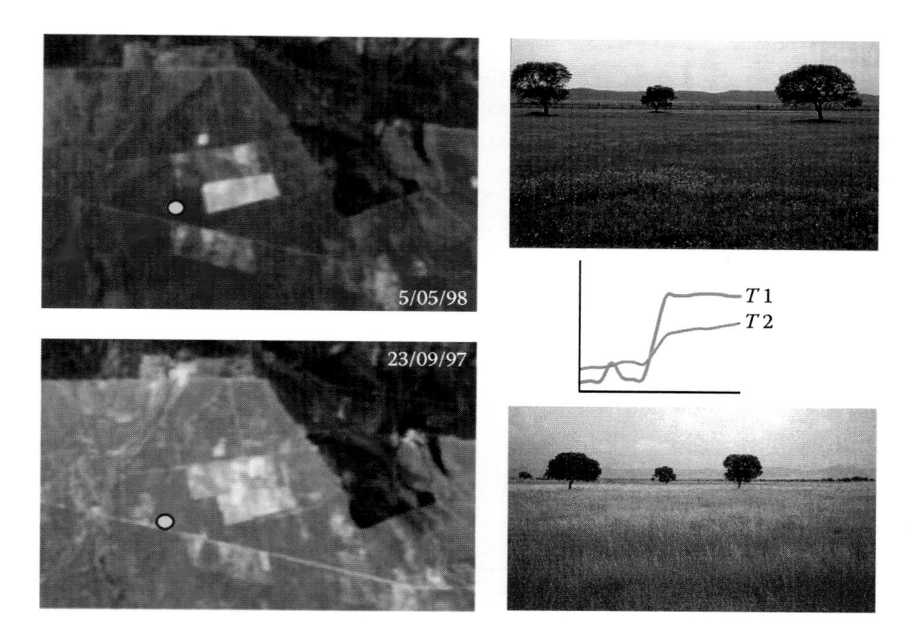


**FIGURE 7.51** Analysis of seasonal changes. Landsat thematic mapper images acquired in May (left) and September (right) over the Cabañeros National Park in Central Spain.

conditions, compare seasonal trends, and extract patterns of interannual variability. Some examples of the potentials of this approach are the monitoring of snow cover (Kelly and Hall 2008), analyzing impacts of climate change on vegetation productivity (Nemani et al. 2003), estimating crop yields (Doraiswamy et al. 2005), and estimating ET (Yebra et al. 2013b). The increasing accessibility to the Landsat historical archive has also made it possible to apply trend analysis to medium-resolution images, improving the detection of smaller features. Specific methods to retrieve changes from temporal trajectories of TM/ETM+/OLI historical images have been proposed in this regard (Cohen et al. 2010; Kennedy et al. 2010).

Sensors with coarse spatial resolution usually provide high temporal observation frequency (1–2 days), and therefore seasonal trends can be estimated by analyzing a large number of images (commonly several hundreds). Consequently, interpretation techniques have been developed to extract short-term indicators of change.

Daily images from these coarse-resolution sensors (such as MODIS or AVHRR) are acquired from different observation and Sun angles and are affected by clouds and cloud shadows or other atmospheric effects. Therefore, the trends of the daily observations may be quite noisy. To reduce this variability, making the selection of significant changes difficult, different techniques may be used, such as smoothing filters, window operations, logistic curve-fitting, and frequency-domain transformations (Sakamoto et al. 2005).

A simple way to characterize temporal trends is calculating descriptive statistics of the time series. Central tendency metrics (such as the mean or the median) would refer to the regular conditions of the measured parameter, while the standard deviation or interquartile range would be associated with the temporal variability. These measures can be computed for the whole time series or, more commonly, for specific periods. An example would be using the NDVI as an indicator of vegetation activity. The average NDVI would be computed as

$$\overline{NDVI}_k = \frac{\sum_{i=1,n} NDVI_{i,k}}{n_k} \tag{7.53}$$

where

$NDVI_{i,k}$ indicates the value of NDVI for a specific year $i$ and period $k$
$n$ is the number of years available for the period considered

This parameter can refer to the whole time series or just to specific periods of time, such as spring or summer biweekly composites. Similarly, the variability (standard deviation, $s_k$), defined as

$$s_k = \sqrt{\frac{\sum_{i=1,n} (NDVI_{i,k} - \overline{NDVI}_k)^2}{n_k}}, \tag{7.54}$$

can be computed for all periods or just for the number of times a specific time period occurs (commonly, the number of years of the time series). Mean and variability indices are useful in detecting anomaly periods, which are episodes with higher or lower than usual values. The anomalies can be expresses as standard scores ($z$-values) (Yool 2001):

$$Z_{NDVI_K} = \frac{NDVI_{i,k} - \overline{NDVI}_k}{s_k} \tag{7.55}$$

The $z$-value provides an estimation of the probability that a specific value is above or below the mean, assuming the NDVI temporal trends are normally distributed. This index would be computed for every pixel of the images, and the time period could be the whole series or just composites of $m$ days. The NDVI anomaly of each pixel for a single period can also be expressed as a percentage of that pixel value and period with respect to the mean NDVI value for the same period:

$$\Delta t_i = \frac{NDVI_{i,k}}{\overline{NDVI}_k} * 100 \tag{7.56}$$

The same or similar metrics can also be used with other vegetation indices, or even with STs. For instance, Figure 7.52 shows a heat wave affecting almost the whole European continent at the beginning of July 2015. The figure was built by comparing land STs of a 10-day period with the average ST of the same period for the

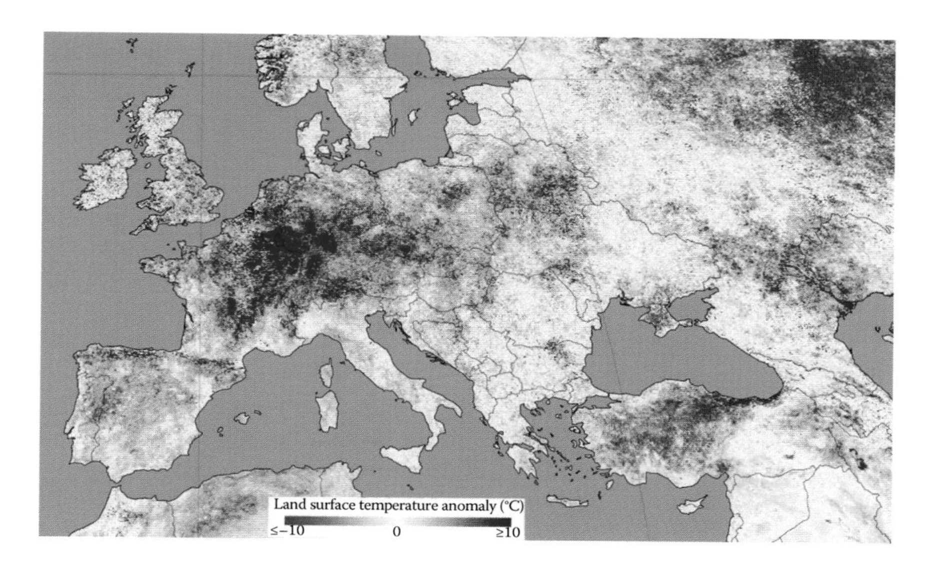


**FIGURE 7.52** Daytime land surface temperature anomalies in Europe from June 30 to July 9, 2015, computed from MODIS images. (From visibleearth.nasa.gov.)

2001–2010 historical series. Red colors imply higher temperatures and blue shades cooler. In early July, 2015, country historical maximum temperatures were registered in the Netherlands, Germany, and Switzerland.

Another approach to a temporal series of images is to use indices of relative change, which attempt to standardize the temporal variation within the spatial variability of the image. They have been frequently used with vegetation indices, most commonly the NDVI. The simplest metric of comparative change is the relative greenness (GR), defined as (Kogan 1990)

$$GR_i = \frac{NDVI_i - NDVI_{\min}}{NDVI_{\max} - NDVI_{\min}} * 100 \tag{7.57}$$

where $NDVI_{\min}$ and $NDVI_{\max}$ are the minimum and maximum NDVI values of each pixel for the whole time series, respectively. RG values have been shown to be more related to seasonal changes than absolute NDVI values since they emphasize the temporal variability within the variation of a single pixel (Figure 7.53). In fact, the absolute NDVI is associated not only with the vegetation seasonal activity but also with vegetation abundance/turgidity. For instance, an NDVI value of 0.4 may imply either a sparsely healthy vegetated area or a densely stressed vegetated area. Therefore, variations caused by weather conditions are more difficult to distinguish in absolute than in relative NDVI values. For this reason, RG has been used to monitor vegetation stress conditions, such as those leading to fire risk (Burgan et al. 1998) and drought susceptibility (Peters et al. 1991).

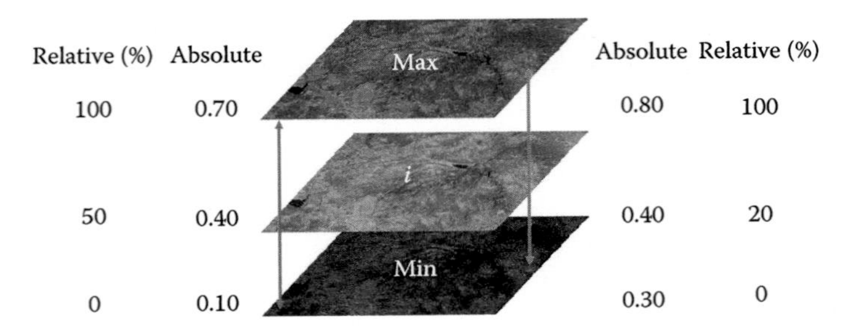


**FIGURE 7.53** Absolute and relative variation of normalized difference vegetation index values.

Other authors have suggested computing the temporal slope of NDVI values to detect sudden changes in vegetation conditions that may indicate either an important increase in productivity (spring greening up) or a rapid decrease (browning), perhaps caused by severe storms or heat waves. The temporal slope is computed as follows (Illera et al. 1996):

$$PT_i = \frac{\sum_{i=1,n} (NDVI(t_i) - NDVI(t_{i-1}))}{t_i - t_{i-1}} \quad (7.58)$$

where

$PT_i$ indicates the slope of the period $i$
$t_i$ is the period of time considered

As a surrogate of vegetation phenology, different methods to extract critical dates in the seasonal trend of VI have been proposed. These methods try to account for the dates when significant changes in temporal trends of VI occur. For instance, Zhang et al. (2003) suggest using a series of logistic functions to estimate VI at a certain time in the year, finding later the dates when the fitted function present significant rates of change.

Detection of long-term changes in vegetation seasonality is quite relevant to better understand the impacts of climate change, particularly in Northern latitudes (Goetz et al. 2005). Several authors have used time series statistical techniques (such as the seasonal Mann–Kendal trend test) to describe VI trends and relate their significance to the impact of disturbing factors, such as fire or climate warming (Alcaraz-Segura et al. 2010).

Finally, PCA has also been used to analyze time series. The input images are the different periods of the index considered, and the output components would indicate average trends (the first ones) or variability aspects (the secondary components). Some of the components may be associated with phenological features and subtle changes in vegetation productivity (Eastman and Fulk 1993).

### 7.3.4 Change Detection Techniques

Change detection techniques have been widely used with remotely sensed images, since they are applicable to a broad range of disciplines (Mouat et al. 1993). They aim to identify what features were modified between two or more dates. A remarkable example of environmental monitoring from satellite images is the *Atlas of Our Changing Environment*, released in 2010 by the UN Environment Program (UNEP), which includes Landsat multitemporal images of areas where striking landscape changes have occurred in the last 30 years. A similar initiative is the Google Earth engine, which incorporates the historical Landsat archive from 1984 to 2012 to iteratively observe changes anywhere on Earth.

As in other phases of image interpretation, the detection of changes usually implies a compromise between omission and commission errors, trying to avoid both missing a real change and misidentifying as change something that is stable, respectively. The importance of one or another type of error for a specific application will define the most convenient technique to use. For some studies, avoiding false alarms (commission errors) will be the main objective, for instance, in the case of applying chemical treatments to crop plots that are not in fact affected by a certain plague. Most commonly, though, the omission errors are more important, especially in the area of natural disasters, when it is critical to provide help to all areas that have been affected by a certain phenomenon.

Changes detected by remotely sensed images are mainly associated with alterations of terrestrial spectral properties (reflectance, temperature, etc.), but spatial features can also be observed, for instance, changes in plot size, shape, or fragmentation, using segmentation and object-oriented classification techniques (Bruzzone and Fernández-Prieto 2000; Hazel 2001; Smits and Annoni 1999).

Several authors have summarized the main techniques used for change detection studies (Eastman et al. 1994; Jensen 1981; Mouat et al. 1993; Singh 1986). In this section, we will review and update those studies with recent developments in this field. This section is organized around two groups of techniques: those that assume a quantitative scale in the input images (interval scale of DLs) and those that are applicable to classified images (nominal scale of DLs). In the first case, several arithmetic techniques can be used: differences, ratios, regression, principal components, etc., while in the second, only category comparisons, such as cross-tabulation, are pertinent. Before describing those techniques in more detail, a color composition technique for change detection will be discussed.

We will illustrate this section with the multitemporal pair of Landsat TM images acquired over the state of Acre, on the Brazilian Amazonia, which shows a striking example of an area affected by severe deforestation processes.

#### 7.3.4.1 Multitemporal Color Composites

As was previously reviewed (Section 5.3.2), a color composition can be created by assigning three spectral bands to the three basic colors: red, green, and blue (RGB). When a single band is displayed, it is usually observed in gray levels, because all RGB color guns have the same amount of brightness. Similarly, when mixing three bands, they would be shown in gray tones if they had the same DLs and in color otherwise.

Following the same logic, if we mixed three dates of the same spectral band in RGB, those areas with similar DLs in all dates would appear in gray levels, while areas affected by change would display in color tones. This technique has been successfully used in several change detection studies (Martin 1989; Sader and Winne 1992).

**FIGURE 7.54** Multitemporal color composite of normalized difference vegetation index values computed from thematic mapper images acquired in the Acre state of Brazil.

The interpretation of the resulting colors will depend on what bands were combined and what the color assignment was (which date was assigned to which color gun). To simplify matters, we can analyze the composition of two dates, the first one displayed in the red tones and the second one in green and blue (Figure 7.54). The grey tones correspond to stable areas, while those pixels that reduced their DLs from the first to the second date would be displayed in red, and those with increased DLs in cyan (green + blue).

In case of comparing NDVI values, as is the case of Figure 7.54, gray tones represent areas that have maintained their vegetation (both dense or sparse), while red tones indicate areas that have lost vegetation cover (lower NDVI in the second date), and cyan those that increased vegetation cover (higher NDVI in the second date). For instance, in areas affected by deforestation, mid-gray tones identify areas that keep forest cover and dark gray zones that are treeless in both dates, while red areas would identify deforested areas and cyan those being reforested.

When using other spectral bands, this color interpretation has to be adapted to the meaning of reflectance changes in the bands compared. Obviously, when three dates are combined, the color combinations become more complex to interpret. If the first date is assigned to red, the second to green, and the third to blue, the resultant color palette could be interpreted as follows:

1. *Red, green, or blue*: A feature with high values in one date and low in the others. This is the case of a feature that changed from one date to the other and was invariant afterward.
2. *Yellow* ($R + G$), *cyan* ($G + B$), *or magenta* ($R + B$): A feature with high values in two dates and low in the other. This situation represents changes from two dates to the third one.
3. *Gray levels* ($R + G + B$): Indicate features that are stable in all dates.

#### 7.3.4.2 Image Differencing

A simple way to identify the changes between two images that are properly calibrated and georectified is to compute the pixel-to-pixel difference in single bands or derived spectral indices. The stable areas will present a value close to zero, while those with high positive or negative values will identify areas with significant change:

$$DL_c = DL_{t2} - DL_{t1} \tag{7.59}$$

where $DL_{t2}$ and $DL_{t1}$ are the values of the second and first date, respectively, being compared. DL can be either reflectance or temperature (Pilon et al. 1988) or cross-calibrated digital values or any kind of spectral indices. In fact, the use of vegetation indices is quite common in vegetation change analysis to detect deforestation, clear-cuts, or fires (Coppin and Bauer 1994; Martín and Chuvieco 1995). Negative values will categorize areas with decreasing values and positive values regions with increasing values. The histogram of the output image will usually have a Gaussian distribution, with the maximum frequencies for the stable areas (in the center of the histogram) and the dynamics occupying the tails of the distribution.

Figure 7.55 includes an example of using temporal differencing to detect deforestation processes. The output image shows dark grays for areas where the decrease of NDVI has been more intense (those being deforested), while the medium grays correspond to areas where grassland, soil, or vegetation was maintained. Along with deforested areas, the image of Figure 7.55 shows also sectors that have been burned, and most probably will be deforested in the following years.

FIGURE 7.55 Multitemporal difference of normalized difference vegetation index values in the Acre state of Brazil.

#### 7.3.4.3 Multitemporal Ratios

Temporal differencing reflects absolute changes between two dates, but this absolute value may not properly reflect the significance of the observed increases or decreases. For instance, a decrease of 0.1 in NDVI may be very critical when the original value is 0.2 or almost irrelevant when the original value is 0.7. To correct this problem, an alternative to the temporal difference is the use of ratios, either absolute or normalized ones, as follows:

$$DL_c = \frac{DL_{t2}}{DL_{t1}} \tag{7.60}$$

$$DL_{i,j,c} = \left( \frac{DL_{t2} - DL_{t1}}{DL_{t2} + DL_{t1}} + 1 \right) \times 127 \tag{7.61}$$

with the same meanings as in Equation 7.59. In the normalized index, the result is scaled between 0 and 255 (8 bits). Multitemporal ratios have been used in agricultural inventories (Lo et al. 1986), deforestation studies (Singh 1986), and detection of areas affected by forest fires (Kasischke et al. 1993).

It should be noted that these ratios are asymmetrical, which means that the result is dependent on what date is in the numerator. An increase of twice the DL (from 50 to 100) would give a value of 2 or 0.5 depending on whether the numerator is the first or the second date. In summary, the ratio is a nonlinear transformation of the temporal relationships. To solve this problem, some authors recommend applying a logarithmic conversion of the result (Eastman et al. 1994).

#### 7.3.4.4 Principal Components

The use of PCA in change detection analysis has a particular scope with respect to the traditional use of this transformation. As previously stated (Section 7.1.2), PCA is generally used to summarize a wide set of variables by preserving the common information found in the input variables. However, when using PCA in change detection, the input bands are acquired at different dates, and the main components are precisely those that refer to invariant information (features that do not change), while secondary components show the transformations ("dynamic" components). Therefore, in this case, we should be interested on the less important vectors, instead of the main ones, which would include the stable information.

PCA has been frequently used in change detection analysis. As for Landsat TM/ETM+, some authors interpret the first two components as portraying stable brightness and greenness, while the third indicates brightness change and the fourth, greenness changes (Fung and LeDrew 1987, 1988; Ingebritsen and Lyon 1985).

Since the PCA is a statistical transformation, the interpretation of the resulting components cannot be done until the actual eigenvectors are computed. Depending on the spectral (and in this case temporal change) characteristics of the target area, the outputs may be quite diverse. We have computed the PCA transformation of the 12 bands of the Brazil multitemporal images (bands 1–7 of both dates, excluding the thermal band). In this case, the interpretation of the multitemporal PCA is not very common, as the area is mostly covered by two categories (forest or grass–soil), and the amount of change between the two images is very high. Consequently, the resulting eigenvectors do not show first the common brightness, but rather the tree-less area, with strong and negative contrast between the NIR and other bands in the 2010 image, while it shows brightness in the 1990 image (Table 7.10). The second component shows vegetation changes between the two images, although the spectral contrast is not very clear, with negative values for the B–R–SWIR bands and positive in the NIR in the most recent image, and vice versa. It should be related to area changes from soil to vegetation. The third component also corresponds to changes, but in this case mainly those related to vegetation loss, while the forth implies moderate reforestation. In terms of the relative importance of each principal component, the first one accounts for 85.5% of variance, the second one 6.23%, the third 4.55%, and the fourth 2.25% (all sum up to 98.54% of the original information of the 12 bands).

Figure 7.56 confirms the interpretation of the eigenvector matrix. The first component shows a similar image to the deforested area at the second period, with higher values for those plots converted to grasslands or crops in both periods. The second one displays in brighter tones those areas that have maintained crops in both dates and darker tones for those areas that have lost more vegetation, particularly those with brighter soils. The third component shows in brighter tones those areas that have been burned (from vegetation to dark signal) with medium tones for those maintaining forest cover. Finally, the fourth component shows with brighter tones agricultural areas of the first date while keeping other areas in mid-gray.

#### 7.3.4.5 Regression Analysis

Linear regression is a well-known technique to estimate values of a dependent variable from the values of another variable when both are strongly correlated. The regression coefficients are computed from a set of common observations of the dependent and independent variables. In the change detection context, these common observations are stable pixels between the two dates. Once the regression model has been fit, the DLs of the second date may be estimated from DLs of the first date, as

$$\widehat{DL}_{t2} = a + b * DL_{t1} \tag{7.62}$$

When the second image has not been altered, the predicted DLs will be similar to the actual ones, but when transformations have occurred, there will be a significant difference between the estimated and real values (Figure 7.57). Therefore, the residuals of the regression can be used as a change detection index:

$$DL_c = \widehat{DL}_{t2} - DL_{t2} \tag{7.63}$$

#### 7.3.4.6 Change Vector Analysis

This technique tries to identify the magnitude and the direction of change by using a bivariate space of transitions between the two dates. The vector that links the location of pixels in that bivariate and multitemporal space will contain information about the intensity and orientation of change. Figure 7.58 represents several

**TABLE 7.10**

**Eigenvectors of the Multitemporal Analysis of the Acre Image (Brazil)**

| | 2010 | | | | | | 1990 | | | | | |
|---|---|---|---|---|---|---|---|---|---|---|---|---|
| | **B1** | **B2** | **B3** | **B4** | **B5** | **B7** | **B1** | **B2** | **B3** | **B4** | **B5** | **B7** |
| PCA1 | 0.168 | 0.136 | 0.266 | −0.029 | 0.841 | 0.408 | 0.023 | 0.016 | 0.024 | 0.018 | 0.074 | 0.032 |
| PCA2 | −0.020 | −0.004 | −0.040 | 0.367 | −0.025 | −0.086 | 0.138 | 0.126 | 0.178 | 0.355 | 0.758 | 0.298 |
| PCA3 | −0.003 | −0.043 | 0.031 | −0.887 | −0.109 | 0.068 | 0.103 | 0.079 | 0.130 | −0.092 | 0.348 | 0.174 |
| PCA4 | 0.028 | 0.040 | 0.020 | 0.230 | 0.003 | −0.047 | 0.060 | 0.020 | 0.090 | −0.924 | 0.215 | 0.171 |

FIGURE 7.56 Principal component analysis of a change detection analysis for the Brazil multitemporal image.

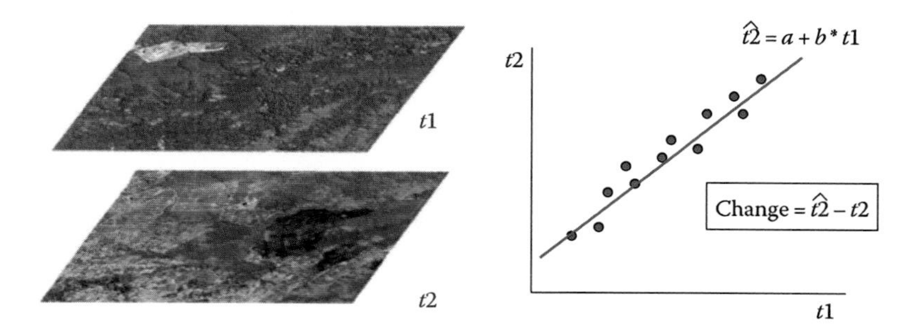


FIGURE 7.57 Change detection by multitemporal regression analysis.

transitions between two images in the R–NIR space. The locations of specific pixel transitions between the two dates are marked. For instance, an increase in NIR reflectance while reducing R reflectance indicates a trend toward increasing vegetation in that pixel. However, a decrease of both reflectances may indicate a trend toward flooding (as water has low reflectance in both bands). Therefore, the angle of

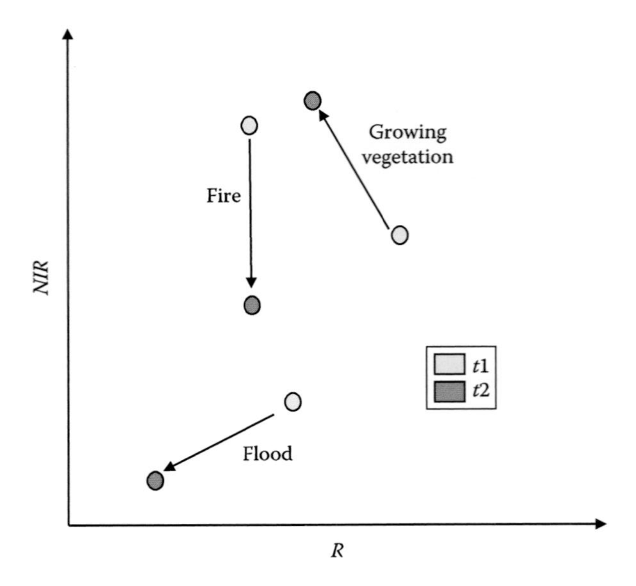


FIGURE 7.58 Example of multitemporal transitions in the red and near-infrared space. Change vectors indicate the magnitude and direction of the change.

the change vector indicates the meaning of the change, while the magnitude of the vector indicates the importance of that particular change.

The components of change vector analysis can be computed from simple geometric rules. The intensity ($I$) of the change vector will be obtained through

$$I_{i,j,c} = \sqrt{\left(DL_{i,t1} - DL_{i,t2}\right)^2 + \left(DL_{j,t1} - DL_{j,t2}\right)^2} \quad (7.64)$$

while the direction will be defined by the angle ($\alpha$), given by

$$\alpha = \arctan\left(\frac{DL_{j,t1} - DL_{j,t2}}{DL_{i,t1} - DL_{i,t2}}\right) \quad (7.65)$$

for the spectral band $i,j$ between the two periods of interest ($t1$ and $t2$). Change vector analysis has been widely used in the last years (Chen et al. 2003), especially in the context of global land cover change (Friedl et al. 2002), and it is the basis of one of the standard products of the MODIS program.

Figure 7.59 shows an example of the change vector analysis computed from the Brazil multitemporal TM pairs. The direction of change shows lower angles for those areas affected by deforestation processes, particularly those where soils are dominating (the burned area does not show such a clear trend). The intensity of change

FIGURE 7.59 Direction and intensity of the change vectors in the Brazil Acre study site.

shows higher values for areas converted from forest to crop soils, but also the burned area (with lower values), while the areas remaining as forested show low values of change, as well as the plots deforested before the first image.

### 7.3.4.7 Defining Change Thresholds

All methods reviewed so far produce continuous images of change. In other words, the output images show a continuous scale of change, from low to high DL values. However, often, the interpreter seeks to generate a categorized image that distinguishes changed areas from stable ones. It is therefore necessary to segment the change images generated by the aforementioned techniques to create significant groups of change.

Theoretically, the histogram of the change image should facilitate that segmentation, by offering three modes corresponding to stable, increased, and decreased pixel values. In practical terms, usually the histograms do not provide such a clear discrimination of change, as a result of the transitional stages of natural changes, which generally do not include sharp contrasts. Therefore, alternative methods should be considered to detect stable and dynamic areas.

Among the statistical approaches, the most common option is to base the thresholds on the mean and the standard deviation, but there is no objective way to fix appropriate threshold values. Some authors recommend gradually changing the multiple of standard deviation around the mean and assessing the most convenient threshold value after analyzing how many pixels of known change are included in the different thresholds (Fung 1992; Fung and LeDrew 1988). Other proposals include establishing thresholds following the sensor signal-to-noise ratio. That factor is estimated from a library of historical images of change. Once the sensor noise is removed, the remaining change can be more easily controlled (Smits and Annoni 2000).

A second approach to establishing change thresholds is based on physical values extracted from external sources. For instance, to detect an active fire, we can establish a threshold of thermal difference between the potential fire pixel and the background, based on the physical measurements of fires in different ecosystems.

A third option focuses on extracting the thresholds from spatial differences of the potential change pixel and the neighboring characteristics. The spatial context is commonly estimated from a moving window of different size, which serves to account for the background trend. When a pixel has a strong contrast with the surrounding areas, it is labeled as potential change. This spatial context has been extensively used for cloud detection (Saunders and Kriebel 1988).

Regardless of which approach is used, it should be emphasized that having more flexible or more rigid thresholds will imply increasing or reducing omission errors, or conversely, commission errors. For this reason, the user should have a clear idea of which error is more relevant for that particular application. For instance, in a burned land mapping exercise, very demanding (hard) thresholds reduce the detected burned areas, while relaxing (soft) thresholds would increase the burned areas at the risk of including unaffected surfaces. For this reason, it might be appropriate to follow a two-step procedure. In the first phase, the goal would be to eliminate commission errors by finding out the most clearly burned pixels, while in the second phase, omission errors are reduced by analyzing the spatial context of only those areas identified as burned in the first phase (Alonso-Canas and Chuvieco 2015; Chuvieco et al. 2008b).

Figure 7.60 shows the histogram of the NDVI difference for the Brazil study site. The mode of the histogram shows values of 0 (no change), but there is a strong dispersion around that value, particularly toward negative values (as most of the changes imply deforestation and therefore decrease in NDVI). Using simple thresholds of −0.1 (deforestation), −0.1 to 0.1 (stable), and >0.1 (increase), we can make a first stratification of the main areas affected by change.

#### 7.3.4.8 Multitemporal Analysis of Classified Images

The identification of changes can also be handled using classification techniques. In this case, it is not required to define a change threshold, since classification already implies transformation of the continuous scale of DLs into a nominal scale of categories. Therefore, the change analysis can be based on analyzing transitions between classes in two or more dates.

The first approach to using classification techniques for change detection analysis is based on classifying two or more dates simultaneously. Using either supervised or unsupervised methods, the interpreter needs to identify both the significant covers and the relevant cover changes. In other words, the resulting classification should include categories that include somehow the multitemporal component. "Stable urban" and "transition from rural to urban" or "from rainfed to irrigated agriculture" are good examples of such categories. Since the classification legend needs to be dynamic when performing supervised classification, the interpreter should select training areas in both stable and dynamic areas to make an appropriate sample of all relevant classes. In the case of unsupervised classification, the automatic algorithm will identify the most significant change/stability groups within the multitemporal images, and later the interpreter should assign those spectrotemporal groups to thematic categories of interest.

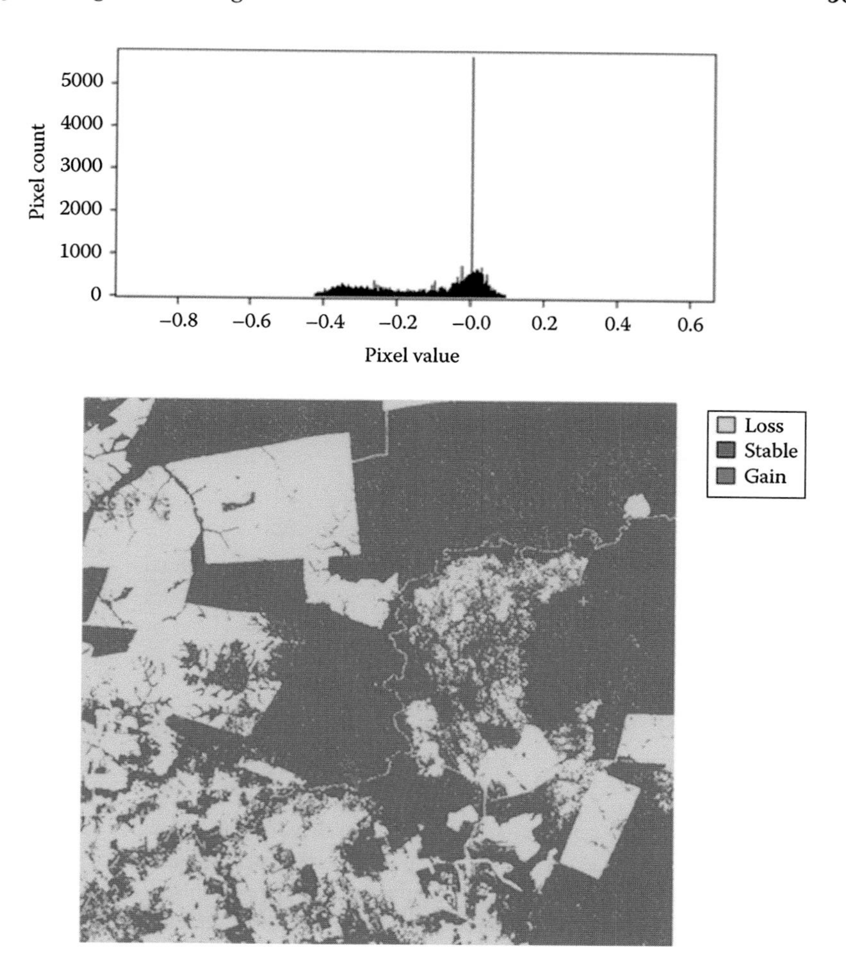


**FIGURE 7.60** Histogram of the change image and image segmentation for the Brazil study site.

Another approach to multitemporal change detection analysis is to classify two dates of the same area and then compare those classifications to detect which areas have maintained or have changed the land cover (Figure 7.61). For doing so, the same categories should be used in both dates; otherwise the comparison will be meaningless. Land cover changes can be obtained by cross-tabulation analysis, in which all transitions between the two dates are clearly presented (Table 7.11). In the diagonal of this table appear the stable pixels (those with the same category in both dates), while the dynamic pixels occupy the other cells of the matrix. The analysis of this matrix provides critical information on the transitions between the two dates. When comparing just the total areas covered by certain categories in the two dates, the net difference provides a first insight into the evolution of that cover (increase or

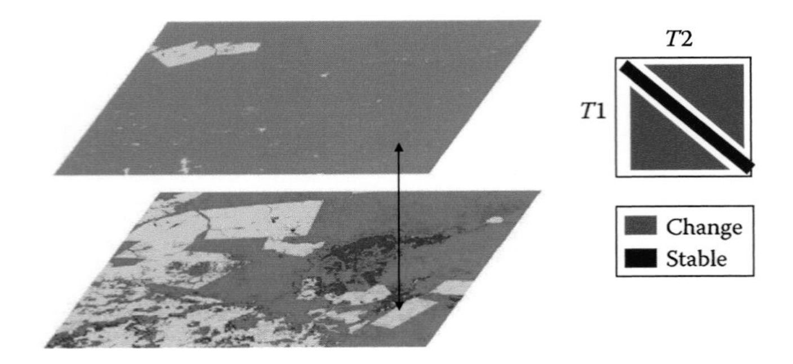


FIGURE 7.61 Multitemporal comparison of classified images.

**TABLE 7.11**
**Cross-Tabulation of the Brazil Images**

| 2010→ 1990↓ | Forest | Crops | Water | Burned | Total |
|---|---|---|---|---|---|
| Forest | 346,888 | 192,349 | 1,745 | 51,048 | 592,030 |
| Crops | 794 | 16,233 | 47 | 260 | 17,334 |
| Water | 371 | 66 | 477 | 44 | 958 |
| Total | 348,053 | 208,648 | 2,269 | 51,352 | 610,322 |

decrease), but the multitemporal table facilitates determination of the categories to which or from which these transitions have occurred. Actually, the total difference in area covered does not reveal the full set of transitions, since in the same area, there might be gains and losses of the same category that are cancelled by considering only the totals (Pontius et al. 2004).

The multitemporal table also provides an excellent framework for computing change metrics. For instance, the Kappa or other agreement indices, which will be discussed in greater detail in Chapter 8, used to measure the strength of the association between two set of categories, could facilitate a quantitative evaluation of temporal stability between two dates. Transitions of each category may be accounted for by computing the initial and final areas covered by that category. For instance, in Table 7.11, the transitions of the sample area between 1990 and 2010 account for 40% of the total area, from which the vast majority was changed from forest to crops (net loss 41%); 32% of original forested area is now crops, while 8.6% is burned. The crop area of 1990 has been mostly maintained (93% remained as crops) with a very low percentage changing to forested area (4.5%). Total crop area of 2010 was mainly forested in 1990 (92%).

Multitemporal transition tables are as confident as the accuracy of individual classification is. If the classes have not been properly discriminated in either of the dates, the temporal comparisons will be erroneous as well, offering poor information on change. In fact, the accuracy of the temporal comparisons is the product of the individual classifications, as errors are propagated when the comparison is done.

A final remark on change detection concerns to the complexity of displaying clearly changes in dynamic maps. Even with a small number of input categories (say, 8 per date), the number of potential transitions (64) exceeds the number of readable classes in a map. Even though some of those transitions will not exist (since some changes are very unlikely or impossible), the cartographic representation of changes is always a challenge. An alternative to representing changes would be to show them in a series of maps, one for each category on the first date, and each showing the output category in the second date. Figure 7.62 illustrates two different approaches to mapping land cover changes. The former relies in simple cross-tabulation, which in this case provides an efficient way of displaying changes, as the number of input categories is small (Figure 7.62a). When this is not the case, it may be preferable to create a series of maps (one for each input category, showing the transitions between the two dates). In this case (Figure 7.62b), three transitional maps have been created, one for each category of the first date, showing in thematic colors the final cover at the second date. With this representation, a clearer view of the relevance of stable and transition covers is perceived.

## 7.4 ANALYSIS OF SPATIAL PROPERTIES

### 7.4.1 Remote Sensing and Landscape Ecology

Landscape ecology emphasizes the importance of considering the impacts of spatial patterns in critical environmental processes, such as the distribution and alteration of species, regeneration trends, and succession cycles (Forman and Godron 1986; Haines-Young et al. 1993). EO images provide critical information for this spatial-oriented ecological approach, since they facilitate a quantitative measure of spatial pattern (Hoffer 1994; Wilkie and Finn 1996), while GIS facilitates the processing tools needed to perform the required spatial analysis (Davis and Goetz 1990; Haines-Young et al. 1993).

A satellite image can be considered a digital representation of the landscape mosaic, and therefore, spatial relationships can be extracted from image information. Several authors have explored the use of satellite images for landscape ecology studies. The topics are very diverse, including habitat suitability (Congalton et al. 1993; Longmire and Stow 2001); the spatial spread of certain alterations, such as those caused by fires or plagues (Hargrove et al. 2000; Muchoki 1988; Sample 1994); relations between landscape pattern and vector diseases (Hay et al. 2000); and biodiversity studies (DeFries 2013; Hansen et al. 2001; Roy and Tomar 2000). In addition, other studies have focused on analyzing spatial structures at different spatial scales and their relations to spatial resolution (Quattrochi and Goodchild 1997; Woodcock and Strahler 1987).

In summary, remote sensing images provide an assessment of spatial aggregation of features at different spatial scales (Figure 7.63), imply a spatial representation of landscape structure, and facilitate an analysis of spatial autocorrelation for different landscape variables and can be used to compute landscape metrics.

The measurement of image spatial properties, in addition to being of interest for landscape ecological studies, is also relevant for image interpretation. As previously

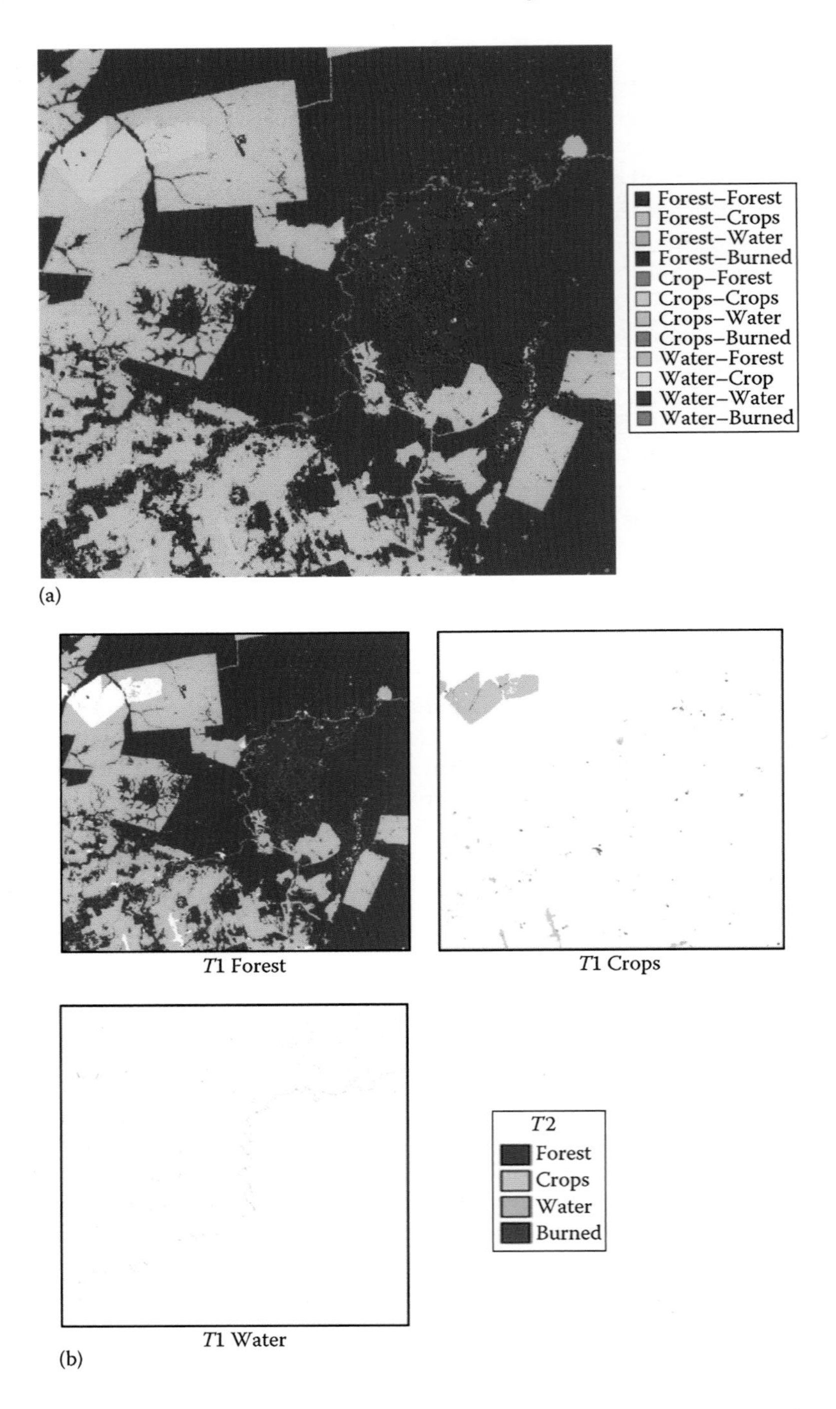


**FIGURE 7.62** Two approaches to land cover change mapping for the Brazil study case: (a) single cross-analysis and (b) transitional land covers.

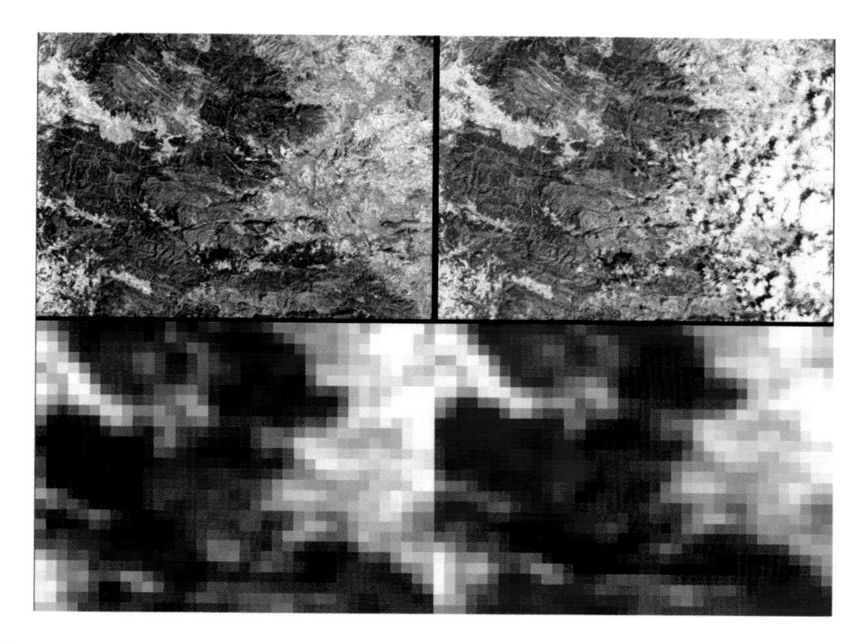

FIGURE 7.63 Landscape structure of an area affected by a large forest fire in Eastern Spain as observed by thematic mapper (top) and Advanced Very High Resolution Radiometer (bottom) images.

stated, the classification of target categories may need to take into consideration spatial features, such as shape, size, or neighbor relations. Therefore, computing spatial metrics is useful to generate additional bands that can be input into digital classification.

The revision of the spatial metrics is grouped in two broad categories. On the one hand, we will comment those methods suitable for continuous interval-scale variables and, on the other hand, those oriented toward nominal or categorical-scale variables. The former are applicable to raw DLs or those resulting from any quantitative transformation (physical transformations, PCA, vegetation indices, etc.), while the latter refer to classified images.

## 7.4.2 Spatial Metrics for Interval-Scale Images

### 7.4.2.1 Global Metrics for Continuous Data

The objective of these techniques is to measure the texture or spatial variability of the image. They try to calculate the spatial complexity of pixels at certain neighboring distances. Some of these metrics operate on the whole image and provide a single measure of the spatial variability, while others calculate the local texture, usually comparing the DL of a pixel with those that surround it. The former are useful in comparing the spatial complexity of different areas, or in analyzing the temporal evolution of landscape patterns for a certain area. The latter output texture images, and they can be used as input for digital classification or in predictive models (Vega-Garcia and Chuvieco 2006; Vega-García et al. 2010).

Fractal analysis is a sound alternative to measure the spatial properties of the whole image, since it provides an estimation of the spatial complexity of the internal shapes that account for that area. In classic geometry, an object's dimension could only be an integer (1 = line; 2 = plane; 3 = volume), but in fractal geometry objects have real-number dimensions (between 1 and 2 or between 2 and 3), depending on their grade of spatial complexity. The more complex the surface, the closer the dimension is to 3, while the more complex the line, the closer it is to 2.

Although fractal geometry has been extensively used in the last few decades (Dimri and Turcotte 2005; Mandelbrot 1982), its calculation is not very evident on interval-scale images, since the objects are not explicitly defined. In a pioneering application of fractal geometry to the analysis of Landsat images, Lam (1990) proposed methods to compute fractal dimension from images, finding higher fractal values for urban than for rural landscapes. The most heterogeneous bands were the visible ones (B, G, R), and the most homogeneous was the TIR band. In all cases, the fractal dimension of a raw image was found to be high, exceeding 2.6 (except for the thermal one, with a value of 2.2). Some studies have used fractal measures to characterize the spatial variation of the tree heights for forest inventory (Drake and Weishampel 2000) and to analyze postfire regeneration patterns from NDVI images (Ricotta et al. 1998).

Another method of measuring the spatial complexity of landscapes is the spatial autocorrelation (SA), proposed some decades ago in the context of understanding the spatial associations of geographical phenomena (Cliff and Ord 1973). Since natural variables are gradually distributed on the Earth's surface, the values for a single location tend to be more similar to those nearby than to those far away. Two indices have been commonly used to measure the strength of this spatial association: the Moran I metric and the semivariogram. The former was proposed by Moran (1948) and is based on the standard correlation index, although in this case it considers the variance of neighboring observations:

$$I = \left(\frac{n}{2A}\right) * \left(\frac{\sum_{i=1,v}\sum_{j=1,n} z_i z_j}{\sum_{i=1,n} z_i^2}\right) \tag{7.66}$$

where

$n$ indicates the number of considered pixels
$A$ is the number of possible unions among those pixels
$z_i$ is the deviation of the pixel $i$ from image mean $(DL_i - DL_x)$
$z_j$ is the deviation from the image mean of the adjacent pixels

The Moran index oscillates between −1 and +1, increasing when the spatial uniformity is greater. It can be calculated for a particular direction (only vertical or horizontal links, for instance) or for all directions. This index has been used to measure the degree of spatial correlation in order to select the most appropriate sampling or to estimate the landscape complexity (Congalton 1988b; Henebry 1993).

The semivariogram is a graph that shows how spatial variance changes at different distances, and it is based in the following function (Chica 1988):

$$\gamma(h) = \frac{1}{2NP(h)} \sum_{i=1,NP(h)} \left(DL(x_i + h) - DL(x_i)\right)^2 \qquad (7.67)$$

where

$\gamma(h)$ indicates the spatial variance for a distance $h$

$NP(h)$ is the number of pixels included in that radius $h$

$DL(x_i + h)$ and $DL(x_i)$ are the values corresponding to the target pixel and those at distance $h$

The semivariogram can be computed for all directions or just for specific ones. The graphic representation of the semivariogram shows how the variance increases as distances become higher, up to a certain threshold when the variance stabilizes (Figure 7.64). That means that the spatial autocorrelation is lost after a certain distance (named *range*), beyond which the new observations would not affect the variability. The value at which the variance stabilizes is named *sill* and should be related to the general complexity of the input image. Range and sill are important in understanding the spatial structure of an image (Woodcock et al. 1988a,b).

Semivariograms of high-resolution images have been used in forest inventory to discriminate between different species (Cohen et al. 1990). These authors observed that the range was related to the crown width, being higher for plots with old growths, while the sill made reference to the stratification of the forest canopy. However, these

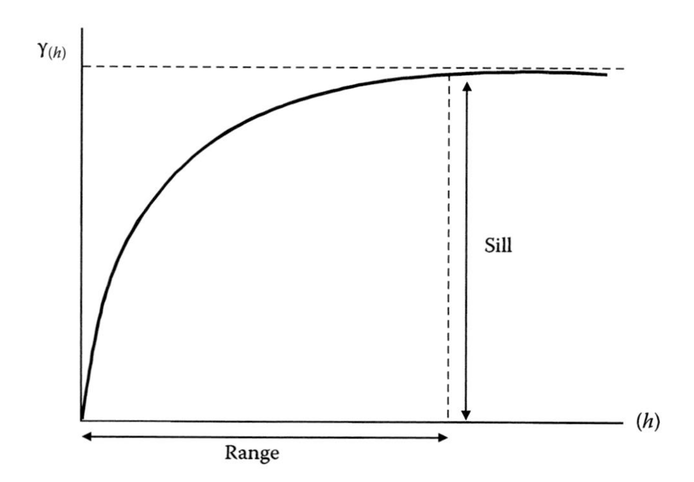


**FIGURE 7.64** Semivariogram.

observations were very dependent on the relation between object size and image resolution, because the results were different when using images with coarser resolution (Cohen et al. 1990). In this sense, the semivariogram has been shown to be of great interest in describing the relationships between the size of the objects and the spatial resolution of the sensor, the maximum of local variance taking place between one-half and three-fourths of the mean size of the objects (Woodcock and Strahler 1987). They have also been used to analyze the impact of a large fire on landscape structure (Chuvieco 1999).

#### 7.4.2.2 Local Metrics for Continuous Data

These metrics estimate the local variation of spatial properties and provide texture bands that can be later use as inputs for classification (Riaño et al. 2001; Smits and Annoni 1999). One of the simplest ways to compute local texture is to calculate the standard deviation of the DLs within a certain window size. Similar to the filtering process previously described, that window moves systematically to all image pixels, and the output is a standard deviation matrix. This first-order texture metric has proved very effective in forest inventory studies, as an indicator of density and homogeneity of vegetation species, providing significant improvements over raw classifications (Strahler et al. 1978).

An alternative to the simple measures of texture is those based on the gray-level co-occurrence matrix (Haralick et al. 1973), which measures, for a specific vicinity window, how many times a certain $DL_i$ is contiguous to another $DL_j$. The co-occurrence matrix is square and theoretically should have the same number of columns and rows as the number of different DLs the image has (the radiometric resolution). However, to simplify the calculation, a previous reduction of the original scale is performed (commonly from 256 to just 16 different DLs). The matrix counts the number of spatial connections between each DL value (Figure 7.65). The connection (angle between the central pixel of the window and the neighbors) can be counted in different directions or be averaged for all possible connections (0°, 45°, 90°, and 135°). The vicinity distance can also be altered.

From the co-occurrence matrix, various second-order metrics of texture can be computed (Gong et al. 1992; Moller-Jensen 1990; Smits and Annoni 1999;

| | | | | |
|---|---|---|---|---|
| 4 | 6 | 8 | 5 | 4 |
| 5 | 5 | 8 | 7 | 6 |
| 6 | 7 | 7 | 7 | 9 |
| 8 | 8 | 4 | 8 | 6 |
| 9 | 8 | 9 | 5 | 6 |

Input image

| DL | 4 | 5 | 6 | 7 | 8 | 9 |
|---|---|---|---|---|---|---|
| 4 | 0 | 1 | 1 | 1 | 0 | 1 |
| 5 | 1 | 0 | 2 | 2 | 1 | 0 |
| 6 | 1 | 2 | 2 | 0 | 1 | 2 |
| 7 | 1 | 2 | 0 | 2 | 3 | 0 |
| 8 | 0 | 1 | 1 | 3 | 4 | 1 |
| 9 | 1 | 0 | 2 | 0 | 1 | 0 |

Co-occurrence matrix

**FIGURE 7.65** Sample image and its corresponding co-occurrence matrix.

Vega-Garcia and Chuvieco 2006). The most common in remote sensing applications have been the following:

$$\text{Homogeneity: } H = \frac{\sum_{i=0,n-1} \sum_{j=0,n-1} p_{i,j}}{(1+(R-C)^2)} \quad (7.68)$$

$$\text{Contrast: } Con = \sum_{i=0,n-1} \sum_{j=0,n-1} (p_{i,j}(R-C)^2) \quad (7.69)$$

$$\text{Dissimilarity: } D = \sum_{i=0,n-1} \sum_{j=0,n-1} \left(p_{i,j} \left|R-C\right|\right) \quad (7.70)$$

$$\text{Mean: } M = \sum_{i=0,n-1} \sum_{j=0,n-1} (Rp_{i,j}) \quad (7.71)$$

$$\text{Standard Deviation: } SD = \sqrt{\sum_{i=0,n-1} \sum_{j=0,n-1} \left(p_{i,j}(R-M_i)^2\right)} \quad (7.72)$$

$$\text{Angular Second Moment: } ASM = \sum_{i=0,n-1} \sum_{j=0,n-1} (p_{i,j})^2 \quad (7.73)$$

$$\text{Entropy: } E = \sum_{i=0,n-1} \sum_{j=0,n-1} \left(-p_{i,j} \log(p_{i,j})\right) \quad (7.74)$$

where

$p_{i,j}$ indicates the relative frequency of co-occurrence for *DL* values *i,j*
*R* and *C* indicate the matrix position in rows and columns, respectively
*n* the number of columns and rows of the matrix

In summary, $p_{i,j}$ measures the probability that a pixel with a DL value *i* is contiguous to another with a value *j*.

Two examples of these metrics over the Tucson image are displayed in Figure 7.66. Homogeneity is higher in volcanic features and natural areas to the east and west, while it is lower in the urbanized areas and nearby transportation routes. The opposite is true for the contrast image, with higher values for the central urban area and lower values for natural areas, gardens, and golf courses.

### 7.4.3 Spatial Metrics for Classified Images

Several authors have proposed metrics to compute the spatial complexity of thematic maps (Frohn 1998; O'Neill et al. 1988; Riitters et al. 1995). As in the case

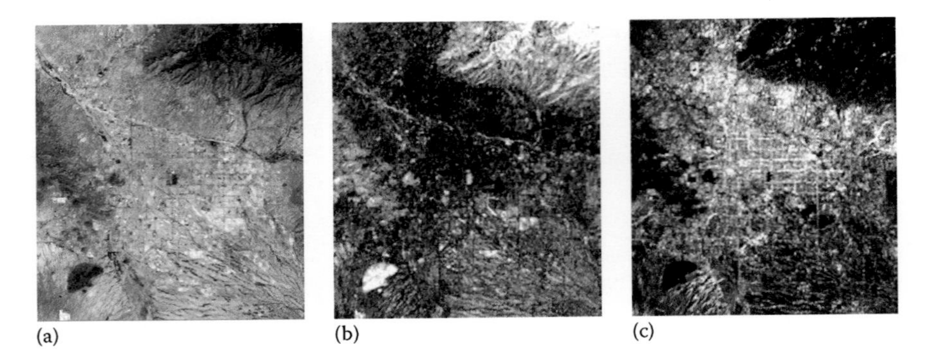

FIGURE 7.66 Texture images of the Tucson image. (a) Original band 4, (b) homogeneity, and (c) contrast computed from the same band 4.

of quantitative images, some of those metrics are computed for the whole image, while others are computed from local moving windows. Both rely heavily on the concept of spatial patch, which is very common in landscape ecology. A patch can be defined as an object on a thematic map that is a homogeneous and contiguous group of cells/pixels. A classified image defines each category by a certain numerical code, but the same code (category) may occur in different areas of the image, not necessarily contiguous. Patches can be obtained from classified images using spatial aggregation algorithms (clumping). These programs label each pixel with the same number if they are included in the same thematic category, sequentially, until pixels from another category are found, after which a new label is issued (Figure 7.67).

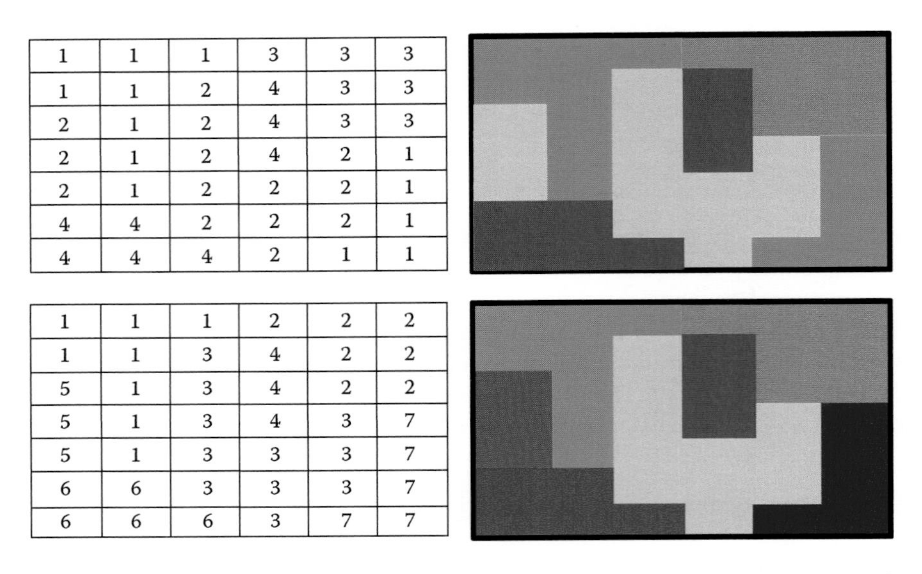

| 1 | 1 | 1 | 3 | 3 | 3 |
|---|---|---|---|---|---|
| 1 | 1 | 2 | 4 | 3 | 3 |
| 2 | 1 | 2 | 4 | 3 | 3 |
| 2 | 1 | 2 | 4 | 2 | 1 |
| 2 | 1 | 2 | 2 | 2 | 1 |
| 4 | 4 | 2 | 2 | 2 | 1 |
| 4 | 4 | 4 | 2 | 1 | 1 |

| 1 | 1 | 1 | 2 | 2 | 2 |
|---|---|---|---|---|---|
| 1 | 1 | 3 | 4 | 2 | 2 |
| 5 | 1 | 3 | 4 | 2 | 2 |
| 5 | 1 | 3 | 4 | 3 | 7 |
| 5 | 1 | 3 | 3 | 3 | 7 |
| 6 | 6 | 3 | 3 | 3 | 7 |
| 6 | 6 | 6 | 3 | 7 | 7 |

FIGURE 7.67 The original classified image (top) is converted to patches (bottom) by segmenting contiguous pixels with the same category.

In summary, clumping algorithms transform the original classified image (with as many DLs as classification categories) into a new file where there are as many DLs as there are different polygons. It is important to note that pixels with the same thematic category will have a different patch number if they are not in the same polygon. Therefore, the new image facilitates additional spatial information to better understand landscape patterns. For instance, an agricultural classification would discriminate between different crops, and therefore estimations of total area under a certain crop can be derived. However, we do not know from such classification whether those crops are in single or different lots, or whether the area has crop concentration or not. Therefore, to analyze agricultural spatial patterns, we should convert first the agricultural map to a patch map.

#### 7.4.3.1 Global Metrics for Classified Data

Once we have a patch map, simple metrics can be used to analyze the landscape pattern of a certain region. For instance, the number and density of patches are two simple indices of spatial complexity. Patch density can be used as a fragmentation index (Dillworth et al. 1994; Ricotta and Retzlaff 2000). An area with a high density of patches will indicate a very fragmented territory. The average size and the size distribution of patches are complementary measures of complexity and provide a view on regional diversities within the study area. Landscape fragmentation is critical to understanding the spatial effects in species mobility (Ripple et al. 1991) and has been commonly associated with studies of vulnerability with respect to the impacts of potential disturbances (climate change, wind storms, fire: Chuvieco et al. 2014).

Patch shape may also be relevant for different studies because they may indicate temporal trends (expansion or regression) or patch function (connectivity will be higher for elongated than circular patches). Several indices of shape have been used, but the most common is compactness, which tries to measure the similarity of a patch to a circular area. Compactness has been defined as the relation between area ($A$) and perimeter ($P$), using the $F$ index (Gulinck et al. 1991):

$$F = \frac{A4\pi}{P^2}100 \tag{7.75}$$

or the circularity ratio ($C$), which relates the area of a patch ($Ap$) to the area of a circle with the same perimeter ($Ac$):

$$C = \sqrt{\frac{Ap}{Ac}} \tag{7.76}$$

When increasing $C$, the polygon would be more compact and less elongated. Shape can also be computed from the fractal dimension, which measures the average degree of complexity of the patches in the study area. A simple method of calculating the image fractal dimension ($D$) is by finding out the relationship between the perimeter and area of every patch (Ripple et al. 1991):

$$P \approx A^{D/2} \tag{7.77}$$

The average value of the image can simply be obtained by averaging $D$ for all patches. Other authors have proposed using the coefficients of the regression line between logs of patch perimeter and area (O'Neill et al. 1988):

$$\log P = a + (D/2) \log A \tag{7.78}$$

where $a$ is a constant of proportionality. Some authors have observed that $D$ is directly associated with landscape complexity. In a wide sector of the eastern United States, it was shown that $D$ offered a numeric assessment of land use patterns: when $D$ was small, the forms were quite geometric, which was a consequence of prevalent human land use, whereas other areas with higher $D$ have more presence of natural landscapes (O'Neill et al. 1988; Ricotta et al. 1998).

A similar analysis of area–perimeter relations has been explored in the theory of self-organizing criticality (SOC: Bak et al. 1987), which explores the scale-invariant properties of many natural phenomena by using a power-law model. This model relates the frequency and area of the distribution of a certain feature, and the exponent of the power-law model indicates the relevance of large or small objects in the target region. For instance, the power-law model can be a good approximation of fire regime characterization, as it accounts for the relation between the frequency of certain fire sizes and the proportion of the area they burn (Hantson et al. 2015).

A third aspect of patch patterns in the image refers to the average distance between patches. This aspect is measured using dispersion indices computed from the centroids of the different patches (Figure 7.68).

#### 7.4.3.2 Local Metrics for Classified Data

In addition to global metrics, the internal spatial variation of patch conditions should also be very valuable in landscape ecology studies. As for continuous data, the analysis is based on moving windows of different size (commonly 3 × 3 pixels).

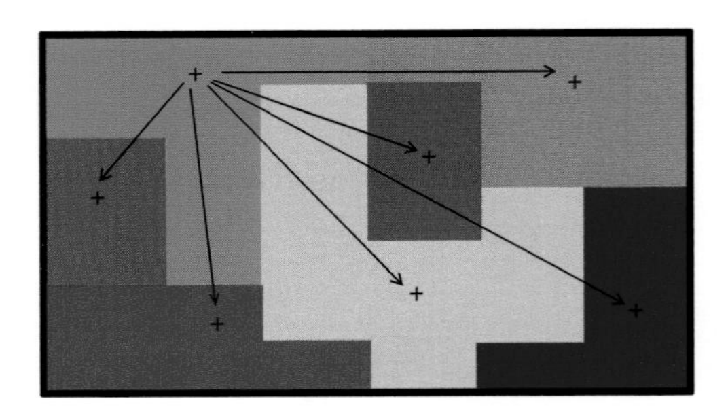

**FIGURE 7.68** Dispersion indices.

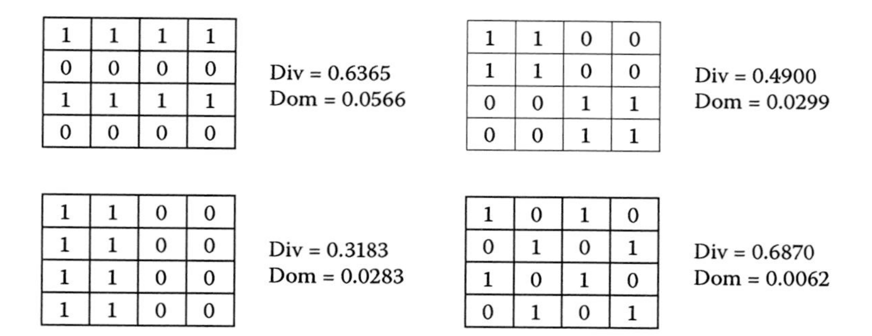


**FIGURE 7.69** Diversity and dominance for test images.

One of the most popular local indices is diversity (D), which attempts to measure the variability of categories within a certain window. It was defined as (Turner 1989)

$$D = -\sum_{k=1,m} p_k \ln(p_k) \tag{7.79}$$

where

$p_k$ is the proportion of category $k$ in the window
$m$ is the number of categories in that window

The more diversity, the more spatial variation in categories' distribution or, to put it in another way, the area will present a more varied mosaic (Figure 7.69). The diversity is directly related to the scale of the images, being smaller when lowering the resolution (Turner et al. 1994).

Complementary to diversity is dominance, which measures the deviation of the estimated value versus the maximum diversity (O'Neill et al. 1988):

$$Do = \ln(m) + \sum_{k=1,m} p_k \ln(p_k) \tag{7.80}$$

with the same significance as Equation 7.79. The higher the dominance, the larger the prevalence of one of the categories over the other ones in that window.

To measure the degree of spatial fragmentation of different patches, different indices have been suggested. A simple one is named *contagion* and measures the degree of adjacency among categories versus the potential maximum (Li and Reynolds 1993; O'Neill et al. 1988):

$$Con = 2\ln(m) + \sum_{i=1,m}\sum_{k=1,m} p_{ik} \ln(p_{ik}) \tag{7.81}$$

where

$p_{ik}$ indicates the proportion of pixels in the window where category $i$ meets category $k$
$m$ is the number of classes

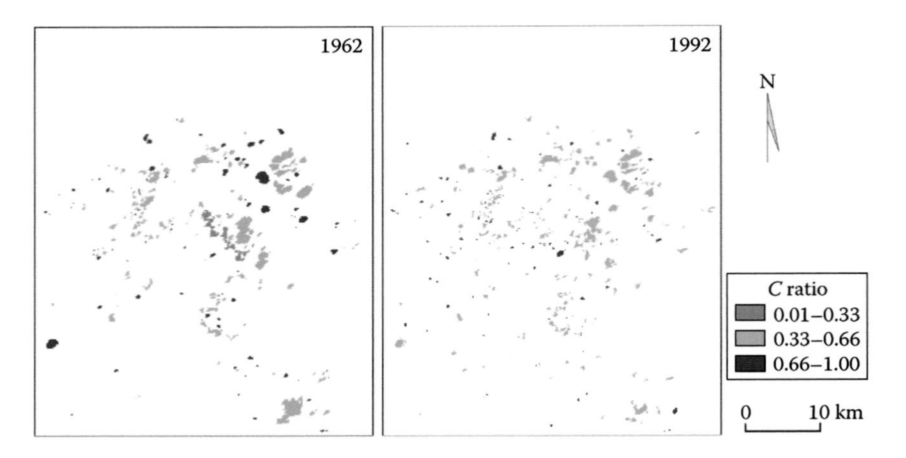


**FIGURE 7.70** Changes in compactness in the central region of San Luis (Argentina). (From Collado, A.D. et al., *J. Arid Environ.*, 52, 121, 2002.)

### 7.4.4 Landscape Structural Dynamics

Landscape metrics can be used to compare various landscapes on a reference date, or to analyze trends of a particular area through time. In this second approach, the spatial impact of certain human activities or natural events may be assessed (Ne'eman et al. 1999; O'Neill et al. 1988; Reinhardt and Ringleb 1990; Turner et al. 1994). For instance, the spatial impacts of a large fire may be analyzed at different scales (Chuvieco 1999), or the spatial arrangements of sand dunes may be monitored (Collado et al. 2002). Figure 7.70 shows changes in compactness associated with soil degradation processes in a semiarid area. The more compact dunes show expansion processes (in this case, as a result of short-term agricultural practices), while the elongated ones display regression trends, as a result of greater water availability.

These studies are complementary to the spectral change detection analysis described in Section 7.3, since environmental impacts of changes are associated not only with alterations in cover (spectral changes) but also with spatial arrangement.

## 7.5 REVIEW QUESTIONS

1. Which one of the following characteristics favors the use of simulation versus empirical models?
   a. Simplicity
   b. Accuracy
   c. Generalization power
   d. None of the above
2. Identify which of these processes does not benefit from the use of simulation models.
   a. Classification
   b. Identification of the most sensitive spectral bands

c. Estimation of a quantitative parameter
d. Reduction of the atmospheric effects

3. The main objective of using spectral vegetation indices is to
   a. Reduce the topographic effect in image interpretation
   b. Facilitate multitemporal comparisons
   c. Estimate the leaf area index
   d. Improve separation of vegetation and soil
4. Which of the following indices should be more suitable to estimate chlorophyll content?
   a. NIR/R
   b. SWIR/NIR
   c. SWIR/B
   d. R/B
5. Which of the following indices improves the separation of water and soil?
   a. NIR/R
   b. SWIR/NIR
   c. B/SWIR
   d. R/B
6. The relationship between NDVI and plant water content is
   a. Linear
   b. Exponential
   c. Quadratic
   d. None of the above
7. Spectral mixture analysis is more convenient for
   a. High spatial resolution images
   b. High radiometric resolution images
   c. High angular resolution images
   d. High spectral resolution images
8. Where can endmembers be derived from?
   a. Input image
   b. Spectral libraries
   c. Simulation models
   d. All of the above
9. Calculation of object heights from lidar data mainly depends on
   a. Estimating the elevation of the ground
   b. Performing atmospheric correction
   c. Retrieving the canopy cover
   d. Estimating the overlapping between lidar flights
10. The supervised training is preferable to unsupervised training when
    a. Classes are heterogeneous
    b. The area has rough terrain
    c. The area is well known
    d. The area is very complex
11. What is the main difference between traditional and object-oriented classifiers?
    a. Use of hyperspectral data
    b. Use of pixels versus polygons

   c. Use of neural network classifiers
   d. Use of high-resolution images
12. Which of the following parameters is not required for maximum likelihood classification?
   a. Normal distribution
   b. Extract training fields
   c. Quantitative input variables
   d. Include textural bands
13. Which of the following algorithms leaves unclassified pixels?
   a. Parallelepiped
   b. Minimum distance
   c. Maximum likelihood
   d. Random Forest
14. Which filters can be used with classified images?
   a. Low pass
   b. High pass
   c. Median filters
   d. Modal filters
15. Multitemporal comparison requires
   a. Using digital images
   b. Applying low-pass filters
   c. Ensuring geometric and radiometric matching
   d. Classifying the images
16. Which of the following techniques can be used to detect changes in classified images?
   a. Temporal color composites
   b. Temporal regression
   c. Temporal cross-tabulation
   d. Temporal vectors
17. Which of the following multitemporal techniques would be more suitable for burn scar mapping?
   a. Subtraction of classified images
   b. Cross-tabulation of classified images
   c. Subtraction of principal component analysis
   d. Subtraction of vegetation indices
18. The analysis of temporal anomalies is especially suitable for
   a. Landsat OLI images
   b. SAR images
   c. MODIS images
   d. SPOT-HRV images
19. Which of the following textural techniques is not suitable for interval-scale images?
   a. Patch size
   b. Co-occurrence matrix
   c. Autocorrelation analysis
   d. Semivariogram

20. To compute patch analysis metrics, we should first
    a. Apply multitemporal matching
    b. Apply atmospheric correction
    c. Clump classifications
    d. Apply spatial filters

# 8 Validation

## 8.1 RELEVANCE OF VALIDATING RESULTS

The final goal of visual or digital interpretation is to provide an estimation of ground truth conditions. Checking whether results are close to the actual ground conditions is a critical part of any EO analysis. "Validation" is defined by the Committee on Earth Observing Satellites Working Group on Calibration and Validation (CEOS-WGCV) as "the process of assessing, by independent means, the quality of the data products derived from the system outputs" (CEOS-WGCV 2012). Validation commonly includes estimating different aspects (Figure 8.1):

1. Accuracy, defined in terms of distance between the estimated and actual measurement. It can also be termed as "error" or "bias".
2. Precision, expressed as the dispersion of estimations around the average estimated value, commonly in units of standard deviation ($\sigma$).
3. Stability, or consistency of accuracy throughout time, measured as temporal variations of accuracy or whether significant trends throughout time exist.

Validation helps in evaluating the utility and limitations of using any particular remote sensing product. For example, when estimating crop yields, validation provides an idea of how close our results are to the actual yields. As is the case of any statistical inference, the estimates are based on probabilistic assumptions that provide a certain confidence level around the estimated value. Depending on the significance of the results, the agricultural manager can make decisions on future actions; for instance, whether to commercialize that crop or not. In addition to testing the significance of the results, accuracy assessment is also critical for evaluating a certain methodology, as it provides an objective criterion to determine whether that method provides (significant) better or worse results than others.

Frequently, accuracy assessment is considered the last step in digital classification. However, this process should also be included in visual interpretation, as well as in other kinds of digital analysis, such as generating biophysical variables (temperature, chlorophyll, leaf area index [LAI], etc.), or multitemporal change detection. For this reason, we have dedicated a separate chapter to this topic, instead of embedding it into the digital classification section.

We should not confuse accuracy assessment with uncertainty characterization. The latter implies estimating how confident the interpreter is about his or her estimation. Uncertainty is estimated by taking into account the errors associated with the different phases of the interpretation processes. It includes modeling errors related to the sensitivity of the sensor system, the geometric and radiometric correction processes, and the classification algorithms. For instance, if we are using a balance to

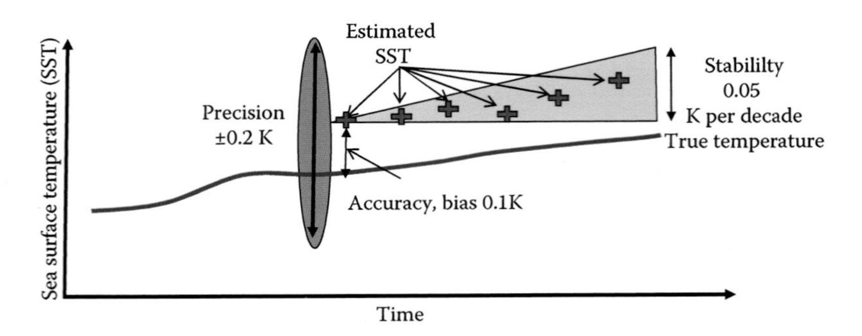


**FIGURE 8.1** Concepts involved in validation.

weigh a leaf sample, the precision of the balance is an important source of error, as the uncertainty of our measurement could never be below the inherent error of the balance. In other words, we cannot weigh a sample with accuracy higher than 1 g if we are using a medium-precision balance with a sensitivity of 10 g. Precision of the final estimation is commonly computed as an error integration of the different sources of potential errors, and it is measured as a dispersion value (commonly in $\pm\sigma$, where $\sigma$ is the standard deviation of errors).

Uncertainty can be characterized without ground measurements of the target variable, but accuracy assessment cannot, as it implies estimating the average deviation between our estimations and the true value. It is commonly expressed as averaged bias and can be positive or negative depending on whether the estimations are above or below the reference. Obviously, the reference measurement should be acquired as close as possible to the ground conditions when the satellite observed the target area. This is particularly important for dynamic variables, such as soil temperature or humidity, which may vary quite rapidly.

Logically, the interpreter will not have those ground conditions for the whole target area, as otherwise, one would not need any remote sensing interpretation, but only a sample area that is assumed to be representative of the ground conditions. In this regard, accuracy assessment relies on sampling techniques that provide a good balance between the information retrieved and low cost. As most assessment exercises are based on 2D results, spatial sampling techniques should be used to select plots that can objectively represent the ground conditions. Once the ground conditions are obtained, they are compared with classification results to compute different metrics of errors.

In summary, the verification of results includes the following phases (Congalton and Green 1999):

1. Sampling design, indicating how many plots must be sampled and how they should be selected.
2. Collecting reference data and results over the selected plots to be sampled.
3. Extracting the interpretation results for the same plots used with the reference data.

4. Comparing reference and interpreted data using appropriate statistical techniques.
5. Analyzing the causes and distribution of errors and, eventually, comparing results from different interpretation techniques.

## 8.2 SOURCES OF ERROR

The reliability of the results depends on different factors of potential error, both related to the acquisition and interpretation phases. They should be considered in order to explain, on one hand, the confidence of the results and, on the other, the accuracy found in the validation process. Actually, the validation process has also potential sources of errors, for instance, those related to the sampling design and reference data collection. In the following sections, we will focus on the former factors and leave the interested reader to consult sampling statistics textbooks (Berry and Baker 1968; Cochran 1977; Congalton and Green 1999) for the latter factors.

### 8.2.1 Sensor Limitations

Various errors are possible in the acquisition of an image, such as those caused by failures in the sensor, platform, or reception systems. Sensor limitations, however, refer to the technical capabilities of the system (spatial, spectral, radiometric, or temporal resolution), which constrain the use of the collected data. When the user is unaware of these limitations, he or she may try to use the images for applications that may exceed the sensor's potential. For instance, if we try to estimate urban growth from MODIS data, we should realize that omission errors will be likely, considering the coarse pixel size of this sensor. Similar errors are also likely when trying to use Landsat images for active fire detection. In this case, the limitation would stem from the temporal resolution.

As mentioned in Chapter 4, it is critical to make a proper selection of the sensor to meet project requirements. Otherwise, the risk of errors will increase. Logically, the selection of the most adequate sensor will be a compromise between the investment needed and the level of acceptable error.

### 8.2.2 Method of Analysis

We have reviewed throughout this book a wide variety of visual and digital techniques to extract thematic information from images. The choice of one or the other will depend on the interpreter's resources and technical preparation. Obviously, some of the errors in the final product should be attributed to the interpretation procedures. For instance, when using RapidEye images to map urban areas, simple classification algorithms may offer significant errors, as many pixels would have mixed covers. However, those errors should be reduced when using contextual or textural classifiers, which take greater advantage of fine spatial resolution than pixel-based classifiers. In the same way, errors in temperature calculation are related not only to the sensor sensitivity but also to the consistency of the methods used for atmospheric or emissivity correction.

In a broad sense, we can include among the deficiencies of the interpretation method those errors that occur because of an incorrect selection of image acquisition dates. In the case of vegetation covers, when seasonal dynamics are not carefully considered, the interpreter may find assignment errors that are otherwise solvable. For example, in rural sites, dry-season images may present confusions between buildings and recently harvested plots. This is caused by the spectral similarity of roof tiles and bare soil. However, when using images from the growing plant season, the cultivated areas will show a greater spectral contrast with built-up areas.

### 8.2.3 Landscape Complexity

Homogeneous landscapes are easier to classify than complex ones, where more borders, and therefore mixed pixels, exist. The spectral contrast between neighboring categories is also related to landscape pattern as well as to the degree of mixture of land cover types.

The effect of patch size and shape is especially evident in regions with a long history of human occupation. These regions frequently have tiny patches, smaller than the pixel size in most medium- and coarse-resolution sensors. The digital levels (DLs) of the border pixels will be a mixture of two or more covers, making its classification quite complex (Figure 8.2). This effect is known as *boundary error*; it is very common in agricultural or land cover mapping in very traditional landscapes.

Closely connected with this problem is the spatial arrangement of cover types. When there is a clear spectral contrast among neighboring covers, they are easier to separate than if they offer similar radiance (Figure 8.3). For example, a rural settlement when surrounded by green vegetation is easier to discriminate than when bounded by bare soils.

Another potential source of error is related to the target categories. Some end users may be interested to include in the final land cover map categories that include two or more covers. For instance, in Mediterranean areas it is very traditional to combine forest and agricultural uses in the so-called Dehesa, where grasslands and evergreen oaks occupy the same plot. They are used for extensive cattle grazing (frequently pigs or bulls). These categories are difficult to classify because they are

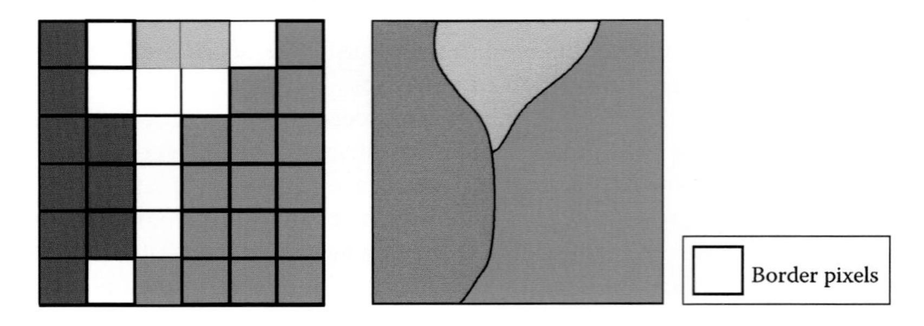


**FIGURE 8.2** Consequences of border errors: the border pixels will register an intermediate radiometric signal between two or more cover types.

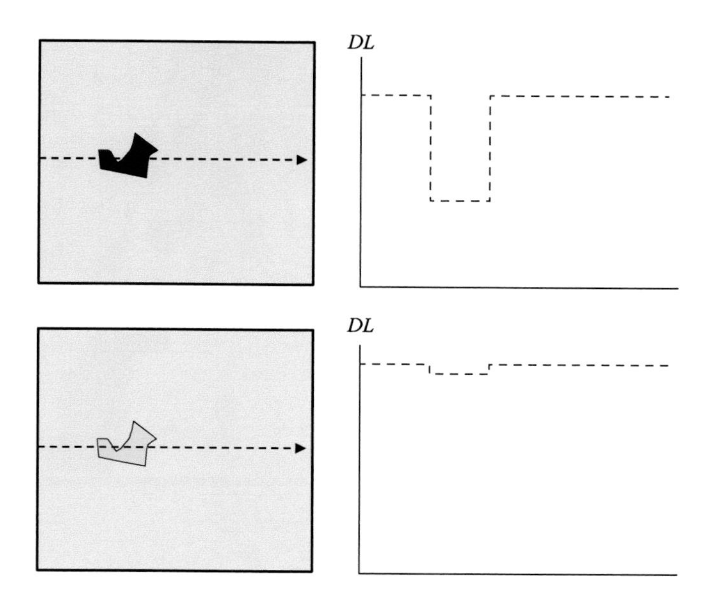


**FIGURE 8.3** Effect of radiometric contrast between neighboring covers: this can be observed in the profiles of areas with high and low contrasts.

intrinsically mixed and their signal is an average of pure signatures. Spectral unmixing or fuzzy classification may be a good alternative to discriminate them.

The influence of landscape pattern in classification errors is evident when studying the spatial distribution of error patterns. These studies have proved that there is a significant spatial autocorrelation in the erroneous pixels (Congalton 1988b). That means that a misclassified pixel has more chances to be close to another erroneous pixel than to a correct one because errors are highly influenced by the landscape structure. This tendency is more evident in those images that show less spatial complexity, that is, images with more systematic patterns of distribution, such as agricultural areas (Figure 8.4). On the other hand, in more complex landscapes, such as urban areas or natural vegetation, the errors are less systematic.

Finally, we mention a group of assignation errors that occur due to the environmental conditions of a given cover. As we have seen, the different orientation, slope, density, and phenological state in which a thematic cover appears imply spectral variability about the typical behavior of that category. Such dispersion hinders the spectral characterization of a class, generating overlap and confusion with similar spectral categories.

### 8.2.4 Verification Process

Some of the errors found in the accuracy assessment process may be caused by the validation process itself (Congalton and Green 1999). Some examples of potential noise factors are related to the mechanisms of data collection, errors in the geometric rectification, and the subjectivity of human samplers.

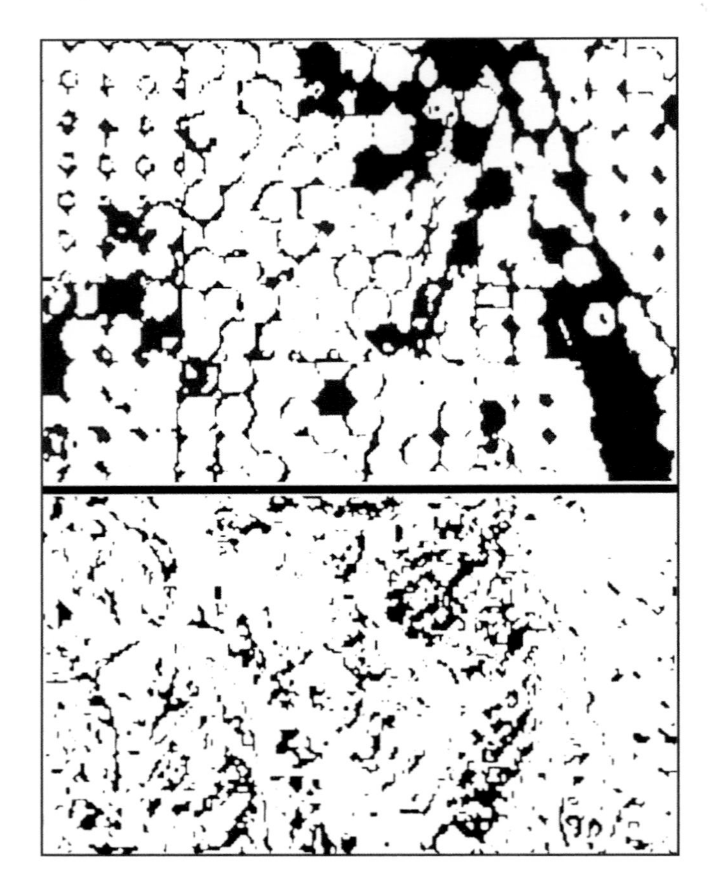

**FIGURE 8.4** Spatial distribution of the error in digital classifications: the upper part corresponds to errors from classifying an agricultural area, and the lower portion from a forested area. (Reprinted from Congalton, R.G., *Photogram. Eng. Remote Sens.*, 54, 587, 1988. With permission from the American Society for Photogrammetry and Remote Sensing.)

The process of ground data collection is commonly named "ground truth," but in fact what is collected cannot be fairly considered "the truth." In some cases, there are also errors associated with the ground measurement, such as those caused by the field equipment (precision of the thermometer of soil moisture meter, for instance), or by the need to simplify some data collection procedures. For example, when estimating LAI on the ground, all leaves of the tree should be cut and scanned to compute the area of all leaves. This is unfeasible, and therefore, either a small sampling of the leaves is taken or an estimation of LAI is done through allometric equations or using field sensors measuring the radiation that penetrates the plant canopy (such as the Licor LAI or fish-eye photography: Bréda 2003). Obviously, all these methods have estimation errors, which introduce uncertainty in their use for validation studies.

Problems in accuracy assessment may also occur when ground data collection is conducted by different people, because they may be inconsistent. For this reason, when several surveyors are involved, it is very important to define precisely the different thematic categories and sampling procedures in order to reduce the margins of error. In a forest inventory, all samplers should have the same understanding of what a tree is (a wood species taller than 2 m, for instance) to make the measurements fully comparable among the different teams involved. In any case, the lack of full correspondence between a ground-measured and satellite-measured biophysical parameter does not necessarily mean errors in the latter.

With regard to thematic errors, the interpreter should be aware of potential problems caused by using documents from different dates. For example, when using aerial photography as the reference data, the acquisition should be from a date close to the satellite imagery; otherwise, errors may be confused with temporal changes. Finally, the reference data might also include geometric or thematic errors. Geometric location is critical when validating results from high-resolution images, since small displacements may severely affect the assessment process. In this regard, a high-precision global positioning systems (GPS) receiver (Figure 8.5) or topographic equipment is required to locate control plots for airborne lidar applications (Riaño et al. 2004a); otherwise, the potential errors may be caused by displacements of target and reference plots.

FIGURE 8.5 High-precision GPS receivers. These are important for ensuring a precise geometrical location in ground-truth surveys, particularly when working with high-spatial-resolution data.

## 8.3 METHODS TO ESTIMATE ACCURACY

The estimation of the quality of results has been approached from different perspectives. Obviously, the simplest way is a pure subjective evaluation coming from the interpreter or end user who knows the study site. We can summarize this procedure as the judgment "It looks good," which is in practical terms a common way of interpreting the results, since this subjective validation is evidently less expensive and faster than a rigorous verification. However, few practical and scientific conclusions can be deduced from this simple appraisal, and therefore, it cannot be accepted as a sound validation procedure.

A more adequate alternative is comparing the results with external statistical sources. For instance, the results of a crop classification can be compared with agricultural statistics to estimate whether the total acreages are similar for the two sources. This assessment is known as *non-site-specific assessment*. It is simple to obtain and provides a numerical estimation of accuracy, but it involves two important problems. First, it is assumed that the statistical source is the ground truth, which is obviously not the case, since it may be based on estimations. The second problem is related to the nonspatial character of the evaluation, since the agreement of the total area does not imply that the class is properly mapped, as omission and commission errors may balance each other (Figure 8.6).

A third approach for accuracy assessment relies on using as reference data the same observations that are used to train the classifier. This option reduces the cost of fieldwork, since no additional validation plots are required, but introduces a bias in the assessment process. Training observations will have a greater probability of being correctly classified than other pixels of the image, since the classification rules are based on those training statistics. In the same way, a regression model should not be validated with the same data used to fit the equation, since the adjustment is generated by minimizing the distances to those points. Additional independent data need to be considered for testing the robustness of the model (Figure 8.7).

Finally, the most reliable assessment involves selecting a series of independent samples not used for producing the results. Logically, the selection and collection of additional samples will raise the cost of the project, and this increase is not used to improve the results but only to validate them. In spite of the problems associated with this process, we should emphasize the need to undertake it with a sound methodology, to obtain a quantitative estimation of the potentials and limitations of our

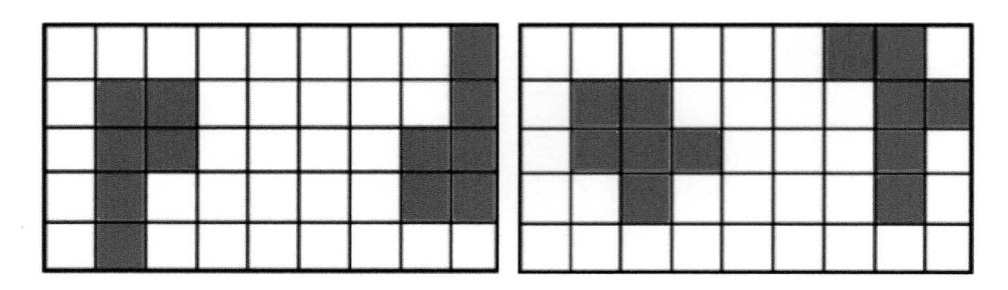

**FIGURE 8.6** Problems caused by non-spatial verification: the total number of gray pixels is the same, but the spatial distribution is totally different.

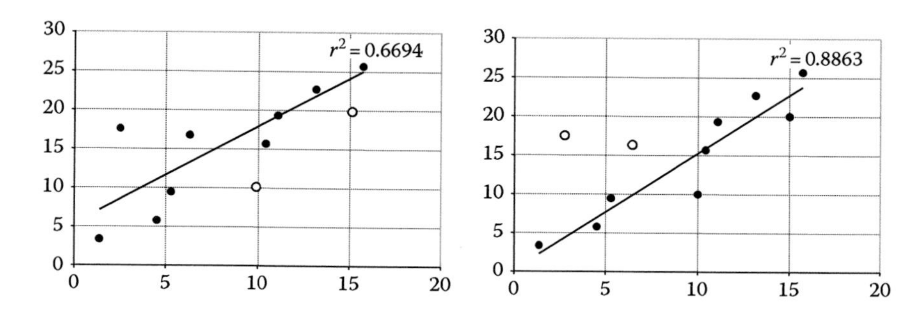


**FIGURE 8.7** Impact of changing the observations to compute the explained variance of a regression model: the white circles are points not used to compute the determination coefficient ($r^2$).

final product. Performing a proper validation implies developing a solid strategy for sample selection, collection of ground data, and comparison of those reference data with the EO results.

The CEOS Land Product Validation (LPV) subgroup has defined four stages in the validation of an EO variable (http://lpvs.gsfc.nasa.gov/, last accessed November 2015), which have increasing levels of confidence in the estimation results:

1. Product accuracy is assessed from a small (typically <30) set of locations and time periods by comparison with in situ or other suitable reference data.
2. Product accuracy is estimated over a significant set of locations and time periods by comparison with reference in situ or other suitable reference data.
3. Uncertainties are characterized in a statistically rigorous way over multiple locations and time periods representing global conditions. Spatial and temporal consistency of the product and with similar products has been evaluated over globally representative locations and periods.
4. Validation results for stage 3 are systematically updated when new product versions are released and as the time series expands.

In summary, validation may be carried out at different levels of spatial and temporal sampling, depending on product maturity and the extent it covers. We will review briefly some principles of sampling design to better adapt the LPV stages to our particular product.

## 8.4 SAMPLING DESIGN

The design and development of sampling is the basis of the assessment process. Spatial sampling helps in making a suitable selection of plots to be surveyed. The two main characteristics of a sound sampling scheme are to provide an unbiased estimation of the target variable (in this case, the error of the results) and, at the same time, to be efficient; that is, it must provide the maximum information at the lowest possible cost. Both properties imply that the sampling scheme should follow a statistical design that is easy to implement and analyze, and provide a sound estimation of the spatial variability of errors. The selection of the type of sampling technique depends on a series of factors (Congalton and Green 1999; Stehman 1999): how the error is distributed in the study area, what the most adequate sampling unit is, how the samples should be selected, and how many samples should be collected.

### 8.4.1 Error Distribution

Any sampling scheme must consider the characteristics of the target variable, that is, whether it is a discrete or continuous variable, and its spatial distribution. In our case, most validation exercises are applied to thematic classifications, and therefore, the interpreter should use statistical techniques adapted to nominal variables. Categorical distributions are the most common in this regard, and their validation requires the use of a binomial distribution (truth/error). When the assessment is intended for continuous variables, such as the temperature or vegetation water content, sampling can be based on continuous distributions, such as the normal or Poisson distribution (Biging et al. 1998; Stehman 1999).

The spatial distribution of errors is a critical consideration when selecting the most appropriate spatial sampling. For instance, the presence of high spatial autocorrelation precludes the use of systematic sampling (Congalton 1988b). The spatial association of an error tends to present diverse patterns based on landscape complexity, which is frequently associated with the natural or man-made character of areas. Agricultural landscapes tend to offer geometrical patterns, while natural areas tend to follow a more random distribution.

### 8.4.2 Sampling Unit

The sampling unit refers to the element in which the reference information will be collected. Most commonly, but not always, it is identified with the pixel. In fact, on many occasions, it is more convenient to use a group of pixels (e.g., a 3 × 3 window) as a sampling unit, which may reduce potential georeferencing errors. The thematic polygons derived from the classification process or the objects in contextual classifiers can also be considered as sampling units.

### 8.4.3 Sampling Strategies

This concept refers to the different ways of selecting the cases that will be part of the validation sample. The most common are the following (Congalton 1988a, 1991; Rosenfeld 1982; Rosenfeld et al. 1982; Stehman 1992, 1999; Figure 8.8):

1. *Simple random*: The elements to be verified are chosen in such a way that all of them have the same probability of being selected, since the selection of one does not influence the selection that follows. Random sampling is very solid statistically, but it presents practical problems, as it can imply high transportation costs. Besides, it may not be suitable for the spatial variation of the classified image because the low-frequency categories may not be represented.

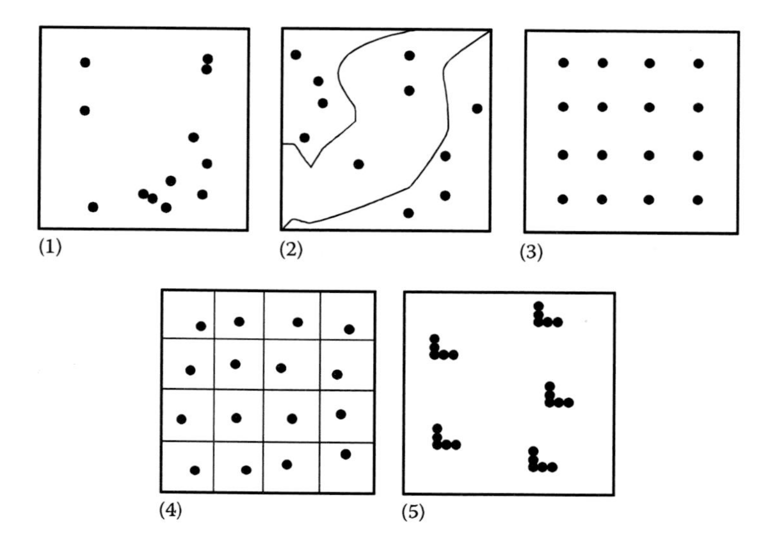


**FIGURE 8.8** Types of sampling schemes applied in verification. (1) Random, (2) stratified random, (3) systematic, (4) unaligned systematic, and (5) cluster.

2. *Stratified random*: The sample is selected by dividing the population into regions or strata, according to an auxiliary variable (e.g., elevation or the same thematic variables used in the classification). This method is more complex to design than the first one, but it gives information on subpopulation categories while reducing the sampling error as long as the stratification variable is closely related to error distribution.
3. *Systematic*: The cases are selected at regular intervals from a starting point randomly selected. It guarantees complete coverage of the target area but can generate erroneous estimations if errors are distributed systematically as well (e.g., geometry of parcels). On the other hand, it does not ensure a probabilistic estimation because there is no randomness in the selection of the observations.
4. *Systematic unaligned*: This modifies the previous scheme by changing randomly—for each line and column of the sampling—one coordinate, keeping the other one fixed. This allows certain randomness in the selection of the samples, while it reduces the bias due to regularity. At the same time, it ensures complete coverage of the territory.
5. *Cluster sampling*: A group of observations, known as a *cluster*, is selected as a sampling unit instead of using single pixels. In other words, on each randomly selected point to be verified, several samples are collected in the neighboring points, following a predetermined pattern. For example, a cluster can consist of five points at a certain distance from the central point and forming an L shape. This method reduces the cost of transportation, but it is more complex and less accurate than the previous ones.

Of these schemes, systematic unaligned sampling has been widely used in classification verification tasks (Dozier and Strahler 1983; Rosenfeld et al. 1982;

Stehman 1992). Other authors recommend alternative methods such as the simple random (Hord and Brooner 1976), stratified sampling by categories (Ginevan 1979), and cluster sampling (Todd et al. 1980). The experimental works by Congalton and collaborators (Congalton 1984, 1988a, 1991) provided a more rigorous evaluation of different sampling schemes. On the basis of errors associated with classification of three different types of landscapes (agricultural lands, forests, and grasslands), simple random sampling was found to be the most effective method, followed by the stratified random sampling. Cluster sampling also showed good results, being especially useful in verifying large areas, where the cost of transportation is critical. The use of systematic or systematic unaligned sampling is not recommended, especially when the image presents a systematic error, as is the case with agricultural zones or pastures.

In a recent study to validate global burned area maps, Padilla et al. (2014) used stratified random sampling to select the set of reference sites. In this case, two criteria were used to select the strata: the biomes and the relevance of fire occurrence. Two categories were discriminated: high and low occurrence based on burned area maps from 2008, resulting in 14 strata (7 biomes × 2 fire occurrence conditions). Sample size for each stratum was selected proportionally to fire occurrence. Sampling units were the Thiessen scene areas (TSAs: Cohen et al. 2010) instead of the Landsat frames, since they do not overlap, and therefore, the selection is not biased toward Northern latitudes with higher repeating overpasses.

### 8.4.4 Sample Size

As in any statistical sampling, the sample size depends on the confidence level to be assigned to the estimation, the tolerable estimation error, and the variability of the target variable itself. The sample size will increase as the level of confidence increases, the margin of error decreases, and the target variable becomes more heterogeneous.

When validating a quantitative variable, such as plant water content, sea surface temperature, or water turbidity, the sample size can be estimated by using the general sampling inference equation as follows:

$$\bar{x} = \hat{x} \pm z \times SE \tag{8.1}$$

where

$\bar{x}$ is the mean of the population (in this case, the actual error in the image)
$\hat{x}$ is the estimated error from the sampling
$z$ is the $x$-axis of the standard normal curve (the level of probability) assigned to the estimation
$SE$ is the sampling error, calculated differently for each type of sampling

The term ($z \times SE$) indicates the confidence interval, that is, the value below and above the mean at which one can expect to find the true value of the variable with a certain level of confidence. To calculate the sample size, we need to fix the confidence level in accordance with the desired precision of the sampling. Assuming that maximum acceptable error is $L$, then for a simple random sampling,

$$L^2 = z^2 \times \frac{s^2}{n}\frac{(N-n)}{N} \tag{8.2}$$

where
$s^2$ is the sampling variance
$N$ is the size of the population
$n$ is the sample size

For large samples ($n > 30$), the last term (known as *correction for finite populations*) can be ignored. From the previous equation, the sample size can be computed as follows:

$$n = \frac{z^2 s^2}{L^2 + (s^2 z^2 / N)} \tag{8.3}$$

The most difficult parameter to estimate in this equation is the sampling variance ($s^2$), since before performing the sampling, the variability of the target variable is unknown. As a first approximation, $s$ can be estimated from previous samplings of the same variable or from a variable range ($r_g$). Different authors have proposed $s = 0.29 \times r_g$ for a regular distribution and $s = 0.21 \times r_g$ for a biased one. For example, if we want to estimate the precision of calculating the temperature over an area of 5000 km$^2$ with a level of probability of 95% ($z$ = 1.96) and an accepted estimated error of 1.5°C, having observed a range in temperatures of 20.69°C (therefore, $s = 0.29 \times 20.69 = 6$), we would have

$$n = \frac{1.96^2 \times 6^2}{1.5^2 + ((6^2 \times 1.96^2)/5000)} = 60.72$$

which means that we have to sample 61 points of 1 km$^2$ area to obtain the expected level of confidence.

In the case of a classified image, where the variable is not quantitative but categorical, it is advised that the bimodal distribution of probability be used. The simplest expression (without considering the correction for finite populations) would be as follows:

$$n = \frac{z^2 pq}{L^2} \tag{8.4}$$

where
$p$ is the estimated percentage of successes
$q$ is the percentage of errors ($q = 1 - p$)
$L$ is the level of error allowed

The values of $p$ and $q$ can be estimated from auxiliary data, or they can be simply made equal to 50.

For example, if we want to know the required number of test points to estimate with a 95% probability the error of a certain map, supposing that the estimated success is 85% and the maximum error allowed is ±5%, we would have

$$n = \frac{1.96^2 \times 85 \times 15}{5^2} = 195.92$$

which indicates that at least 196 samples would have to be taken to achieve the accuracy levels established. This size can be very small if we refer to the total number of pixels in the image. We have to consider that in this application, we need not follow the classic rules of statistics because the reference number of pixels (total area classified) is very high. As a result, some authors suggest a minimum number of samples to obtain a reliable estimation of the error. This threshold is located at 50 pixels per thematic class (Congalton 1991; Hay 1979). It is also recommended to carry out the sampling separately for each class, starting with the class having the smallest size. This will mark the proportion of area to be sampled for the rest of the classes.

## 8.5 GATHERING INFORMATION

Once the sampling design is set, the next step is acquiring the data from the ground or from a reference document (typically, a higher-resolution image). Ground measurement or direct observation requires fieldwork to be done simultaneously with the collection of the image. This is especially critical for very dynamic variables (e.g., surface temperature or soil moisture), while in other cases (e.g., vegetation species), a temporal discrepancy can be accepted.

When measuring biophysical variables, the ground sensors should be well calibrated to avoid errors in estimation. Additionally, the protocol for data measurement should be standardized when several sampling teams are working on the same project. For instance, direct measurement of LAI on the ground is very complex and costly, and therefore, ground methods are based on sampling. Indirect methods rely on measuring the light passing through the canopy and generate a model of leaf distribution to estimate the number of leaves present (Figure 8.9). Since this is an estimation, the comparison with satellite-estimated LAI is not always fully reliable (Bréda 2003; Wang et al. 2004).

Another problem is the spatial uncertainty of locating the exact area observed by the sensor's instantaneous field of view (IFOV). In other words, we do not know exactly the limits of the area observed by the sensor, and therefore, it is difficult to measure the same area included in the image. An alternative is to use field radiometers and take simultaneous measurements of reflectance and the target parameter on the ground.

In the case of validating classified maps, the interpreter on the field has to apply the same definition of thematic categories used to classify the image. As previously mentioned, this is crucial when several people work on the ground (Congalton and

**FIGURE 8.9** Field sensors for collecting LAI information.

Green 1999). An adequate coordination can avoid subjectivities in assigning classes, especially when they are heterogeneous: crop mosaics, pasture-scrub, low-dense vegetation, etc. These terms are often too vague to collect adequate ground information.

An alternative to fieldwork, particularly for continental or global studies, is the extraction of reference data from higher-resolution images. In the case of quantitative variables, it is important that the sensors be well calibrated and have good temporal resolution, in order to extract the data closest to the time of image acquisition.

Regarding which image value should be compared with the reference value, many authors recommend the use of average or median values of a window around the sample point (usually, a 3 × 3 window) to reduce the impact of small georeferencing errors in the image. In the case of classified images, the extraction is done from the sampling point, but it is more common to extract the entire polygon observed in the field and not just single pixels, which are more difficult to locate.

## 8.6 VALIDATING INTERVAL-SCALE VARIABLES

Once the reference data are obtained, we can compute several statistical metrics to estimate the error of our results. For interval-scale variables, the most common index to estimate errors is the root-mean-square error (*RMSE*), defined as

$$RMSE = \sqrt{\frac{\sum_{i=1,n} \left(\hat{x}_i - x_i\right)^2}{n}} \tag{8.5}$$

where

$\hat{x}_i$ is the estimated value from the image
$x_i$ is the reference value

As previously mentioned, *RMSE* will really be an estimation of the error present in a variable since it has been obtained by sampling. We should apply the sampling inference equation (Equation 8.1) to estimate the confidence interval around the estimated error.

A rigorous estimation of the error requires points that were not used in calibrating the model. In some cases, only a few ground observations are available; therefore, splitting the observations into one set for validation and another set for calibration may imply having few cases in both sets. In these situations, a cross-validation procedure can be implemented. The empirical function is calibrated several times, leaving out one observation at a time ("jackknife" is the term commonly used for this resampling procedure). This case is used for validation of the fitted model. The process is repeated as many times as there are observations minus one (McGwire 1996). An average error is then computed as follows:

$$RMSE^{*} = \sqrt{\frac{\sum_{i=1,n}\left(f(\hat{x}_i) - x_i\right)^2}{n}} \tag{8.6}$$

where $f(\hat{x}_i)$ is the fitting function without using case $i$.

## 8.7 VALIDATING CLASSIFIED IMAGES

### 8.7.1 CONFUSION MATRIX

In the case of evaluating a classified image, the result of collecting the ground data will be a list of observations with their classified (satellite) and reference (ground) categories. From this list, we can create the "confusion matrix," which is a cross-tabulation of assignments done from the image versus the real assignments of the same classes. The confusion matrix reflects the agreements and disagreements between the classification and reference sources. Usually, the columns indicate the reference classes, and the rows the classification categories. Logically, both will have the same number of observations and meaning; it is a square matrix: $n \times n$, where $n$ is the number of categories (Figure 8.10). The diagonal of this matrix represents the sampled points at which there is agreement between both sources (image outputs and reference), whereas the marginal ones represent classification errors.

The relationship between the number of pixels correctly classified and the total number of pixels sampled expresses the global accuracy of the map. The column residuals indicate true assignments that were not identified in the map, while the row residuals indicate classified pixels that do not correspond to reality. They represent the omission and commission errors, respectively (Aronoff 1982; Story and Congalton 1986). The confusion matrix shows not only the general accuracy of the process but also the accuracy of each category and the conflicts between categories.

Tables 8.1 and 8.2 include two confusion matrices derived from a fuel-type classification using two different algorithms. The area covers around 500 km$^2$ in Central

| | | Reference | | | | | | |
|---|---|---|---|---|---|---|---|---|
| | | Class 1 | Class 2 | Class 3 | Class *n* | Total | User's accuracy | Commission error |
| Classification | Class 1 | $X_{11}$ | | | | $X_{1+}$ | $X_{11}/X_{1+}$ | $1-X_{11}/X_{1+}$ |
| | Class 2 | | $X_{22}$ | | | $X_{2+}$ | $X_{22}/X_{2+}$ | $1-X_{22}/X_{2+}$ |
| | Class 3 | | | $X_{33}$ | | $X_{3+}$ | $X_{33}/X_{3+}$ | $1-X_{33}/X_{3+}$ |
| | Class *n* | | | | $X_{nn}$ | $X_{n+}$ | $X_{nn}/X_{n+}$ | $1-X_{nn}/X_{n+}$ |
| | Total | $X_{+1}$ | $X_{+2}$ | $X_{+3}$ | $X_{+n}$ | $\Sigma X_{ij}$ | | |
| | Producer's accuracy | $X_{11}/X_{+1}$ | $X_{22}/X_{+2}$ | $X_{33}/X_{+3}$ | $X_{nn}/X_{+n}$ | | | |
| | Omission error | $1-X_{11}/X_{+1}$ | $1-X_{22}/X_{+2}$ | $1-X_{33}/X_{+3}$ | $1-X_{nn}/X_{+n}$ | | | |

**FIGURE 8.10** Structure of a confusion matrix.

Spain, and the fieldwork was based on 144 sample plots. More than 9500 pixels were included in the validation process, collecting a minimum of 50 pixels per category. Fuel maps were generated from a digital classification of a Landsat TM image in the context of a fire-risk mapping project (Chuvieco and Salas 1996). The fieldwork was carried out within a few months of the satellite acquisition. Fuel types refer to groups of vegetation species classified according to their fire propagation potential. The numbers refer to the fuel models developed by the U.S. National Forest Fire Laboratory in Missoula, Montana (Burgan and Rothermel 1984). The main confusions observed in Tables 8.1 and 8.2 are the overlapping of different types of scrub (fuel models 4, 5, and 6 are separable only by their height; it is a complex task to discriminate them in Landsat images); the difficult discrimination of the forested scrubland (model 7, only present in the classification Table 8.1); and the confusion between scrub, pasture, and the mixture between forest and high dense scrub.

## 8.7.2 Global Accuracy

A series of accuracy metrics can be derived from the confusion matrix. The simplest one is a ratio of the diagonal cells (those samples well classified) to the total number of validation samples. The index is called the *overall accuracy* (*OA*), defined as

$$\widehat{OA} = \frac{\sum_{i=1,n} x_{ii}}{\sum_{i=1,n} \sum_{j=1,n} x_{ij}} \tag{8.7}$$

where

$x_{ii}$ is the diagonal of each column
$x_{ij}$ is any cell in the confusion matrix

**TABLE 8.1**
**Confusion Matrix for a Fuel-Type Classification Map Using Nonsupervised Algorithms**

| | Reference | | | | | | | | | | | |
|---|---|---|---|---|---|---|---|---|---|---|---|---|
| **Classification** | **1** | **2** | **4** | **5** | **6** | **7** | **9** | **A** | **B** | **Total** | **User's Accuracy** | **Commission Error** |
| **1** | 551 | 4 | | 17 | 12 | 48 | 23 | 128 | | 783 | 70.4 | 29.6 |
| **2** | 71 | 543 | 190 | | 193 | 24 | 75 | 75 | | 1171 | 46.4 | 53.6 |
| **4** | 57 | 75 | 209 | 3 | 67 | 55 | 145 | 17 | | 628 | 33.3 | 66.7 |
| **5** | 7 | 3 | 2 | 525 | 49 | 2 | 19 | 9 | | 616 | 85.2 | 14.8 |
| **6** | 207 | 31 | 57 | | 273 | 3 | 5 | | | 576 | 47.4 | 52.6 |
| **7** | | | | | | | | | | | 0.0 | 100 |
| **9** | | 116 | 751 | 46 | 107 | 393 | 2507 | 9 | | 3929 | 63.8 | 37.2 |
| **A** | 255 | 61 | 31 | 1 | 31 | 57 | 84 | 1145 | | 1665 | 68.7 | 31.3 |
| **B** | | | | | | | | | 147 | 147 | 100 | 0 |
| **Total** | 1148 | 833 | 1240 | 592 | 732 | 582 | 2858 | 1383 | 147 | 9515 | | |
| Producer's Accuracy | 48.0 | 65.2 | 16.9 | 88.7 | 37.3 | 0.0 | 87.7 | 82.8 | 100.0 | | | |
| Omission Error | 52.0 | 34.8 | 83.1 | 11.3 | 62.7 | 100.0 | 12.3 | 17.2 | 0.0 | | | |

*Note:* Numbers 1–9 refer to the fuel models defined by the Spanish Forest Service: 1, pastures; 2, wooded pastures; 4, dense and tall scrubs; 5, thin scrubs; 6, medium height scrubs; 7, wooded low scrubs; 9, tree cover with clean underbrush. Class A identifies the agricultural and urban zones, and B water bodies.

**TABLE 8.2**

**Confusion Matrix for a Fuel-Type Classification Map Using Supervised Algorithms**

| | Reference | | | | | | | | | | | |
|---|---|---|---|---|---|---|---|---|---|---|---|---|
| **Classification** | **1** | **2** | **4** | **5** | **6** | **7** | **9** | **A** | **B** | **Total** | **User's Accuracy** | **Commission Error** |
| **1** | 927 | 49 | 63 | | 42 | | 5 | 87 | | 1173 | 79 | 21 |
| **2** | 35 | 575 | 74 | | 70 | 6 | 34 | 58 | | 852 | 67.5 | 32.5 |
| **4** | 6 | 7 | 812 | | 39 | 104 | 145 | | | 1113 | 73 | 27 |
| **5** | | | | 590 | 19 | | 1 | | | 610 | 96.7 | 3.3 |
| **6** | 11 | 192 | 77 | | 489 | 1 | 6 | | | 776 | 63 | 37 |
| **7** | | 2 | 11 | | | 93 | 51 | | | 157 | 59.2 | 40.8 |
| **9** | 99 | 5 | 181 | | 59 | 329 | 2273 | | | 2945 | 77.2 | 22.8 |
| **A** | 70 | 3 | 22 | 2 | 14 | 49 | 343 | 1238 | | 1741 | 71.1 | 28.9 |
| **B** | | | | | | | | | 147 | 147 | 100 | 0 |
| **Total** | 1148 | 833 | 1240 | 592 | 732 | 582 | 2858 | 1383 | 147 | 9515 | | |
| Producer's Accuracy | 80.7 | 69.0 | 65.5 | 99.7 | 66.8 | 16.0 | 79.5 | 89.5 | 100.0 | | | |
| Omission Error | 19.3 | 31.0 | 34.5 | 0.3 | 33.2 | 84.0 | 20.6 | 10.5 | 0.0 | | | |

For the data included in our example confusion matrices, the estimated global accuracy is 62% (Table 8.1) and 75% (Table 8.2). To estimate the actual *OA*, the confidence levels need to be computed. From Equation 8.1, we have

$$OA = \widehat{OA} \pm z \times SE \tag{8.8}$$

The standard error for this estimation can be generated from the simple random sampling equation:

$$SE = \sqrt{\frac{pq}{n}} \tag{8.9}$$

For the matrix in Table 8.2, the number of pixels correctly assigned was 5900 out of 9515 in the sample. This indicates an estimated global accuracy of 62%. The standard error would be as follows:

$$SE = \sqrt{\frac{62 \times 38}{9515}} = 0.497$$

With a level of significance of 0.05 (95% probability), the confidence interval can be calculated as follows:

$$A_g = 62 \pm 1.96 \times 0.497 = 62 \pm 0.975(\%)$$

Thus, the actual global accuracy would be between 61.02% and 62.98% with a 95% probability. In case of using other sampling strategies, the formulas to compute the *SE* would differ. For instance, in the stratified random sample, the relative importance of each stratum in both the sample and the population need to be taken into account (Padilla et al. 2014).

### 8.7.3 User and Producer Accuracy

Overall accuracy is an initial estimation of the performance of results, but it can obscure important differences between categories. A more rigorous analysis should also consider the marginal cells of the matrix. In the case of columns, the margins indicate the number of pixels that, in spite of belonging to a given category, were not assigned to that category. The proportion of these unassigned pixels versus the total number of pixels that are in that category defines the omission error (*OE*). For each class, it is calculated as follows:

$$OE_i = \frac{X_{+i} - X_{ii}}{X_{+i}} \tag{8.10}$$

where $X_{+i}$ represents the total of column *i*.

Similarly, the marginal cells in the rows express the commission errors (*CE*), that is, pixels that are included in a given category when they are not in fact part of that category:

$$CE_i = \frac{X_{i+} - X_{ii}}{X_{i+}} \tag{8.11}$$

where $X_{i+}$ represents the total of row *i*.

*OE* and *CE* express two approaches to the same problem. The former refers to an underestimation of a particular category. The latter is caused by an overestimation. In terms of accuracies, these two types of error can be related to two different types of accuracies: *user accuracy* (*UA*) and *producer accuracy* (*PA*) (Aronoff 1982; Story and Congalton 1986). The first is inversely related to commission errors, whereas the second is inversely related to omission errors:

$$PA_i = \frac{X_{ii}}{X_{+i}} \tag{8.12}$$

and

$$UA_{u,i} = \frac{X_{ii}}{X_{i+}} \tag{8.13}$$

for each of the *n* categories of the classification. The importance of this distinction can be shown with a simple example. If the omission error for a particular forest species is low, there is a high probability that the area actually covered by that species is included as such in the map. This was the main objective of the person who produced the map. However, simultaneously a high commission error for the same category may occur, which implies that many areas mapped as having that cover will in fact be covered by another category. Since the user of a map expects that everything included is well mapped, high commission errors imply low user accuracy.

In the case of the confusion matrix in Table 8.1, model 9 (tree cover with clean underbrush) offers a producer's accuracy of 87.7%. This means that approximately 9 out of 10 times the actual cover was correctly included in the classification. However, the user's accuracy is just over 60%, which means that only three-fifths of the area classified as such cover are really so. Here, the omission error is low but the commission error is medium high. Both measurements are complementary and very important. The first one is important from the perspective of the map producer and the second one from the user's perspective.

### 8.7.4 Kappa Statistic

This index measures the agreement between map and reference categories, removing the effects of random factors. The estimation of κ can be computed from the following equation (Hudson and Ramn 1987):

$$\hat{\kappa} = \frac{n\sum_{i=1,n} X_{ii} - \sum_{i=i,n} X_{i+}X_{+i}}{n^2 - \sum_{i=i,n} X_{i+}X_{+i}} \tag{8.14}$$

where

- $n$ is the sample size
- $X_{ii}$ indicates the observed agreement
- the product of the total row and column ($X_{i+}$, $X_{+i}$) estimates the expected agreement for each category $i$

In the case of the confusion matrices we have been analyzing in this section, the estimated values of κ are 0.53 (Table 8.1) and 0.70 (Table 8.2), which indicate that the classifications are 53% and 70% better than chance agreement, respectively. A value of $\kappa = 1$ indicates total agreement between map and reference, whereas values close to 0 suggest that the agreements observed are due to random effects. A negative value also indicates a poor classification due to other external factors, such as the size of the sample.

The kappa index can also be computed for specific classes (named *conditional* κ) or can weigh the importance of specific errors in the confusion matrix. This option may be useful for specific applications. For instance, in a forest inventory analysis, confusion between water and urban areas is much less relevant than that between water and coniferous species. The new κ value will be computed as follows:

$$C(\hat{\kappa}) = \frac{\sum_{i=1,n} w_{ii}X_{ii}/n - \sum_{i=i,n} w_{ij}X_{i+}X_{+i}/n}{1 - \sum_{i=i,n} w_{ij}X_{i+}X_{+i}/n} \tag{8.15}$$

where $w_{ij}$ indicates the weights assigned to the confusion of classes $i$ and $j$.

One of the most obvious applications of the κ index is to compare classifications obtained by different methods in order to test whether they differ significantly (Fitzegarld and Lees 1994). In this case, we can use a normal distribution to estimate the confidence intervals (Skidmore 1989):

$$z = \frac{\left|\hat{\kappa}_1 - \hat{\kappa}_2\right|}{\sqrt{\sigma_{\kappa 1}^2 + \sigma_{\kappa 1}^2}} \tag{8.16}$$

The variance of κ can be calculated from the following equation (Congalton and Green 1999):

$$\sigma_{\hat{\kappa}}^2 = \left(\frac{1}{n}\right)\left\{\frac{\theta_1(1-\theta_1)}{(1-\theta_2)^2} + \frac{2(1-\theta_1)(2\theta_1\theta_2 - \theta_3)}{(1-\theta_2)^3} + \frac{(1-\theta_2)^2(\theta_4 - 4\theta_2^2)}{(1-\theta_2)^4}\right\} \tag{8.17}$$

where $\theta_1$, $\theta_2$, $\theta_3$, and $\theta_4$ are calculated as follows:

$$\theta_1 = \left(\frac{1}{n}\right)\sum_{1=1,n} X_{ii} \tag{8.18}$$

$$\theta_2 = \left(\frac{1}{n^2}\right)\sum_{1=1,n} X_{i+}X_{+i} \tag{8.19}$$

$$\theta_3 = \left(\frac{1}{n^2}\right)\sum_{1=1,n} X_{ii}(X_{i+}+X_{+i}) \tag{8.20}$$

$$\theta_4 = \left(\frac{1}{n^3}\right)\sum_{1=1,n}\sum_{j=1,n} X_{ij}(X_{j+}+X_{+i})^2 \tag{8.21}$$

This test can be used to contrast the effective validity of different classification methods (Chuvieco and Congalton 1988; Fitzegarld and Lees 1994). For example, in the case of the two classification methods mentioned earlier (Tables 8.1 and 8.2), the estimated values of κ are 0.53 and 0.70, and the variances are 0.000034 and 0.0000282, respectively. Therefore, the $z$ value after applying Equation 8.16 will be 20.40. This value exceeds the one expected for a random scenario with a high confidence level (the threshold for 99% would be 2.56). Therefore, we can conclude that both classifications are significantly different, or, in other words, the second one is better than the first one.

Similar approaches can be computed for other metrics of accuracy, such as *OA*, *OE*, or *CE*, but the actual computation of metrics variance will be different depending on the sampling strategy and the actual metrics used (see Padilla et al. 2015 for an example of testing significant differences in the accuracy of global burned-area products).

### 8.7.5 Normalizing the Confusion Matrix

When comparing two confusion matrices—generated from different assignation methods, different images, or different individuals—κ can indicate which of them is the most effective as a whole. However, when comparing the reliability of two maps of different sample sizes, the statistical κ does not offer an adequate comparison because it depends on the sample size. In this context, Congalton (1984) proposed the application of a multivariate procedure to standardize a square matrix (Bishop et al. 1975). It is an iterative method to adjust the total number of rows and columns to a fixed value (+1) through successive increments or reductions of the cells of the matrix. The process stops when the margins of each row and column add up to +1.00, or close to it. Tables 8.3 and 8.4 show the normalized matrices of the previous analysis.

**TABLE 8.3**
**Normalization Matrix Corresponding to Table 8.1**

| | Reference | | | | | | | | |
|---|---|---|---|---|---|---|---|---|---|
| **Classification** | **1** | **2** | **4** | **5** | **6** | **7** | **9** | **A** | **B** |
| **1** | 0.528 | 0.010 | 0.001 | 0.029 | 0.021 | 0.194 | 0.038 | 0.180 | 0.000 |
| **2** | 0.030 | 0.527 | 0.166 | 0.000 | 0.137 | 0.042 | 0.052 | 0.046 | 0.000 |
| **4** | 0.044 | 0.137 | 0.340 | 0.005 | 0.089 | 0.179 | 0.187 | 0.020 | 0.000 |
| **5** | 0.007 | 0.008 | 0.005 | 0.847 | 0.080 | 0.010 | 0.030 | 0.013 | 0.000 |
| **6** | 0.231 | 0.082 | 0.135 | 0.001 | 0.523 | 0.016 | 0.010 | 0.001 | 0.001 |
| **7** | 0.058 | 0.137 | 0.123 | 0.101 | 0.100 | 0.244 | 0.097 | 0.085 | 0.054 |
| **9** | 0.000 | 0.034 | 0.199 | 0.010 | 0.023 | 0.207 | 0.525 | 0.002 | 0.000 |
| **A** | 0.099 | 0.056 | 0.026 | 0.001 | 0.021 | 0.093 | 0.055 | 0.649 | 0.000 |
| **B** | 0.003 | 0.008 | 0.007 | 0.006 | 0.006 | 0.014 | 0.006 | 0.005 | 0.944 |

**TABLE 8.4**
**Normalization Matrix Corresponding to Table 8.2**

| | Reference | | | | | | | | |
|---|---|---|---|---|---|---|---|---|---|
| **Classification** | **1** | **2** | **4** | **5** | **6** | **7** | **9** | **A** | **B** |
| **1** | 0.804 | 0.040 | 0.031 | 0.001 | 0.042 | 0.000 | 0.002 | 0.080 | 0.001 |
| **2** | 0.046 | 0.692 | 0.054 | 0.001 | 0.103 | 0.006 | 0.016 | 0.080 | 0.001 |
| **4** | 0.010 | 0.011 | 0.708 | 0.002 | 0.069 | 0.115 | 0.083 | 0.001 | 0.001 |
| **5** | 0.000 | 0.000 | 0.000 | 0.978 | 0.019 | 0.000 | 0.001 | 0.001 | 0.001 |
| **6** | 0.015 | 0.225 | 0.055 | 0.001 | 0.698 | 0.001 | 0.003 | 0.001 | 0.001 |
| **7** | 0.005 | 0.024 | 0.066 | 0.010 | 0.006 | 0.680 | 0.194 | 0.005 | 0.009 |
| **9** | 0.074 | 0.004 | 0.076 | 0.001 | 0.050 | 0.174 | 0.621 | 0.000 | 0.001 |
| **A** | 0.045 | 0.002 | 0.008 | 0.003 | 0.010 | 0.022 | 0.080 | 0.830 | 0.001 |
| **B** | 0.002 | 0.002 | 0.001 | 0.004 | 0.002 | 0.001 | 0.001 | 0.002 | 0.986 |

This process offers a new measure of global reliability. It is sufficient to calculate the mean value of the elements in the diagonal still indicating the agreement between rows and columns. Obviously, the ideal case would be to have the diagonal values equal to 1. This would indicate a perfect agreement between reference and map, whereas very low values in the diagonal would indicate a poor classification.

In this case, the average value of the diagonal offers an average percentage of accuracy between 57% and 78%. The values contributed by the original tables are better clarified by classes, especially in the case of perfectly estimated agreements, for example, agricultural areas that range from 100% accuracy to values close to 94%.

It is convenient to consider that these measurements represent a low estimation of reliability because of the characteristics of the normalization process itself. It is

**TABLE 8.5**
**Comparison of the Validation Metrics of Several Digital Classifications of Fuel Types**

| Classification | *OA* | κ Index | Normalized Accuracy |
|---|---|---|---|
| Nonsupervised (TM + ancillary data) | 62 | 53 | 57 |
| Nonsupervised (TM) | 60 | 50 | 56 |
| Supervised | 62 | 54 | 60 |
| Mixed | 75 | 70 | 78 |

*Source:* Salas, F.J. and Chuvieco, E., *Revista de Teledetección*, 5, 18, 1995.

necessary to view the confusion matrix as a very particular case of a contingency table, since it is common to find cells with very low or zero values because there are very unlikely confusions between categories. The normalization process adds a very small amount to each cell to avoid zeros; thus, the cells that express success are underrated with respect to the rest. This is the case of the fuel model 5, with 99.7% reliability in the original table (mixed classification test) but reduces to 97% in the normalized version, as a consequence of the small size of the sample.

For this problem, the normalization process offers a complementary perspective to the one obtained in the original matrix. It provides a simultaneous estimation of global and category accuracies and makes possible the comparisons of matrices with different numbers of samples.

Table 8.5 shows a summary of the reliability values, according to the criteria discussed in this chapter for analyzed matrices.

### 8.7.6 Validation of Binary Classes

In the case of assessing the results of classifying just two categories (forested/unforested, burned/unburned, cloudy/cloud free, etc.), additional analysis can complement the measures of confusion matrices. When using a high-resolution image to assess the accuracy of a product generated with a sensor with coarser resolution, the confusion matrix includes the effects of the different pixel sizes (Boschetti et al. 2006; Morisette et al. 2005). Some of the areas classified as errors may be, in fact, the product of different generalizations that imply various spatial resolutions (Figure 8.11). To reduce this effect, an alternative method of assessing a binary product is based on comparing the area proportions covered by the target category in the high- and low-resolution document. The total study area is divided into grids of 5–10 times the size of the coarse-resolution image. Then, we register for each cell the proportion covered by the two sources. Finally, a scatter graph is created with these two proportions in all cells, and a regression model can be computed to find out whether there is a consistent over- or underestimation of the target variable (Figure 8.12).

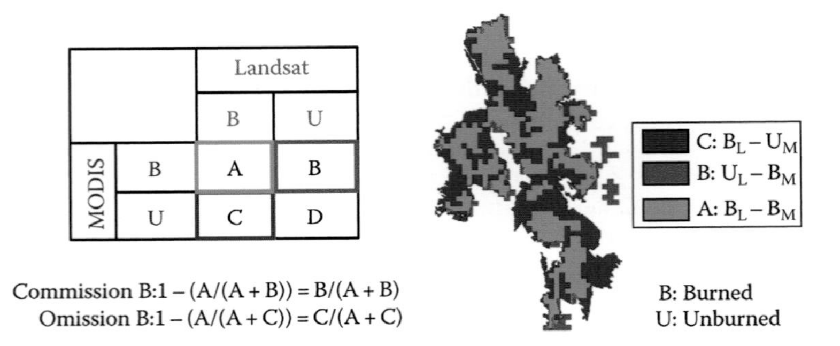


**FIGURE 8.11** Example of cross-tabulation of a burned land map produced by two images with different spatial resolutions. (Courtesy of Patricia Oliva.)

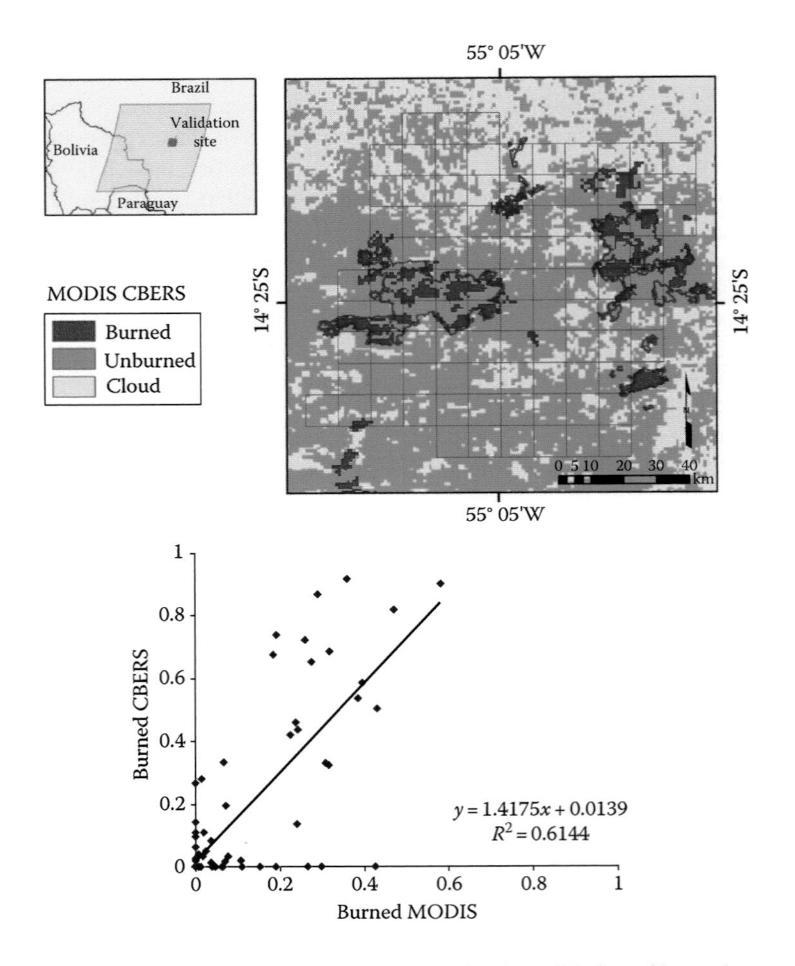


**FIGURE 8.12** Example of accuracy scatter graph for the validation of burned area. (From Chuvieco, E. et al., *Ecol. Appl.*, 18, 64, 2008b.)

### 8.7.7 Verification in Multitemporal Analysis

The validation of multitemporal analysis is especially challenging, as it complicates the comparison of reference and classified data with an additional time dimension. This implies errors associated with proper overlapping between the two images, as well as the difficulty of gathering reference information for historical periods. Summarizing these aspects, we can name the main differences in verification applied to change detection studies:

1. Increase in the number of categories in the error matrix. The potential number of categories ("stable," "from," and "to") for two dates is the square of the number of categories in one date. All these classes must be considered during error evaluation, and at the end the confusion matrix will have $n^4$ cells, $n$ being the starting number of classes for one date (Table 8.6).
2. More complexity in the sampling scheme, due to the great increase in number of categories to be verified. Since some of the potential changes will be uncommon or impossible (e.g., two urban areas to water bodies), we have to considerably increase the sample size to verify these intersections. In these cases, stratified sampling or one based on auxiliary information can be a better alternative (Biging et al. 1998).
3. Difficulty in obtaining the reference information for past dates. In most cases, we will not have detailed information of terrestrial covers for distant past dates, which complicates the assessment of the oldest image. Aerial photography or cartographic documents of similar dates can solve this problem, although they may not be always reliable.

**TABLE 8.6**
**Confusion Matrix for Change Detection Analysis**

| | | Reference | | | | | | | | |
|---|---|---|---|---|---|---|---|---|---|---|
| | | Stable | | | Change | | | | | |
| **Classification** | | AA | BB | CC | AB | AC | BA | BC | CA | CB |
| **Stable** | AA | 1 | 3 | 3 | 5 | 5 | 5 | 5 | 5 | 5 |
| | BB | 3 | 1 | 3 | 5 | 5 | 5 | 5 | 5 | 5 |
| | CC | 3 | 3 | 1 | 5 | 5 | 5 | 5 | 5 | 5 |
| **Change** | AB | 4 | 4 | 4 | 2 | 6 | 6 | 6 | 6 | 6 |
| | AC | 4 | 4 | 4 | 6 | 2 | 6 | 6 | 6 | 6 |
| | BA | 4 | 4 | 4 | 6 | 6 | 2 | 6 | 6 | 6 |
| | BC | 4 | 4 | 4 | 6 | 6 | 6 | 2 | 6 | 6 |
| | CA | 4 | 4 | 4 | 6 | 6 | 6 | 6 | 2 | 6 |
| | CB | 4 | 4 | 4 | 6 | 6 | 6 | 6 | 6 | 2 |

*Source:* Adapted from Congalton, R.G. and Green, K., *Assessing the Accuracy of Remotely Sensed Data: Principles and Applications*, Lewis Publishers, Boca Raton, FL, 1999; Biging, G. et al., *Remote Sensing Change Detection. Environmental Monitoring Methods and Applications*, Ann Arbor Press, Chelsea, MI, 1998.

*Note:* A, B, C are different thematic categories. See text for explanation.

When we attempt to verify change detection, the confusion matrix gets notably complicated. To illustrate this process, we will use a simple example of validation: a change detection product (taken from Biging et al. 1998). Let us assume that we have three classes in two different dates. This would imply nine possible categories of change. To verify these transitions, we would need a confusion matrix of 81 cells (the 9 true × the 9 estimated changes), which includes both the permanent classes (those with the same category between dates) as well as the dynamic classes. All these potential transitions/permanencies require verification. Therefore, the confusion matrix will include the tabulation of pixels, which our map of changes presents as "from class X to class Y," while in reality they may be that or some other kind of transition. As in the case of simple confusion matrices, the diagonal will include the true detections of change (those sampled pixels that were correctly detected as stable or as affected by a real change), and outside the diagonal the incorrect changes will be present.

By being more specific, we can better explain the structure of a confusion matrix using the notation described by Biging et al. (1998) in Table 8.6. The cells annotated with 1 are estimated stable areas that really are constant, while the cells with 2 are accurately classified as changes. The addition of 1 and 2 indicates the global accuracy of the change detection performed by the interpreter. The rest are errors due to several possibilities: 3 indicates stable areas incorrectly classified to other stable categories; 4 indicates changes when the cells maintained their category (i.e., commission errors, identified as a change for something did not); 5 marks real changes that were not detected as such (i.e., omission errors); and 6 denotes inaccurately identified changes (detected changes, but assigned to an another class, i.e., omission or commission errors, depending on the class considered).

## 8.8 REVIEW QUESTIONS

1. Which one of the following options would you consider more reliable for assessing a digital classification of crops?
   a. Comparison with crops statistics
   b. Comparison with training fields
   c. Comparison with test fields
   d. Comparison with previous crop maps
2. Which one of the following sampling strategies is more appropriate to assess the classification of a remote area?
   a. Cluster sampling
   b. Systematic sampling
   c. Stratified sampling
   d. Random sampling
3. Which one of the following sampling strategies is more appropriate to assess the classification of an easily accessible area?
   a. Cluster sampling
   b. Systematic sampling
   c. Stratified random sampling
   d. Simple random sampling

4. Which one of the following metrics is not used with classified images?
   a. Global accuracy
   b. Residual mean square
   c. User's accuracy
   d. Producer's accuracy
5. User's accuracy is complementary to
   a. Omission errors
   b. Commission errors
   c. Global errors
   d. None of the above
6. Producer's accuracy is complementary to
   a. Omission errors
   b. Commission errors
   c. Global errors
   d. None of the above
7. To test whether two classifications have significant differences, we can compute
   a. The *RMSE*
   b. The normalized accuracy
   c. The global accuracy
   d. The variance of kappa
8. Normalized error matrices are useful to
   a. Compare classifications
   b. Reduce omission errors
   c. Compare classes
   d. None of the above
9. The assessment of a change detection map requires
   a. Evaluation of errors of dynamic areas
   b. Evaluation of errors of static areas
   c. Evaluation of errors of both
   d. Evaluation of errors in the first date
10. The main obstacle in multitemporal accuracy assessment is
    a. Estimating errors for the first date
    b. Complexity
    c. Matrix size
    d. Estimating errors for the second date

# 9 Remote Sensing and Geographic Information Systems

The final objective of most users of satellite remote sensing (RS) is to combine their results with other geographic variables for different purposes: forest inventory, agricultural suitability, natural hazard assessment, analysis of climate patterns, etc. In this regard, image interpretation is not the final goal of the user but just a source of input information in the framework of an integrated spatial analysis. This integration is carried out with geographical information systems (GIS) technologies. GIS can be defined as a set of programs that store, manage, manipulate, and represent data with some kind of spatial component (Bolstad 2008; Burrough and McDonell 1998; Longley et al. 2005). This geographically referenced information includes maps, statistics, or climatic data (Figure 9.1). With a common location, all these variables can be mutually related in diverse ways. Since the information is stored in digital format, a GIS takes advantage of a very diverse range of computer analytical capabilities, facilitating multiple operations that are impossible to perform by conventional means: cartographic generalization, path analysis, variable overlaying, slope, aspect, or visibility calculations, neighbor analysis, etc.

The wide availability of GIS manuals (Aronoff 1989; Bolstad 2008; Burrough and McDonell 1998; Clarke 2003; IDRISI 1992; Jensen and Jensen 2012; Juppenlatz and Tian 1996; Longley et al. 2005; Maguire et al. 1991; Star and Estes 1990; Tomlin 1990) makes it unnecessary to detail the fundamentals of this technique. Therefore, we will concentrate on presenting the relationships between GIS and RS technologies, focusing on the process of transferring the information from one to the other.

## 9.1 TRENDS IN GIS AND REMOTE SENSING DEVELOPMENT

RS and GIS followed, to some extent, a parallel trajectory during their early stages of development. Both techniques appeared as independent disciplines, connected only by their environmental approach. The first GIS, designed in Canada in the 1960s, tried to facilitate environmental conservation by focusing on the integration of spatial data from different sources. However, how those input data were generated was considered a different field.

On the other hand, satellite RS interpretation was growing during the 1960s as a discipline oriented toward generating environmental information. It aimed to assist in the mapping and inventory of critical phenomena such as natural disasters or agricultural inventories. The development of digital interpretation techniques, especially

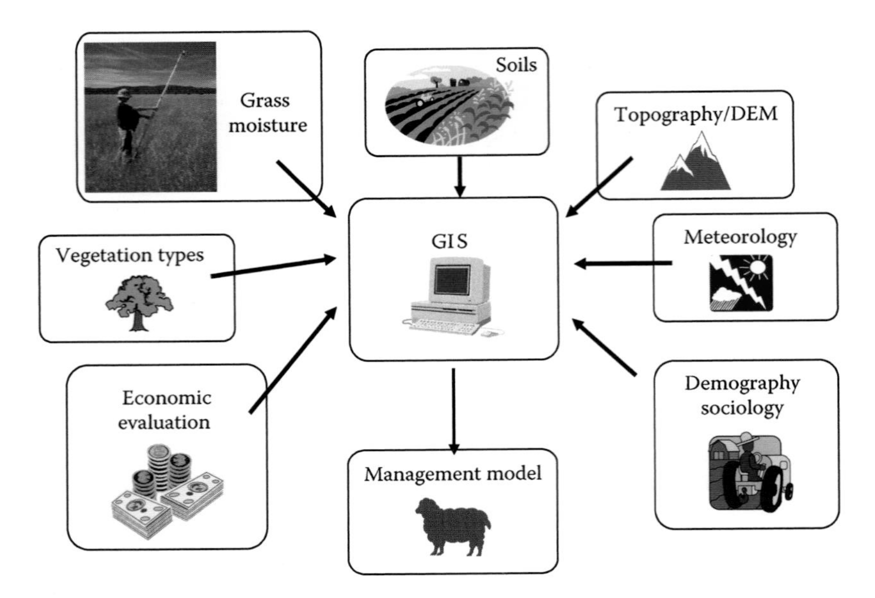


**FIGURE 9.1** Management model of spatial information for a grazing assessment.

after the launch of the first Landsat satellites in 1972, extended this approach and increased the connection of satellite data with other geographic variables. Initially, the integration of both techniques intended to support digital classification through the use of auxiliary data (Strahler et al. 1978). During the next few decades, the quick propagation of GIS in planning and academic institutions led to centering the management of spatial data around this technology. In the 1990s, RS was starting to be considered as an auxiliary technique of GIS, providing input variables to the spatial database. In those years, GIS textbooks started including chapters on RS essentials, although those focused more on urban or socioeconomic applications were skeptical about using satellite data within that context. The growing availability of high-spatial-resolution sensors at the beginning of this century has further blurred the borders between these technologies.

These days, GIS is considered to be the backbone of geographic information analysis since it can integrate all sources of spatial data. RS is considered in this context as just a source of information for a GIS, along with other techniques (global positioning systems [GPS], ground sensors, census, etc.). The multiple relations between all these techniques and the applications to which they are directed have reshaped the initial concept of GIS toward a more integrated view, which also includes the models and processes applied to geographical data. In this context, some authors favor renaming GIS as *geographic information science* (Longley et al. 2005; McMaster and Usery 2004). The increasing convergence between RS and GIS is justified, among others, for the following reasons:

1. Both have a clear territorial interest.
2. They have similar hardware and software requirements.

3. Many professionals commonly use both technologies.
4. They have similar research demands, as in the case of error analysis, the structure of and access to the data, the development of computer hardware, and the interpretation methodology (Star et al. 1991).

The integration of GIS and RS also has problems, mainly related to data availability (costs, level of spatial, spectral, and temporal resolution), education, and hardware costs (Lauer et al. 1991), but a trend toward the convergence of both technologies is evident.

GIS and RS interpretation require hardware and software that have many similarities, especially when GIS programs are designed for raster structures. Vector GIS processes information in a different way and requires specific input units (digitizer) that are unusual in digital image processing although most input data come from external databases or spatial data infrastructures (Granell et al. 2007).

## 9.2 GIS AS INPUT FOR RS INTERPRETATION

Throughout this book, we have emphasized the importance of auxiliary information to correct potential errors associated with image acquisition as well as to improve the classification of different categories. Digital elevation models (DEMs), soil or climatic maps, and property or administrative boundaries are examples of cartographic variables that can significantly benefit the digital analysis of images. This nonspectral information can be incorporated before, during, or after image classification, thereby greatly benefiting the discrimination in problematic categories.

Prior to classification, the auxiliary data can be valuable for geometric and radiometric preprocessing of images. For instance, a DEM is critical to correct topographic shades, which impact reflectance in rough terrains (see Section 6.7.3.3). Auxiliary data can also be beneficial to stratify the image in more homogeneous sectors (following bioclimatic or terrain characteristics). In this way, automatic classification may avoid potential errors by reducing the natural variability of some of the thematic categories (Rollins et al. 2004; Ustin and Xiao 2001).

During the classification stage, auxiliary data can be used to refine the location of training sites or to help the thematic labeling of unsupervised groups, particularly for global land cover mapping (Loveland et al. 2000). Auxiliary variables are valuable as new input bands for the classification, providing additional non-spectral information to discriminate certain classes. An example is the use of elevation, slope, or aspect to classify vegetation types (Figure 9.2), thereby reducing the effect of the local topography (Riaño et al. 2002a). In conventional classifiers, including topographic variables is straightforward, because they are measured on an interval scale. Other geographic variables, such as soils or biogeographic regions, are more complex to incorporate in statistical classifications. Alternative methods, such as decision trees or neural networks, should be used in those cases.

After the classification, auxiliary data may be very helpful for accuracy assessment or to refine some assignments that may not have been correctly achieved with spectral information (Hutchinson 1982). For example, some crops or forest species with similar spectral behavior could be discriminated on the basis of the types of soils on which they grow. Another interesting aspect refers to the boundary error, mentioned in Chapter 8. It is very common in transition areas where the definition

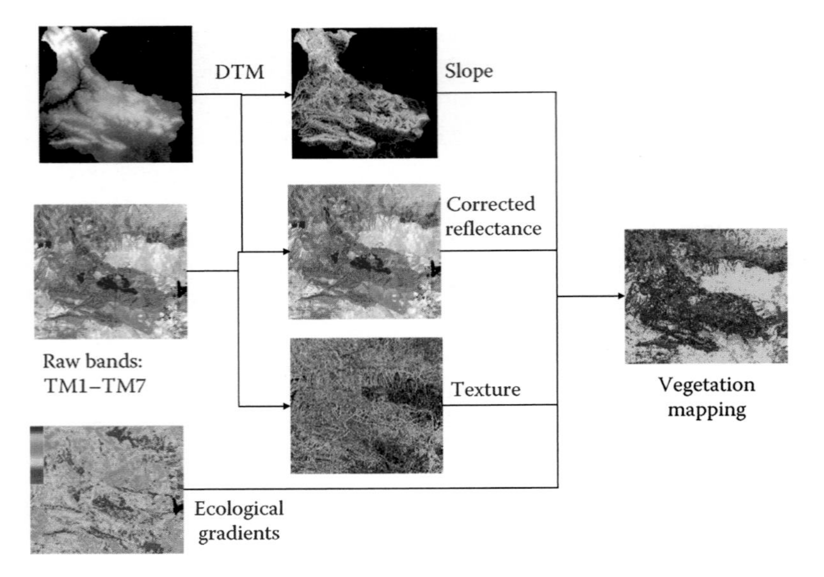


**FIGURE 9.2** Integrated classification with RS images and GIS variables for vegetation mapping. (Courtesy of David Riaño.)

of a pixel can be ambiguous since it is in fact the average reflectance of two or more covers. A solution to reduce the influence of this error would be to include the plot boundaries from a cadastral database, determining from it the edge pixels and modifying the assignment of those pixels to any of the boundary covers.

Finally, the display of results is a clear field of integrating auxiliary data and RS. The use of 3D perspectives or flight simulations provides a realistic view of landscape characteristics (Figure 9.3).

## 9.3 RS AS INPUT FOR GIS

### 9.3.1 Availability of Geographic Information

Within the acronym GIS, the second letter, denoting *information*, is the most important, since *geographic* only qualifies it and *system* can also refer to many other computer applications. Several authors have shown that the cost of generating and maintaining geographic information is the highest in establishing a corporate GIS (Lachowski et al. 1992) and certainly is the most relevant. When a GIS has unreliable, outdated, or incomplete information, it is useless in practical terms, regardless of the quality of the hardware and software it comprises. A GIS without high-quality information cannot provide good solutions to real problems. Albert Einstein used to say, “Information is not knowledge,” which is certainly true, but it is also true that good knowledge requires good information.

Satellite RS is not the only source of geographic information for a GIS, but it is one of the most beneficial, as it provides updated standardized data in digital format from a wide range of geographical variables. Given the high cost of generating

FIGURE 9.3 Three-dimensional view of Tucson and the Catalina mountains integrating satellite images and DEM.

good-quality geographic information, the cost of satellite images can be justified, particularly when high spatial resolution is required. We were recently asked to help the design of a national agricultural census for a developing African country. Without good cartography or aerial sensors, the recourse to satellite high-resolution images was the most logical, both in terms of cost and operation. The growing trend toward low-cost or free access to satellite data will promote the use of RS to generate GIS variables for a wide set of applications.

### 9.3.2 Generation of Input Variables

RS can be an ideal tool to map certain variables: sea surface temperature, chlorophyll content, snow or ice cover, crop yields, landslides, flooding areas, etc., and the list grows with the availability of new sensors with higher spatial, temporal, and spectral resolution.

In some cases, RS satellites are the only reasonable way to retrieve environmental variables, which are very costly or unfeasible to obtain using ground sampling. Detection of oil spills, mapping water quality, and monitoring ozone, ice, or soil moisture are routinely done with satellite information. Ground sampling is important to calibrate or validate EO-based estimations, but generating those variables from ground survey would not be feasible and would require interpolation/extrapolation techniques, which always incur a certain range of errors (Burrough and McDonell 1998; Chapters 5 and 6). Satellite information is also particularly critical for remote areas where other ground or aerial sensors may not be available.

Even coarse-resolution satellite images provide a more complete spatial sampling than those sparse ground observations and therefore can give more accurate estimations of biophysical variables such as evapotranspiration (Choudhury 1994; Nieto et al. 2010; Yebra et al. 2013b), vegetation water content (Jurdao et al. 2013; Yebra et al. 2013a), or biomass spatial distribution (Anaya et al. 2009; Steininger 2000).

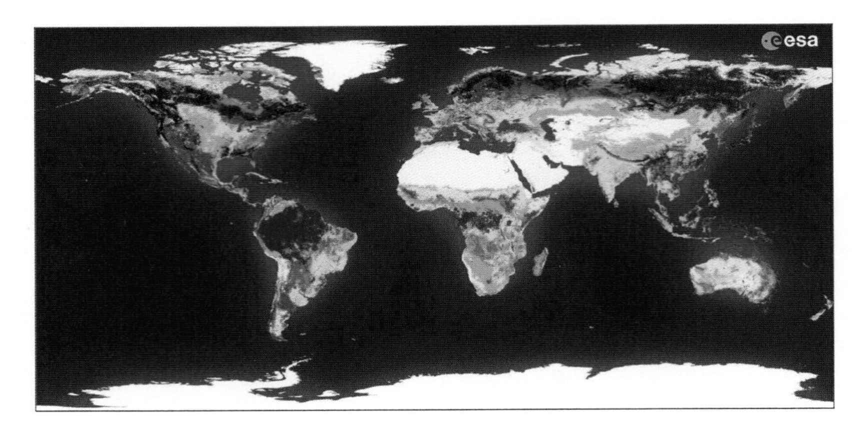


FIGURE 9.4 Global land cover map derived from Envisat-MERIS observations. (Courtesy of Pierre Defourny, ESA Landcover_cci project.)

Another clear contribution of RS to GIS is the generation of global databases. As is well known, global databases are very difficult to create when they come from national sources because of thematic and geometric inconsistencies (Unninayar 1988). Satellite images cover almost the whole earth, with the same or a similar sensor, at regular time intervals. This facilitates the generation of global coverage, which is especially valuable with regard to essential climate variables (Hollmann et al. 2013). For this reason, the Global Climate Observing System (GCOS) strongly recommends the use of satellite information as a basic component of the climate monitoring system (Bojinski et al. 2014). Many examples are available on the use of RS information to monitor global ice or snow resources (Kelly and Hall 2008; Shepherd et al. 2012), soil moisture (Wagner et al. 2013); land cover (Bontemps et al. 2012; Figure 9.4), sea height (Ablain et al. 2015), fire activity (Mouillot et al. 2014), and water conditions (Oki and Kanae 2006).

### 9.3.3 Updating the Information

As previously commented, the practical utility of a GIS greatly depends on having accurate and up-to-date information. This implies keeping track of environmental changes with a reasonable periodicity depending on the variable's dynamism. When the original inventory is accurate but the information has not been updated, the practical use of that information becomes riskier.

Satellite images offer great advantages for updating geographic data because of the systematic acquisition of images taken under comparable conditions in space and time (as seen in Section 7.3). This facilitates an updated view of the territory as well as the temporal monitoring of certain dynamic phenomena: urban growth, crop status, and natural hazards (floods, volcanic eruptions, etc.). Since the images are transmitted in real time, the end user may receive the interpretation very close in time to the acquisition of the image; thus, quick evaluation of the phenomenon is possible. This factor is very important in the management of natural hazards

(Nourbakhsh et al. 2006). In fact, the urgency of data access in these emergencies has led to the creation of the International Charter of Space and Major Disasters, an international treaty among space agencies to help quickly disseminate RS data in case of natural or man-made hazards.

## 9.4 INTEGRATION OF SATELLITE IMAGES AND GIS

The importation of satellite data to a GIS can be done in several ways. The easiest procedure is using the satellite image as a background for data already stored in the system. The new features may be digitized on the screen (Figure 9.5). This simple visual overlay facilitates the updating of attributes that are difficult to define from automatic algorithms, such as urban features (new buildings, roads, streets, etc.), parcel limits, highways, and dams. To avoid positional errors, it is critical that the image be well georectified before manual digitizing.

When importing an RS-derived variable to a GIS, the user has several options, which would depend on whether the interpretation was done visually or digitally (Figure 9.6). The former requires prior digitizing of the limits extracted from visual interpretation, either using a semi-manual procedure or a line-search algorithm. Then, if the GIS has a vector structure, the integration finishes, while if the GIS is raster, a rasterization process needs to be carried out, using the same pixel size as other variables in the database.

When the interpretation is done digitally, the export of the results is much simpler. As we have seen in Chapter 7, digital classification outputs a numerical matrix in which each pixel value indicates the assigned category. Therefore, the resulting

FIGURE 9.5 Satellite image used as a background source to update geographic information by manual digitizing.

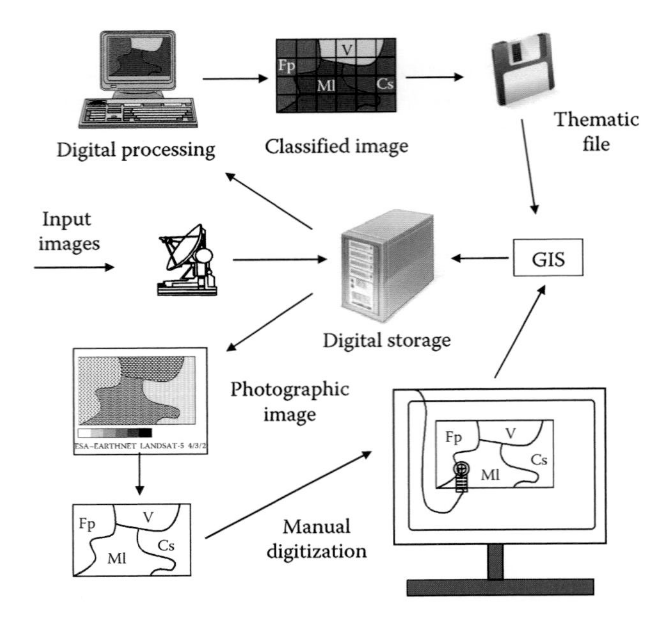


FIGURE 9.6 Methods of connection between RS and GIS.

file can be exported to the GIS, taking into account the conversion of formats and structures if needed (Figure 9.7). When the GIS is raster, the structure is the same as in the image, and the interpreter only needs to georeference the image with the same cartographic projection, covering the same area, and with the same pixel size of the other variables in the target GIS database.

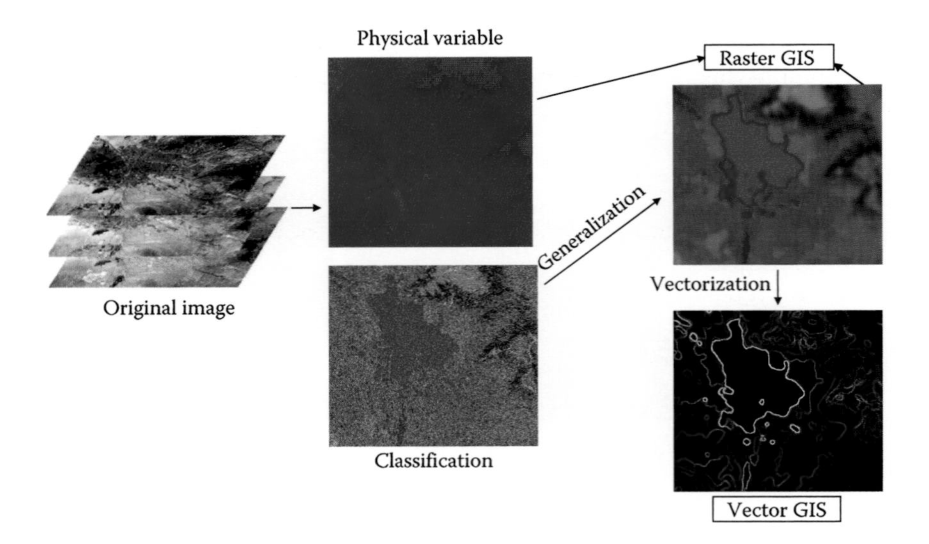


FIGURE 9.7 Integration of digital images in a GIS.

When the image is not previously georectified, a full geometric correction process needs to be applied (see Section 6.6). When the image is already georectified, it may occur that it has a different cartographic projection to the target GIS database. In this case, the user should make a coordinate projection transformation. As most GIS programs include automatic projection conversions, this step should be fairly simple. In the process of resampling to the new projection system, the user should select exactly the same coordinate limits ($X_{min}$, $X_{max}$, $Y_{min}$, $Y_{max}$) and pixel size to the other variables stored in the target GIS, so the resulting files have the same size and occupy the same geographical area. When resampling the pixel size, the interpreter should use the nearest-neighbor algorithm, as a digital classification is a categorical variable and other interpolation algorithms would produce incorrect results.

When the results of RS interpretation overlap cartographically with the other GIS variables, the process is done as in the case of working with a raster-based GIS. When working with a vector-based GIS, image results need to be converted to the vector format. Different algorithms are available for this task (Bolstad 2008). They commonly extract the borders between thematic categories and then create the topological structure of the resulting polygons. These algorithms are automatic, but they are not error free, and therefore a close supervision by the interpreter is recommended, especially in areas of multiple intersections in the image. To perform this task satisfactorily, it is essential to previously smooth the image using modal filters or similar techniques.

Independently of the problems associated with the integration of satellite images and GIS, the interaction between these technologies is evident and should be fostered (Figure 9.7). It offers enormous possibilities for environmental studies, evaluation of resources, environmental impact analysis, allocation models, etc. With the support of information technologies, environmental studies can focus on analyzing rather than obtaining information. Much effort is still needed to understand environmental systems, the impacts of different factors, the relevance of data accuracy, and the assessment of explicative models, but these challenges will, in any case, benefit from having broader perspectives and access to data from many different sources.

## 9.5 REVIEW QUESTIONS

1. Which of the following remote sensing processes does not require auxiliary geographic information?
   a. Topographic correction
   b. Stratification
   c. Sensor calibration
   d. Assessment
2. Which of the following applications should be most favored by using geographic information external to RS?
   a. Multitemporal comparison
   b. Classification
   c. Filtering
   d. Spatial metrics

3. Which GIS application field is more demanding of RS data?
   a. Crop monitoring
   b. Water pollution
   c. Urban growth
   d. Global change
4. Connection of GIS and RS data requires which of the following?
   a. Digital raw data
   b. Digital interpreted data
   c. Visually interpreted data
   d. All of the above
5. What is the main strength of using satellite data for a GIS?
   a. Information is updated
   b. Input information is digital
   c. Images are multiband
   d. Scaling-up processes are feasible
6. When exporting digital classification outputs to a GIS, we should primarily be concerned with
   a. Generalizing the results
   b. Performing raster to vector conversion
   c. Assuring geometric consistency
   d. Resolving the images
7. When the raster image has the same projection as the GIS, then
   a. The user should adjust frame limits and pixel size
   b. The user should adjust classification categories
   c. The user should adjust geographical coordinates
   d. All of the above
8. The main advantage of using GIS information in the classification phase is
   a. Use of spectral information
   b. Use of spatial context
   c. Use of radiometric resolution
   d. Include nonspectral bands
9. GIS information is useful in post-classification to
   a. Reduce errors of border pixels
   b. Reduce the spatial desegregation
   c. Fuse similar classes
   d. Eliminate small polygons

# Appendix

## A.1 ACRONYMS USED IN THIS TEXTBOOK

| | |
|---|---|
| AATSR | Advanced along track scanning radiometer |
| ALI | Advanced land imager |
| ALOS | Advanced land observing system |
| AMI | Active microwave instrument |
| ANN | Artificial neural network |
| ARVI | Atmospherically resistant vegetation index |
| ASAR | Advanced synthetic aperture radar |
| ASTER | Advanced spaceborne thermal emission and reflection radiometer |
| ATS | Applications technology satellite |
| ATSR | Along-track scanning radiometer |
| AVHRR | Advanced very high resolution radiometer |
| AVIRIS | Airborne visible/infrared imaging spectrometer |
| BIL | Band interleaved by line |
| BIP | Band interleaved by pixel |
| BRDF | Bidirectional reflectance distribution factor |
| BSQ | Band sequential |
| BV | Brightness value |
| CASI | Compact airborne spectrographic imager |
| CCD | Charge coupled devices |
| CERES | Clouds and the Earth's Radiant Energy System |
| CLUT | Color lookup table |
| CNES | Centre National d'Etudes Spatiales, France |
| CZCS | Coastal zone color scanner |
| DAAC | Distributed Active Archive Center |
| DAIS | Digital airborne imaging spectrometer |
| DEM | Digital elevation model |
| DL | Digital level; each pixel's numeric value |
| DLR | Deutschen Zentrum für Luft-un Raumfahrt, Germany |
| DMSP | Defense Meteorological Satellite Program |
| EIFOV | Effective instantaneous field of view |
| EO-1 | Earth Observing-1 |
| EOS | Earth Observing System |
| EOSDIS | Earth Observation System Data and Information System |
| ERBE | Earth radiation budget experiment |
| ERE | Effective resolution element |
| ERS | European Remote Sensing Satellite |
| ESA | European Space Agency |
| ESC | Electronic still camera |
| ETC | Earth terrain camera |

| | |
|---|---|
| ETM | Enhanced thematic mapper |
| EWT | Equivalent water thickness |
| FBA | Foreground/background analysis |
| FC | Filter coefficients |
| FMC | Fuel moisture content |
| FTP | File transfer protocol |
| GAC | Global area coverage |
| GARP | Global Atmospheric Research Programme |
| Gb | Gigabyte; $10^9$ bytes |
| GCP | GCPWorks |
| GEMI | Global environment monitoring index |
| GHz | Gigahertz; $10^9$ Hz |
| GIF | Graphics interchange format |
| GIS | Geographic information systems |
| GMT | Greenwich Mean Time |
| GOES | Geostationary Operational Environmental Satellite |
| GOME | Global ozone monitoring experiment |
| GOMOS | Global ozone monitoring by occultation of stars |
| GPS | Global positioning system |
| GVI | Global vegetation index |
| HCMR | Heat capacity mapping radiometer |
| HH | Horizontal-horizontal |
| HIRIS | High-resolution imaging spectrometer |
| HIRS/3 | High-resolution infrared radiation sounder |
| HRPT | High-resolution picture transmission |
| HRV | Haute resolution visible |
| HV | Horizontal-vertical |
| Hz | Hertz |
| IFOV | Instantaneous field of view |
| IGBP | International Geosphere Biosphere Program |
| IHS | Intensity, hue, saturation |
| INPE | Instituto Nacional de Pesquisas Espaciais, Brasil |
| IPS | Image processing software |
| IRSS | Indian Remote Sensing Satellite |
| JPEG | Joint Photographic Experts Group |
| JPL | Jet Propulsion Laboratory |
| kb | Kilobyte; $10^3$ bytes |
| LAC | Local area coverage |
| LAD | Leaf angle distribution |
| LAI | Leaf area index |
| LFC | Large format camera |
| LISS | Linear imaging self-scanning |
| Mb | Megabyte; $10^6$ bytes |
| MERIS | Medium-resolution imaging spectrometer |
| MISR | Multi-angle imaging spectroradiometer |
| MNF | Minimum noise fraction |

| | |
|---|---|
| MODIS | Moderate-resolution imaging spectroradiometer |
| MOPITT | Measurements of pollution in the troposphere |
| MOS | Marine Observation Satellite |
| MSG | Meteosat second generation |
| MSI | Multispectral Scanner Instrument (Sentinel-2) |
| MSS | Multispectral scanner |
| MSU | Microwave sounding unit |
| NASA | National Aeronautics and Space Administration |
| NASDA | National Space Development Agency, Japan |
| NDII | Normalized difference infrared index |
| NDVI | Normalized difference vegetation index |
| NGDC | National Geophysical Data Center, United States |
| NIR | Near infrared band, from 0.7 to 1.2 μm |
| NOAA | National Oceanic and Atmospheric Administration Satellite |
| NPOESS | National Polar-Orbiting Operational Environmental Satellite System |
| NRSA | National Remote Sensing Agency, India |
| OLI | Operational line imager |
| OLS | Operational Linescan System |
| PCA | Principal component analysis |
| PDUS | Primary data user stations |
| PNG | Portable network graphics |
| POAM | Polar ozone and aerosol measurement |
| POLDER | Polarization and directionality of the Earth's reflectance |
| PPI | Pixel purity index |
| PPU | Patch per unit |
| RBV | Return beam vidicon |
| RGB | Red, green, blue |
| RMSE | Root-mean-square error |
| SAIL | Scattered by arbitrarily inclined leaves |
| SAM | Spectral angle mapper |
| SAR | Synthetic aperture radar |
| SAVI | Soil-adjusted vegetation index |
| SBUV/2 | Solar backscatter ultraviolet spectral radiometer |
| SEVIRI | Spinning enhanced visible and infrared imager |
| SIR | Shuttle imaging radar |
| SLA | Specific leaf area |
| SLAR | Side-looking airborne radar |
| SMA | Spectral Mixing Analysis |
| SMMR | Scanning multichannel microwave radiometer |
| SPOT | Satellite Pour l'Observation de la Terre |
| SRTM | Shuttle Radar Topography Mission |
| SSM/I | Special sensor microwave imager |
| SSU | Stratospheric sounding unit |
| ST | Surface Temperature |
| SWIR | Short wave infrared band; 1.2–2.5 μm |
| TCT | Tasseled cap transformation |

| | |
|---|---|
| TIR | Thermal infrared band; 8–14 μm |
| TM | Thematic mapper |
| TOVS | Tiros operational vertical sounder |
| TRMM | Tropical Rainfall Measuring Mission |
| VCL | Vegetation canopy lidar |
| VH | Vertical-horizontal |
| VI | Vegetation Indices |
| VIIRS | Visible/Infrared Imager/Radiometer Suite |
| VIS | Visible spectrum; 0.4–0.7 μm |
| VV | Vertical-vertical |

## A.2 ANSWERS TO REVIEW QUESTIONS

Chapter 1: 1b 2c 3d 4b 5a
Chapter 2: 1d 2a 3c 4b 5c 6b 7a 8c 9b 10b
Chapter 3: 1a 2b 3c 4c 5b 6a 7b 8b 9b 10a
Chapter 4: 1a 2b 3b 4b 5d
Chapter 5: 1d 2a 3c 4b 5d 6a 7c 8d 9a 10d
Chapter 6: 1b 2d 3a 4c 5a 6a 7a 8c 9c 10c 11d 12b 13a 14a 15b 16b
Chapter 7: 1c 2a 3d 4a 5c 6d 7d 8d 9a 10c 11b 12d 13a 14d 15c 16c 17d 18c 19a 20c
Chapter 8: 1c 2a 3d 4b 5b 6a 7d 8c 9c 10a
Chapter 9: 1c 2b 3d 4d 5a 6c 7a 8d 9a

# References


Ablain, M., Cazenave, A., Larnicol, G., Balmaseda, M., Cipollini, P., Faugère, Y., Fernandes, M.J. et al. (2015). Improved sea level record over the satellite altimetry era (1993–2010) from the Climate Change Initiative project. *Ocean Science*, 11, 67–82.

Achard, F. and Hansen, M.C. (2012). *Global Forest Monitoring from Earth Observation*. Boca Raton, FL: CRC Press.

Ackermann, F. (1999). Airborne laser scanning—Present status and future expectations. ISPRS *Journal of Photogrammetry and Remote Sensing*, 54, 64–67.

Adams, J.B., Sabol, D.E., Kapos, V., Almeida, R., Roberts, D.A., Smith, M.O., and Gillespie, A.R. (1995). Classification of multispectral images based on fractions of endmembers: Application to land-cover change in the Brazilian Amazon. *Remote Sensing of Environment*, 52, 137–154.

Adeniyi, P.O. (1980). Land-use change analysis using sequential aerial photography and computer techniques. *Photogrammetric Engineering and Remote Sensing*, 46, 1447–1464.

Adeniyi, P.O. (1987). Using remotely sensed data for census surveys and population estimation in developing countries: Examples from Nigeria. *Geocarto International*, 2, 11–32.

Ahern, F.J., Goldammer, J.G., and Justice, C.O. (Eds.) (2001). *Global and Regional Vegetation Fire Monitoring from Space: Planning a Coordinated International Effort*. The Hague, the Netherlands: SPB Academic Publishing.

Alcaraz-Segura, D., Chuvieco, E., Epstein, H.E., Kasischke, E.S., and Trishchenko, A. (2010). Debating the greening vs. browning of the North American boreal forest: Differences between satellite datasets. *Global Change Biology*, 16, 760–770.

Allan, T.D. (Ed.) (1983). *Satellite Microwave Remote Sensing*. New York: Ellis Howard Ltd.

Allen, R.G., Jensen, M.E., Wright, J.L., and Burman, R.D. (1989). Operational estimates of reference evapotranspiration. *Agronomy Journal*, 81, 650–662.

Alonso-Canas, I. and Chuvieco, E. (2015). Global Burned Area Mapping from ENVISAT-MERIS data. *Remote Sensing of Environment*, 163, 140–152.

Al-Yaari, A., Wigneron, J.-P., Ducharne, A., Kerr, Y., Wagner, W., De Lannoy, G., Reichle, R., Al Bitar, A., Dorigo, W., and Richaume, P. (2014). Global-scale comparison of passive (SMOS) and active (ASCAT) satellite based microwave soil moisture retrievals with soil moisture simulations (MERRA-Land). *Remote Sensing of Environment*, 152, 614–626.

Amatulli, G., Rodrigues, M.J., Trombetti, M., and Lovreglio, R. (2006). Assessing long-term fire risk at local scale by means of decision tree technique. *Journal of Geophysical Research—Biosciences*, 111, G04S05, doi:10.1029/2005JG000133.

Anaya, J.A., Chuvieco, E., and Palacios-Orueta, A. (2009). Aboveground biomass assessment in Colombia: A remote sensing approach. *Forest Ecology and Management*, 257, 1237–1246.

Anderson, J.R., Hardy, E.E., Roach, J.T., and Witmer, R.E. (1976). *A Land Use and Land Cover Classification System for Use with Remote Sensor Data*. Washington, DC: U.S. Geological Survey.

Aronoff, S. (1982). The map accuracy report: A user's view. *Photogrammetric Engineering and Remote Sensing*, 48, 1309–1312.

Aronoff, S. (1989). *Geographic Information Systems: A Management Perspective*. Ottawa, Ontario, Canada: WDL Publications.

Asner, G.P. (1998). Biophysical and biochemical sources of variability in canopy reflectance. *Remote Sensing of Environment*, 64, 234–253.

Asner, G.P. (2000). Contributions of multi-view angle remote sensing to land-surface and biogeochemical research. *Remote Sensing Reviews*, 18, 137–162.

Asner, G.P., Wessman, C.A., Bateson, C.A., and Privette, J.L. (2000). Impact of tissue, canopy, and landscape factors on the hyperspectral reflectance variability of arid ecosystems. *Remote Sensing of Environment*, 74, 69–84.

Asrar, G. (Ed.) (1989). *Theory and Applications of Optical Remote Sensing*. New York: Wiley.

Asrar, G., Fuchs, M., Kanemasu, E.T., and Hatfield, S.L. (1984). Estimating absorbed photosynthetic radiation and leaf area index from spectral reflectance in wheat. *Agronomy Journal*, 76, 300–306.

Asrar, G., Myneni, R.B., and Choudhury, B.J. (1992). Spatial heterogenity in vegetation canopies and remote sensing of absorbed photosynthetically active radiation: A modeling study. *Remote Sensing of Environment*, 41, 85–103.

Astaras, T. (1984). Land complex classification of the Mula area, Murcia Province, Southeast Spain. *Geografiska Annaler*, 66A, 307–325.

Au, K.N. (Ed.) (1993). *Cities of the World as Seen From Space*. Hong Kong, China: Geocarto International Centre.

Bachmann, M. and Bendix, J. (1992). An improved algorithm for NOAA-AVHRR image referencing. *International Journal of Remote Sensing*, 13, 3205–3215.

Bak, P., Tang, C., and Wiesenfeld, K. (1987). Self-organized criticality: An explanation of the 1/f noise. *Physical Review Letters*, 59, 381.

Baker, J.C. (2001). New users and established experts: Bridging the knowledge gap in interpreting commercial satellite imagery. In J.C. Baker, K.M. O'Connell, and R.A. Williamson (Eds.), *Commercial Observation Satellites. At the Leading Edge of Global Transparency* (pp. 533–557). Santa Monica, CA: RAND—ASPRS.

Baret, F. (1995). Use of spectral reflectance variation to retrieve canopy biophysical characteristics. In F.M. Danson and S.E. Plummer (Eds.), *Advances in Environmental Remote Sensing* (pp. 33–51). Chichester, U.K.: John Wiley & Sons.

Baret, F. and Guyot, G. (1991). Potentials and limits of vegetation indices for LAI and APAR assessment. *Remote Sensing of Environment*, 35, 161–173.

Baret, F., Guyot, G., and Major, D. (1989). TSAVI: A vegetation index which minimizes soil brightness effects on LAI and APAR estimation. In *12th Canadian Symposium on Remote Sensing and IGARSS'89* (pp. 1355–1358). Vancouver, British Columbia, Canadá.

Barr, S. and Barnsley, M.J. (2000). Reducing structural clutter in land cover classifications of high spatial resolution remotely-sensed images for urban land use mapping. *Computers and Geosciences*, 26, 433–449.

Barret, E.C. (1974). *Climatology from Satellites*. London, U.K.: Methuen.

Barret, E.C. and Curtis, L.F. (Eds.) (1999). Calibration, evaluation and validation in remote sensing. In *Introduction to Environmental Remote Sensing* (pp. 117–134). Cheltenham, U.K.: Stanley Thornes Publishers Ltd.

Bartholome, J.E. and Belward, A.S. (2005). GLC2000: A new approach to global land cover mapping from Earth observation data. *International Journal of Remote Sensing*, 26, 1959–1977.

Bastarrika, A., Chuvieco, E., and Martín, M.P. (2011). Mapping burned areas from Landsat TM/ETM+ data with a two-phase algorithm: Balancing omission and commission errors. *Remote Sensing of Environment*, 115, 1003–1012.

Baudot, Y. (1990). Integration of high-quality remote sensing images with low cost geographical information systems. In *EGIS'90* (pp. 40–46). Utrecht, the Netherlands.

Ben-Dor, E., Taylor, R.G., Hill, J., Dematt, J.A.M., Whiting, M.L., Chabrillat, S., and Sommer, S. (2008). Imaging spectrometry for soil applications. *Advances in Agronomy*, 97, 321–392.

Benediktsson, J.A., Swain, P.H., and Ersoy, O.K. (1990). Neural network approaches versus statistical methods in classification of multisource remote sensing data. *IEEE Transactions on Geoscience and Remote Sensing*, 28, 540–552.

Beraldin, J.A., Blais, F., and Lohr, U. (2010). Laser Scanning Technology. In G. Vosselman and H.-G. Maas (Eds.), *Airborne and Terrestrial Laser Scanning* (pp. 1–42). Dunbeath, U.K.: Whittles Publishing.

Berk, A., Anderson, G.P., Acharya, P.K., Bernstein, L.S., Muratov, L., Lee, J., Fox, M., Adler-Golden, S.M., Chetwynd Jr, J.H., and Hoke, M.L. (2006). MODTRAN5: 2006 update. In *Defense and Security Symposium* (pp. 62331F–62338F). Orlando, FL: International Society for Optics and Photonics.

Bernstein, R., Lotspiech, J.B., Myers, J., Kolskoy, H.G., and Lees, R.D. (1984). Analysis and processing of Landsat-4 sensor data using advanced image processing techniques and technologies. *IEEE Transactions on Geoscience and Remote Sensing*, GE-22, 192–221.

Berry, B.J. and Baker, A.M. (1968). Geographic sampling. In B.J. Berry and D.F. Marble (Eds.), *Spatial Analysis* (pp. 91–100). Englewood Cliffs, NJ: Prentice Hall.

Bhargava, D.S. and Mariam, D.W. (1990). Spectral reflectance relationships to turbidity generated by different clay materials. *Photogrammetric Engineering and Remote Sensing*, 56, 225–229.

Biging, G., Colby, D.R., and Congalton, R.G. (1998). Sampling systems for change detection accuracy assessment. In R.S. Luneta and C.D. Elvidge (Eds.), *Remote Sensing Change Detection. Environmental Monitoring Methods and Applications* (pp. 281–308). Chelsea, MI: Ann Arbor Press.

Birkett, C.M. (1998). Contribution of the TOPEX NASA radar altimeter to the global monitoring of large rivers and wetlands. *Water Resources Research*, 34, 1223–1239.

Bischof, H., Schneider, W., and Pinz, A.J. (1992). Multispectral classification of Landsat images using neural networks. *IEEE Transactions on Geoscience and Remote Sensing*, 30, 482–489.

Bishop, Y., Fienberg, S., and Holland, P. (1975). *Discrete Multivariate Analysis: Theory and Practice*. Cambridge, MA: MIT Press.

Blackburn, G.A. (2002). Remote sensing of forest pigments using airborne imaging spectrometer and LIDAR imagery. *Remote Sensing of Environment*, 82, 311–321.

Blom, R.G. and Daily, M. (1982). Radar image processing for rock-type discrimination. *IEEE Transactions on Geoscience Electronics*, GE20, 343–351.

Bo Cai, G. and Goetz, A.F.H. (1995). Retrieval of equivalent water thickness and information related to biochemical components of vegetation canopies from AVIRIS data. *Remote Sensing of Environment*, 52, 155–162.

Bojinski, S., Verstraete, M., Peterson, T.C., Richter, C., Simmons, A., and Zemp, M. (2014). The concept of essential climate variables in support of climate research, applications, and policy. *Bulletin of the American Meteorological Society*, 95, 1431–1443.

Bolstad, P.V. (2008). *GIS Fundamentals: A First Textbook on Geographic Information Systems*. White Bear Lake, MN: Eider Press.

Bontemps, S., Herold, M., Kooistra, L., Van Groenestijn, A., Hartley, A., Arino, O., Moreau, I., and Defourny, P. (2012). Revisiting land cover observation to address the needs of the climate modeling community. *Biogeosciences*, 9, 2145–2157.

Booth, D.J. and Oldfield, R.B. (1989). A comparison of classification algorithms in terms of speed and accuracy after the applications of a post-classification modal filter. *International Journal of Remote Sensing*, 10, 1271–1276.

Bordes, P., Brunel, P., and Marsovin, A. (1992). Automatic adjustment of AVHRR navigation. *Journal of Atmospheric and Oceanic Technology*, 9, 15–27.

Boschetti, L., Brivio, P.A., Eva, H.D., Gallego, J., Baraldi, A., and Gregoire, J.M. (2006). A sampling method for the retrospective validation of Global Burned Area products. *IEEE Transactions on Geoscience and Remote Sensing*, 44, 1765–1773.

Boschetti, L., Roy, D.P., Justice, C.O., and Humber, M.L. (2015). MODIS–Landsat fusion for large area 30m burned area mapping. *Remote Sensing of Environment*, 161, 27–42.

Bréda, N.J. (2003). Ground-based measurements of leaf area index: A review of methods, instruments and current controversies. *Journal of Experimental Botany*, 54, 2403–2417.

Breiman, L. (2001). Random forests. *Machine Learning*, 45, 5–32.

Broge, N.H. and Leblanc, E. (2001). Comparing prediction power and stability of broadband and hyperspectral vegetation indices for estimation of green leaf area index and canopy chlorophyll density. *Remote Sensing of Environment*, 76, 156–172.

Brookes, A.J. (1975). *Photo Reconnaissance: The Operational History*. London, U.K.: Ian Allan.

Bruce, B. and Hormsby, J.K. (1987). A Canadian perspective on the application of satellite remote sensing to regional geobotany. *Geocarto International*, 1, 53–59.

Bruzzone, L. and Fernández-Prieto, D. (2000). An adaptive parcel-based technique for unsupervised change detection. *International Journal of Remote Sensing*, 21, 817–822.

Bruzzone, L. and Serpico, S.B. (1998). Fusion of multisensor and multitemporal data in remote sensing image analysis. *Geoscience and Remote Sensing Symposium Proceedings, IGARSS '98* (pp. 162–164).

Burgan, R.E., Klaver, R.W., and Klaver, J.M. (1998). Fuel models and fire potential from satellite and surface observations. *International Journal of Wildland Fire*, 8, 159–170.

Burgan, R.E. and Rothermel, R.C. (1984). *BEHAVE: Fire Behavior Prediction and Fuel Modeling System. Fuel Subsystem*. Ogden, UT: USDA Forest Service.

Burrough, P.A. and McDonell, R.A. (1998). *Principles of Geographical Information Systems*. Oxford, U.K.: Oxford University Press.

Büttner, G., Feranec, J., and Jaffrain, G. (2000a). *Corine Land Cover Technical Guide. Addendum 2000*. Copenhagen, Denmark: European Environmental Agency.

Büttner, G., Feranec, J., and Jaffrain, G. (2000b). *Corine Land Cover Update 2000. Technical Guidelines*. Copenhagen, Denmark: European Environmental Agency.

Caetano, M.S., Mertes, L., Cadete, L., and Pereira, J.M.C. (1996). Assessment of AVHRR data for characterising burned areas and post-fire vegetation recovery. *EARSeL Advances in Remote Sensing*, 4, 124–134.

Calle, A., Casanova, J.L., and Romo, A. (2006). Fire detection and monitoring using MSG Spinning Enhanced Visible and Infrared Imager (SEVIRI) data. *Journal of Geophysical Research—Biosciences*, 111, 2156–2202.

Campbell, J.B. (1996). *Introduction to remote sensing*. New York: The Guilford Press.

Campbell, J.B. and Wyme, R.H. (2011). *Introduction to Remote Sensing* (5th edn.). New York: The Guilford Press.

Carleton, A.M. (1991). *Satellite Remote Sensing in Climatology*. London, U.K.: Belhaven Press.

Carlson, T.N. and Ripley, D.A. (1997). On the relation between NDVI, Fractional Vegetation Cover and Leaf Area Index. *Remote Sensing of Environment*, 62, 241–252.

Carper, W.J., Lillesand, T.M., and Kiefer, R.W. (1990). The use of Intensity-Hue-Saturation transformation for merging SPOT pancromatic and multispectral image data. *Photogrammetric Engineering and Remote Sensing*, 56, 459–467.

Caselles, V. and Sobrino, J.A. (1989). Determination of frosts in orange groves from NOAA-9 AVHRR Data. *Remote Sensing of Environment*, 29, 135–146.

Cazenave, A. and Cozannet, G.L. (2014). Sea level rise and its coastal impacts. *Earth's Future*, 2, 15–34.

Ceccato, P., Flasse, S., and Gregoire, J.M. (2002). Designing a spectral index to estimate vegetation water content from remote sensing data: Part 2. Validation and applications. *Remote Sensing of Environment*, 82, 198–207.

CEOS-WGCV (2012). Working Group on Calibration and Validation—Land Product Validation Subgroup. http://lpvs.gsfc.nasa.gov/, last accessed in November 2015.

Chander, G., Helder, D.L., Markham, B.L., Dewald, J.D., Kaita, E., Thome, K.J., Micijevic, E., and Ruggles, T.A. (2004). Landsat-5 TM reflective-band absolute radiometric calibration. *IEEE Transactions on Geoscience and Remote Sensing*, 42, 2747–2760.

Chavez, P.S. (1975). Atmospheric, solar, and MTF corrections for ERTS digital imagery. In *Proceedings of the American Society of Photogrammetry Fall Technical Meeting* (p. 69). Phoenix, AZ.

Chavez, P.S. (1988). An improved dark-object subtraction technique for atmospheric scattering correction of multispectral data. *Remote Sensing of Environment*, 24, 459–479.

Chavez, P.S. (1996). Image-based atmospheric corrections. Revisited and improved. *Photogrammetric Engineering and Remote Sensing*, 62, 1025–1036.

Chavez, P.S. and Kwarteng, A.Y. (1989). Extracting spectral contrast in Landsat Thematic Mapper image data using selective principal component analysis. *Photogrammetric Engineering and Remote Sensing*, 55, 339–348.

Chavez, P.S., Sides, S.C., and Anderson, J.A. (1991). Comparison of three different methods to merge multiresolution and multispectral data: Landsat TM and SPOT Panchromatic. *Photogrammetric Engineering and Remote Sensing*, 57, 295–303.

Chen, H.S. (1985). *Space Remote Sensing Systems: An Introduction*. Orlando, FL: Academic Press.

Chen, J., Gong, P., He, C., Pu, R., and Shi, P. (2003). Land use/land cover change detection using improved change-vector analysis. *Photogrammetric Engineering and Remote Sensing*, 69, 369–379.

Chica, M. (1988). *Análisis Geoestadístico en el Estudio de la Explotación de los Recursos Minerales*. Granada, Spain: Universidad de Granada.

Choudhury, B.J. (1994). Synergism of multispectral satellite observations for estimating regional land surface evaporation. *Remote Sensing of Environment*, 49, 264–274.

Chuvieco, E. (1999). Measuring changes in landscape pattern from satellite images: Short-term effects of fire on spatial diversity. *International Journal of Remote Sensing*, 20, 2331–2347.

Chuvieco, E. (Ed.) (2003). *Wildland Fire Danger Estimation and Mapping: The Role of Remote Sensing Data*. Singapore: World Scientific Publishing.

Chuvieco, E. (Ed.) (2008). *Earth Observation of Global Change: The Role of Satellite Remote Sensing in Monitoring the Global Environment*. Berlin, Germany: Springer.

Chuvieco, E. (Ed.) (2009). *Earth Observation of Wildland Fires in Mediterranean Ecosystems*. Berlin, Germany: Springer.

Chuvieco, E., Cocero, D., Riaño, D., Martín, M.P., Martínez-Vega, J., de la Riva, J., and Pérez, F. (2004). Combining NDVI and Surface Temperature for the estimation of live fuel moisture content in forest fire danger rating. *Remote Sensing of Environment*, 92, 322–331.

Chuvieco, E. and Congalton, R.G. (1988). Using cluster analysis to improve the selection of training statistics in classifying remotely sensed data. *Photogrammetric Engineering and Remote Sensing*, 54, 1275–1281.

Chuvieco, E., De Santis, A., Riaño, D., and Halligan, K. (2007). Simulation approaches for burn severity estimation using remotely sensed images. *Journal of Fire Ecology*, 3, 129–150.

Chuvieco, E., Englefield, P., Trishchenko, A.P., and Luo, Y. (2008a). Generation of long time series of burn area maps of the boreal forest from NOAA–AVHRR composite data. *Remote Sensing of Environment*, 112, 2381–2396.

Chuvieco, E., Li, J., and Yang, X. (Eds.) (2010). *Advances in Earth Observation of Global Change*. London, U.K.: Springer.

Chuvieco, E., Martín, M.P., and Palacios, A. (2002). Assessment of different spectral indices in the red-near-infrared spectral domain for burned land discrimination. *International Journal of Remote Sensing*, 23, 5103–5110.

Chuvieco, E., Martinez, S., Roman, M.V., Hantson, S., and Pettinari, L. (2014). Integration of ecological and socio-economic factors to assess global wildfire vulnerability. *Global Ecology and Biogeography*, 23, 245–258.

Chuvieco, E. and Martínez Vega, J. (1990). Visual versus digital analysis for vegetation mapping: Some examples on Central Spain. *Geocarto International*, 3, 21–29.

Chuvieco, E., Opazo, S., Sione, W., Del Valle, H., Anaya, J., Di Bella, C., Cruz, I. et al. (2008b). Global burned land estimation in Latin America using MODIS composite data. *Ecological Applications*, 18, 64–79.

Chuvieco, E. and Salas, F.J. (1996). Mapping the spatial distribution of forest fire danger using GIS. *International Journal of Geographical Information Systems*, 10, 333–345.

Civco, D.L. (1989). Topographic normalization of landsat Thematic Mapper digital imagery. *Photogrammetric Engineering and Remote Sensing*, 55, 1303–1309.

Civco, D.L. (1993). Artificial neural networks for land-cover classification and mapping. *International Journal of Geographical Information Systems*, 7, 173–186.

Clarke, K.C. (2003). *Getting Started with Geographic Information Systems*. Upper Saddle River, NJ: Prentice-Hall.

Cliff, A.D. and Ord, J.K. (1973). *Spatial Autocorrelation*. London, U.K.: Pion.

Cochran, W.G. (1977). *Sampling Techniques*. New York: John Wiley & Sons.

Cohen, W.B. and Spies, T.A. (1992). Estimating structural attributes of Douglas-fir/Western Hemlock forest stands from Landsat and SPOT imagery. *Remote Sensing of Environment*, 41, 1–17.

Cohen, W.P., Spies, T.A., and Bradshaw, G.A. (1990). Semivariograms of digital imagery for analysis of conifer canopy structure. *Remote Sensing of Environment*, 34, 167–178.

Cohen, W.B., Yang, Z., and Kennedy, R. (2010). Detecting trends in forest disturbance and recovery using yearly Landsat time series: 2. TimeSync—Tools for calibration and validation. *Remote Sensing of Environment*, 114, 2911–2924.

Colby, J.D. (1991). Topographic normalization in rugged terrain. *Photogrammetric Engineering and Remote Sensing*, 57, 531–537.

Coleman, A. (1968). *Land Use Survey Handbook*. Isle of Thanet, U.K.: Geographical Association.

Coll, C. and Caselles, V. (1997). A global split-window algorithm for land surface temperature from AVHRR data: Validation and algorithm comparison. *Journal of Geophysical Research*, 102B14, 16697–16713.

Coll, C., Caselles, V., Sobrino, J.A., and Valor, E. (1994). On the atmospheric dependence of the split winclow equation for land surface temperature. *International Journal of Remote Sensing*, 15, 105–122.

Collado, A.D., Chuvieco, E., and Camarasa, A. (2002). Satellite remote sensing analysis to monitor desertification processes in the crop-rangeland boundary of Argentina. *Journal of Arid Environments*, 52, 121–133.

Colwell, J.E. (1974). Vegetation canopy reflectance. *Remote Sensing of Environment*, 3, 175–183.

Combal, B., Baret, F., Weiss, M., Trubuil, A., Mace, D., Pragnère, A., Myneni, R., Knyazikhin, Y., and Wang, L. (2002). Retrieval of canopy biophysical variables from bidirectional reflectance using prior information to solve the ill-posed inverse problem. *Remote Sensing of Environment*, 84, 1–15.

Conese, C., Gilabert, M.A., Maselli, F., and Bottai, L. (1993). Topographic normalization of TM scenes through the use of an atmospheric correction method and digital terrain models. *Photogrammetric Engineering and Remote Sensing*, 59, 1745–1753.

Congalton, R.B. (1984). *A comparison of Five Sampling Schemes Used in Assessing the Accuracy of Land Cover/Land Use Maps Derived from Remotely Sensed Data*. Blacksburg, VA: Virginia Polytechnic Institute.

Congalton, R.G. (1988a). A comparison of sampling schemes used in generating error matrices for assessing the accuracy of maps generated from remotely sensed data. *Photogrammetric Engineering and Remote Sensing*, 54, 593–600.

Congalton, R.G. (1988b). Using spatial autocorrelation analysis to explore the errors in maps generated from remotely sensed data. *Photogrammetric Engineering and Remote Sensing*, 54, 587–592.

Congalton, R.G. (1991). A review of assessing the accuracy of classifications of remotely sensed data. *Remote Sensing of Environment*, 37, 35–46.

Congalton, R.G. and Green, K. (1999). *Assessing the Accuracy of Remotely Sensed Data: Principles and Applications*. Boca Raton, FL: Lewis Publishers.

Congalton, R.G., Stenback, J.M., and Barret, R.H. (1993). Mapping deer habitat suitability using remote sensing and geographic information systems. *Geocarto International*, 8, 23–33.

Conway, E.D. (1997). *An Introduction to Satellite Image Interpretation*. Baltimore, MD: John Hopkins University Press.

Coppin, P.R. and Bauer, M.E. (1994). Processing of multitemporal Landsat TM imagery to optimize extraction of forest cover change features. *IEEE Transactions on Geoscience and Remote Sensing*, 32, 918–927.

Couzy, A. (1981). *La Télédétection*. Paris, France: Presses Universitaires de France.

Cracknell, A.P. (1997). *The Advanced Very High Resolution Radiometer (AVHRR)*. London, U.K.: Taylor & Francis Group.

Cracknell, A.P. and Hayes, L.W.B. (1991). *Introduction to Remote Sensing*. London, U.K.: Taylor & Francis Group.

Crist, E.P. and Cicone, R.C. (1984a). Application of the tasseled cap concept to simulated Thematic Mapper data. *Photogrammetric Engineering of Remote Sensing*, 50, 343–352.

Crist, E.P. and Cicone, R.C. (1984b). A physically-based transformation of Thematic Mapper data the TM tasseled cap. *IEEE Transactions on Geoscience and Remote Sensing*, GE22, 256–263.

Crist, E.P. and Kauth, R.J. (1986). The tasseled cap de-mystified. *Photogrammetric Engineering and Remote Sensing*, 52, 81–86.

Curran, P.J. (1985). *Principles of Remote Sensing*. London, U.K.: Longman.

Curran, P.J. and Foody, G.M. (1994). The use of remote sensing to characterise the regenerative states of tropical forests. In G.M. Foody and P.J. Curran (Eds.), *Environmental Remote Sensing from Regional to Global Scales* (pp. 44–83). Chichester, U.K.: John Wiley & Sons.

Curran, P.J., Foody, G.M., Kondratyev, K.Y., Kozoderov, W., and Fedchenko, P.P. (1990). *Remote Sensing of Soils and Vegetation in the USSR*. London, U.K.: Taylor & Francis.

Cushnie, J.L. (1987). The interactive effect of spatial resolution and degree of internal variability within land-cover types on classification accuracies. *International Journal of Remote Sensing*, 8, 15–29.

Cushnie, J.L. and Atkinson, P. (1985). The effect of spatial filtering on scene noise and boundary detail in Thematic Mapper imagery. *Photogrammetric Engineering and Remote Sensing*, 51, 1483.

Dai, X. and Khorram, S. (1998). The effects of image misregistration on the accuracy of remotely sensed change detection. *IEEE Transactions on Geoscience and Remote Sensing*, 36, 1566–1577.

Danjoy, W.A. and Sadowski, F.G. (1978). Use of Landsat in the study of forest classification in the tropical jungle. In *12th International Symposium on Remote Sensing of Environment* (pp. 947–951). Ann Arbor, MI.

Danson, F.M. and Plummer, S.E. (Eds.) (1995). *Advances in Environmental Remote Sensing*. Chichester, U.K.: John Wiley & Sons.

Danson, F.M., Plummer, S.E., and Briggs, S.A. (1995). Remote Sensing and the information extraction problem. In F.M. Danson and S.E. Plummer (Eds.), *Advances in Environmental Remote Sensing* (pp. 171–177). Chichester, U.K.: John Wiley & Sons.

Davis, F.W. and Goetz, S. (1990). Modeling vegetation pattern using digital terrain data. *Landscape Ecology*, 4, 69–80.

Day, D.A., Logsdon, J.M., and Latell, B. (1998). *Eye in the Sky: The Story of the Corona Spy Satellites.* Washington, DC: Smithsonian Institution Press.

Deane, G.C. (1980). Preliminary evaluation of Seasat-1 SAR data for land use mapping. *Geographical Journal*, 146, 408–418.

De'ath, G. and Fabricius, K.E. (2000). Classification and regression trees: A powerful yet simple technique for ecological data analysis. *Ecology*, 81, 3178–3192.

DeFries, R. (2013). Remote sensing and image processing. In S.A. Levin (Ed.), *Encyclopedia of Biodiversity* (2nd edn.) (pp. 389–399). Waltham, MA: Academic Press.

D'Errico, M., Mocha, A., and Vetrella, S. (1995). High-frequency observation of Natural disasters by SAR interferometry. *Photogrammetric Engineering and Remote Sensing*, 61, 891–898.

de Leeuw, G., Holzer-Popp, T., Bevan, S., Davies, W.H., Descloitres, J., Grainger, R.G., Griesfeller, J., Heckel, A., Kinne, S., and Klüser, L. (2013). Evaluation of seven European aerosol optical depth retrieval algorithms for climate analysis. *Remote Sensing of Environment*, 62, 295–315.

De Santis, A. and Chuvieco, E. (2007). Burn severity estimation from remotely sensed data: Performance of simulation versus empirical models. *Remote Sensing of Environment*, 108, 422–435.

Di Gregorio, A. and Jansen, L.J.M. (2000). *Land Cover Classification System (LCCS). Classification Concepts and User Manual.* Rome, Italy: FAO.

Dillworth, M.E., Whistler, J.L., and Merchant, J.W. (1994). Measuring landscape structure using geographic and geometric windows. *Photogrammetric Engineering and Remote Sensing*, 60, 1215–1224.

Dils, B., Buchwitz, M., Reuter, M., Schneising, O., Boesch, H., Parker, R., Guerlet, S. et al. (2014). The Greenhouse Gas Climate Change Initiative (GHG-CCI): Comparative validation of GHG-CCI SCIAMACHY/ENVISAT and TANSO-FTS/GOSAT $CO_2$ and $CH_4$ retrieval algorithm products with measurements from the TCCON. *Atmospheric Measurement Techniques*, 7, 1723–1744.

Dimri, V. and Turcotte, D.L. (2005). *Fractal Behaviour of the Earth System.* Berlin, Germany: Springer-Verlag.

Diner, D.J., Asner, G.P., Davies, R., Knyazikhin, Y., Muller, J.P., Nolin, A.W., Pinty, B., Schaaf, C., and Stroeve, J. (1999). New directions in Earth observing: Scientific applications of multiangle remote sensing. *Bulletin of the American Meteorological Society*, 80, 2209–2228.

Doraiswamy, P.C., Sinclair, T.R., Hollinger, S., Akhmedov, B., Stern, A., and Prueger, J. (2005). Application of MODIS derived parameters for regional crop yield assessment. *Remote Sensing of Environment*, 97, 197–202.

Dozier, J. (1989). Spectral signature of alpine snow cover from Landsat Thematic Mapper. *Remote Sensing of Environment*, 28, 9–22.

Dozier, J. and Strahler, A.H. (1983). Ground investigations in support of remote sensing. In R.N. Colwell (Ed.), *Manual of Remote Sensing* (pp. 959–986). Falls Church, VA: American Society of Photogrammetry.

Drake, J.B., Dubayah, R.O., Clark, D.B., Knox, R.G., Blair, J.B., Hofton, M.A., Chazdon, R.L., Weishampel, J.F., and Prince, S.D. (2002). Estimation of tropical forest structural characteristics using large-footprint lidar. *Remote Sensing of Environment*, 79, 305–319.

Drake, J.B. and Weishampel, J.F. (2000). Multifractal analysis of canopy height measures in a longleaf pine savanna. *Forest Ecology and Management*, 128, 121–127.

Drury, S.A. (1998). *Images of the Earth: A Guide to Remote Sensing.* Oxford, U.K.: Oxford University Press.

Dubayah, R.O. and Drake, J.B. (2000). Lidar remote sensing of forestry. *Journal of Forestry*, 98, 44–46.

Duda, R.D. and Hart, P.E. (1973). *Pattern Classification and Scene Analysis.* New York: John Wiley & Sons.

Duguay, C.R. and LeDrew, E.F. (1992). Estimating surface reflectance and albedo from Landsat-5 TM over rugged terrain. *Photogrammetric Engineering and Remote Sensing*, 58, 551–558.

Dymond, J.R., Shepherd, J.D., and Qi, J. (2001). A simply physical model of vegetation reflectance for standardising optical satellite imagery. *Remote Sensing of Environment*, 77, 230–239.

Eastman, J.R., McKendry, J., and Fulk, M. (1994). *UNITAR Explorations in GIS Technology*, Vol. 1: *Change and Time Series Analysis*. Geneva, Switzerland: U.N. Institute for Training and Research.

Eastman, R. and Fulk, M. (1993). Long sequence time series evaluation using standarized principal components. *Photogrammetric Engineering and Remote Sensing*, 59, 1307–1312.

Ehleringer, J.R. and Field, C.B. (Eds.) (1993). *Scaling Physiological Processes. Leaf to Globe*. San Diego, CA: Academic Press.

Eidenshink, J. (2006). A 16-year time series of 1 km AVHRR satellite data of the conterminous United States and Alaska. *Photogrammetric Engineering and Remote Sensing*, 72, 1027–1035.

Elachi, C. (1987). *Introduction to the Physics and Techniques of Remote Sensing*. New York: John Wiley & Sons.

Emery, W.J., Brown, W.J., and Nowak, Z.P. (1989). AVHRR image navigation: Summary and review. *Photogrammetric Engineering and Remote Sensing*, 55, 1175–1183.

Estes, J.E. and Lenger, L.W. (Eds.) (1974). *Remote Sensing: Techniques for Environmental Analysis*. Santa Barbara, CA: Hamilton Publishing Co.

Estes, J.E. and Simonett, D.S. (1975). Fundamentals of image interpretation. In R.G. Reeves (Ed.), *Manual of Remote Sensing* (pp. 869–1076). Falls Church, VA: American Society of Photogrammetry.

Eugenio, F. and Marqués, F. (2003). Automatic satellite image georeferencing using a contour-matching approach. *IEEE Transactions on Geoscience and Remote Sensing*, 41, 2869–2880.

European Commission (1993). *Corine Land Cover: Guide Technique*. Luxembourg, Europe: Office for Official Publications of the European Union.

Eva, H. and Lambin, E.F. (1998). Remote sensing of biomass burning in tropical regions: Sampling issues and multisensor approach. *Remote Sensing of Environment*, 64, 292–315.

Fang, H. and Liang, S. (2003). Retrieving leaf area index with a neural network method: Simulation and validation. *IEEE Transactions on Geoscience and Remote Sensing*, 41, 2052–2062.

Fernández, A., Illera, P., and Casanova, J.L. (1997). Automatic mapping of surfaces affected by forest fires in Spain using AVHRR NDVI composite image data. *Remote Sensing of Environment*, 60, 153–162.

Ferreira, N.C., Ferreira, L., Huete, A.R., and Ferreira, M.E. (2007). An operational deforestation mapping system using MODIS data and spatial context analysis. *International Journal of Remote Sensing*, 28, 47–62.

Ferris, J. and Congalton, R.G. (1989). Satellite and Geographic Information System estimates of Colorado River Basin snowpack. *Photogrammetric Engineering and Remote Sensing*, 55, 1629–1635.

Ferro, C.J.S. and Warner, T.A. (2002). Scale and texture in digital image classification. *Photogrammetric Engineering and Remote Sensing*, 68, 51–63.

Fisher, P.F. and Pathirana, S. (1990). The evaluation of fuzzy membership of land cover classes in the suburban zone. *Remote Sensing of Environment*, 34, 121–132.

Fitzegarld, R.W. and Lees, B.G. (1994). Assessing the classification accuracy of multisource remote sensing data. *Remote Sensing of Environment*, 47, 362–368.

Flasse, S.P. and Ceccato, P. (1996). A contextual algorithm for AVHRR fire detection. *International Journal of Remote Sensing*, 17, 419–424.

Florini, A.M. and Dehqanzada, Y.A. (2001). The global politics of commercial observation satellites. In J.C. Baker, K.M. O'Connell and R.A. Williamson (Eds.), *Commercial Observation Satellites: At the Leading Edge of Global Transparency* (pp. 433–448). Santa Monica, CA: RAND—ASPRS.

Folk, M., McGrath, R.E., and Yeager, N. (1999). HDF: An update and future directions. In *IEEE 1999 International Geoscience and Remote Sensing Symposium, 1999. IGARSS'99 Proceedings* (pp. 273–275). Hamburg: IEEE.

Foody, G.M. (1992). A fuzzy sets approach to the representation of vegetation continua from remotely sensed data: An example from lowland heath. *Photogrammetric Engineering and Remote Sensing*, 58, 221–225.

Foody, G.M. (1995). Using prior knowledge in artificial neural network classification with a minimal training set. *International Journal of Remote Sensing*, 16, 301–312.

Foody, G.M. (2001). Monitoring the magnitude of land-cover change around the Southern limits of the Sahara. *Photogrammetric Engineering and Remote Sensing*, 67, 841–847.

Foody, G.M. and Cox, D.P. (1994). Sub-pixel land-cover composition estimation using a linear mixture model and fuzzy membership functions. *International Journal of Remote Sensing*, 15, 619–631.

Forman, R.T.T. and Godron, M. (1986). *Landscape Ecology*. New York: John Wiley & Sons.

Fourty, T. and Baret, F. (1997). Vegetation water and dry matter contents estimated from top-of-the atmosphere reflectance data: a simulation study. *Remote Sensing of Environment*, 61, 34–45.

Francis, P. and Jones, P. (1984). *Images of Earth*. London, U.K.: George Phillip & Son Ltd.

Frankenberg, C., Fisher, J.B., Worden, J., Badgley, G., Saatchi, S.S., Lee, J.E., Toon, G.C., Butz, A., Jung, M., and Kuze, A. (2011). New global observations of the terrestrial carbon cycle from GOSAT: Patterns of plant fluorescence with gross primary productivity. *Geophysical Research Letters*, 38, L17706.

Franklin, S.E. (Ed.) (2001). *Remote Sensing for Sustainable Forest Management*. Boca Raton, FL: Lewis Publishers.

Fraser, R.H., Fernandes, R., and Latifovic, R. (2003). Multi-temporal mapping of burned forest over Canada using satellite-based change metrics. *Geocarto International*, 18, 37–47.

Friedl, M.A. and Brodley, C.E. (1997). Decision tree classification of land cover from remotely sensed data. *Remote Sensing of Environment*, 61, 399–409.

Friedl, M.A., Brodley, C.E., and Strahler, A.H. (1999). Maximizing land cover classification accuracies produced by decision trees at continental to global scales. *IEEE Transactions on Geoscience and Remote Sensing*, 37, 969–977.

Friedl, M.A., McIver, D.K., Hodges, J.C.F., Zhang, X.Y., Muchoney, D., Strahler, A.H., Woodcock, C.E. et al. (2002). Global land cover mapping from MODIS: Algorithms and early results. *Remote Sensing of Environment*, 83, 287–302.

Frohn, R.C. (1998). *Remote Sensing for Landscape Ecology: New Metric Indicators for Monitoring, Modeling and Assessment of Ecosystems*. Boca Raton, FL: Lewis Publishers.

Fujii, T. and Fukuchi, T. (2005). *Laser Remote Sensing*. Boca Raton, FL: CRC Press.

Fung, T. (1992). Land use and land cover change detection with Landsat MSS and SPOT-HRV data in Hong Kong. *Geocarto International*, 7, 33–40.

Fung, T. and LeDrew, E. (1987). Application of principal components analysis to change detection. *Photogrammetric Engineering and Remote Sensing*, 53, 1649–1658.

Fung, T. and LeDrew, E. (1988). The determination of optimal threshold levels for change detection using various accuracy indices. *Photogrammetric Engineering and Remote Sensing*, 54, 1449–1454.

Gallo, K.P. and Eidenshink, J.C. (1988). Differences in visible and near-IR responses, and derived vegetation indices, for the NOAA-9 and NOAA-10 AVHRRs: A case study. *Photogrammetric Engineering and Remote Sensing*, 54, 485–490.

Gao, F., Masek, J., and Wolfe, R.E. (2009). Automated registration and orthorectification package for Landsat and Landsat-like data processing. *Journal of Applied Remote Sensing*, 3, 033515–033520.

Gao, F., Schaaf, C.B., Strahler, A.H., Jin, Y., and Li, X. (2003). Detecting vegetation Structure using a kernel-based BRDF model. *Remote Sensing of Environment*, 86, 198–205.

Garcia, M., Chuvieco, E., and Danson, F.M. (2010). Estimating biomass carbon stocks for a Mediterranean forest in central Spain using LiDAR height and intensity data. *Remote Sensing of Environment*, 114, 816–830.

Garcia, M., Danson, F.M., Riano, D., Chuvieco, E., Ramirez, F.A., and Bandugula, V. (2011a). Terrestrial laser scanning to estimate plot-level forest canopy fuel properties. *International Journal of Applied Earth Observation and Geoinformation*, 13, 636–645.

García, M., Popescu, S., Riaño, D., Zhao, K., Neuenschwander, A., Agca, M., and Chuvieco, E. (2012). Characterization of canopy fuels using ICESat/GLAS data. *Remote Sensing of Environment*, 123, 81–89.

Garcia, M., Riano, D., Chuvieco, E., Salas, J., and Danson, F.M. (2011b). Multispectral and LiDAR data fusion for fuel type mapping using Support Vector Machine and decision rules. *Remote Sensing of Environment*, 115, 1369–1379.

Gardner, A.S., Moholdt, G., Cogley, J.G., Wouters, B., Arendt, A.A., Wahr, J., Berthier, E., Hock, R., Pfeffer, W.T., and Kaser, G. (2013). A reconciled estimate of glacier contributions to sea level rise: 2003 to 2009. *Science*, 340, 852–857.

Garguet-Duport, B. (1997). Wavemerg: A multiresolution software for merging SPOT panchromatic and SPOT multispectral data. *Environmental Modeling and Software*, 12, 85–92.

Gates, D.M., Keegan, H.J., Schleter, J.C., and Weidner, V.R. (1965). Spectral properties of plants. *Applied Optics*, 4, 11–20.

Gibson, P. and Power, C.H. (2000a). *Introductory Remote Sensing: Principles and Concepts*. London, U.K.: Routledge.

Gibson, P. and Power, C.H. (2000b). *Introductory Remote Sensing: Digital Image Processing and Applications*. London, U.K.: Routledge.

Giglio, L., Descloitres, J., Justice, C.O., and Kaufman, Y.J. (2003). An enhanced contextual fire detection algorithm for MODIS. *Remote Sensing of Environment*, 87, 273–282.

Giglio, L., Randerson, J.T., van der Werf, G.R., Kasibhatla, P.S., Collatz, G.J., Morton, D.C., and DeFries, R.S. (2010). Assessing variability and long-term trends in burned area by merging multiple satellite fire products. *Biogeosciences Discussion*, 7, 1171–1186, doi:1110.5194/bg-1177-1171-2010.

Gilabert, M.A., Conese, C., and Maselli, F. (1994). An atmospheric correction method for the automatic retrieval of surface reflectances from TM images. *International Journal of Remote Sensing*, 15, 2065–2086.

Gilabert, M.A., González Piqueras, J., García-Haro, F.J., and Meliá, J. (2002). A generalized soil-adjusted vegetation index. *Remote Sensing of Environment*, 82, 303–310.

Ginevan, M.E. (1979). Testing land-use map accuracy: Another look. *Photogrammetric Engineering and Remote Sensing*, 45, 1371–1377.

Gislason, P.O., Benediktsson, J.A., and Sveinsson, J.R. (2006). Random forests for land cover classification. *Pattern Recognition Letters*, 27, 294–300.

Gitas, I.Z., Mitri, G.H., and Ventura, G. (2004). Object-based image classification for burned area mapping of Creus Cape, Spain, using NOAA-AVHRR imagery. *Remote Sensing of Environment*, 92, 409–413.

Gitelson, A., Kaufmam, J.Y., Stark, R., and Rundquist, D. (2002). Novel algorithms for remote estimation of vegetation fraction. *Remote Sensing of Environment*, 80, 76–87.

Gitelson, A.A. (2004). Wide dynamic range vegetation index for remote quantification of biophysical characteristics of vegetation. *Journal of Plant Physiology*, 161, 165–173.

Gitelson, A.A., Kogan, F., Zakarin, E., Spivak, L., and Lebed, L. (1998). Using AVHRR data for quantitative estimation of vegetation conditions: Calibration and validation. *Advances in Space Research*, 22, 673–676.

Gobron, N., Pinty, B., Verstraete, M.M., and Widlowski, J.L. (2000). Advanced vegetation indices optimized for up-coming sensors: Design, performance, and applications. *IEEE Transactions on Geoscience and Remote Sensing*, 38, 2489–2505.

Goel, N.S. (1988). Models of vegetation canopy reflectance and their use in estimation of biophysical parameters from reflectance data. *Remote Sensing Reviews*, 4, 1–212.

Goetz, S., Bunn, A., Fiske, G., and Houghton, R.A. (2005). Satellite-observed photosynthetic trends across boreal North America associated with climate and fire disturbance. *Proceedings of National Academy of Sciences of the United States of America*, 102, 13521–13525.

Goetz, S., Fiske, G., and Bunn, A. (2006). Using satellite time-series data sets to analyze fire disturbance and forest recovery across Canada. *Remote Sensing of Environment*, 92, 411–423.

Gogineni, S., Ampe, J., and Budihardjo, A. (1991). Radar estimates of soil moisture over the Konza Prairie. *International Journal of Remote Sensing*, 12, 2425–2432.

Goldshleger, N., Ben-Dor, E., Lugassi, R., and Eshel, G. (2010). Soil degradation monitoring by remote sensing: Examples with three degradation processes. *Soil Science Society of America Journal*, 74, 1433–1445.

Gong, P. and Howarth, P.J. (1990a). An assessment of some factors influencing multispectral land-cover classification. *Photogrammetric Engineering and Remote Sensing*, 56, 597–603.

Gong, P. and Howarth, P.J. (1990b). The use of structural information for improving land-cover classification accuracies at the rural-urban fringe. *Remote Sensing of Environment*, 56, 67–73.

Gong, P., Marceau, O.J., and Howarth, P.J. (1992). A comparison of spatial feature extraction algorithms for land-use classification with SPOT-HRV data. *Remote Sensing of Environment*, 40, 137–151.

Gong, P., Miller, J.R., and Spanner, M. (1994). Forest canopy closure from classification and spectral unmixing of scene components. Multisensor evaluation of an open canopy. *IEEE Transactions on Geoscience and Remote Sensing*, 32, 1067–1079.

González, F. and López, S. (1991). Using contextual information to improve land use classification of satellite images in Central Spain. *International Journal of Remote Sensing*, 12, 2227–2235.

Gonzalez, R.C. and Wintz, P. (1977). *Digital Image Processing*. Reading, MA: Addison & Wesley.

Goodenough, D.G., Dyk, A., Niemann, K.O., Pearlman, J.S., Chen, H., Han, T., Murdoch, M., and West, C. (2003). Processing hyperion and ALI for forest classification. *IEEE Transactions on Geoscience and Remote Sensing*, 41, 1321–1331.

Gopal, S. and Woodcock, C. (1996). Remote sensing of forest change using artificial neural networks. *IEEE Transactions on Geoscience and Remote Sensing*, 34, 398–404.

Gordon, D.K. and Philipson, R. (1986). A texture-enhancement procedure for separating orchard from forest in Thematic Mapper data. *International Journal of Remote Sensing*, 7, 301–304.

Gordon, S.I. (1980). Utilizing Landsat imagery to monitor land use change: A case study in Ohio. *Remote Sensing of Environment*, 9, 189–196.

Goward, S.N. (2001). Landsat—30 years and counting. *Remote Sensing of Environment*, 78, 1–2.

Goward, S.N. and Dye, D.G. (1987). Evaluating North American net productivity with satellite observations. *Advances in Space Research*, 7, 165–174.

Goward, S.N. and Huemmrich, K.F. (1992). Vegetation canopy PAR absorptance and the normalized difference vegetation index: An assessment using the SAIL model. *Remote Sensing of Environment*, 39, 119–139.

Goward, S.N., Markham, B., Dye, D.G., Dulaney, W., and Yang, I. (1991). Normalized difference vegetation index measurements from the advanced very high resolution radiometer. *Remote Sensing of Environment*, 35, 257–277.

Gower, J.F.R. (Ed.) (1994). *Oceanography from Space*. New York: Plenum Press.

Granell, C., Gould, M., Bernabé, M., and Manso, M. (2007). Spatial data infrastructures. In H.A. Karimi (Ed.), *Handbook of Research on Geoinformatics* (pp. 36–42). Hershey, PA: Information Science Reference.

Green, A.A., Berman, M., Switzer, P., and Craig, M.D. (1988). A transformation for ordering multispectral data in terms of image quality with implications for noise removal. *IEEE Transactions on Geoscience and Remote Sensing*, 26, 65–74.

Green, K. (2000). Selecting and interpreting high-resolution images. *Journal of Forestry*, 98, 37–39.

Gulinck, H., Walpot, D., Janssens, P., and Dries, I. (1991). The visualization of corridors in the landscape using SPOT data. In D. Saunders and R.J. Hobbs (Eds.), *Nature Conservation 2: The Role of Corridors* (pp. 9–17). Surrey, U.K.: Beatty & Sons.

Gurney, C.M. and Townshend, J.R.G. (1983). The use of contextual information in the classification of remotely sensed data. *Photogrammetric Engineering and Remote Sensing*, 49, 55–64.

Gutman, G.G. (1991). Vegetation indices from AVHRR: an update and future prospects. *Remote Sensing of Environment*, 35, 121–136.

Haack, B., Bryant, N., and Adams, S. (1987). An assessment of Landsat MSS and TM data for urban and near-urban land-cover digital classification. *Remote Sensing of Environment*, 21, 201–213.

Haines-Young, R., Green, D.R., and Cousins, S.H. (Eds.) (1993). *Landscape Ecology and Geographic Information Systems*. London, U.K.: Taylor & Francis Group.

Hall, D.K. (1988). Assessment of Polar climate change using satellite technology. *Reviews of Geophysics*, 26, 26–39.

Hall, D.K. and Martinec, J. (1985). *Remote Sensing of Ice and Snow*. London, U.K.: Chapman & Hall.

Hall, D.K. and Riggs, G.A. (2007). Accuracy assessment of the MODIS snow products. *Hydrological Processes*, 21, 1534–1547.

Hall, D.K., Riggs, G.A., and Salomonson, V.V. (1995). Development of methods for mapping global snow cover using moderate resolution imaging spectroradiometer data. *Remote Sensing of Environment*, 54, 127–140.

Hall, F.G., Botkin, D.B., Strebel, D.E., Woods, K.D., and Goetz, S.J. (1991a). Large-scale patterns of forest succession as determined by remote sensing. *Ecology*, 72, 628–640.

Hall, F.G., Strebel, D.E., Nickeson, E., and Goetz, S.J. (1991b). Radiometric rectification: Toward a common radiometric response among multidate multisensor images. *Remote Sensing of Environment*, 35, 11–27.

Hall, R.J., Crown, P.H., and Titus, S.J. (1984). Change detection methodology for aspen defoliation with Landsat MSS digital data. *Canadian Journal of Remote Sensing*, 10, 135–142.

Hansen, M., Dubayah, R., and Defries, R. (1996). Classification trees: An alternative to traditional land cover classifiers. *International Journal of Remote Sensing*, 17, 1075–1081.

Hansen, M.C., Defries, R.S., Townshend, J.R.G., and Sohlberg, R. (2000). Global land cover classification at 1 Km spatial resolution using a classification tree approach. *International Journal of Remote Sensing*, 21, 1331–1364.

Hansen, M.C., DeFries, R.S., Townshend, J.R.G., Sohlberg, R., Dimiceli, C., and Carroll, M. (2002). Towards an operational MODIS continuous field of percent tree cover algorithm: Examples using AVHRR and MODIS data. *Remote Sensing of Environment*, 83, 303–319.

Hansen, M.J., Franklin, S.E., Woudsma, C.G., and Peterson, M. (2001). Caribou habitat mapping and fragmentation analysis using Landsat MSS, TM, and GIS data in the North Columbia Mountains, British Columbia, Canada. *Remote Sensing of Environment*, 77, 50–65.

Hantson, S. and Chuvieco, E. (2011). Evaluation of different topographic correction methods for Landsat imagery. *International Journal of Applied Earth Observation and Geoinformation*, 13, 691–700.

Hantson, S., Pueyo, S., and Chuvieco, E. (2015). Global fire size distribution is driven by human impact and climate. *Global Ecology and Biogeography*, 24, 77–86.

Haralick, R.M., Shanmugan, K., and Dinstein, I. (1973). Textural features for image classification. *IEEE Transactions on Systems, Man and Cybernetics*, SMC-3, 610–621.

Hargrove, W.W., Gardner, R.H., Turner, M.G., Romme, W.H., and Despain, D.G. (2000). Simulating fire patterns in heterogeneous landscapes. *Ecological Modelling*, 135, 243–263.

Harnapp, V. (1978). Landsat imagery: A tool for updating land use in Gulf Coast Mexico. *Journal of Geography*, 78, 141–144.

Harper, D. (1983). *Eye in the Sky. Introduction to Remote Sensing.* Montreal, Quебec, Canada: Multiscience Publications Ltd.

Harris, R. (1987). *Satellite Remote Sensing. An Introduction.* London, U.K.: Routledge & Kegan Paul.

Hay, A.M. (1979). Sampling designs to test land-use map accuracy. *Photogrammetric Engineering and Remote Sensing*, 45, 529–533.

Hay, S.I., Randolph, S.E., and David, J.R. (2000). *Remote Sensing and Geographical Information Systems in Epidemiology.* New York: Academic Press.

Hazel, G.G. (2001). Object-level change detection in spectral imagery. *IEEE Transactions on Geoscience and Remote Sensing*, 39, 553–561.

Henderson, F.M. and Lewis, A.J. (Eds.) (1998). *Principals and Applications of Imaging Radar.* New York: John Wiley & Sons, Inc.

Henebry, G.M. (1993). Detecting change in grasslands using measures of spatial dependence with Landsat-TM data. *Remote Sensing of Environment*, 46, 223–234.

Heo, J. and Fitzhugh, T.W. (2000). A standardized radiometric normalization method for change detection using remotely sensed imagery. *Photogrammetric Engineering and Remote Sensing*, 66, 173–181.

Hepner, G., Logan, T., Ritter, N., and Bryant, N. (1990). Artificial neural network classification using a minimal training set: Comparison to conventional supervised classification. *Photogrammetric Engineering and Remote Sensing*, 56, 469–473.

Hermosilla, T., Wulder, M.A., White, J.C., Coops, N.C., and Hobart, G.W. (2015). An integrated Landsat time series protocol for change detection and generation of annual gap-free surface reflectance composites. *Remote Sensing of Environment*, 158, 220–234.

Hill, J., and Sturm, B. (1991). Radiometric correction of multitemporal Thematic Mapper data for use in agricultural land-cover classification and vegetation monitoring. *International Journal of Remote Sensing*, 12, 1471–1491.

Hilwig, F.W. (1980). Visual interpretation of multitemporal Landsat data for inventories of natural resources. *ITC Journal*, 2, 297–327.

Hixson, M., Scholz, D., Fuhs, N., and Akiyama, T. (1980). Evaluation of several schemes for classification of remotely sensed data. *Photogrammetric Engineering and Remote Sensing*, 46, 1547–1553.

Hlavka, C.A. and Spanner, M.A. (1995). Unmixing AVHRR imagery to assess clearcuts and forest regrowth in Oregon. *IEEE Transactions on Geoscience and Remote Sensing*, 33, 788–795.

Ho, D. and Asem, A. (1986). NOAA AVHRR image referencing. *International Journal of Remote Sensing*, 7, 895–904.

Hobbs, R.J. and Mooney, H.A. (1990). *Remote Sensing of Biosphere Functioning*. New York: Springer-Verlag.

Hoffer, R.M. (1994). Challenges in developing and applying remote sensing to ecosystem management. In V.A. Sample (Ed.), *Remote Sensing and GIS in Ecosystem Management* (pp. 25–40). Washington, DC: Island Press.

Hoffer, R.M. and Swain, P.H. (1980). Computer processing of satellite data for assessing agriculture, forest, and rangeland resources. In *14th International Society of Photography* (pp. 437–446). Hamburg, Germany.

Holben, B. and Justice, C. (1981). An examination of spectral band ratioing to reduce the topographic effect on remotely sensed data. *International Journal of Remote Sensing*, 2, 115–133.

Holben, B.N. and Shimabukuro, Y.E. (1993). Linear mixing model applied to coarse spatial resolution data from multispectral satellite sensors. *International Journal of Remote Sensing*, 14, 2231–2240.

Hollmann, R., Merchant, C.J., Saunders, R.W., Downy, C., Buchwitz, M., Cazenave, A., Chuvieco, E. et al. (2013). The ESA Climate Change Initiative: Satellite data records for essential climate variables. *Bulletin of the American Meteorological Society*, 94, 1541–1552.

Holz, R.K. (Ed.) (1973). *The Surveillant Science: Remote Sensing of the Environment*. Boston, MA: Houghton Mifflin Company.

Hord, R.M. (1986). *Remote Sensing: Methods and Applications*. New York: John Wiley & Sons.

Hord, R.M. and Brooner, W. (1976). Land use map accuracy criteria. *Photogrammetric Engineering and Remote Sensing*, 42, 671–677.

Howard, J.A. (1991). *Remote Sensing of Forest Resources*. London, U.K.: Chapman & Hall.

Hudson, W.D. and Ramn, C.W. (1987). Correct formulation of the kappa coefficient of agreement. *Photogrammetric Engineering and Remote Sensing*, 53, 421–422.

Huete, A., Didan, K., Miura, T., Rodriguez, E.P., Gao, X., and Ferreira, L.G. (2002). Overview of the radiometric and biophysical performance of the MODIS vegetation indices. *Remote Sensing of Environment*, 83, 195–213.

Huete, A.R. (1988). A soil-adjusted vegetation index (SAVI). *Remote Sensing of Environment*, 25, 295–309.

Huete, A.R., Didan, K., Shimabukuro, Y.E., Ratana, P., Saleska, S.R., Hutyra, L.R., Yang, W., Nemani, R.R., and Myneni, R. (2006). Amazon rainforests green-up with sunlight in dry season. *Geophysical Research Letters*, 33, L06405.

Huete, A.R., Jackson, R.D., and Post, D.F. (1985). Spectral response of a plant canopy with different soil backgrounds. *Remote Sensing of Environment*, 17, 37–53.

Huete, A., Justice, C., and Liu, H. (1994). Development of vegetation and soil indices for MODIS-EOS. *Remote Sensing of Environment*, 49, 224–234.

Huete, A., Liu, H.Q., Batchily, K., and Van Leeuwen, W. (1997). A Comparison of Vegetation Indices over a Global Set of TM Images for EOS-MODIS. *Remote Sensing of Environment*, 59, 440–451.

Hutchinson, C.F. (1982). Techniques for combining Landsat and ancillary data for digital classification improvement. *Photogrammetric Engineering and Remote Sensing*, 48, 123–130.

IDRISI (1992). *IDRISI Technical Reference Manual*. Worcester, MA: Clark University.

Illera, P., Fernández, A., and Delgado, J.A. (1996). Temporal evolution of the NDVI as an indicator of forest fire danger. *International Journal of Remote Sensing*, 17, 1093–1105.

Ingebritsen, S.E. and Lyon, R.J.P. (1985). Principal components analysis of multitemporal image pairs. *International Journal of Remote Sensing*, 6, 687–695.

Irons, J.R., Dwyer, J.L., and Barsi, J.A. (2012). The next Landsat satellite: The Landsat data continuity mission. *Remote Sensing of Environment*, 122, 11–21.

Irons, J.R. and Kennard, R.L. (1986). The utility of Thematic Mapper sensor characteristics for surface mine monitoring. *Photogrammetric Engineering and Remote Sensing*, 52, 389–396.

Itten, K.I. and Meyer, P. (1993). Geometric and radiometric correction of TM data of mountainous forested areas. *IEEE Transactions on Geoscience and Remote Sensing*, 31, 764–770.

Iverson, L.R., Cook, E.A., and Graham, R.L. (1989). A technique for extrapolating and validating forest cover across large regions. Calibrating AVHRR data with TM data. *International Journal of Remote Sensing*, 10, 1805–1812.

Jackson, R.D., Idso, S.B., Reginato, R.J., and Pinter, P.J. (1981). Canopy temperature as a crop water stress indicator. *Water Resources Research*, 17, 1133–1138.

Jacquemoud, S. (1990). PROSPECT: A model to leaf optical properties spectra. *Remote Sensing of Environment*, 34, 74–91.

Jacquemoud, S., Baret, F., Andrieu, B., Danson, F.M., and Jaggard, K. (1995). Extraction of vegetation biophysical parameters by inversion of the PROSPECT+SAIL models on sugar beet canopy reflectance data. Application to TM and AVIRIS sensors. *Remote Sensing of Environment*, 52, 163–172.

Jensen, J.R. (1981). Urban change detection mapping using Landsat digital data. *The American Cartographer*, 8, 127–147.

Jensen, J.R. (1996). *Introductory Digital Image Processing: A Remote Sensing Perspective.* Upper Saddle River, NJ: Prentice-Hall.

Jensen, J.R. (2000). *Remote Sensing of the Environment: An Earth Resource Perspective.* Upper Saddle River, NJ: Prentice-Hall.

Jensen, J.R. and Jensen, R.R. (2012). *Introductory Geographic Information Systems.* Boston, MA: Prentice Hall.

Ji, W., Civco, D.L., and Kennard, W.C. (1992). Satellite remote bathymetry: A new mechanism for modelling. *Photogrammetric Engineering and Remote Sensing*, 58, 545–549.

Jiang, Z., Huete, A.R., Chen, J., Chen, Y., Li, J., Yan, G., and Zhang, X. (2006). Analysis of NDVI and scaled difference vegetation index retrievals of vegetation fraction. *Remote Sensing of Environment*, 101, 366–378.

Jimenez-Munoz, J.C., Sobrino, J., Skokovic, D., Mattar, C., and Cristobal, J. (2014). Land surface temperature retrieval methods from Landsat-8 thermal infrared sensor data. *IEEE Geoscience and Remote Sensing Letters*, 11, 1840–1843.

Jong, S.M.d. (1993). An application of spatial filtering techniques for land cover mapping using TM-images. *Geocarto International*, 1, 43–49.

Jordan, C.F. (1969). Derivation of leaf area index from quality of light on the forest floor. *Ecology*, 50, 663–666.

Juppenlatz, M. and Tian, X. (1996). *Geographic Information Systems and Remote Sensing: Guidelines for Use by Planners and Decision Makers.* Roseville, New South Wales, Australia: McGraw-Hill.

Jurdao, S., Yebra, M., Guerschman, J.P., and Chuvieco, E. (2013). Regional estimation of woodland moisture content by inverting Radiative Transfer Models. *Remote Sensing of Environment*, 132, 59–70.

Justice, C.O. and Townshend, J.R.G. (1981). The use of Landsat data for land cover inventories of Mediterranean lands. In J.R.G. Townshend (Ed.), *Terrain Analysis and Remote Sensing* (pp. 135–153). London, U.K.: George Allen & Unwin.

Kasischke, E.S., Bourgeau-Chavez, L.L., and Johnstone, J.F. (2007). Assessing spatial and temporal variations in surface soil moisture in fire-disturbed black spruce forests in Interior Alaska using spaceborne synthetic aperture radar imagery—Implications for post-fire tree recruitment. *Remote Sensing of Environment*, 108, 42–58.

Kasischke, E.S., French, N.H.F., Harrell, P., Christensen, N.L., Ustin, S.L., and Barry, D. (1993). Monitoring of wildfires in Boreal Forests using large area AVHRR NDVI composite image data. *Remote Sensing of Environment*, 45, 61–71.

Kaufman, Y. and Tanré, D. (1992). Atmospherically Resistant Vegetation Index (ARVI) for EOS-MODIS. *IEEE Transactions on Geoscience and Remote Sensing*, 30, 261–270.

Kauth, R.J. and Thomas, G.S. (1976). The Tasseled Cap. A graphic description of the spectral-temporal development of agricultural crops as seen by Landsat. In *Symposium on Machine Processing of Remotely Sensed Data* (pp. 41–51). West Lafayette, IN: Purdue.

Kelly, R. and Hall, D.K. (2008). Remote sensing of terrestrial snow and ice for global change studies. In E. Chuvieco (Ed.), *Earth Observation and Global Change* (pp. 189–220). New York: Springer.

Kennedy, R.E., Yang, Z., and Cohen, W.B. (2010). Detecting trends in forest disturbance and recovery using yearly Landsat time series: 1. LandTrendr—Temporal segmentation algorithms. *Remote Sensing of Environment*, 114, 2897–2910.

Kerdiles, H. and Grondona, M.O. (1995). NOAA-AVHRR NDVI decomposition and subpixel classification using linear mixing in the Argentinean Pampa. *International Journal of Remote Sensing*, 95, 1303–1325.

Kerr, J.T. and Ostrovsky, M. (2003). From space to species: Ecological applications for remote sensing. *TRENDS in Ecology and Evolution*, 18, 299–305.

Ketting, R.L. and Landgrebe, D.A. (1975). Classification of multispectral image data by extraction and classification of homogeneous objects. In *Symposium on Machine Classification of Remotely Sensed Data* (pp. 1–11). West Lafayette, IN.

Khlopenkov, K.V. and Trishchenko, A.P. (2007). SPARC: New cloud, snow, and cloud shadow detection scheme for historical 1-km AVHHR data over Canada. *Journal of Atmospheric and Oceanic Technology*, 24, 322–343.

Kidwell, K.B. (1991). *NOAA Polar Orbiter Data (TIROS-N, NOAA-6, NOAA-7, NOAA-8, NOAA-9, NOAA-10, NOAA-11, and NOAA-12). Users Guide*. Washington, DC: NOAA/NESDIS.

King, M.D., Kaufman, Y.J., Tanré, D., and Nakajima, T. (1999). Remote sensing of troposheric aerosols from space: Past, present and future. *Bulletin of the American Meteorological Society*, 80, 2229–2259.

Knipling, E.B. (1970). Physical and Physiological basis for the reflectance of visible and near-infrared radiation from vegetation. *Remote Sensing of Environment*, 1, 155–159.

Kogan, F.N. (1990). Remote sensing of weather impacts on vegetation in non-homogeneous areas. *International Journal of Remote Sensing*, 11, 1405–1419.

Konecny, G. (1986). First results of the European spacelab photogrammetric camera mission. In K.H. Szekielda (Ed.), *Satellite Remote Sensing for Resources Development* (pp. 115–121). London, U.K.: Graham & Trotman Ltd.

Kotchenova, S.Y., Vermote, E.F., Levy, R., and Lyapustin, A. (2008). Radiative transfer codes for atmospheric correction and aerosol retrieval: Intercomparison study. *Applied Optics*, 47, 2215–2226.

Koutsias, N. and Karteris, M. (1998). Logistic regression modelling of multitemporal Thematic Mapper data for burned area mapping. *International Journal of Remote Sensing*, 19, 3499–3514.

Koutsias, N., Karteris, M., and Chuvieco, E. (2000). The use of intensity-hue-saturation transformation of Landsat-5 Thematic Mapper data for burned land mapping. *Photogrammetric Engineering and Remote Sensing*, 66, 829–839.

Kramer, H.J. (2002). *Observation of the Earth and Its Environment: Survey and Missions and Sensors*. Berlin, Germany: Springer-Verlag.

Krasnopolsky, V.M. and Breaker, L.C. (1994). The problem of AVHRR image navigation revisited. *International Journal of Remote Sensing*, 15, 979–1008.

Kraus, K. and Pfeifer, N. (1998). Determination of terrain models in wooded areas with airborne laser scanner data. *ISPRS Journal of Photogrammetry and Remote Sensing*, 53, 193–203.

Kruse, F.A., Lefkoff, A.B., Boardman, J.B., Heidebrecht, K.B., Shapiro, A.T., Barloon, P.J., and Goetz, A.F.H. (1993a). The spectral image processing (SIPS)—Interactive visualization and analysis of imaging specrometer data. *Remote Sensing of Environment*, 44, 145–163.

Kruse, F.A., Lefkoff, A.B., and Dietz, J.B. (1993b). Expert system-based mineral mapping in Northern Death Valley, California/Nevada, using the airborne visible/infrared imaging spectrometer (AVIRIS). *Remote Sensing of Environment*, 44, 309–336.

Kumar, R. and Silva, L.F. (1977). Separability of agricultural cover types by remote sensing in the visible and infrared wavelength regions. *IEEE Transactions on Geoscience Electronics*, 15, 49–59.

Lachowski, H., Maus, P., and Platt, B. (1992). Integrating remote sensing with GIS. Procedures from the forest service. *Journal of Forestry*, 90, 16–21.

Lam, N.S. (1990). Description and measurement of Landsat-TM images using fractals. *Photogrammetric Engineering and Remote Sensing*, 56, 187–195.

Lasaponara, R. and Masini, N. (2012). *Satellite Remote Sensing: A New Tool for Archaeology.* New York: Springer Science.

Latifovic, R., Trishchenko, A.P., Chen, J., Park, W.B., Khlopenkov, K.V., Fernandes, R., Pouliot, D. et al. (2005). Generating historical AVHRR 1 km baseline satellite data records over Canada suitable for climate change studies. *Canadian Journal of Remote Sensing*, 31, 324–346.

Latifovic, R., Zhu, Z.L., Cihlar, J., Giri, C., and Olthof, I. (2004). Land cover mapping of north and central America—Global Land Cover 2000. *Remote Sensing of Environment*, 89, 116–127.

Lauer, D.T., Estes, J.E., Jensen, J.R., and Greenlee, D.D. (1991). Institutional issues affecting the integration and use of remotely sensed data and geographic information systems. *Photogrammetric Engineering and Remote Sensing*, 57, 647–654.

Lawrence, R.L., Wood, S.D., and Sheley, R.L. (2006). Mapping invasive plants using hyperspectral imagery and Breiman Cutler classifications (RandomForest). *Remote Sensing of Environment*, 100, 356–362.

Lawrence, R.L. and Wright, A. (2001). Rule-based classification systems using classification and regression tree (CART) analysis. *Photogrammetric Engineering and Remote Sensing*, 67, 1137–1142.

Leblon, B., Kasischke, E.S., Alexander, M.E., Doyle, M., and Abbott, M. (2002). Fire danger monitoring using ERS-1 SAR images in the case of northern boreal forests. *Natural Hazards*, 27, 231–255.

Leckie, D.G. (1990). Synergism of synthetic aperture radar and visible/infrared data for forest type discrimination. *Photogrammetric Engineering and Remote Sensing*, 56, 1237–1246.

Lefsky, M.A., Harding, D., Cohen, W.B., Parker, G., and Shugart, H.H. (1999). Surface lidar remote sensing of basal area and biomass in deciduous forests of eastern Maryland, USA. *Remote Sensing of Environment*, 67, 83–98.

Leghorn, R.S. and Herken, G. (2001). The origins and evolution of openness in overhead global observations. In J.C. Baker, K.M. O'Connell and R.A. Williamson (Eds.), *Commercial Observation Satellites: At the Leading Edge of Global Transparency* (pp. 17–36). Santa Monica, CA: RAND—ASPRS.

Leprieur, C., Durand, J.M., and Peyron, J.L. (1988). Influence of topography on forest reflectance using Landsat Thematic Mapper and digital terrain data. *Photogrammetric Engineering and Remote Sensing*, 54, 491–496.

Lewis, A.J., Henderson, F.M., and Holcomb, D.W. (1998). Radar fundamentals: The Geoscience perspective. In F.M. Henderson and A.J. Lewis (Eds.), *Principals and Applications of Imaging Radar* (pp. 131–182). New York: John Wiley & Sons, Inc.

Li, F. and Lyons, T.J. (2002). Remote estimation of regional evapotranspiration. *Environmental Modelling and Software*, 17, 61–75.

Li, H. and Reynolds, J.F. (1993). A new contagion index to quantify spatial patterns of landscapes. *Landscape Ecology*, 8, 155–162.

Li, Z.-L., Tang, B.-H., Wu, H., Ren, H., Yan, G., Wan, Z., Trigo, I.F., and Sobrino, J.A. (2013). Satellite-derived land surface temperature: Current status and perspectives. *Remote Sensing of Environment*, 131, 14–37.

Liang, S. (2004). *Quantitative Remote Sensing for Land Surface Characterization*. Hoboken, NJ: Wiley.

Lillesand, T.M. and Kiefer, R.W. (2000). *Remote Sensing and Image Interpretation*. New York: John Wiley & Sons.

Lillo Saavedra, M. and Gonzalo, C. (2006). Spectral or spatial quality for fused satellite imagery? A trade-off solution using the wavelet à trous algorithm. *International Journal of Remote Sensing*, 27, 1453–1464.

Linden, D.S. (2000). Videography for foresters. *Journal of Forestry*, 98, 25–27.

Links, H.F. (1976). Land-use mapping from Skylab S190B photography. *Photogrammetric Engineering and Remote Sensing*, 42, 301–307.

Linnerooth-Bayer, J., Mechler, R., and Pflug, G. (2005). Refocusing disaster aid. *Science*, 309, 1044.

Lira, J. (1987). *La percepción remota. Nuestros ojos desde el espacio*. México: Fondo de Cultura Económica.

Liu, G.-R., Chen, A.J., Lin, T.-H., and Kuo, T.-H. (2002). Applying SPOT data to estimate the aerosol optical depth and air quality. *Environmental Modelling & Software*, 17, 3–9.

Liu, H.q. and Huete, A. (1995). A feedback based modification of the NDVI to minimize canopy background and atmospheric noise. *IEEE Transactions on Geoscience and Remote Sensing*, 33, 457–465.

Liu, X. and Lathrop, R.G. (2002). Urban change detection based on an artificial neural network. *International Journal of Remote Sensing*, 23, 2513–2518.

Lo, C.P. (1986). *Applied Remote Sensing*. New York: Longman Scientific and Technical.

Lo, T.H.C., Scarpace, F.L., and Lillesand, T.M. (1986). Use of multitemporal spectral profiles in agricultural land-cover classification. *Photogrammetric Engineering and Remote Sensing*, 52, 535–544.

Lobo, A. (1997). Image segmentation and discriminant analysis for the identification of land cover units in ecology. *IEEE Transactions on Geoscience and Remote Sensing*, 35, 1136–1145.

Lobser, S. and Cohen, W. (2007). MODIS tasselled cap: Land cover characteristics expressed through transformed MODIS data. *International Journal of Remote Sensing*, 28, 5079–5101.

Longley, P., Goodchild, M.F., Maguire, D.J., and Rhind, D.W. (2005). *Geographic Information Systems and Science*. New York: John Wiley & Sons.

Longmire, P. and Stow, D. (2001). Use of very high spatial resolution remotely sensed imagery for assessing land-cover changes in shrub habitat preserves of Southern California. *Geocarto International*, 16, 47–57.

Loveland, T.R., Reed, B.C., Brown, J.F., Ohlen, D.O., Zhu, J., Yang, L., and Merchant, J.W. (2000). Development of a global land cover characteristics database and IGBP DISCover from 1-km AVHRR data. *International Journal of Remote Sensing*, 21, 1303–1330.

Lucas, R.M., Honzak, M., Foody, G.M., Curran, P.J., and Corves, C. (1993). Characterizing tropical secondary forests using multi-temporal Landsat sensor imagery. *International Journal of Remote Sensing*, 14, 3016–3067.

Lulla, K. (1993). Space shuttle earth observations database for global urban applications. In K.N. Au (Ed.), *Cities of the World as Seen from Space* (pp. 15–19). Hong Kong, China: Geocarto International.

Lulla, K. and Holland, S.D. (1993). NASA electronic still camera (ESC) system used to image the Kamchatka volcanoes from the space shuttle. *International Journal Remote Sensing*, 14, 2745–2746.

Lulla, K.P. and Dessinov, L.V. (Eds.) (2000). *Dynamic Earth Environments. Remote Sensing Observations from Shuttle—Mir Missions*. New York: John Wiley & Sons.

Lusch, R. (1989). Spectral characteristics of land surfaces. In M.D. Nellis, R. Lougeay, and K. Lulla (Eds.), *Current Trends in Remote Sensing Education* (pp. 81–86). Hong Kong, China: Geocarto International Centre.

Madsen, S.N. and Zebker, H.A. (1998). Imaging Radar Interferometry. In F.M. Henderson and A.J. Lewis (Eds.), *Principals and Applications of Imaging Radar* (pp. 359–380). New York: John Wiley & Sons, Inc.

Magnussen, S. and Boudewyn, P. (1998). Derivations of stand heights from airborne laser scanner data with canopy-based quantile estimators. *Canadian Journal of Forest Research*, 28, 1016–1031.

Maguire, D.J., Goodchild, M.F., and Rhind, D.W. (Eds.) (1991). *Geographical Information Systems: Principles and Applications*. London, U.K.: Longman.

Maillard, P. (2003). Comparing texture analysis methods through classification. *Photogrammetric Engineering and Remote Sensing*, 69, 357–367.

Major, D.J., Baret, F., and Guyot, G. (1990). A ratio vegetation index adjusted for soil brightness. *International Journal of Remote Sensing*, 11, 727–740.

Mandelbrot, B.B. (1982). *The Fractal Geometry of Nature*. New York: WH Freeman.

Markham, G.L. and Townshend, J.R.G. (1981). Land cover classification accuracy as a function of sensor spatial resolution. In *15th International Symposium on Remote Sensing of Environment* (pp. 1075–1090). Ann Arbor, MI.

Marsh, S.E., Switzer, P., Kowalick, W.S., and Lyon, R.J.P. (1980). Resolving the percentage of component terrains within single resolution elements. *Photogrammetric Engineering and Remote Sensing*, 46, 1079–1086.

Martin, L.R.G. (1989). Accuracy assessment of Landsat-based visual change detection methods applied to the rural-urban fringe. *Photogrammetric Engineering and Remote Sensing*, 55, 209–215.

Martín, M.P. and Chuvieco, E. (1995). Mapping and evaluation of burned land from multitemporal analysis of AVHRR NDVI images. *EARSeL Advances in Remote Sensing*, 4(3), 7–13.

Martín, M.P. and Chuvieco, E. (1998). Cartografía de grandes incendios forestales en la Península Ibérica a partir de imágenes NOAA-AVHRR. *Serie Geográfica*, 7, 109–128.

Martín, M.P. and Chuvieco, E. (2001). Propuesta de un nuevo índice para cartografía de áreas quemadas: Aplicación a imágenes NOAA-AVHRR y Landsat-TM. *Revista de Teledetección*, 16, 57–64.

Martínez Vega, J. (1996). Una revisión sobre las imágenes espaciales como fuentes cartográficas. *Revista Española de Teledetección*, 6, 37–50.

Maselli, F., Conese, C., De Filippis, T., and Norcini, S. (1995). Estimation of forest parameters through fuzzy classification of TM data. *IEEE Transactions on Geoscience and Remote Sensing*, 33, 77–84.

Maselli, F., Petkov, L., Maracchi, G., and Conese, C. (1996). Eco-climatic classification of Tuscany through NOAA-AVHRR data. *International Journal of Remote Sensing*, 17, 2369–2384.

Massonet, D., Rossi, M., Carmona, C., Adragna, F., Peltzer, G., Feigl, K., and Rabaute, T. (1993). The displacement field of the Landers Earthquake mapped by radar interferometry. *Nature*, 364, 138–142.

Mather, P.M. (1998). *Computer Processing of Remotely Sensed Images*. Chichester, U.K.: John Wiley & Sons.

Mather, P.M. (1999). Land cover classification revisited. In P.M. Atkinson and N.J. Tate (eds.), *Advances in Remote Sensing and GIS Analysis* (pp. 7–16). Chichester, U.K.: John Wiley & Sons.

Mather, P.M. and Coch, M. (2011). *Computer Processing of Remotely Sensed Images* (4th edn.). Chichester, U.K.: John Wiley & Sons.

Mazer, A.S., Martin, M., Lee, M., and Solomon, J.E. (1988). Image Processing software for Imaging Spectrometry Data Analysis. *Remote Sensing of Environment*, 24, 201–210.

McCloy, K.R. (1995). *Resource Management Information Systems: Process and Practice*. London, U.K.: Taylor & Francis Group.

McGwire, K.C. (1996). Cross-validated assessment of geometric accuracy. *Photogrammetric Engineering and Remote Sensing*, 62, 1179–1187.

McMaster, R.B. and Usery, E.L. (2004). *A Research Agenda for Geographic Information Science*. Boca Raton, FL: CRC Press.

Merchant, C.J. (2013). Thermal remote sensing of sea surface temperature. In C. Kuenzer and S. Dech (Eds.), *Thermal Infrared Remote Sensing* (pp. 287–313). Dordrecht, the Netherlands: Springer.

Meroni, M., Rossini, M., Guanter, L., Alonso, L., Rascher, U., Colombo, R., and Moreno, J. (2009). Remote sensing of solar-induced chlorophyll fluorescence: Review of methods and applications. *Remote Sensing of Environment*, 113, 2037–2051.

Metternicht, G. and Zinck, A. (2008). *Remote sensing of soil salinization: Impact on land management*. Boca raton: CRC Press.

Meyer, P., Itten, K.I., Kellenbenberger, T., Sandmeier, S., and Sandmeier, R. (1993). Radiometric corrections of topographically induced effects on Landsat TM data in an alpine environment. *ISPRS Journal of Photogrammetry and Remote Sensing*, 48, 17–28.

Middelkoop, H. and Janssen, L.L. (1991). Implementation of temporal relationships in knowledge based classification of satellite images. *Photogrammetric Engineering and Remote Sensing*, 57, 937–945.

Millington, A.C., Wellens, J., Settle, J.J., and Saull, R.J. (1994). Explaining and monitoring land-cover dynamics in drylands using multi-temporal analysis of NOAA-AVHRR imagery. In G.M. Foody and P.J. Curran (Eds.), *Environmental Remote Sensing from Regional to Global Scales* (pp. 16–43). Chichester, U.K.: John Wiley & Sons.

Milton, E.J., Rollin, E.M., and Emery, D.R. (1995). Advances in field spectroscopy. In F.M. Danson and S.E. Plummer (Eds.), *Advances in Environmental Remote Sensing* (pp. 9–32). Chichester, U.K.: John Wiley & Sons.

Minnaert, M. (1941). The reciprocity principle in lunar photometry. *The Astrophysical Journal*, 93, 403–410.

Moller-Jensen, L. (1990). Knowledge-based classification of an urban area using texture and context information in Landsat TM imagery. *Photogrammetric Engineering and Remote Sensing*, 56, 899–904.

Morain, S.A. and Klankamsorn, B. (1978). Forest mapping and inventory techniques through visual analysis of Landsat imagery: Examples from Thailand. In *12th International Symposium on Remote Sensing of Environment* (pp. 417–426). Ann Arbor, MI.

Moran, M.S., Clarke, T.R., Inoue, Y., and Vidal, A. (1994). Estimating crop water deficit using the relation between surface-air temperature and spectral vegetation index. *Remote Sensing of Environment*, 49, 246–263.

Moran, P. (1948). The interpretation of statistical maps. *Journal of the Royal Statistical Society*, 108, 243–251.

Morisette, J.T., Giglio, L., Csiszar, I., and Justice, C.O. (2005). Validation of the MODIS active fire product over Southern Africa with ASTER data. *International Journal of Remote Sensing*, 26, 4239–4264.

Morsdorf, F., Meier, E., Kotz, B., Itten, K.I., Dobbertin, M., and Allgower, B. (2004). LIDAR-based geometric reconstruction of boreal type forest stands at single tree level for forest and wildland fire management. *Remote Sensing of Environment*, 92, 353–362.

Mouat, D.A., Mattin, G.G., and Lancaster, J. (1993). Remote sensing techniques in the analysis of change detection. *Geocarto International*, 8, 39–50.

Mouillot, F., Schultz, M.G., Yue, C., Cadule, P., Tansey, K., Ciais, P., and Chuvieco, E. (2014). Ten years of global burned area products from spaceborne remote sensing—A review: Analysis of user needs and recommendations for future developments. *International Journal of Applied Earth Observation and Geoinformation*, 26, 64–79.

Muchoki, C.H.K. (1988). Remotely sensed relationships between wooded patch habitats and agricultural landscape type: A basis for ecological planning. In M.R. Moss (Ed.), *Landscape Ecology and Management* (pp. 85–95). Montreal, Québec, Canada: Polyscience Publications.

Mulders, M.A. (1987). *Remote Sensing in Soil Science*. Amsterdam, the Netherlands: Elsevier.

Muller, J.P. (1988a). Computing issues in digital image processing in remote sensing. In J.P. Muller (Ed.), *Digital Image Processing in Remote Sensing* (pp. 1–20). London, U.K.: Taylor & Francis.

Muller, J.P. (Ed.) (1988b). *Digital Image Processing in Remote Sensing*. London, U.K.: Taylor & Francis Group.

Murphy, R.E., Ardanuy, P., DeLuccia, F.J., Clement, J.E., and Schueler, C.F. (2006). The visible infrared imaging radiometer suite. In J.J. Qu, G. Wei, M. Kafatos, R.E. Murphy, and V.V. Salomonson (Eds.), *Earth Science Satellite Remote Sensing*. Vol. 1: *Science and Instruments* (pp. 33–49). Beijing, China: Tsinghua University Press.

Murtha, P.A. (1978). Remote sensing and vegetation damage: A theory for detection and assessment. *Photogrammetric Engineering and Remote Sensing*, 44, 1147–1158.

Myneni, R.B., Hall, F.G., Sellers, P.J., and Marshak, A.L. (1995). The interpretation of spectral vegetation indexes. *IEEE Transactions on Geoscience and Remote Sensing*, 33, 481–486.

Myneni, R.B., Hoffman, S., Knyazikhin, Y., Privette, J.L., Glassy, J., Tian, Y., Wang, Y. et al. (2002). Global products of vegetation leaf area and fraction absorbed PAR from year one of MODIS data. *Remote Sensing of Environment*, 83, 214–231.

Nadeau, L.B. and Englefield, P. (2006). Fine-resolution mapping of wildfire fuel types for Canada: Fuzzy logic modeling for an Alberta pilot area. *Environmental Monitoring and Assessment*, 120, 127–152.

Naesset, E. and Okland, T. (2002). Estimating tree height and tree crown properties using airborne scanning laser in a boreal nature reserve. *Remote Sensing of Environment*, 79, 105–115.

NASA (1974). *Skylab. Earth Resources Data Catalog*. Washington, DC: NASA, JSC 09016.

NASA (1976). *Mission to Earth. Landsat Views of the World*. Greenbelt, MD: Goodard Space Flight Center.

National Geographic (1999). *Satelllite Atlas of the World*. Washington, DC: National Geographic Society.

Ne'eman, G., Fotheringham, C.J., and Keely, J.E. (1999). Patch to landscape patterns in post fire recruitment of a serotinous conifer. *Plant Ecology*, 145, 235–242.

Nelson, R. and Grebowsky, G. (1982). Evaluation of temporal registration of Landsat scenes. *International Journal of Remote Sensing*, 3, 45–50.

Nelson, R., Jimenez, J., Schnell, C.E., Hartshorn, G.S., Gregoire, T.G., and Oderwald, R. (2000). Canopy height models and airborne lasers to estimate forest biomass: Two problems. *International Journal of Remote Sensing*, 21, 2153–2162.

Nemani, R.R., Keeling, C.D., Hashimoto, H., Jolly, W.M., Piper, S.C., Tucker, C.J., Myneni, R.B., and Running, S.W. (2003). Climate-driven increases in global terrestrial net primary production from 1982 to 1999. *Science*, 300, 1560–1563.

Nemani, R.R., Pierce, L., Running, S.W., and Edward, S. (1993). Developing satellite-derived estimates of surface moisture status. *Journal of Applied Meteorology*, 32, 548–557.

Niclòs, R., Valor, E., Caselles, V., Coll, C., and Sánchez, J.M. (2005). In situ angular measurements of thermal infrared sea surface emissivity—Validation of models. *Remote Sensing of Environment*, 94, 83–93.

Nieto, H., Aguado, I., Chuvieco, E., and Sandholt, I. (2010). Dead fuel moisture estimation with MSG–SEVIRI data. Retrieval of meteorological data for the calculation of the equilibrium moisture content. *Agricultural and Forest Meteorology*, 150, 861–870.

Nieto, H., Sandholt, I., Aguado, I., Chuvieco, E., and Stisen, S. (2011). Air temperature estimation with MSG-SEVIRI data: Calibration and validation of the TVX algorithm for the Iberian Peninsula. *Remote Sensing of Environment*, 115, 107–116.

Nourbakhsh, I., Sargent, R., Wright, A., Cramer, K., McClendon, B., and Jones, M. (2006). Mapping disaster zones. *Nature*, 439, 787–788.

O.T.A. (1984). *Remote Sensing and the Private Sector: Issues for Discussion*. A Technical Memorandum. Washington, DC: U.S. Congress, Office of Technology Assessment.

Oki, T. and Kanae, S. (2006). Global hydrological cycles and world water resources. *Science*, 313, 1068–1072.

Okin, G.S., Roberts, D.A., Murray, B., and Okin, W.J. (2001). Practical limits on hyperspectral vegetation discrimination in arid and semiarid environments. *Remote Sensing of Environment*, 77, 212–225.

Olorunfemi, J.F. (1987). Identification and measurement of the areal extent settlements from Landsat. *International Journal of Remote Sensing*, 8, 1839–1843.

O'Neill, R.V., Krummel, J.R., Gardner, R.H., Sugihara, G., and Jackson, B. (1988). Indices of landscape pattern. *Landscape Ecology*, 1, 153–162.

Ouaidrari, H. and Vermote, E.F. (1999). Operational atmospheric correction of Landsat TM data. *Remote Sensing of Environment*, 70, 4–15.

Padilla, M., Stehman, S.V., and Chuvieco, E. (2014). Validation of the 2008 MODIS-MCD45 global burned area product using stratified random sampling. *Remote Sensing of Environment*, 144, 187–196.

Padilla, M., Stehman, S.V., Hantson, S., Oliva, P., Alonso-Canas, I., Bradley, A., Tansey, K., Mota, B., Pereira, J.M., and Chuvieco, E. (2015). Comparing the accuracies of remote sensing global burned area products using stratified random sampling and estimation. *Remote Sensing of Environment*, 160, 114–121.

Palà, V. and Pons, X. (1995). Incorporation of relief in polynomial-based geometric corrections. *Photogrammetric Engineering and Remote Sensing*, 61, 935–944.

Palacios-Orueta, A. and Ustin, S.L. (1996). Multivariate statistical classification of soil spectra. *Remote Sensing of Environment*, 57, 108–118.

Palacios-Orueta, A. and Ustin, S.L. (1998). Remote sensing of soil properties in the Santa Monica Mountains I. Spectral analysis. *Remote Sensing of Environment*, 65, 170–183.

Parkinson, C.L., Ward, A., and King, M.D. (Eds.) (2006). *Earth Science Reference Handbook. A Guide to NASA's Earth Science Program and Earth Observing Satellite Missions*. Washington, DC: National Aeronautics and Space Administration.

Pellemans, A.H., Jordans, R.W., and Allewijn, R. (1993). Merging multispectral and pancromatic SPOT images with respect to the radiometric properties of the sensor. *Photogrammetric Engineering and Remote Sensing*, 59, 81–87.

Pereira, J.M.C. (1999). A comparative evaluation of NOAA/AVHRR vegetation indexes for burned surface detection and mapping. *IEEE Transactions on Geoscience and Remote Sensing*, 37, 217–226.

Perry, C.R. and Lautenschlager, L.F. (1984). Functional equivalence of spectral vegetation indices. *Remote Sensing of Environment*, 14, 169–182.

Peters, A., Rundquist, D.C., and Wilhite, P.A. (1991). Satellite detection of the geographic core of the 1988 Nebraska drought. *Agricultural and Forest Meteorology*, 57, 35–47.

Peters, A.J., Walter-Shea, E.A., Ji, L., Viña, A., Hayes, M., and Svodoba, M.D. (2002). Drought monitoring with NDVI-based standardized vegetation index. *Photogrammetric Engineering and Remote Sensing*, 62, 71–75.

Pettorelli, N., Vik, J.O., Mysterud, A., Gaillard, J.-M., Tucker, C.J., and Stenseth, N.C. (2005). Using the satellite-derived NDVI to assess ecological responses to environmental change. *Trends in Ecology and Evolution*, 20, 503–510.

Pfeifer, N., Reiter, T., Briese, C., and Rieger, W. (1999). Interpolation of high quality ground models from laser scanner data in forested areas. In *ISPRS Workshop: Mapping Forest Structure and Topography by Airborne and Spaceborne Lasers*. La Jolla, CA.

Pilon, P., Howarth, P.J., Bullock, R.A., and Adeniyi, P.O. (1988). An enhanced classification approach to change detection in semi-arid environments. *Photogrammetric Engineering and Remote Sensing*, 54, 1709–1716.

Pinty, B. and Verstraete, M.M. (1992). GEMI: A non-linear index to monitor global vegetation from satellites. *Vegetatio*, 101, 15–20.

Pinzón, J.E., Ustin, S.L., Castañeda, C.M., and Smith, M.O. (1998). Investigation of Leaf biochemistry by Hierarchical Foreground/Background Analysis. *IEEE Transactions on Geoscience and Remote Sensing*, 36, 1913–1927.

Pohl, C. and Van Genderen, J.L. (1998). Review article multisensor image fusion in remote sensing: Concepts, methods and applications. *International Journal of Remote Sensing*, 19, 823–854.

Pons, X. and Solé-Sugrañes, L. (1994). A Simple radiometric correction model to improve automatic mapping of vegetation from multispectral satellite data. *Remote Sensing of Environment*, 48, 191–204.

Pontius, J., Robert G., Shusas, E., and McEachern, M. (2004). Detecting important categorical land changes while accounting for persistence. *Agriculture, Ecosystems and Environment*, 101, 251–268.

Price, J.C. (1987). Calibration of satellite radiometers and the comparison of vegetation indices. *Remote Sensing of Environment*, 21, 15–27.

Privette, J.L., Fowler, C., Wick, G.A., Baldwin, D., and Emery, W.J. (1995). Effects of orbital drift on advanced very high resolution radiometer products: Normalized difference vegetation index and sea surface temperature. *Remote Sensing of Environment*, 53, 164–171.

Pulliainen, J., Kallio, K., Eloheimo, K., Koponen, S., Servomaa, H., Hannonen, H., Tauriainen, S., and Hallikainen, M. (2001). A semi-operative approach to lake water quality retrieval from remote sensing data. *The Science of the Total Environment*, 268, 79–93.

Purkis, S.J. and Klemas, V.V. (2011). *Remote Sensing and Global Environmental Change*. Chichester, U.K.: John Wiley & Sons.

Qi, J., Chehbouni, A., Huete, A.R., Kerr, Y.H., and Sorooshian, S. (1994). A modified soil adjusted vegetation index. *Remote Sensing of Environment*, 47, 1–25.

Quarmby, N.A., Townshend, J.R.G., Settle, J.J., White, K.H., Milnes, M., Hindle, T.L., and Silleos, N. (1992). Linear mixture modelling applied to AVHRR data for crop area estimation. *International Journal of Remote Sensing*, 13, 415–425.

Quattrochi, D.A. and Goodchild, M.F. (Eds.) (1997). *Scale in Remote Sensing and GIS*. Boca Raton, FL: CRC Press.

Quegan, S. (1995). Recent advances in understanding SAR imagery. In F.M. Danson and S.E. Plummer (Eds.), *Advances in Environment Remote Sensing* (pp. 89–104). Chichester, U.K.: John Wiley & Sons.

Rees, G. (1999). *The Remote Sensing Data Book*. Cambridge, MA: Cambridge University Press.

Reginato, R.J., Idso, S.B., Vedder, J.F., Jackson, R.D., Blanchard, M.B., and Goettelman, R. (1976). Soil water content and evaporation determined by thermal parameters obtained from ground based and remote measurements. *Journal of Geophysical Research*, 81, 1617–1620.

Reginato, R.J., Vedder, J.F., Idso, S.B., Jackson, R.D., and Blanchard, M.B. (1977). An evaluation of total solar reflectance and spectral band ratioing techniques for estimating soil water content. *Journal of Geophysical Research*, 82, 2101–2104.

Reinhardt, E.D. and Ringleb, R.V. (1990). Analysis of changes in patterns of a forested landscape following wildfire using Landsat data and landscape ecology methodology. Proceedings *Resource Technology 90* (pp. 83–93). Washington, DC: American Society of Photogrammetry and Remote Sensing.

Rencz, A. and Ryerson, R.A. (Eds.) (1999). *Remote Sensing for the Earth Sciences*. New York: John Wiley & Sons.

Riaño, D., Chuvieco, E., Condés, S., González-Matesanz, J., and Ustin, S.L. (2004a). Generation of crown bulk density for *Pinus sylvestris* L. from lidar. *Remote Sensing of Environment*, 92, 345–352.

Riaño, D., Chuvieco, E., Salas, F.J., and Aguado, I. (2003a). Assessment of different topographic corrections in Landsat-TM data for mapping vegetation types. *IEEE Transactions on Geoscience and Remote Sensing*, 41, 1056–1061.

Riaño, D., Chuvieco, E., Salas, J., Palacios-Orueta, A., and Bastarrica, A. (2002a). Generation of fuel type maps from Landsat TM images and ancillary data in Mediterranean ecosystems. *Canadian Journal of Forest Research*, 32, 1301–1315.

Riaño, D., Chuvieco, E., Ustin, S.L., Salas, J., Rodríguez-Pérez, J.R., Ribeiro, L.M., Viegas, D.X., Moreno, J.M., and Fernández, H. (2007a). Estimation of shrub height for fuel type mapping combining airborne LiDAR and simultaneous color infrared ortho image. *International Journal of Wildland Fire*, 16, 341–348.

Riaño, D., Chuvieco, E., Ustin, S.L., Zomer, R., Dennison, P., Roberts, D., and Salas, J. (2002b). Assessment of vegetation regeneration after fire through multitemporal analysis of AVIRIS images in the Santa Monica Mountains. *Remote Sensing of Environment*, 79, 60–71.

Riaño, D., Meier, E., Allgöwer, B., Chuvieco, E., and Ustin, S.L. (2003b). Modeling airborne laser scanning data for the spatial generation of critical forest parameters in fire behavior modeling. *Remote Sensing of Environment*, 86, 177–186.

Riaño, D., Ruiz, J.A.M., Isidoro, D., Ustin, S.L., and Riaño, D. (2007b). Global spatial patterns and temporal trends of burned area between 1981 and 2000 using NOAA-NASA Pathfinder. *Global Change Biology*, 13, 40–50, doi: 10.1111/j.1365-2486.2006.01268.

Riaño, D., Salas, J., and Chuvieco, E. (2001). Cartografía de modelos de combustible con teledetección: Aportaciones a un desarrollo ambiental sostenible. *Estudios Geográficos*, 62, 309–333.

Riaño, D., Valladares, F., Condés, S., and Chuvieco, E. (2004b). Estimation of leaf area index and covered ground from airborne laser scanner (Lidar) in two contrasting forests. *Agricultural and Forest Meteorology*, 124, 269–275.

Riaño, D., Vaughan, P., Chuvieco, E., Zarco-Tejada, P., and Ustin, S.L. (2005). Estimation of Fuel Moisture Content by Inversion of Radiative Transfer Models to Simulate Equivalent Water Thickness and Dry Matter Content: Analysis at Leaf and Canopy Level. *IEEE Transactions on Geoscience and Remote Sensing*, 43, 819–826.

Richards, J.A. (1993). *Remote Sensing Digital Image Analysis: An Introduction*. Berlin, Germany: Springer-Verlag.

Richards, J.A. and Xia, X. (1999). *Remote Sensing Digital Image Analysis: An Introduction*. Berlin, Germany: Springer-Verlag.

Richardson, A.J. and Wiegand, C.L. (1977). Distinguishing vegetation from soil background information. *Photogrammetric Engineering and Remote Sensing*, 43, 1541–1552.

Richter, R. (1996). Atmospheric correction of satellite data with haze removal including a haze/clear transition region. *Computers and Geosciences*, 22, 675–681.

Richter, R. (1997). Correction of atmospheric and topographic effects for high spatial resolution satellite imagery. *International Journal of Remote Sensing*, 18, 1099–1111.

Ricotta, C., Avena, G.C., Olsen, E.R., Ramsey, R.D., and Winn, D.S. (1998). Monitoring the landscape stability of the Mediterranean vegetation in relation to fire with a fractal algorithm. *International Journal of Remote Sensing*, 19, 871–881.

Ricotta, C., and Retzlaff, R. (2000). Self-similar spatial clustering of wildland fires: The example of a large wildfire in Spain. *International Journal of Remote Sensing*, 21, 2113–2118.

Riitters, K.H., O'Neill, R.V., Hunsaker, C.T., Wickham, J.D., Yankee, D.H., and Timmins, S.P. (1995). A factor analysis of landscape pattern and structure metrics. *Landscape Ecology*, 10, 23–39.

Ripple, W.J., Bradshaw, G.A., and Spies, T.A. (1991). Measuring forest landscape patterns in the cascade range of Oregon. *Biological Conservation*, 57, 73–88.

Roberts, D., Smith, G.M., and Adams, M.L. (1993). Green Vegetation, Non photosynthetic vegetation and soils in AVIRIS data. *Remote Sensing of Environment*, 44, 255–269.

Roberts, D.A., Gardner, M., Church, R., Ustin, S., Scheer, G., and Green, R.O. (1998). Mapping chaparral in the Santa Monica Mountains using multiple endmembers spectral mixture models. *Remote Sensing of Environment*, 65, 267–279.

Roberts, D.A., Gardner, M., Church, R., Ustin, S., Scheer, G., and Green, R.O. (2003). Evaluation of the Potential of hyperion for fire danger assessment by comparison to the airborne visible/infrared imaging spectrometer. *IEEE Transactions on Geoscience and Remote Sensing*, 4, 1297–1310.

Roberts, D.A., Green, R.O., and Adams, J.B. (1997). Temporal and spatial patterns in vegetation and atmospheric properties from AVIRIS. *Remote Sensing of Environment*, 62, 223–240.

Robin, M. (1998). *La Télédétection.* Paris, France: Nathan.

Robinson, I.S. (1985). *Satellite Oceanography: An Introduction for Oceanographers and Remote Sensing Scientist.* Chichester, U.K.: John Wiley & Sons.

Rocchio, L. (2011). Chronicling the landsat legacy. *The Earth Observer*, 23, 4–10.

Rock, B.N., Vogelmann, J.E., Williams, D.L., Vogelmann, A.F., and Hoshizaki, T. (1986). Remote detection of forest damage. *BioScience*, 36, 439–445.

Rodriguez-Galiano, V.F., Ghimire, B., Rogan, J., Chica-Olmo, M., and Rigol-Sanchez, J.P. (2012). An assessment of the effectiveness of a random forest classifier for land-cover classification. *ISPRS Journal of Photogrammetry and Remote Sensing*, 67, 93–104.

Rollins, M.G., Keane, R.E., and Parsons, R.A. (2004). Mapping fuels and fire regimes using remote sensing, ecosystem simulation, and gradient modeling. *Ecological Applications*, 14, 75–95.

Román, M.O., Schaaf, C.B., Woodcock, C.E., Strahler, A.H., Yang, X., Braswell, R.H., Curtis, P.S., Davis, K.J., Dragoni, D., and Goulden, M.L. (2009). The MODIS (Collection V005) BRDF/albedo product: Assessment of spatial representativeness over forested landscapes. *Remote Sensing of Environment*, 113, 2476–2498.

Rosborough, G.W., Baldwin, D., and Emery, W.J. (1994). Precise AVHRR image navigation. *IEEE Transactions on Geoscience and Remote Sensing*, 32, 644–657.

Rosenfeld, G.H. (1982). Sample design for estimating change in land use and land cover. *Photogrammetric Engineering and Remote Sensing*, 48, 793–801.

Rosenfeld, G.H., Fitzpatrick-Lins, K., and Ling, H.S. (1982). Sampling for thematic map accuracy testing. *Photogrammetric Engineering and Remote Sensing*, 48, 131–137.

Roy, D. (2000). The impact of misregistration upon composited wide field of view satellite data and implications for change detection. *IEEE Transactions on Geoscience and Remote Sensing*, 38, 2017–2032.

Roy, P.S. and Tomar, S. (2000). Biodiversity characterization at landscape level using geospatial modelling technique. *Biological Conservation*, 95, 95–109.

Rumelhart, D.E., Hinton, G.E., and Williams, R.J. (1986). Learning representations by back-propagating errors. *Nature*, 323, 533–535.

Saatchi, S.S., Halligan, K., Despain, D.G., and Crabtree, R.L. (2007). Estimation of forest fuel load from radar remote sensing. *IEEE Transactions on Geoscience and Remote Sensing*, 45, 1726–1740.

Sadar, S.A., Linden, D.S., and McGuire, M. (1982). Fuels mapping from Landsat imagery and digital terrain data for fire suppression decisions. In *ACSM-ASP Congress* (pp. 345–351). Fort Lauderdale, FL.

Sader, S.A. and Winne, J.C. (1992). RGB-NDVI colour composites for visualizing forest change dynamics. *International Journal of Remote Sensing*, 13, 3055–3067.

Sakamoto, T., Yokozawa, M., Toritani, H., Shibayama, M., Ishitsuka, N., and Ohno, H. (2005). A crop phenology detection method using time-series MODIS data. *Remote Sensing of Environment*, 96, 366–374.

Salas, F.J. and Chuvieco, E. (1995). Aplicación de imágenes Landsat-TM a la cartografía de modelos combustibles. *Revista de Teledetección*, 5, 18–28.

Sample, V.A. (Ed.) (1994). *Remote Sensing and GIS in Ecosystem Management*. Washington, DC: The Island Press.

Sandholt, I., Rasmussen, K., and Andersen, J. (2002). A simple interpretation of the surface temperature/vegetation index space for assessment of surface moisture status. *Remote Sensing of Environment*, 79, 213–224.

Sandmeier, S.R. and Itten, K.I. (1999). A Field Goniometer System (FIGOS) for Acquisition of Hyperspectral BRDF data. *IEEE Transactions on Geoscience and Remote Sensing*, 37, 978–986.

Santos, A.M.P. (2000). Fisheries oceanography using satellite and airborne remote sensing methods: A review. *Fisheries Research*, 49, 1–20.

Saunders, R.W. and Kriebel, K.T. (1988). An improved method for detecting clear sky and cloudy radiances from AVHRR data. *International Journal of Remote Sensing*, 9, 123–150.

Schaaf, C.B., Gao, F., Strahler, A.H., Lucht, W., Li, X., Tsang, T., Strugnell, N.C. et al. (2002). First operational BRDF, albedo nadir reflectance products from MODIS. *Remote Sensing of Environment*, 83, 135–148.

Schanda, E. (Ed.) (1976). *Remote Sensing for Environment Sciences*. New York: Springer-Verlag.

Schiewe, J., Tufte, L., and Ehlers, M. (2001). Potential and problems of multi-scale segmentation methods in remote sensing. *GIS*, 06, 34–39.

Schowengerdt, R.A. (1983). *Techniques for Image Processing and Classification in Remote Sensing*. New York: Academic Press.

Schowengerdt, R.A. (2007). *Remote Sensing, Models, and Methods for Image Processing*. Burlington, MA: Elsevier Academic Press.

Schriever, J.R. and Congalton, R.G. (1995). Evaluating seasonal variability as an aid to cover-type mapping from Landsat Thematic Mapper data in the Northeast. *Photogrammetric Engineering and Remote Sensing*, 61, 321–327.

Schroeder, W., Prins, E., Giglio, L., Csiszar, I., Schmidt, C., Morisette, J., and Morton, D. (2008). Validation of GOES and MODIS active fire detection products using ASTER and ETM+ data. *Remote Sensing of Environment*, 112, 2711–2726.

Sellers, P.J. (1989). Vegetation-canopy spectral reflectance and biophysical processes. In G. Asrar (Ed.), *Theory and Applications of Optical Remote Sensing* (pp. 297–335). New York: Wiley.

Serrano, L., Ustin, S.L., Roberts, D.A., Gamon, J.A., and Peñuelas, J. (2000). Deriving Water Content of Chaparral Vegetation from AVIRIS Data. *Remote Sensing of Environment*, 74, 570–581.

Settle, J.J. and Drake, N.A. (1993). Linear mixing and the estimation of ground cover proportions. *International Journal of Remote Sensing*, 14, 1159–1177.

Shaw, G. and Wheeler, D. (1985). *Statistical Techniques in Geographical Analysis*. Chichester, U.K.: Wiley.

Sheffield, C. (1981). *Earth Watch: A Survey of the World from Space*. London, U.K.: Sidgwick & Jackson Ltd.

Sheffield, C. (1983). *Man on Earth. The Marks of Man. A Survey from Space*. London, U.K.: Sidgwick & Jackson Ltd.

Sheffield, C. (1985). Selecting band combinations from multispectral data. *Photogrammetric Engineering and Remote Sensing*, 51, 681–687.

Sheng, Y. (2008). Quantifying the size of a lidar footprint: A set of generalized equations. *IEEE Geoscience and Remote Sensing Letters*, 5, 419–422.

Shepherd, A., Ivins, E.R., Geruo, A., Barletta, V.R., Bentley, M.J., Bettadpur, S., Briggs, K.H., Bromwich, D.H., Forsberg, R., and Galin, N. (2012). A reconciled estimate of ice-sheet mass balance. *Science*, 338, 1183–1189.

Shepherd, J.D. and Dymond, J.R. (2000). BRDF correction of vegetation in AVHRR imagery. *Remote Sensing of Environment*, 74, 397–408.

Shepherd, K.D. and Walsh, M.G. (2002). Development of reflectance spectral libraries for characterization of soil properties. *Soil Science Society of America Journal*, 66, 988–998.

Shimabukuro, Y.E., Holben, B.N., and Tucker, C.J. (1994). Fraction images derived from NOAA-AVHRR data for studying the deforestation in the Brazilian Amazon. *International Journal of Remote Sensing*, 15, 517–520.

Shimabukuro, Y.E. and Smith, J.A. (1991). The least-squares mixing models to generate fraction images derived from remote sensing multispectral data. *IEEE Transactions on Geosciences and Remote Sensing*, 29, 16–20.

Short, N.M. (1982). *The Landsat Tutorial Workbook: Basics of Satellite Remote Sensing*. Washington, DC: NASA Scientific and Technical Information Branch.

Short, N.M. and Blair, R.W. (Eds.) (1986). *Geomorphology from Space*. Washington, DC: NASA, Scientific and Technical Information Branch.

Short, N.M. and Stuart, L.M. (1982). *The Heat Capacity Mapping Mission (HCMM) Anthology*. Washington D.C.: NASA Scientific and Technical Information Branch.

Siljeström, P. and Moreno, A. (1995). Monitoring burnt areas by principal components analysis of multi-temporal TM data. *International Journal of Remote Sensing*, 16, 1577–1587.

Silva, J., Sá, A., and Pereira, J.M.C. (2005). Comparison of burned area estimation derived from SPOT-VEGETATION and Landsat ETM+ data in Africa: Influencia de spatial pattern and vegetation type. *Remote Sensing of Environment*, 96, 188–201.

Silva, J.M.N., Cadima, J.F.C.L., Pereira, J.M.C., and Gregoire, J.M. (2004). Assessing the feasibility of a global model for multi-temporal burned area mapping using SPOT-VEGETATION data. *International Journal of Remote Sensing*, 25, 4889–4913.

Simard, M., Pinto, N., Fisher, J.B., and Baccini, A. (2011). Mapping forest canopy height globally with spaceborne lidar. *Journal of Geophysical Research*, 116(G0421), 1–12.

Simard, M., Rivera-Monroy, V.H., Mancera-Pineda, J.E., Castaneda-Moya, E., and Twilley, R.R. (2008). A systematic method for 3D mapping of mangrove forests based on Shuttle Radar Topography Mission elevation data, ICEsat/GLAS waveforms and field data: Application to Cienaga Grande de Santa Marta, Colombia. *Remote Sensing of Environment*, 112, 2131–2144.

Sims, D.A., Rahman, A.F., Cordova, V.D., El-Masri, B.Z., Baldocchi, D.D., Flanagan, L.B., Goldstein, A.H., Hollinger, D.Y., Misson, L., and Monson, R.K. (2006). On the use of MODIS EVI to assess gross primary productivity of North American ecosystems. *Journal of Geophysical Research: Biogeosciences* (2005–2012), 111.

Singh, A. (1986). Change detection in the tropical forest environment of Northeastern India using Landsat. In M.J. Eden and J.T. Parry (Eds.), *Remote Sensing and Tropical Land Management* (pp. 237–254). Chichester, U.K.: John Wiley & Sons.

Skidmore, A.K. (1989). An expert system classifies eucalypt forest types using Thematic Mapper data and digital terrain model. *Photogrammetric Engineering and Remote Sensing*, 55, 1149–1464.

Slater, P.N. (1980). *Remote Sensing, Optics and Optical Systems*. Reading, MA: Addison-Wesley Publishing Co.

Slatton, K.C., Crawford, M.M., and Evans, B.L. (2001). Fusing interferometric radar and laser altimeter data to estimate surface topography and vegetation heights. *IEEE Transactions on Geoscience and Remote Sensing*, 39, 2470–2482.

Smith, E.A., Asrar, G., Furuhama, Y., Ginati, A., Mugnai, A., Nakamura, K., Adler, R.F., Chou, M.-D., Desbois, M., et al. (2007). International global precipitation measurement (GPM) program and mission: An overview. In Levizzani, V., Bauer, and Turk, F. J. (Eds.), *Measuring Precipitation from Space* (pp. 611–653). Berlin, Germany: Springer.

Smith, M.O., Ustin, S.L., Adams, J.B., and Gillespie, A.R. (1990). Vegetation in deserts: II. Environmental influences on regional vegetation. *Remote Sensing of Environment*, 31, 27–52.

Smits, P. and Annoni, A. (1999). Updating land-cover maps by using texture information from very high-resolution space-borne imagery. *IEEE Transactions on Geoscience and Remote Sensing*, 37, 1244–1254.

Smits, P. and Annoni, A. (2000). Toward specification-driven change detection. *IEEE Transactions on Geoscience and Remote Sensing*, 38, 1484–1488.

Snedecor, G.W. and Cochran, W.G. (1980). *Statistical Methods*. Ames, IA: The Iowa University Press.

Sobrino, J.A. (Ed.) (2000). *Teledetección*. Valencia, Spain: Servicio de Publicaciones, Universidad de Valencia.

Sobrino, J.A., Jiménez-Muñoz, J.C., and Paolini, L. (2004). Land surface temperature retrieval from LANDSAT TM 5. *Remote Sensing of Environment*, 90, 434–440.

Soja, A.J., Sukhinin, A.I., Cahoon, D.R., Shugart, H.H., and Stackhouse, P.W. (2004). AVHRR-derived fire frequency, distribution and area burned in Siberia. *International Journal of Remote Sensing*, 25, 1939–1960.

Souza, C.M., Roberts, D.A., and Cochrane, M.A. (2005). Combining spectral and spatial information to map canopy damage from selective logging and forest fires. *Remote Sensing of Environment*, 98, 329–343.

Star, J.L. and Estes, J.E. (1990). *Geographic Information Systems*. Englewood Cliffs, NJ: Prentice Hall.

Star, J.L., Estes, J.E., and Davis, F. (1991). Improved integration of remote sensing and Geographic Information Systems: A background to NCGIA initiative 12. *Photogrammetric Engineering and Remote Sensing*, 57, 643–645.

Stehman, S.V. (1992). Comparison of systematic and random sampling for estimating the accuracy of maps generated from remotely sensed data. *Photogrammetric Engineering and Remote Sensing*, 58, 1343–1350.

Stehman, S.V. (1999). Basic probability sampling designs for thematic map accuracy assessment. *International Journal of Remote Sensing*, 20, 2423–2441.

Stehman, S.V., Arora, M., Kasetkasem, T., and Varshney, P. (2007). Estimation of Fuzzy error matrix accuracy measures under stratified random sampling. *Photogrammetric Engineering and Remote Sensing*, 73, 165–174.

Steininger, M.K. (2000). Satellite estimation of tropical secondary forest above-ground biomass: Data from Brazil and Bolivia. *International Journal of Remote Sensing*, 21, 1139–1157.

Stoner, E.R. and Baumgardner, M.F. (1981). Characteristic variations in reflectance of surface soils. *Soil Science Society of America Journal*, 45, 1161–1165.

Story, M. and Congalton, R.G. (1986). Accuracy assessment: A user's perspective. *Photogrammetric Engineering and Remote Sensing*, 52, 397–399.

Story, M.H. and Campbell, J.B. (1986). The effect of training data on classification accuracy. In *ACSM-ASPRS Convention*. Technical Papers (pp. 370–379). Anchorage, AK: American Society for Photogrammetry & Remote Sensing.

Strahler, A.H. (1980). The use of prior probabilities in maximum likelihood classification of remotely sensed data. *Remote Sensing of Environment*, 10, 135–163.

Strahler, A.H., Logan, T.L., and Bryant, N.A. (1978). Improving forest cover classification accuracy from Landsat by incorporating topographic information. In *12th International Symposium on Remote Sensing of Environment* (pp. 927–941). Ann Arbor, MI.

Strahler, A.H., Woodcock, C.E., and Smith, J.A. (1986). On the nature of models in remote sensing. *Remote Sensing of Environment*, 20, 121–140.

Strain, P. and Engle, F. (1993). *Looking at Earth*. Atlanta, GA: Turner Publishing Inc.

Stroppiana, D., Grégoire, J.M., and Pereira, J.M.C. (2003). The use of SPOT VEGETATION data in a classification tree approach for burnt area mapping in Australian savanna. *International Journal of Remote Sensing*, 24, 2131–2151.

Sun, G., Ranson, K., Kimes, D., Blair, J., and Kovacs, K. (2008). Forest vertical structure from GLAS: An evaluation using LVIS and SRTM data. *Remote Sensing of Environment*, 112, 107–117.

Swain, P.H. and Davis, S.M. (Eds.) (1978). *Remote Sensing: The Quantitative Approach*. New York: McGraw-Hill.

Szekielda, K.H. (1988). *Satellite Monitoring of the Earth*. New York: John Wiley & Sons.

Tachikawa, T., Hato, M., Kaku, M., and Iwasaki, A. (2011). Characteristics of ASTER GDEM version 2. In *2011 IEEE International Geoscience and Remote Sensing Symposium (IGARSS)* (pp. 3657–3660). Vancouver, Canada: IEEE International.

Tanase, M.A., Santoro, M., de la Riva, J., Perez-Cabello, F., and Le Toan, T. (2010a). Sensitivity of X-, C-, and L-band SAR backscatter to burn severity in Mediterranean pine forests. *IEEE Transactions on Geoscience and Remote Sensing*, 48, 3663–3675.

Tanase, M.A., Santoro, M., Wegmuller, U., de la Riva, J., and Perez-Cabello, F. (2010b). Properties of X-, C- and L-band repeat-pass interferometric SAR coherence in Mediterranean pine forests affected by fires. *Remote Sensing of Environment*, 114, 2182–2194.

Tansey, K., Gregorie, J., Binaghi, E., Boschetti, L., Brivio, P.A., Ershov, D., Flasse, E. et al. (2004a). A global inventory of burned areas at 1 km. resolution for the year 2000 derived from SPOT VEGETATION data. *Climatic Change*, 67, 345–377.

Tansey, K., Grégoire, J.M., Stroppiana, D., Sousa, A., Silva, J., Pereira, J.M., Boschetti, L. et al. (2004b). Vegetation burning in the year 2000: Global burned area estimates from SPOT VEGETATION data. *Journal of Geophysical Research—Atmospheres*, 109, D14S03, doi:10.1029/2002JD003598, 2-22.

Teillet, P.M., Guindon, B., and Goodeonugh, D.G. (1982). On the slope-aspect correction of multispectral scanner data. *Canadian Journal of Remote Sensing*, 8, 84–106.

Thomas, I.L. (1980). Spatial post-processing of spectrally-classified Landsat data. *Photogrammetric Engineering and Remote Sensing*, 46, 1201–1206.

Thomas, I.L., Benning, V.M., and Ching, N.P. (1987a). *Classification of Remotely Sensed Images*. Bristol, CT: Adam Hilger.

Thomas, I.L., Ching, N.P., Benning, V.M., and D'Aguanno, J.A. (1987b). A review of multichannel indices of class separability. *International Journal of Remote Sensing*, 8, 331–350.

Tian, Q., Tong, Q., Pu, R., Guo, X., and Zhao, C. (2001). Spectroscopic determination of wheat water status using 1650–1850 nm spectral absorption features. *International Journal of Remote Sensing*, 22, 2329–2338.

Tian, Y., Woodcock, C.E., Wang, Y., Privette, J.L., Shabanov, N.V., Zhou, L., Zhang, Y. et al. (2002). Multiscale analysis and validation of the MODIS LAI product. II. Sampling strategy. *Remote Sensing of Environment*, 83, 431–441.

Tindal, M.A. (1978). *Educator's Guide for Mission to Earth: Landsat Views the World*. Greenbelt, MD: NASA, Goddard Space Flight Center.

Todd, W.J., Gehring, D.G., and Haman, J.F. (1980). Landsat wildland mapping accuracy. *Photogrammetric Engineering and Remote Sensing*, 46, 509–520.

Togliatti, G. and Moriondo, A. (1986). LFC: The second generation photogrammetric camera for space photography. In *ESA/EaRSEL Symposium on Europe from Space* (pp. 15–18). Lyngby, Denmark: European Space Agency.

Toll, D.L. (1984). An evaluation of simulated Thematic Mapper data and Landsat MSS data for discriminating suburban and regional land use and land cover. *Photogrammetric Engineering and Remote Sensing*, 50, 1713–1724.

Tomlin, D. (1990). *Geographic Information Systems and Cartographic Modeling*. Englewood Cliffs, NJ: Prentice Hall.

Townsend, P.A. (2000). A quantitative fuzzy approach to assess mapped vegetation classifications for ecological applications. *Remote Sensing of Environment*, 72, 253–267.

Townshend, J., Justice, C., Li, W., Gurney, C., and McManus, J. (1991). Global land cover classification by remote sensing: Present capabilities and future possibilities. *Remote Sensing of Environment*, 35, 243–255.

Townshend, J.R.G. (1980). *The Spatial Resolving Power of Earth Resources Satellites: A Review*. Greenbelt, MD: NASA, Goddard Spaceflight Center.

Townshend, J.R.G. (Ed.) (1981). *Terrain Analysis and Remote Sensing*. London, U.K.: George Allen & Unwin Ltd.

Townshend, J.R.G., Justice, C.O., Gurney, C., and McManus, J. (1992). The impact of misregistration on change detection. *IEEE Transactions on Geoscience and Remote Sensing*, 30, 1054–1060.

Trevett, J.W. (1986). *Imaging Radar for Resources Surveys*. London, U.K.: Chapman & Hall.

Trishchenko, A.P. (2006). Solar irradiance and brightness temperature for SWIR channels of AVHRR and GOES imagers. *Journal of Atmospheric and Oceanic Technology*, 23, 198–210.

Trishchenko, A.P., Cihlar, J., and Li, Z.Q. (2002). Effects of spectral response function on surface reflectance and NDVI measured with moderate resolution satellite sensors. *Remote Sensing of Environment*, 81, 1–18.

Trolier, L.J. and Philipson, W.R. (1986). Visual analysis of Landsat Thematic Mapper images for hydrologic land use and cover. *Photogrammetric Engineering and Remote Sensing*, 52, 1531–1538.

Trombetti, M., Riano, D., Rubio, M.A., Cheng, Y.B., and Ustin, S.L. (2008). Multi-temporal vegetation canopy water content retrieval and interpretation using artificial neural networks for the continental USA. *Remote Sensing of Environment*, 112, 203–215.

Tsai, F., and Philpot, W.D. (2002). A derivative-aided hyperspectral image analysis system for land-cover classification. *IEEE Transactions on Geoscience and Remote Sensing*, 40, 416–425.

Tso, B. and Mather, P. (2001). *Classification Methods for Remote Sensed Data*. London, U.K.: Taylor & Francis Group.

Tsuchiya, K., Arai, K., and Igarashi, T. (1987). Marine observation satellite. *Remote Sensing Reviews*, 3, 59–103.

Tucker, C.J. (1979). Red and photographic infrared linear combinations for monitoring vegetation. *Remote Sensing of Environment*, 8, 127–150.

Tucker, C.J., Dregne, H.E., and Newcomb, W.W. (1991). Expansion and contraction of the Sahara desert from 1980 to 1990. *Science*, 253, 299–301.

Tucker, C.J., Pinzon, J.E., Brown, M.E., Slayback, D.A., Pak, E.W., Mahoney, R., Vermote, E.F., and El Saleous, N. (2005). An extended AVHRR 8-km NDVI dataset compatible with MODIS and SPOT vegetation NDVI data. *International Journal of Remote Sensing*, 26, 4485–4498.

Tucker, C.J., Yung, I.Y., Keeling, C.D., and Gammon, R.H. (1986). Relationship between atmospheric $CO_2$ variations and a satellite-derived vegetation index. *Nature*, 319, 195–199.

Turner, M.G. (1989). Landscape ecology: The effect of pattern on process. *Annual Review of Ecology and Systematics*, 20, 171–197.

Turner, M.G., Hargrove, W.W., Gardner, R.H., and Romme, W.H. (1994). Effects of fire on landscape heterogeneity in Yellowstone-National-Park, Wyoming. *Journal of Vegetation Science*, 5, 731–742.

Unninayar, S. (1988). The global system, observing and monitoring change. Data problems, data management and databases. In H. Mounsey (Ed.), *Building Databases for Global Science* (pp. 357–377). London, U.K.: Taylor & Francis Group.

Ustin, S.L. and Trabucco, A. (2000). Using hyperspectral data to assess forest structure. *Journal of Forestry*, 98, 47–49.

Ustin, S.L. and Xiao, Q.F. (2001). Mapping successional boreal forests in interior central Alaska. *International Journal of Remote Sensing*, 22, 1779–1797.

Valor, E. and Caselles, V. (1996). Mapping land surface emissivity from NDVI: Application to European, African, and South American Areas. *Remote Sensing of Environment*, 57, 167–184.

van Leeuwen, W.J.D., Huete, A.R., Duncan, J., and Franklin, J. (1994). Radiative transfer in shrub savanna sites in Niger: Preliminary results from HAPEX-Sahel. 3. Optical dynamics and vegetation index sensitivity to biomass and plant cover. *Agricultural and Forest Meteorology*, 69, 267–288.

Vega-Garcia, C. and Chuvieco, E. (2006). Applying local measures of spatial heterogeneity to Landsat-TM images for predicting wildfire occurrence in Mediterranean landscapes. *Landscape Ecology*, 21, 595–605.

Vega-García, C., Lee, B., and Wooddard, T. (1996). Applying neural network technology to human-caused wildfire occurrence prediction. *AI Applications*, 10, 9–18.

Vega-García, C., Tatay-Nieto, J., Blanco, R., and Chuvieco, E. (2010). Evaluation of the Influence of Local Fuel Homogeneity on Fire Hazard through Landsat-5 TM Texture Measures. *Photogrammetric Engineering and Remote Sensing*, 76, 853–864.

Verbyla, D. (1995). *Satellite Remote Sensing of Natural Resources*. Boca Raton, FL: Lewis Publishers.

Verhoef, W. (1984). Light scattering by leaf layers with application to canopy reflectance modeling: The SAIL model. *Remote Sensing of Environment*, 16, 125–141.

Verhoef, W. and Bach, H. (2003). Simulation of hyperspectral and directional radiance images using coupled biophysical and atmospheric radiative transfer models. *Remote Sensing of Environment*, 87, 23–41.

Vermote, E., Tanré, D., Deuzé, J.L., Herman, M., and Morisette, J.J. (1997). *Second Simulation of the Satellite Signal in the Solar Spectrum (6S)*.Greenbelt, MD: NASA Goddard Space Flight Center Code 923, p. 54.

Vosselman, G. (2000). Slope based filtering of laser altimetry data. *International Archives of Photogrammetry and Remote Sensing* (pp. 958–964). Amsterdam, the Netherlands: ISPRS Workshop.

Vosselman, G. and Klein, R. (2010). Visualisation and structuring of point clouds. In G. Vosselman and H.-G. Maas (Eds.), *Airborne and Terrestrial Laser Scanning* (pp. 45–81). Dunbeath, U.K.: Whittles Publishing.

Wagner, W., Dorigo, W., and Paulik, C. (2013a). The use of Earth observation satellites for soil moisture monitoring. WMO statement on the status of the global climate in 2012 (pp. 32–33). Geneva, Switzerland: World Meteorological Organization.

Wagner, W., Hahn, S., Kidd, R., Melzer, T., Bartalis, Z., Hasenauer, S., Figa-Saldaña, J., de Rosnay, P., Jann, A., and Schneider, S. (2013b). The ASCAT soil moisture product: A review of its specifications, validation results, and emerging applications. *Meteorologische Zeitschrift*, 22, 5–33.

Wan, Z., Zhang, Y., Zhang, Q., and Li, Z.-L. (2002). Validation of the land-surface temperature products retrieved from Terra Moderate Resolution Imaging Spectroradiometer data. *Remote Sensing of Environment*, 83, 163–180.

Wang, D.E., Woodcock, C., Wolfgang, B., Stenber, P., Voipio, P., Smolanderc, H., Häme, T. et al. (2004). Evaluation of the MODIS LAI algorithm at a coniferous forest site in Finland. *Remote Sensing of Environment*, 91, 114–127.

Wang, F. (1990a). Fuzzy supervised classification of remote sensing images. *IEEE Transactions on Geoscience and Remote Sensing*, 28, 194–201.

Wang, F. (1990b). Improving remote sensing image analysis through fuzzy information representation. *Photogrammetric Engineering and Remote Sensing*, 56, 1163–1169.

Webster, T.L., Forbes, D.L., MacKinnon, E., and Roberts, D. (2006). Flood-risk mapping for storm-surge events and sea-level rise using lidar for southeast New Brunswick. *Canadian Journal of Remote Sensing*, 32, 194–211.

Wehr, A. and Lohr, U. (1999). Airborne laser scanning—An introduction and overview. *ISPRS Journal of Photogrammetry and Remote Sensing*, 54, 68–82.

Welch, R. and Usery, E.L. (1984). Cartographic accuracy of Landsat-4 MSS and TM image data. *IEEE Transactions on Geoscience and Remote Sensing*, GE22, 281–288.

Weng, Q. (2012). *An Introduction to Contemporary Remote sensing*. New York: McGraw Hill.

Westman, C.A., Aber, J.D., Peterson, D.L., and Melillo, J.M. (1988). Remote sensing of canopy chemistry and nitrogen cycling in temperate forest ecosystems. *Nature*, 335, 154–156.

Westman, W.E. and Price, C.V. (1988). Spectral changes in conifers subjected to air pollution and water stress: Experimental studies. *IEEE Transactions on Geoscience and Remote Sensing*, 26, 11–20.

Wilkie, D.S. and Finn, J.T. (1996). *Remote Sensing Imagery for Natural Resources Monitoring*. New York: Columbia University Press.

Williamson, R.A. (2001). Remote sensing policy and the development of commercial remote sensing. In J.C. Baker, K.M. O'Connell and R.A. Williamson (Eds.), *Commercial Observation Satellites: At the Leading Edge of Global Transparency* (pp. 37–52). Santa Monica, CA: RAND—ASPRS.

Willis, K.S. (2015). Remote sensing change detection for ecological monitoring in United States protected areas. *Biological Conservation*, 182, 233–242.

Woldai, T. (1983). Landsat and SIR-A interpretation of the Kalpin Chol and Chong Korum mountains of China. *ITC Journal*, 83, 250–252.

Woodcock, C.E., Macomber, S.A., Pax-Lenney, M., and Cohen, W.B. (2001). Monitoring large areas for forest change using Landsat: Generalization across space, time and Landsat sensors. *Remote Sensing of Environment*, 78, 194–203.

Woodcock, C.E. and Strahler, A.H. (1987). The factor of scale in remote sensing. *Remote Sensing of Environment*, 21, 311–332.

Woodcock, C.E., Strahler, A.H., and Jupp, D.L. (1988a). The use of variograms in remote sensing: I. Scene models and simulated images. *Remote Sensing of Environment*, 25, 323–348.

Woodcock, C.E., Strahler, A.H., and Jupp, D.L. (1988b). The use of variograms in remote sensing: II. Real digital images. *Remote Sensing of Environment*, 25, 349–379.

Wulder, M.A., Masek, J.G., Cohen, W.B., Loveland, T.R., Woodcock, C.E., Irons, J.R., Dwyer, J.L., Barsi, J.A., Markham, B.L., and Helder, D.L. (2012). Landsat Legacy Special Issue. *Remote Sensing of Environment*, 122, 1–202.

Wulder, M.A., White, J.C., Alvarez, F., Han, T., Rogan, J., and Hawkes, B. (2009). Characterizing boreal forest wildfire with multi-temporal Landsat and LIDAR data. *Remote Sensing of Environment*, 113, 1540–1555.

Xiao, X., Braswell, B., Zhang, Q., Boles, S., Frolking, S.E., and Moore III, B. (2003). Sensitivity of vegetation indices to atmospheric aerosols: Continental-scale observations in Northern Asia. *Remote Sensing of Environment*, 84, 385–392.

Yang, F., Ichii, K., White, M.A., Hashimoto, H., Michaelis, A.R., Votava, P., Zhu, A.-X., Huete, A., Running, S.W., and Nemani, R.R. (2007). Developing a continental-scale measure of gross primary production by combining MODIS and AmeriFlux data through Support Vector Machine approach. *Remote Sensing of Environment*, 110, 109–122.

Yebra, M., Chuvieco, E., and Riaño, D. (2008). Estimation of live Fuel Moisture Content from MODIS images for fire risk assessment. *Agricultural and Forest Meteorology*, 148, 523–536.

Yebra, M., Dennison, P., Chuvieco, E., Riaño, D., Zylstra, P., Hunt, E.R., Danson, F.M., Qi, Y., and Jurdao, S. (2013a). A global review of remote sensing of live fuel moisture content for fire danger assessment: Moving towards operational products *Remote Sensing of Environment*, 136, 455–468.

Yebra, M., Van Dijk, A., Leuning, R., Huete, A., and Guerschman, J.P. (2013b). Evaluation of optical remote sensing to estimate actual evapotranspiration and canopy conductance. *Remote Sensing of Environment*, 129, 250–261.

Yocky, D.A. (1996). Multiresolution wavelet decomposition image merger of Landsat Thematic Mapper and SPOT pancromatic data. *Photogrammetric Engineering and Remote Sensing*, 62, 1067–1084.

Yool, S.R. (2001). Enhancing fire scar anomalies in AVHRR NDVI time-series data. *Geocarto International*, 16, 5–12.

Zarco-Tejada, P., Miller, J.R., Harron, J., Hu, B., Noland, T.L., Goel, N., Mohammed, G.H., and Sampson, P.H. (2004). Needle chlorophyll content estimation through model inversion using hyperspectral data from boreal conifer forest canopies. *Remote Sensing of Environment*, 89, 189–199.

Zarco-Tejada, P.J., Morales, A., Testi, L., and Villalobos, F. (2013). Spatio-temporal patterns of chlorophyll fluorescence and physiological and structural indices acquired from hyperspectral imagery as compared with carbon fluxes measured with eddy covariance. *Remote Sensing of Environment*, 133, 102–115.

Zarco-Tejada, P.J., Rueda, C.A., and Ustin, S.L. (2003). Water content estimation in vegetation with MODIS reflectance data and model inversion methods. *Remote Sensing of Environment*, 85, 109–124.

Zender, C.S. (2008). Analysis of self-describing gridded geoscience data with netCDF Operators (NCO). *Environmental Modelling and Software*, 23, 1338–1342.

Zeng, C., Shen, H., and Zhang, L. (2013). Recovering missing pixels for Landsat ETM+ SLC-off imagery using multi-temporal regression analysis and a regularization method. *Remote Sensing of Environment*, 131, 182–194.

Zha, Y., Gao, J., Ni, S., and Shen, N. (2005). Temporal filtering of successive MODIS data in monitoring a locust outbreak. *International Journal of Remote Sensing*, 26, 5665–5674.

Zhan, X., Sohlberg, R.A., Townshend, J.R.G., DiMiceli, C., Carroll, M.L., Eastman, J.C., Hansen, M.C., and DeFries, R.S. (2002). Detection of land cover changes using MODIS 250 m data. *Remote Sensing of Environment*, 83, 336–350.

Zhang, Q.F., Pavlic, G., Chen, W.J., Fraser, R., Leblanc, S., and Cihlar, J. (2005). A semi-automatic segmentation procedure for feature extraction in remotely sensed imagery. *Computers and Geosciences*, 31, 289–296.

Zhang, X., Friedl, M.A., Schaaf, C.B., Strahler, A.H., Hodges, J.C.F., Gao, F., Reed, B.C., and Huete, A. (2003). Monitoring vegetation phenology using MODIS. *Remote Sensing of Environment*, 84, 471–475.

Zhang, Z. and Moore, J.C. (2015). Chapter 4—Remote sensing. In Z. Zhang and J.C. Moore (Eds.), *Mathematical and Physical Fundamentals of Climate Change* (pp. 111–124). Boston, MA: Elsevier.

Zhou, G., Song, C., Simmers, J., and Cheng, P. (2004). Urban 3D GIS from LiDAR and digital aerial images. *Computers and Geosciences*, 30, 345–353.

# Index

## D



## E

## F

## G



## H



## I

## J



## K



## L



## M

## S

## T